Current Drug Synthesis

1st Edition

Edited by

Jie Jack Li

GenHouse Bio

WILEY

This edition first published 2023

Registered Office
John Wiley & Sons, Inc., 111 River Street, Hoboken, NJ 07030, USA

Editorial Office
111 River Street, Hoboken, NJ 07030, USA

For details of our global editorial offices, customer services, and more information about Wiley products visit us at www.wiley.com.

Wiley also publishes its books in a variety of electronic formats and by print-on-demand. Some content that appears in standard print versions of this book may not be available in other formats.

Library of Congress Cataloging-in-Publication Data Applied for:
Hardback ISBN: 9781119847250

Cover design by Wiley
Cover image: Courtesy of Brian Lanman

Set in 10/13pt TimesNewRoman by Straive, Chennai, India

Current Drug Synthesis

Dedicated to Dr. Lew Pennington

Contents

Preface

Our first four installments *Wiley's Drug Synthesis Series*, *Contemporary Drug Synthesis*, *The Art of Drug Synthesis*, *Modern Drug Synthesis*, and *Innovative Drug Synthesis* were published in 2004, 2007, 2010, and 2015, respectively. They have been warmly received by the drug discovery community. The current title, *Current Drug Synthesis*, is our fifth installment of this series.

This book has four sections, reviewing total of 18 drugs. Section I, "Infectious Disease Drugs," covers six drugs; Section II, "Cancer Drugs," reviews seven drugs; Section III, "CNS Drugs," covers two drugs; Section IV, "Miscellaneous Drugs," covers three additional drugs.

Each chapter is divided into seven sections:

1. Background
2. Pharmacology
3. Structure–activity relationship
4. Pharmacokinetics and drug metabolism
5. Efficacy and safety
6. Syntheses
7. References

I am very much indebted to all contributing authors from both industry and academia. Many of them are veterans and well-known experts in medicinal chemistry. Some of them discovered the drugs that they reviewed. As a consequence, their work tremendously elevated the quality of this book as a teaching tool.

Meanwhile, I welcome your critique and suggestions so we can make this *Wiley's Drug Synthesis Series* even more useful to the drug discovery community.

Jack Li

Ann Arbor, Michigan

December 1, 2021

Contributing Authors

Dr. Nadia M. Ahmad
Charles River Laboratories
7-9 The Spire Green Centre
Harlow, CM19 5TR
United Kingdom

Dr. Narendra Ambhaikar
Neuland Laboratories Limited
R&D Centre, Bonthapally Village
Gummadidala Mandal
Sangareddy District (near Hyderabad)
Telangana 502313, India

Dr. Yvonne M. Angell
GenEdit, Inc.
681 Gateway Blvd., Suite 313
South San Francisco, CA 94080
United States

Ruby M. Aaron
Department of Chemistry and Biochemistry
Colorado College
14 East Cache La Poudre St.
Colorado Springs, CO 80903
United States

Dr. Richard Beresis
ChemPartner
280 Utah Avenue, Suite 100
South San Francisco, CA 94080
United States

Dr. Brett C. Bookser
ChemPartner
280 Utah Avenue, Suite 100
South San Francisco, CA 94080
United States

Dr. Serge H. Boyer
Qpex Biopharma, Inc.
6275 Nancy Ridge Dr., Suite 100
San Diego, CA 92121
United States

Dr. Dao-Qian Chen
HEC R&D Center
Pharmaceutical Science
Dongguan, GuangDong
P. R. China

Dr. Jinxia Nancy Deng
ChemPartner
280 Utah Avenue, Suite 100
South San Francisco, CA 94080
United States

Dr. Daniel A. Dickman
ChemPartner
280 Utah Avenue, Suite 100
South San Francisco, CA 94080
United States

Prof. Ke Ding
Guangzhou City Key Laboratory of Precision Chemical Drug Development,
School of Pharmacy
Jinan University
601 West Huangpu Avenue
Guangzhou 510632, P. R. China

Prof. Amy B. Dounay
Department of Chemistry and Biochemistry
Colorado College
14 East Cache La Poudre St.
Colorado Springs, CO 80903
United States

Dr. Wendy J. Hartsock
CEM Corporation
3100 Smith Farm Road
Matthews, NC 28104
United States

Dr. Scott J. Hecker
Qpex Biopharma, Inc.
6275 Nancy Ridge Dr., Suite 100
San Diego, CA 92121
United States

Dr. Brian A. Lanman
Amgen, Inc.
1 Amgen Center Drive
Thousand Oaks, CA 91320
United States

Hayden K. Low
Department of Chemistry and Biochemistry
Colorado College
14 East Cache La Poudre St.
Colorado Springs, CO 80903
United States

Dr. Jie Jack Li
ChemPartner
280 Utah Avenue, Suite 100
South San Francisco, CA 94080
United States

Dr. John Mancuso
NuChem Sciences
2350 Cohen St. Suite 201
Ville St-Laurent, QC
H4R 2N6, Canada

Dr. Raymond Ng
Medicinal Chemistry
Olema Oncology
780 Brannan St.
South San Francisco, CA 94103
United States

Dr. Andrew T. Parsons
Amgen, Inc.
1 Amgen Center Drive
Thousand Oaks, CA 91320
United States

Dr. K. Raja Reddy
Qpex Biopharma, Inc.
6275 Nancy Ridge Dr., Suite 100
San Diego, CA 92121
United States

Timothy M. Reichart
Department of Chemistry
Hampden-Sydney College
Hampden-Sydney, VA 23943
United States

Benjamin T. Sokol
Department of Chemistry and Biochemistry
Colorado College
14 East Cache La Poudre St.
Colorado Springs, CO 80903
United States

Dr. Yan Wang
ChemPartner
280 Utah Avenue, Suite 100
South San Francisco, CA 94080
United States

Dr. Yong-Jin Wu
Small Molecule Drug Discovery
Bristol Myers Squibb Research and Early Development
100 Binney St. Cambridge, MA 02142
United States

Dr. Dexi Yang
Merck Research Lab
Merck & Company, Inc.
Kenilworth, NJ 07033
United States

Dr. Ji Zhang
HEC R&D Center
Pharmaceutical Science
Dongguan, GuangDong
P. R. China

Fengtao Zhou
Guangzhou City Key Laboratory of
Precision Chemical Drug Development
School of Pharmacy
Jinan University
601 West Huangpu Avenue
Guangzhou 510632, P. R. China

I

INFECTIOUS DISEASE DRUGS

1

Relebactam (Recarbrio), A β-Lactamase Inhibitor for the Treatment of cIAI/cUTI/HABP/VABP

Dexi Yang

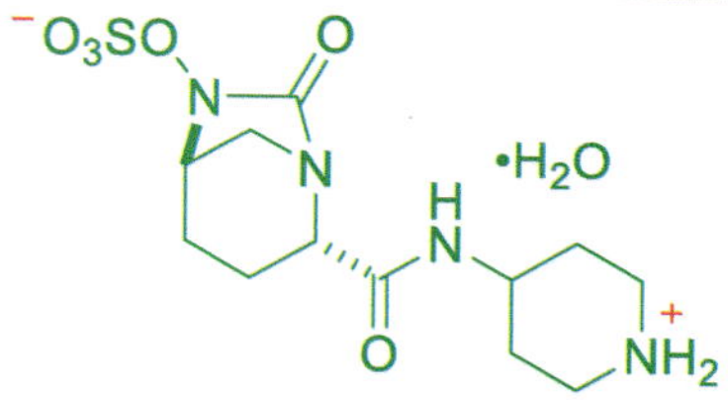

USAN: Relebactam (**1**)
Trade name: Recarbrio
Merck Co.
Launched: 2019

1. Background

The discovery of antibiotics is revolutionary in chemotherapy against infectious diseases in modern medicine history. Unfortunately, after its golden era from the 1950s to 1970s, antimicrobial resistance among common bacterial pathogens became a new threat to public health. Recently, WHO enlisted antibiotic resistance in the top three public health threats. Infections caused by multidrug-resistant organism became a new economic burden in health-care system. In the United States alone, it costs over 20 billion dollars per year, and more than 23,000 people died of infection with antibiotic-resistant annually. With this continuing, CDC estimated that victims will culminate to more than 300 million globally with loss of over 100 trillion dollars by 2050. The stake is high enough to draw more attention to invent new therapeutics to treat infected patients.[1]

Of all known antimicrobial resistance, carbapenem resistance in gram-negative pathogens is the most critical. Clinically, carbapenems are considered the most active and potent agents against MDR gram-negative pathogens. They are the last silver bullets to kill superbugs. However, according to the global priority list of antibiotic-resistant bacteria published by WHO in 2017, three of the top four pathogens critical for

Current Drug Synthesis, First Edition. Edited by Jie Jack Li.

developing antibiotics are carbapenem-resistant, and they are *Enterobacteriaceae* (CRE), *Pseudomonas aeruginosa*, and *Acinetobacter baumannii*.[2,3] In the 1990s, research on MDR revealed that antibiotic resistance of gram-negative bacteria is mainly caused by three mechanisms. The major mechanism of resistance to carbapenems is the production of β-lactamase. It has been identified that MDR gram-negative pathogens produces at least four classes of β-lactamases: A, B, C, and D.[4] They all can degrade antibacterial agents and make them ineffective. In addition to these enzymes, these pathogens also developed other mechanisms to make antibiotics less efficient. One is porin mutation caused by porin expression. This renders the outer bacterial membrane impermeable to antibacterial. The other is efflux pump upregulation. *Via* efflux pump, antibiotics are pumped out of the membrane of bacteria and lose their therapeutic efficacy.[3,4] Based on all these three mechanisms, new generations of β-lactamase inhibitors should not only have high potency to help carbapenem restore potency, but also possess appropriate physicochemical properties to increase permeability and decrease efflux rate.

Guided by this strategy, in the past decade, several combinations of antibiotic with β-lactamase inhibitors, such as polymyxins, ceftazidime-avibactam, ceftolozane–tazobactam, metropenem–veborbactam, etc., have been discovered and approved by FDA.[5–7] But most of them are only effective against a small portion of pathogens, and with no surprise, gradually loose efficacy due to evolved resistance. As a result, new therapeutic options against gram-negative organisms with resistance are still urgently needed.

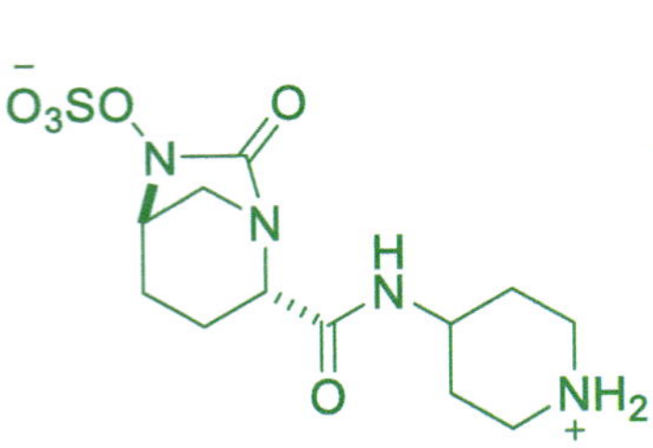

Relebactam (**1**)

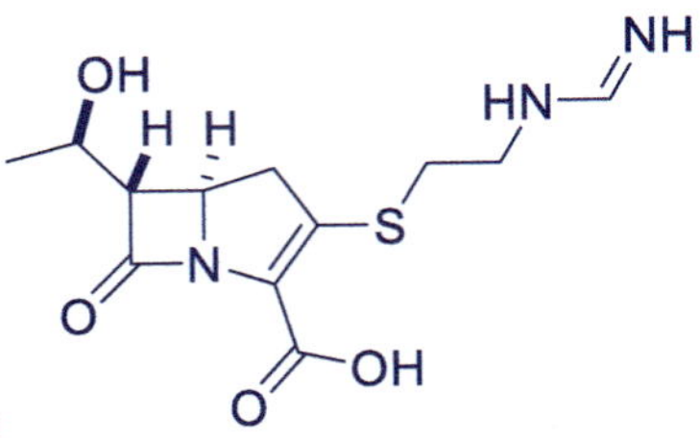

Imipenem (**2**)

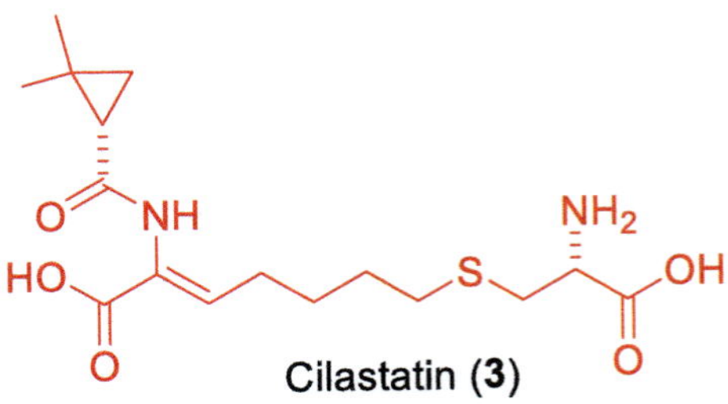

Cilastatin (**3**)

Recarbrio was approved in July 2019 as an alternative treatment option of adults with complicated urinary tract infections (cUTI), including pyelonephritis, and complicated intra-abdominal infections (cIAI) caused by susceptible gram-negative bacteria.[8,9] Recarbrio is a three-drug combination injection containing imipenem, cilastatin, and relebactam (**1**). Imipenem is a carbapenem that inhibits cross-linking of peptidoglycan during cell wall synthesis by deactivating penicillin binding proteins. It is co-administered with cilastatin, a dehydro-peptidase-I inhibitor to reduce renal metabolism of imipenem. Cilastatin itself does not have antibacterial activity. Relebactam (**1**) is a novel β-lactamase inhibitor. It alone has no antibacterial activity either. Its function is to protect imipenem by inhibiting Ambler class A (e.g., KPCs), class C (e.g., AmpC) β-lactamases, and PDC. *In vitro*, the addition of relebactam (**1**) significantly improves the antibacterial activity of imipenem by lowering the minimum inhibitory concentration of imipenem by 2- to 128-folds against ESBL or KPC producing *enterobacterales*, as well as MDR or imipenem-resistant isolates.[10]

In June 2020, FDA further approved a supplemental new drug application (sNDA) for Recarbrio for the treatment of patients 18 years of age and older with hospital-acquired bacterial pneumonia and ventilator-associated bacterial pneumonia (HABP/VABP), caused by a group of susceptible gram-negative microorganism.[9]

It worth mentioning that relebactam (**1**) is inactive against class B metallo-β-lactamases (e.g., NDM, VIM, and IMP) and class D oxacillinases (e.g., OXA-48). This leaves space for further development of novel BLIs with expanded coverage of class B metallo-β-lactamase and class D β-lactamase to secure efficacy of new antibiotics.

2. Pharmacology

Imipenem (trade name Primaxin) is an intravenous β-lactam antibiotic discovered by Merck scientists Burton Christensen, William Leanza, and Kenneth Wildonger in the mid-1970s.[10] As a carbapenem, it is highly resistant to the β-lactamases produced by MDR gram-negative bacteria. It inhibits cross-linking of peptidoglycan during cell wall synthesis by deactivating penicillin binding proteins, thereby causing bacterial cell lysis and death.[11] Since it rapidly degraded by the renal enzyme dehydropeptidase-I when administered alone, imipenem is always co-administered with cilastatin, a dehydropeptidase-I inhibitor to reduce renal metabolism. Ever since its approval, it played a key role in treating infections caused by susceptible strains when other antibiotics failed. However, since more and more bacteria developed drug-resistance *via* production of β-lactamase, imipenem became less effective in some patients. It is necessary to invent new β-lactamase inhibitors as additive to restore the antibacterial activity of imipenem.

Relebactam (MK-7655) is a novel β-lactamase inhibitor discovered by Merck scientists as part of a drug discovery program aimed at novel BLI in 2008. Over several lead series, a bridged class showed broad spectrum of activities. Its cyclic urea can open and bind covalently to an active site serine within Ambler class A, C, and D β-lactamases. To be specific, the constrained five-membered urea bridge facilitates acylation reaction between the C-7 carbonyl and a serine residue within the β-lactamase active site. Modeling studies suggested that the *N,O*-oxysulfate group can further stabilize the ring-opened acyl-β-lactamase intermediate *via* hydrogen-bond formation with neighboring catalytic site residues. As a result, the covalent bond blocks the active site of the β-lactamase, stops hydrolysis of imipenem, and restores its bactericidal activity.[12]

So far, it is active in inhibiting Ambler class A (e.g., KPCs) and class C (e.g., AmpC) actamases and PDC, but inactive against class B metallo-β-lactamases (e.g., NDM, VIM and IMP) and class D oxacillinases (e.g., OXA-48).[13] *In vitro*, the addition of relebactam (**1**) significantly improves the antibacterial activity of imipenem by lowering the minimum inhibitory concentration of imipenem by 2- to 128-folds against ESBL or KPC producing *enterobacterales*, or imipenem-resistant isolates. Neither imipenem nor relebactam is subject to efflux, which is an advantage against strains that overexpress efflux pumps.

3. Structure–Activity Relationship (SAR)

The lead compound for the relebactam (**1**) project has two sources: one is MK-8712, a monobactam β-lactamase inhibitor only effective against class C β-lactamase; the other is avibactam sodium (NXL-104), a covalent and reversible non-β-lactam β-lactamase inhibitor to β-lactamase TEM-1 and CTX-M-15.[14] In order to expand the coverage of novel BLI to both class C (PDC) *P. aeruginosa* and class A (KPC) *enterobacterales* class A and class C β-lactamases, Merck scientists took a hybrid approach by incorporating novel heterocyclic amide side chains for MK-8712 with the bridged core of NXL-104. After overcoming many synthetic difficulties, a series of bridged bicyclic urea with basic heterocyclic side chains have been prepared for SAR study.

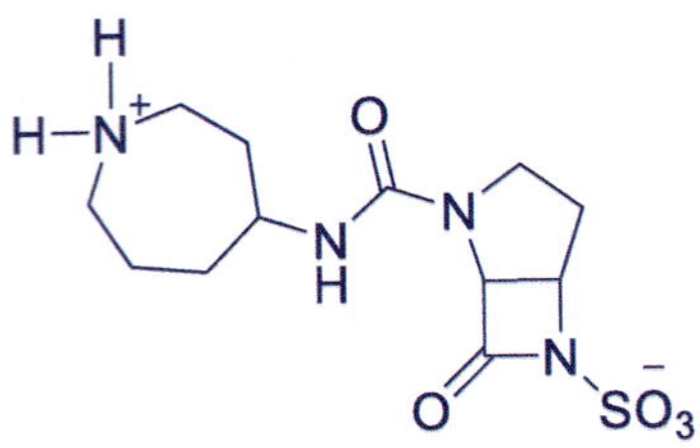

MK-8712 (**4**)

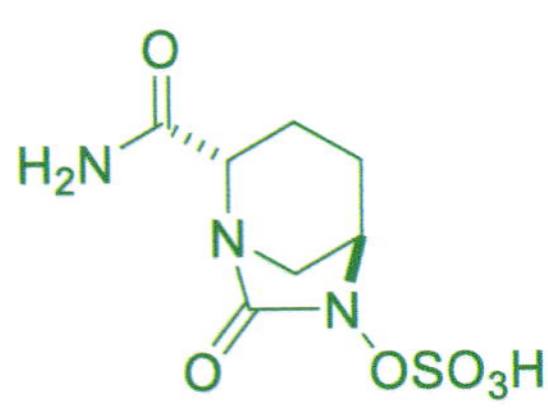

NXL-104 (**5**)

As shown in Table 1, selected compounds were evaluated in enzyme inhibition assays and *in vitro* synergy assays.[12] The enzyme inhibition assays measured each compound's ability to inhibit the hydrolysis of nitrocefin by four β-lactamase: one class A BL, two class C BLs and one class D BL. The synergy data reported the concentration of each compound to reduce the MIC of imipenem to 4 μg/mL against the strains of *Pseudomonas*, *Klebsiella*, and *Acinetobacter*.

Let us discuss the result from the piperidine analog **1**, which was later developed as Merck's clinical compound relebactam (**1**). From enzyme inhibition assay we can see it inhibits KPC-2, which is the key class A BL responsible for carbapenem resistance in *Klebsiella pneumoniae*, as well as AmpC, the BL responsible for the resistance in *Pseudomonas*. However, it shows weak inhibition against AmpC, and no effect against the class D enzyme Oxa-40. Fortunately, the latter two enzymes account for very little portion of clinical cases. It is delighting to see analog **1** also effectively synergized imipenem against *Klebsiella* and *Pseudomonas*, with concentrations of 12.5 and 4.7 μM, to reduce the imipenem MIC to 4 μg/mL. Just like the piperidine analog **1**, the seven membered azepine analog **6** and five membered pyrrolidine analog **7** all showed similar activities in the enzyme and synergy assays. Alkylation of the piperidine nitrogen as in analog **8** has nearly no effect on enzyme activity and synergy. From the X-ray crystal structure (Figure 1) of **1** bound in the active site of AmpC (*Pseudomonas*), it can be easily seen that hydrogen bonding between piperidine free N and backbone of AmpC is a key interaction. Other than that, the amide and the *N,O*-oxysulfate are also important for potency and should be considered for future target design.

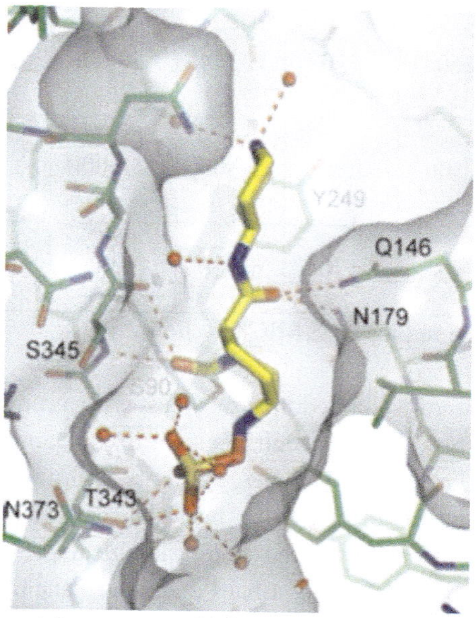

Figure 1. X-ray crystal structure of **1** in AmpC from *Pseudomonas*.

Table 1. SAR of bridged bicyclic urea β-lactamase inhibitors

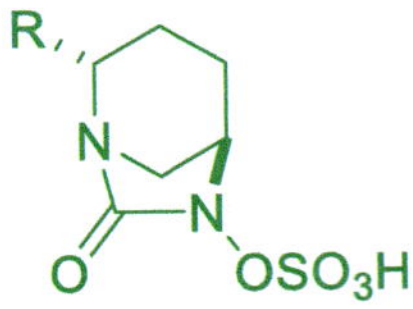

R	Enzyme inhibition (IC_{50}, nM)				*In vitro* synergy (BLi μM)[a]		
	KPC-2 *K.pneum*	AmpC *P.aerug*	AmpC *A.baum*	Oxa40 *A.baum*	CL6569 *K.pneum*	CL5701 *P.aerug*	CL6188 *A.baum*
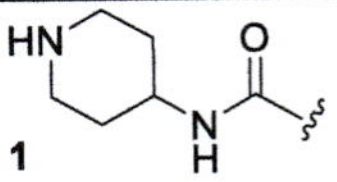 1	210	465	4,100	>50,000	12.5	4.7	>100
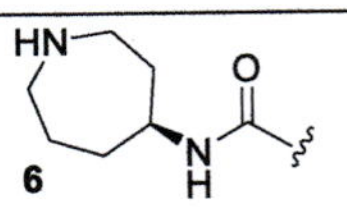 6	245	69	1,700	>50,000	12.5	6.25	>100
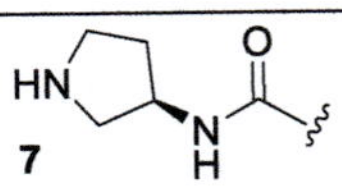 7	260	1,900	27,000	>50,000	6.25	6.25	>100
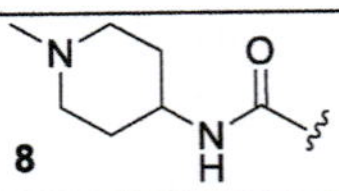 8	240	500	2,700	>50,000	12.5	12.5	>100
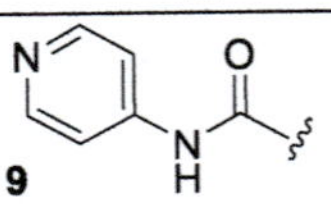 9	54	6	1,050	20,000	50	25	>100
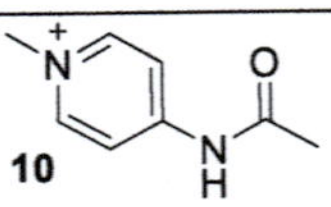 10	260	210	12,000	>50,000	25	25	>100
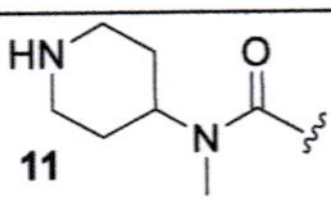 11	2,500	290	2,100	>50,000	50	25	>100
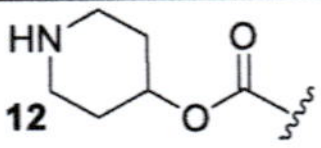 12	9	170	1,100	13,000	6.25	6.25	>100

[a] BLI (μM) to reduce IPM MIC to 4 μg/mL.

Replacement of saturated heterocycles with aromatic heterocycles was also considered. Results shown obvious enhancement in β-lactamase enzyme inhibition (as in analog **9**), but a substantial loss in synergy activity with imipenem is disappointing. It is mainly due to increased efflux from bacterial cells as the nitrogen in aromatic heterocycles would not be protonated at physiological pH. For comparison, a positive charge was introduced by preparing a hard-quaternary analog **10**. It is not surprised that the compound is substantially less reactive than corresponding aromatic analog **9** in enzyme inhibition. It can be concluded that saturated heterocycles are the best with considering both enzyme inhibition and *in vitro* synergy.

As for the amide linker, methylation on the nitrogen was prepared, and it caused substantial loss in enzyme inhibition as shown in analog **11**. Replacement of the amide linker with an ester was also considered, although analog **12** showed comparable enzyme inhibition and similar activity in the synergy assay, further stability study and efflux studies proved it inferior to analog **1**.

In general, most heterocyclic side chains were tolerated and the resulting compounds had activity against both class A and class C β-lactamases. No inhibition against class D was observed. After balancing all pros and cons, not only these assay data, but also efflux and comprehensive ADMET properties, analog **1** is superior to others and selected for further development.

4. Pharmacokinetics and Drug Metabolism

The pharmacokinetic properties of imipenem/cilastatin are not affected by relebactam. C_{max} and AUC of imipenem, cilastatin, and relebactam (**1**) are all increased proportionally with dose. After administered by intravenous infusion over 30 min every 6 hour at a dose of 1,250 mg (imipenem/cilastatin 500/500 mg plus relebactam (**1**) 250 mg) every 6 hour, steady-state C_{max} of imipenem and relebactam reached 88.9 μM and 58.5 μM, respectively, with AUC from 0 to 24 hour 500 μM·h for imipenem and 390.5 μM·h for relebactam (**1**). Their PPBs are 20% for imipenem, 40% for cialastin, and 20% for relebactam (**1**). Their V_{ds} at steady state are 24.3 L for imipenem, 13.8 L for cialastin, and 19.0 L for relebactam.

In Recarbrio, imipenem is exclusively metabolized in the kidney by dehydropeptidase-I. To decrease renal metabolism, cilastatin, a dehydropeptidase-I inhibitor, is co-administered. Although relebactam (**1**) is minimally metabolized by renal, it is mainly excreted renally. After multiple dose administrations in healthy individuals, 63% of imipenem, 77% of cilastatin, and > 90% of relebactam doses were recovered in human urine. Half-lives of imipenem/cilastatin and relebactam (**1**) are 1.0, 1.2 hour respectively. Since no ingredient is an inhibitor or inducer of the cytochrome P450 enzyme system, Recarbrio has no obvious drug–drug interactions.[15]

Sex, race, age, and weight have no impact on PK. Since it mainly excreted renally, hepatic impairment has no effect, while renal impairments have substantial influence. Exposure in patients with mild, moderate, or severe renal impairment was 1.22–2.01 times for imipenem, and 1.38–3.05 times for relebactam (**1**). Dose adjustment therefore is required.[8,9]

5. Efficacy and Safety

In Table 1, we have already seen imipenem/relebactam showed broad-spectrum *in vitro* inhibitory activity on both class A (*K. pneumoniae*) and class C (*P. aeruginosa*) β-lactamase enzymes. Actually, of over 6,000 *P. aeruginosa* isolates collected in 2016 in a surveillance study, 90% were fully susceptible at the approved FDA imipenem/REL breakpoint of 2 mg/L, while only 67% of the isolates were susceptible to imipenem without REL. Additionally, the percent imipenem and imipenem/REL susceptibility for 145 molecularly-sequenced KPC isolates was 9.9% and 98.0%, respectively, demonstrating the effectiveness of REL in restoring antibacterial activity of imipenem in KPC-producing clinical isolates.[16a]

The *in vivo* efficacy of Recarbrio was demonstrated in mouse spleen (*P. aeruginosa* and *K. pneumoniae*) and lung (*P. aeruginosa*) models of infection and afforded greater than a 3-log reduction in colony forming units (CFU) following intravenous (IV) administration. Based on these clinical studies, the pivotal phase III trial RESTORE-IMI 1 was conducted. Recarbrio was administered *via* intravenous infusion in two trials, one each for cUTI and cIAI. The cUTI trial included 298 adult patients with 99 treated with Recarbrio. The cIAI trail included 347 patients with 117 treated with Recarbrio. And later, more assessments were conducted in the phase III trials of adults (age ≥ 18 years) with Hospital-Acquired or Ventilator-Associated Bacterial Pneumonia (HABP/VABP). All treatments achieved favorable overall response. In all these clinical trials, imipenem/cilastatin/relebactam showed good tolerance in patients with cUTI, cIAI, or HABP/VABP, and their safety profiles were the same as those established for imipenem/cilastatin alone.[16]

Observed adverse effects are mainly from imipenem-cilastation (Primaxin). Since generalized seizures have been reported when ganciclovir was co-administered with imipenem/cilastatin, Recarbrio should be avoided being taken with the anticonvulsant valproic acid, divalprox sodium, or the antiviral ganciclovir, unless the potential benefits outweigh the risks. It may reduce the level of corresponding API and increase the risk of seizure in the patients taking these drugs to control seizure. In addition, there were also reported CNS adverse reactions with patients with preexisting CNS disorders and/or those with compromised renal function. For these patients, neurological evaluation should be conducted to determine before taking Recarbrio.[17]

6. Syntheses

6.1. Early Process Synthesis

The compound has two features made its synthesis difficult: one is the bicyclo[3.2.1]urea bridge, and the other is its existence as a zwitterionic salt. Due to low solubility and high reactivity of the final product, the urea bridge had to be constructed as late as possible, so did the formation of Zwitterionic salt. The evolution in its synthesis can be clearly tracked from the several generations of synthesis routes published.[18] Here we will focus on two synthetic routes from process chemists. The earlies route from medicinal chemists took up to 17 steps can be found in the patents.[18a,b]

In one earlier process chemistry route, the piperidine ring was prepared by ring expansion of the known chiral (*S*)-*N*-Bocpyroglutamic acid **13**.[18e] The lactam was first opened by sulfur ylide generated *in situ* from trimethylsulfoxonium iodide and potassium *tert*-butoxide. The ring opening intermediate **14** was carried forward without further workup. Sequential treatment of **14** with benzyloxyammonium chloride and LiCl in DMSO rendered α-chlorooxime **15**. DMSO is critical to this reaction as it can suppress undesired displacement of the chloride by residual iodide in the reaction mixture carried from trimethylsulfoxonium iodide. With no surprise, ring closure was swiftly achieved by treating the α-chlorooxime **15** with potassium *tert*-butoxide in DMF at 0 °C, and the resulting intermediate **16** was deprotected by TMSBr and *N*,*O*-bis(trimethylsilyl)-acetamide to give piperidine **17**.

The oxime in **17** existed as a mixture of (*E*)- and (*Z*)-isomers. After screening reduction conditions for C=N bond, sodium borohydride/triglyme was selected to deliver the desired product **18** with high diastereoselectivity, mainly directed by the carboxylic acid. Conversion of the hydroxylamine to sulfuric acid salt enabled isolation of crystalline form with 99.2:0.8 *dr*. Due to low solubility of sulfuric acid salt, it was converted to trifluoroacetate first, followed by amidation with amino piperidine to give amide **19**. The TFA salt was further converted to tosylate salt for easy crystallization, and compound **20** with high purity (> 99.5%) was separated as a salt.

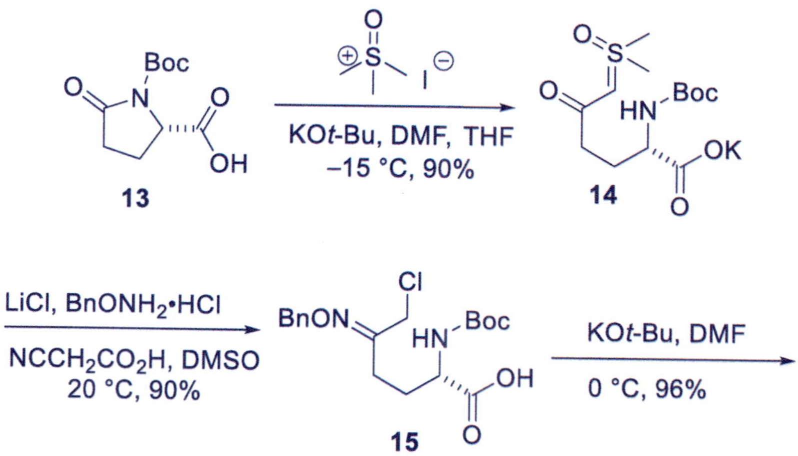

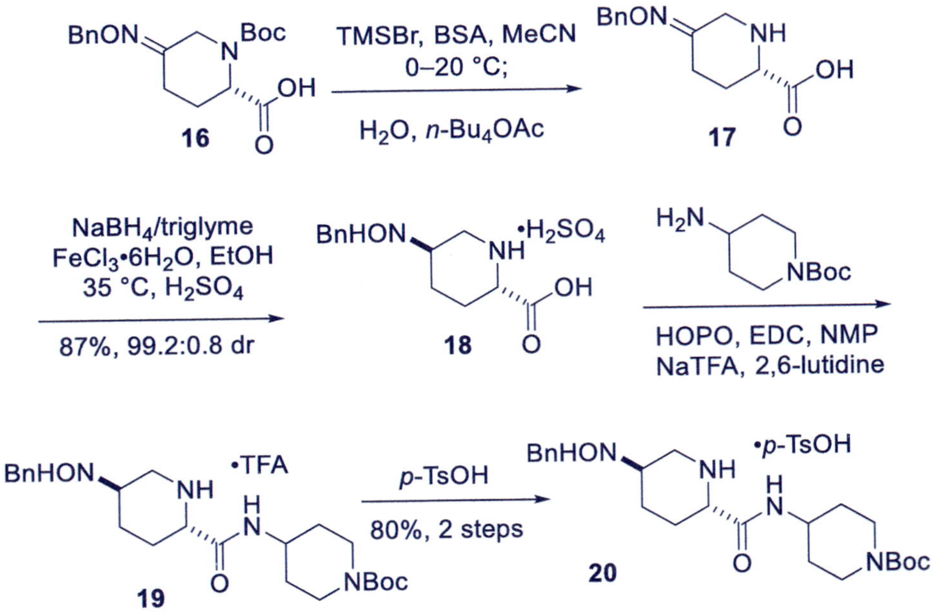

Construction of the reactive bicyclo[3.2.1]urea bridge from diamine **20** proved to be very challenging. Most carbonyl sources are added preferentially at the piperidine nitrogen, without further cyclization to make the bridged urea. After trials and errors, the cyclization finally succeeded with triphosgene in the presence of excess Hünig's base, followed by dilute aqueous phosphoric acid. Urea **21** was crystalized with 87% isolated yield. Deprotection of benzyl group in THF gave free *N*-hydroxylamine, which was sulfated with SO_3•pyridine to give desired product **22**. Finally, TMSI in acetonitrile was able to deprotect Boc and give the desired final product **1**.

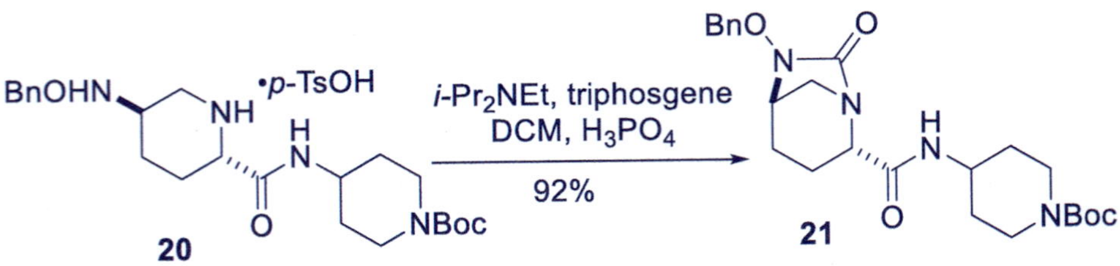

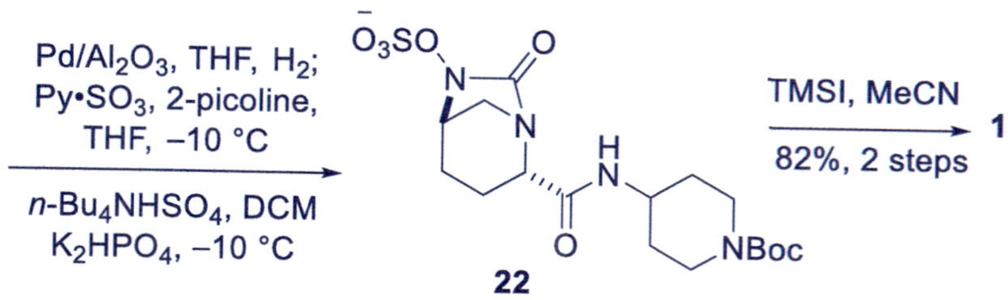

6.2 Later Process Synthesis

After further optimization, Merck process chemists reported several new routes, and the following one is the most efficient and economical for scalable multikilogram synthesis.[18f]

This route started from optically pure *cis*-5-hydroxypipecolic acid **23** available from enzymatic oxidation of pipecolic acid, avoiding more expensive earlier approach. Upon treatment with 2 equiv of 2-NsCl, one for N-protection, the other to active the carboxylic acid to facilitate cyclization to give lactone **24** in > 99% purity. The lactone was then opened readily with commercially Boc-protected aminopiperidine in THF, and the resulting intermediate was treated with another 2 equiv of 2-NsCl in one pot to give desired sulfonate ester **25** in 98% yield for 2 steps from **24**. Subsequent displacement of 2-NsO with protected –OBn hydroxylamine, followed by removal of all 2-Ns gave neutral form of compound **20** in 70% yield with > 97.5% HPLC purity. From **20**, the same chemistry as mentioned before was able to furnish the synthesis of **1**. This new route provided a more efficient way for the synthesis in only eight steps and 42% overall yield.

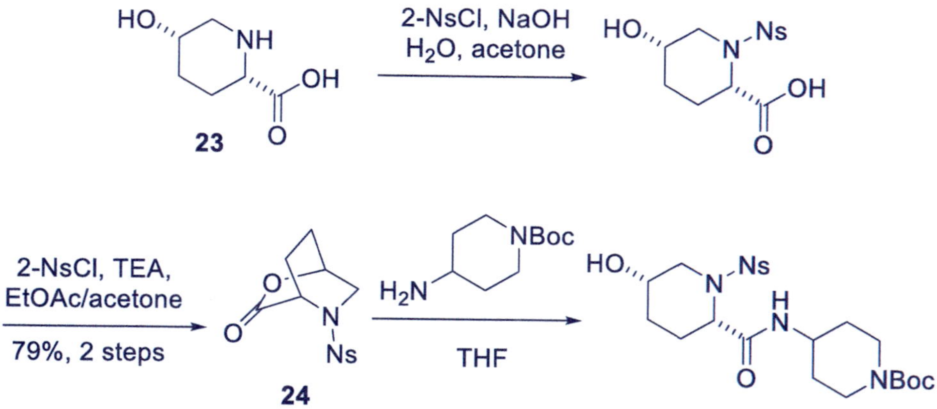

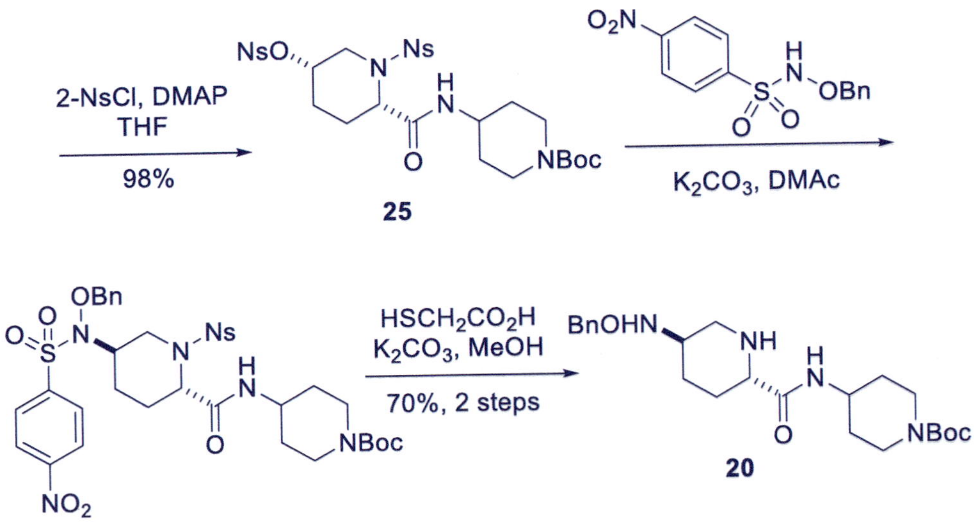

7. Summary

In summary, Recarbrio is a fixed dose combination of imipenem, cilastatin, and relebactam (**1**). It was approved by FDA in July 2019 for the treatment of adults with limited treatment options for cUTI and cIAI caused by a range of susceptible gram-negative bacteria, including but not limited to *Enterobacter cloacae, Escherichia coli, Klebsiella aerogenes*, *K. pneumoniae*, and *P. aeruginosa.* In June 2020, it was further approved to treat hospital-acquired bacterial pneumonia and ventilator-associated bacterial pneumonia (HABP/VABP) in patients 18 years of age and older. Recarbrio is administered by intravenous infusion over 30 minutes every 6 hours at a recommended dose of 1.25 g (imipenem 500 mg, cilastatin 500 mg, and relebactam 250 mg), with dose adjustments for patients with renal impairment. In addition to its pharmacology, PK, metabolism, safety, and SAR, two of its process syntheses have also been discussed in this chapter. As the last treatment options for carbapenem-resistant gram-negative infections, antimicrobial stewardship should be set up to ensure appropriate use of this new therapeutic agent.

8. References

1. Global Report on Surveillance 2014. World Health Organization; 2014. Downloaded from https://apps.who.int/iris/handle/10665/112642, last accessed on Oct 23, 2021; Global Antimicrobial Resistance and Use Surveillance System (GLASS) Report: 2021. Downloaded from https://www.who.int/publications/i/item/9789240027336, last accessed on Oct 23, 2021

2. Zhanel, G. G.; Wiebe, R.; Dilay, L.; Thomson, K.; Rubinstein, E.; Hoban, D. J.; Noreddin, A. M.; Karlowsky, J. A. *Drugs* **2007**, *7*, 1027–1052.
3. Smith, J. R.; Rybak, J. M.; Claeys, K. C. *Pharmacotherapy* **2020**, *40*, 343–356.
4. Resistance Mechanism: (a) Bush, K.; Jacoby, G. A. *Antimicrob. Agents Chemother.* **2010**, *3*, 969–976. (b) Nordmann, P.; Poirel, L. *Clin. Infect. Dis.* **2019**, *69*, 521–528.
5. Doi, Y. *Clin. Infect. Dis.* **2019**, *69*, S565–S575.
6. van Duin, D., Bonomo, R. A. *Clin. Infect. Dis.* **2016**, *2*, 234–241.
7. Hecker, S. J.; Reddy, K. R.; Totrov M.; et al. *J. Med. Chem.* **2015**, *9*, 3682–3692.
8. Merck Sharp & Dohme. Recarbrio® (imipenem, cilastatin, and relebactam): US prescribing information. 2020. https ://www.fda.gov. Accessed 24 Oct 2021.
9. Merck Sharp & Dohme. Recarbrio: EU summary of product characteristics. 2021. https ://www.ema.europa.eu. Accessed 24 Oct 2021.
10. Imipenem reviews: (a) Christensen, B. G.; Leanza, W. J.; Wildonger, K. J. US4194047. (b) Clissold, S. P.; Todd, P. A.; Campoli-Richards, D. M. *Drugs* **1987**, *33*: 183–241. (c) Vardakas, K. Z.; Tansarli, G. S.; Rafailidis, P. I.; Falagas, M. E. *J. Antimicrob. Chemother*. **2012**, *67*, 2793–2803.
11. Imipenem mechanism: (a) Hashizume, T.; Ishino, F.; Nakagawa, J.; Tamaki, S.; Matsuhashi, M. *J. Antibiot.* **1984**, *4*, 394–400. (b) Mitsuhashi, S. *J. Antimicrob. Chemother.* **1983**, *12(Suppl D)*, 53–64.
12. Blizzard, T. A.; Chen, H.; Kim, S.; et al. *Bioorg. Med. Chem. Lett.* **2014**, *3*, 780–785.
13. Canver, M. C.; Satlin, M. J.; Westblade, L. F.; et al. *Antimicrob. Agents. Chemother.* **2019**; *63*, e00672-19.
14. NXL-104: (a) Stachyra, T.; Pechereau, M. C.; Bruneau, J. M.; et al. *Antimicrob. Agents. Chemother.* **2010**, *54*, 5132–5138. (b) Lahiri, S. D.; Mangani, S.; Durand-Reville, T.; et al. *Antimicrob. Agents Chemother.* **2013**, *57*, 2496–2505.
15. PK/PD (a) Bhagunde, P.; Zhang, Z.; Racine, F.; Carr, D.; Wu, J.; Young, K.; Rizk, M. L. A *Int. J. Infect. Dis.* **2019**, *89*, 55–61. (b) Kaushik, A.; Ammerman, N. C.; Lee, J.; Martins, O.; Kreiswirth, B. N.; Lamichhane, G.; Parrish, N. M.; Nuermberger, E. L. *Antimicrob. Agents. Chemother.* **2019**, *63*, e02623-18. (c) McCarthy, M. W. *Ther. Clin. Pharmacokinet.* **2020**, *59*, 567–573.
16. Efficacy and safety: (a) Young, K.; Painter, R.; Raghoobar, S. L.; Hairston, N. N.; Racine, F.; Wisniewski, D.; Balibar, C. J.; Villafania, A.; Zhang, R.; Sahm, D. F.; Blizzard, T.; Murgolo, N.; Hammond, M. L.; Motyl, M. R. *BMC Microbiol.* **2019**, *19*, 150. (b) Sims, M.; Mariyanovski, V.; McLeroth, P.; et al. *J. Antimicrob. Chemother.* **2017**, *72*, 2616–2626. (c) Motsch, J.; Murta de Oliveira, C.; Stus, V.; et al. *Clin. Infect. Dis.* **2020**, *70*, 1799–1808.
17. Cannon, J. P.; Lee, T. A.; Clark, N. M.; Setlak, P.; Grim, S. A. *J. Antimicrob. Chemother.* **2014**, *69*, 2043–2055.
18. Synthesis: (a) Blizzard, T. A.; Chen, H.; Gude, C.; Hermes, J. D.; Imbriglio, J. E.; Kim, S.; Wu, J. Y.; Ha, S.; Mortko, C. J.; Mangion, I.; Rivera, N.; Ruck, R. T.; Shevlin, M. WO 2009091856 A2, (2009). (b) Miller, S. P.; Limanto, J.; Zhong, Y.-L.; Yasuda, N.; Liu, Z. WO 2014200786 A1, (2014). (c) Chung, J. Y. L.; Meng, D.; Shevlin, M.; Gudipati, V.; Chen, Q.; Liu, Y.; Lam, Y.-H.; Dumas,

A.; Scott, J.; Tu, Q.; Xu, F. *J. Org. Chem.* **2020**, *85*, 994−1000. (d) Liu, Z.; Yasuda, N.; Simeone, M.; Reamer, R. A. *J. Org. Chem.* **2014**, *79*, 11792−11796. (e) Mangion, I. K.; Ruck, R. T.; Rivera, N.; Huffman, M. A.; Shevlin, M. *Org. Lett.* **2011**, *13*, 5480−5483. (f) Miller, S.P., Zhong, Y., Liu, Z., Simeone, M., Yasuda, N., Limanto, J., Chen, Z., Lynch, J. E., Capodanno, V.R. *Org. Lett.* **2014**, *16*, 174−177.

2

Vaborbactam (in Combination with Meropenem as Vabomere), a Non-β-Lactam β-Lactamase Inhibitor for Treatment of Complicated Urinary Tract Infections and Pyelonephritis

Brett C. Bookser, K. Raja Reddy,
Serge H. Boyer and Scott J. Hecker

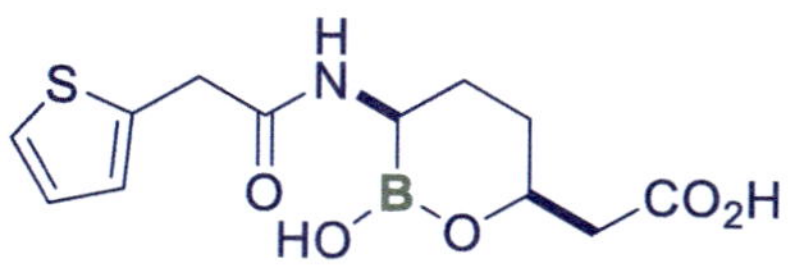

USAN: Vaborbactam **1**
Tradename: Vabomere for Varborbactam/Meropenem combination
Rempex Pharmaceuticals
Launched: 2017

1 Background: Evolution from irreversible β-lactam to irreversible non-β-lactam to reversible boron-containing non-β-lactam β-lactamase inhibitors

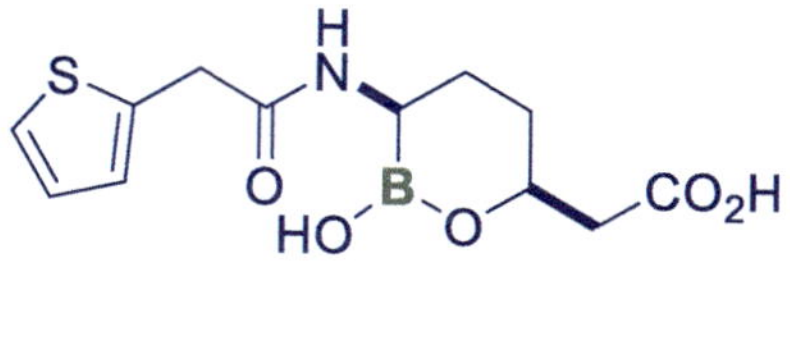

1, Vaborbactam

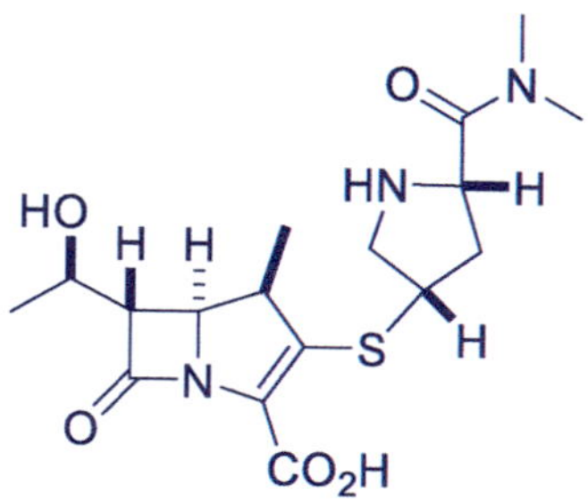

2, Meropenem

Current Drug Synthesis, First Edition. Edited by Jie Jack Li.

Vaborbactam (**1**), a non-β-lactam β-lactamase inhibitor, was approved in 2017 in combination with the carbapenem β-lactam antibiotic meropenem (**2**) as Vabomere for the treatment of complicated urinary tract infections and pyelonephritis, kidney inflammation resulting from bacterial infection.[1,2]

β-Lactam antibiotics have been a mainstay as anti-infectives since the discovery of penicillin in 1928 by Fleming and subsequent use of penicillin F in 1943 for the treatment of streptococcal meningitis.[3] After the ensuing decades of constant application of the drugs, bacteria evolved serine β-lactamases to hydrolyze the β-lactam of the drug and render it ineffective. In the 1980s and 1990s a new strategy developed in the Pharma industry to couple a β-lactamase inhibitor with the active antibiotic to protect the drug from this metabolism.[4] Notable such combinations include clavulanic acid (**3**) as the β-lactamase inhibitor in combination with amoxicillin (**4**, 1984); followed by sulbactam (**5**)/ampicillin (**6**, 1987), and tazobactam (**7**)/piperacillin (**8**, 1993). These compounds **3**, **5**, and **7**, each contained a β-lactam and functioned by out-competing the drug to form a covalent bond with the serine β-lactamase.

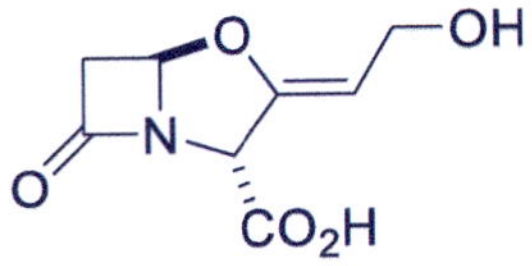

3, Clavulanic acid

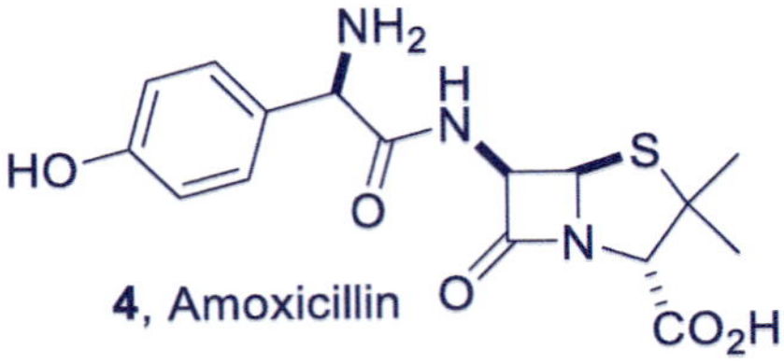

4, Amoxicillin

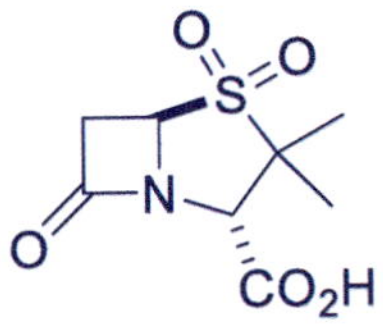

5, Sulbactam

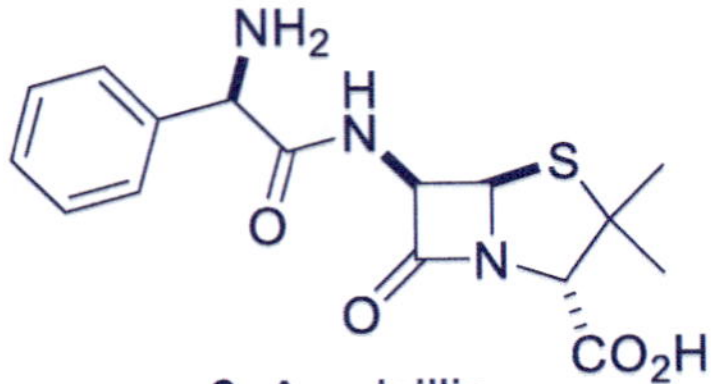

6, Ampicillin

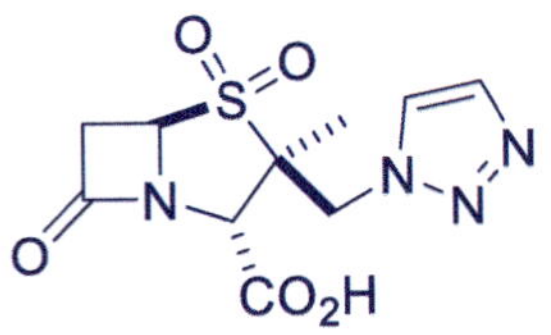

7, Tazobactam

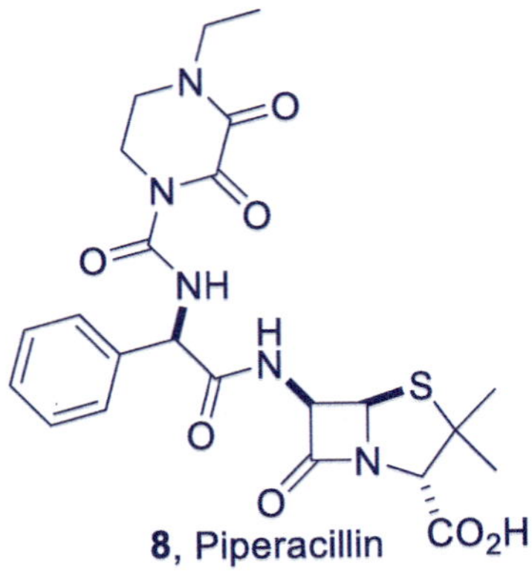

8, Piperacillin

While this strategy worked to protect the cephalosporin class of β-lactams for some time, eventually the bacteria evolved additional β-lactamases that evaded the inhibitors **3**, **5** and **7**. Notably, the antibiotics reserved for multidrug resistant (MDR) gram-negative bacterial (GNB) infection, the carbapenems, were not well protected, particularly since they are susceptible to the *Klebsiella pneumoniae* carbapenemase (KPC) type of β-lactamase. MDR infections from carbapenem-resistant *Enterobacteriaceae* (CRE) were recognized by the WHO and other public health authorities as a serious public health threat.[5] Hence, more recent efforts resulted in another class of β-lactamase inhibitors which have overcome some of these deficiencies: the diazabicyclooctanes (DBOs), avibactam (**9**) and relebactam (**10**). These compounds function similarly to the first-generation compounds, **3**, **5**, and **7**, by terminating the β-lactamase function with covalent bond formation between the β-lactamase serine and the urea carbonyl of the DBO inhibitor. These were approved when administered with ceftazidime (**11**, 2015) and imipenem (**12**)/cilastatin (**13**, 2019), respectively.[4]

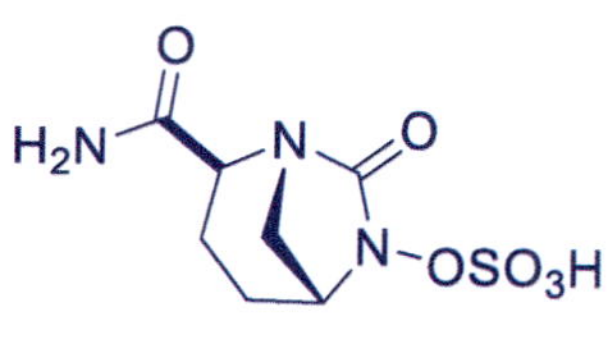

9, Avibactam

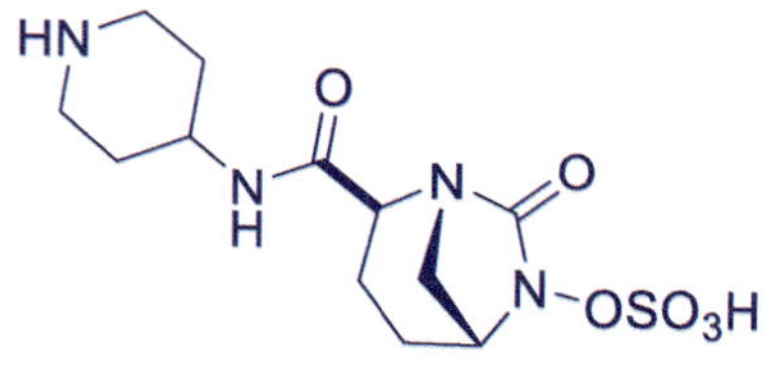

10, Relebactam

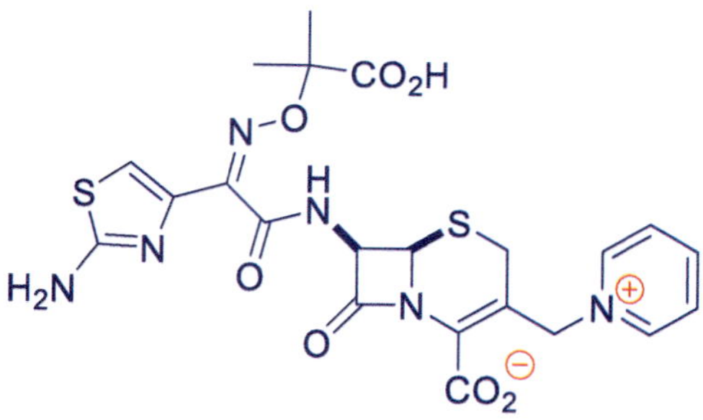

11, Ceftazidime

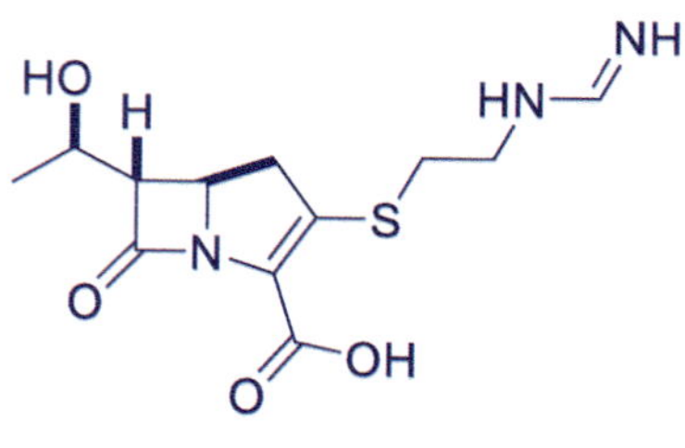

12, Imipenem

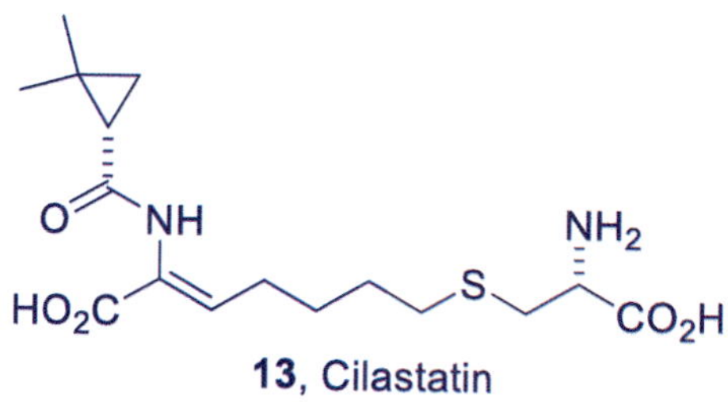

13, Cilastatin

Vaborbactam (**1**) is the first boron-containing β-lactamase inhibitor. This molecule joins a group of other boron-containing drugs and investigational compounds, see **14**–**20**,[6–12] which benefit from the sp^2/sp^3 hybridization interconversions possible at the boron atom. In contrast to the irreversible β-lactamase inhibitors from earlier strategies (see **3**, **5**, **7**, **9** and **10**), the use of this boron-hybridization flexibility allows these types of molecules to reversibly form a covalent bond with and inhibit hydrolase enzymes as transition state mimic inhibitors, see Figure 1.[13,14] Representative of the continued value placed on this boron strategy are the three more recent members of this group, **18**–**20**,[10–12] which are in development as β-lactamase inhibitors with active carbapenem antibiotics.

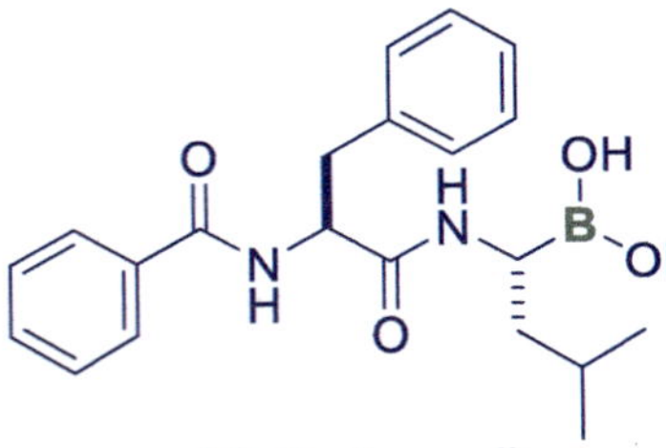

14, Bortezomib
2003 Anticancer

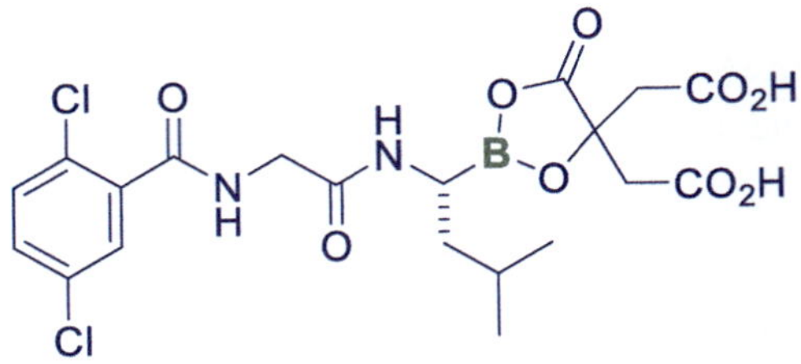

15, Ixazomib citrate
2015 for multiple myeloma

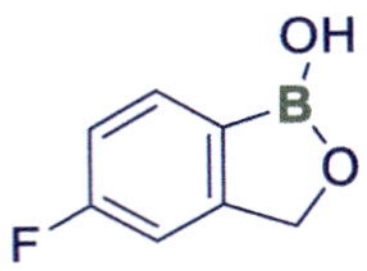

16, Tavaborole
2014 for fungal infections

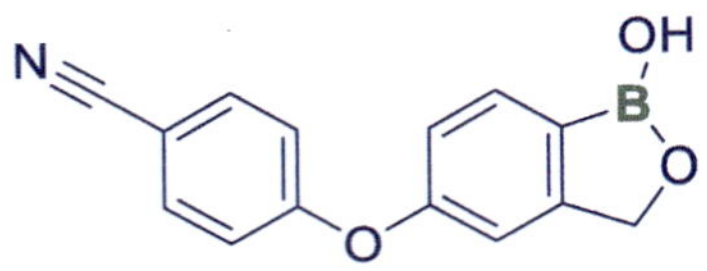

17, Crisaborole
2016 for atopic dermatitis

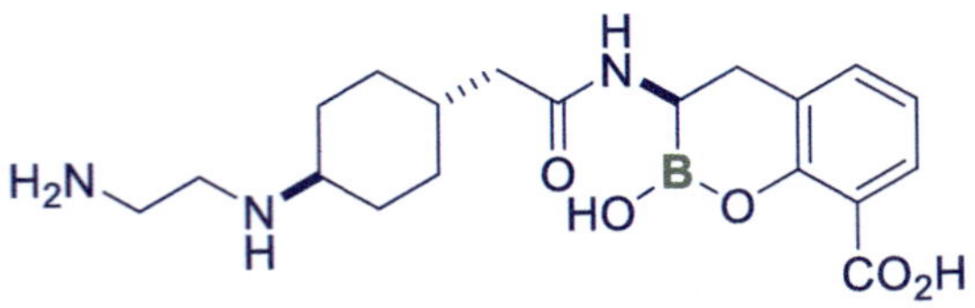

18, Taniborbactam
2019 PIII clinical trials
urinary tract infections

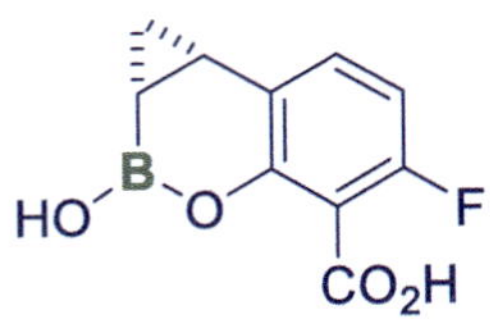

19, Xeruborbactam (**QPX7728**)
2021 PI clinical trials
urinary tract infections

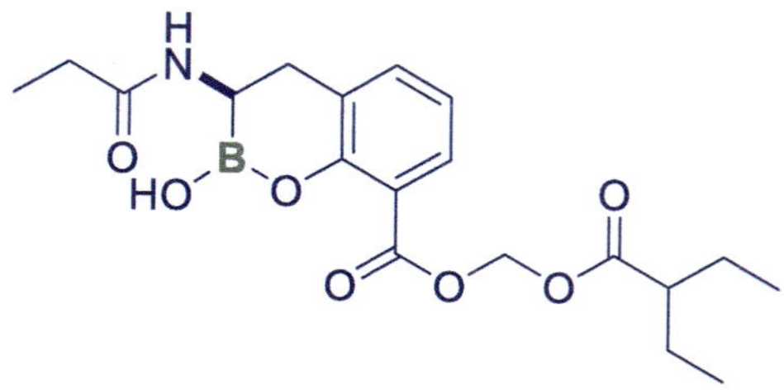

20, Ledaborbactam etzadroxil (**VNRX-7145/VNRX-5236**)
2021 PI Clinical Trials, urinary tract infections

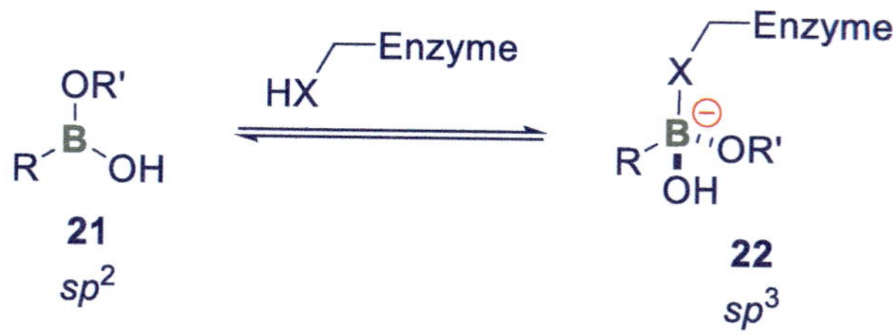

Hydrolysis transition state mimic

Figure 1. Mechanism of boronate transition state mimetic covalent-reversible inhibition.

2 Discovery Medicinal Chemistry

With no effective β-lactamase inhibitor combinations with carbapenems yet available, new research into novel molecule β-lactamase inhibitors was warranted. This resulted in consideration of the early testing of boronic acids as β-lactamase inhibitors. This work originated in academic labs with compounds which included **23**,[15] **24**,[16,17] **25**,[18] and **26**.[19] Acknowledging back to the discovery that borate ion inhibited β-lactamase I,[20] phenylboronic acid (**23**) was determined to inhibit it as well. Publication of the X-ray crystal structure of the class A RTEM-1 β-lactamase[21] inspired the sequential rational design of inhibitors **24** and **25** which had resulting IC_{50} values of 110 and 13 nM, respectively. The latter compound **25** revealed an important potency boost realized by the polar phenolic OH group. Further structure-based design revealed the importance of an additional hydrophobic residue, the thiophene ring of compound **26**. Yet despite these discoveries of potent β-lactamase inhibitors, there were no publications detailing testing of these compounds in an animal infection model. Thus, the road was clear to investigate if this strategy would work to inhibit the problematic serine carbapenemases of CRE and hence protect an anti-infective carbapenem from degradation, enabling efficacy.[22]

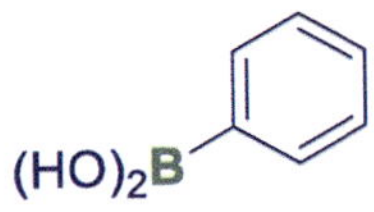

23

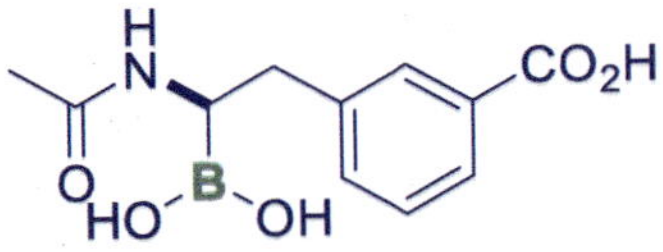

24

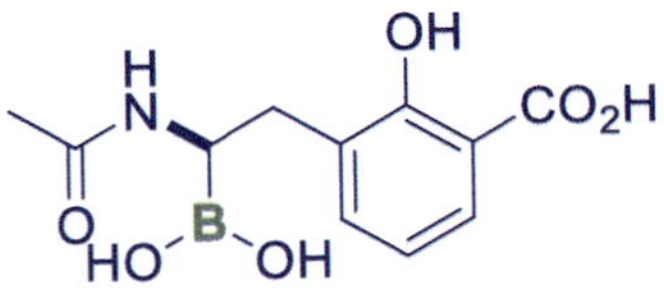

25

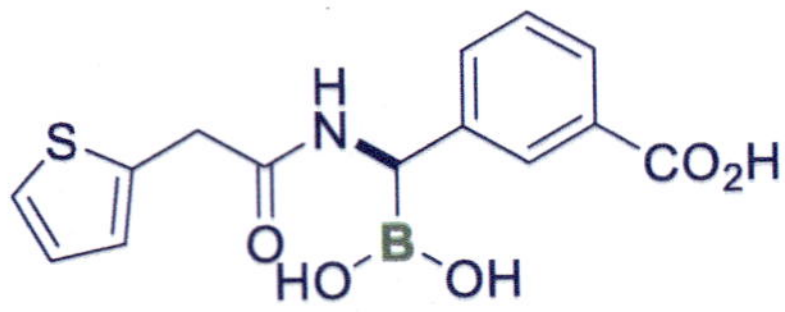

26

While the report describing **25** considered the equilibrium to the cyclic derivative **27** to be potentially important to inhibition of β-lactamase TEM-1, it was considered less likely when models indicated the acyclic version **25** seemed to fit better.

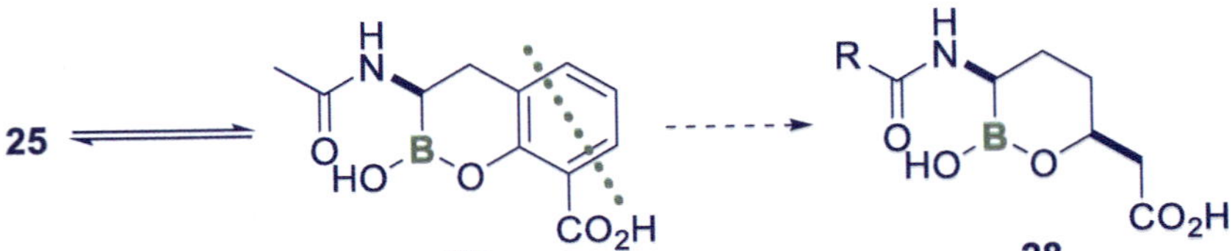

27 28

Figure 2. Equilibrium inspires cyclic active structure design.

While seeking to be selective away from serine hydrolases that favored linear substrates with their smaller binding cavities, research towards vaborbactam (**1**) was directed towards cyclic structures such as **28** in order to specifically target β-lactamases, see Figure 2. Modeling of the cyclic boronate **28** into the active site of AmpC (S64G mutant) revealed all key interactions could be obtained with pre-covalent and covalent bound forms of the prospective compound, see Figure 3.[2]

Early analogues of the compound **28** series were tested in their ability to potentiate the activity of biapenem against a KPC-producing strain of *Klebsiella pneumoniae* (KP1004) as shown in Table 1. Potency is expressed as MPC1, defined as the minimum concentration of the β-lactamase inhibitor required to reduce the biapenem MIC from 32 μg/mL (no inhibitor) to 1 μg/mL. The 2-thienylacetyl analogue **28e** (reminiscent of marketed cephalosporins cephalothin and cefoxitin) proved remarkably potent with MPC1 = 0.02 μg/mL. Few other analogues in this series approached that potency and **28e** was then designated as **RPX7009** (vaborbactam, **1**).[2]

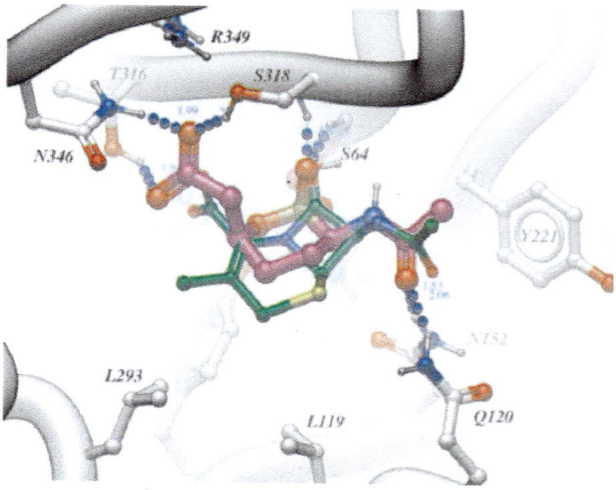

Figure 3. Model of **28** bound at the active site of class C β-lactamase (magenta). X-ray structure of cephalothin (core portion, green) bound to AmpC (enzyme is superimposed) is shown in green. Blue balls represent hydrogen bonds. (Adapted with permission from Reference 2. Copyright 2015 American Chemical Society)

Table 1. Potentiation of Biapenem by compounds **28**.

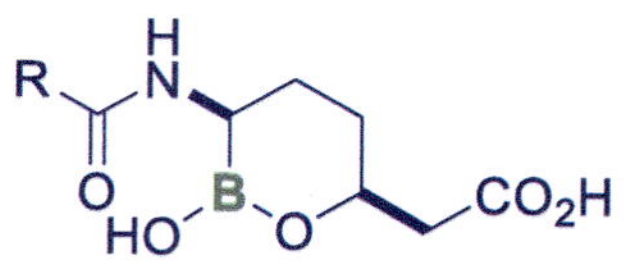

28

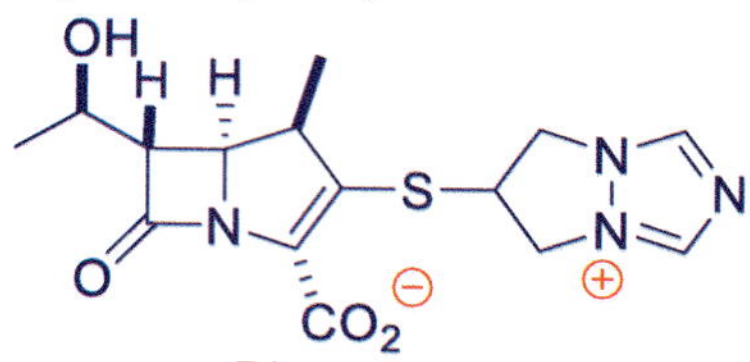

Biapenem

Cmpd	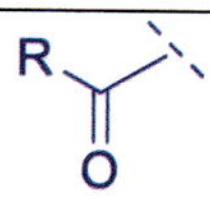	MPC_1 (μg/mL)
28a	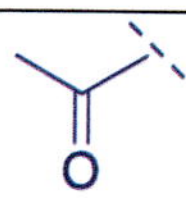	0.30
28b	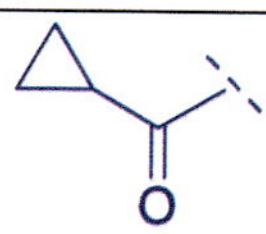	0.15
28c	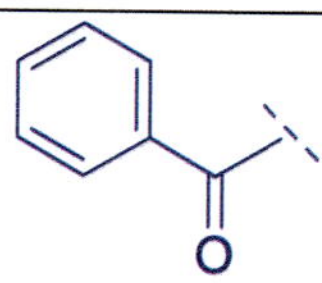	0.30

28d	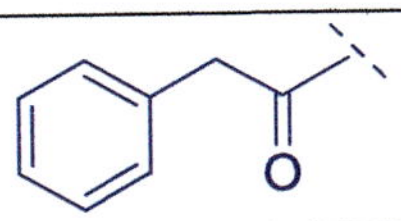	0.15
28e	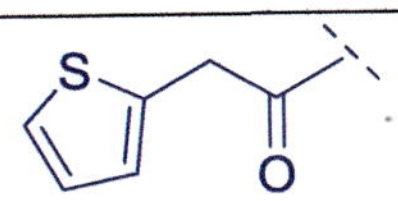	0.020
28f	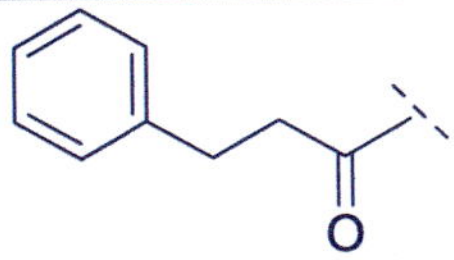	0.30

RPX7009 (vaborbactam, **1**) is considerably more potent than FDA-approved β-lactamase drugs clavulanic acid (**3**) and tazobactam (**7**) against class A carbapenemase KPC-2 as well as class C enzymes P99 and CMY-2.[2] **RPX7009** (vaborbactam, **1**) potentiated the effects of cefepime in whole cell models demonstrating enhanced protection against clinical strains with serine β-lactamases from classes A, C, and D, see Table 2. Importantly, this compound potentiated the protective effect of the carbapenems biapenem, meropenem, ertapenem and imipenem in clinical isolates of *Eterobacteriaceae* expressing class A carbapenemases KPCs-2, and -3; SHVs-1, -11, -30; TEM-1; OXA-2; as well as CTX-M-15.[2] Lastly, in a demonstration of selectivity, **RPX7009** (vaborbactam, **1**) was completely inactive ($IC_{50} \geq 1000$ μM) against eleven common mammalian serine proteases.[2]

Table 2. Potentiation of Cefepime by **RPX7009** (vaborbactam, **1**).

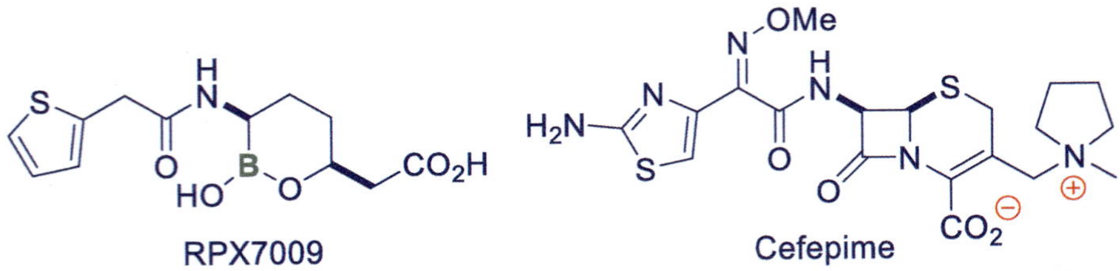

organism	strain	enzyme	Cefepime MIC (μg/mL)	Cefepime MIC (w/**RPX7009** 4 μg/mL)
Escherichia coli	EC1008	CTX-M-3	>64	4
Klebsiella pneumoniae	KP1005	CTX-M-14	64	4
Klebsiella pneumoniae	KP1009	CTX-M-15	>64	2

Klebsiella pneumoniae	KP1011	SHV-5	64	0.25
Klebsiella pneumoniae	KP1010	SHV-12	2	0.25
Escherichia coli	EC1009	TEM-10	8	4
Escherichia coli	EC1011	TEM-26	8	2
Enterobacter cloacae	ECL1003	TEM, SHV	32	8
Enterobacter cloacae	ECL1002	Hyper AmpC	16	0.5
Enterobacter cloacae	ECL1061	Hyper AmpC, KPC-3	>64	2
Enterobacter cloacae	EC1010	CMY-6	>64	4
Klebsiella oxytoca	KX1001	OXA-2	4	0.5
Enterobacter aerogenes	EA1028	OXA-30	>64	0.5

Crystallographic structures of **RPX7009** bound to class A (CTX-M-15) and class C (AmpC) lactamases reveals two chair flipped orientations for the ligand respective to the enzyme, see Figures 4 and 5. This molecular flexibility provides reasoning for why the compound can so effectively inhibit a broad range of these bacterial enzymes. Consistent with the covalent nature of the inhibition, in each enzyme, the key active site serine forms a bond to the boron atom of the inhibitor.

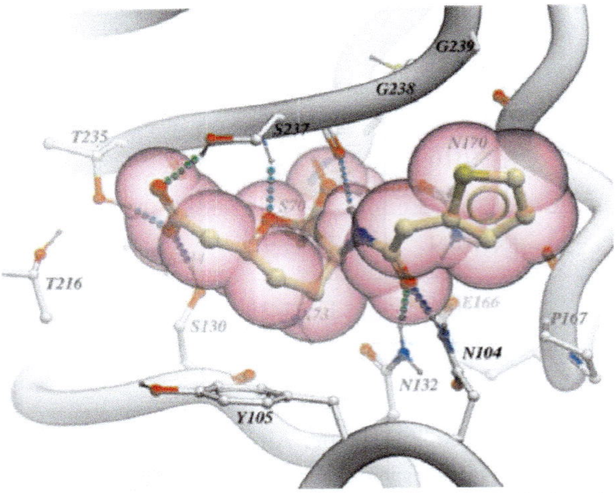

Figure 4. **RPX7009** bound to CTX-M-15. (Adapted with permission from Reference 2. Copyright 2015 American Chemical Society)

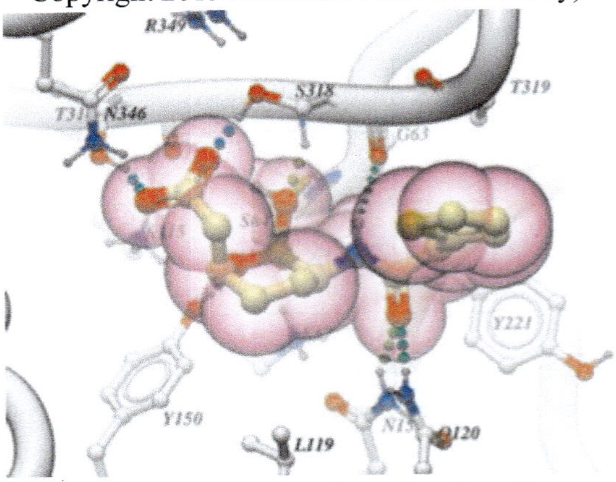

Figure 5. **RPX7009** bound to AmpC.
(Adapted with permission from Reference 2. Copyright 2015 ACS)

Proof of concept *in vivo* model.

The pharmacokinetics of **RPX7009** in rat at multiple doses were similar to those of most β-lactam antibiotics. At 100 mpk intravenous infusion, it had a high C_{max} (231 mg/L), and $AUC_{0\text{-}\infty}$ (64 h•mg/L), a short half-life (0.42 h) and a low volume of distribution (0.97 L/kg).[2] The compound **RPX7009** (vaborbactam, **1**) enhanced the effects of biapenem and meropenem against a KPC-producing strain of *Klebsiella pneumoniae* in a neutropenic mouse lung infection model whereas the carbapenems alone were ineffective (Figure 6).[2]

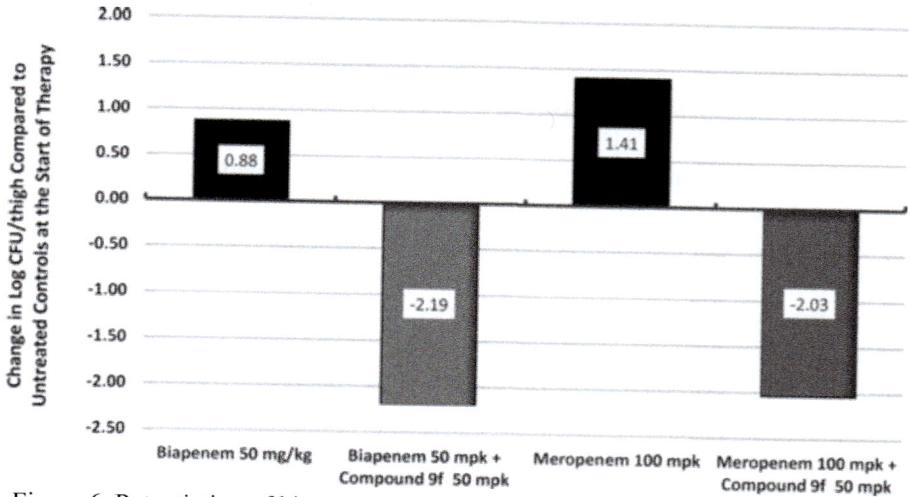

Figure 6. Potentiation of biapenem and meropenem activity by **RPX7009** against a carbapenem-resistant strain of *K. pneeumoniae* in the neutropenic mouse lung infection model. (Adapted with permission from Reference 2. Copyright 2015 American Chemical Society)

3 Vaborbactam/Vabomere Clinical Trials

Safety studies in rats at doses up to 1000 mg kg^{-1} day^{-1} with **RPX7009** (**1**) revealed no discernable toxicity in the standard battery of safety pharmacology, repeat-dose toxicology, genotoxicity, and reproductive and developmental toxicity studies. The compound was designated vaborbactam (**1**) as it entered human clinical trials in combination with meropenem.

The human clinical trials of meropenem/vaborbactam for the *i.v.* treatment of complicated urinary tract infections (cUTI) including pyelonephritis caused by designated susceptible bacteria including *Escherichia coli, Klebsiella pneumoniae* and *Enterobacter cloacae* species have been summarized.[23] Of particular importance is the ability of vaborbactam (**1**) to inhibit the *Klebsiella pneumoniae* carbapenemases (KPCs). The ability to clinically inhibit KPCs was critical and thus extended the treatment capability of the carbapenem. Phase I single dose and multiple dose studies up to 2000 mg/dose vaborbactam (**1**) by *i.v.* infusion every 8 h for 7 days showed no differences in safety findings in clinical laboratory tests (hematology, biochemistry, coagulation, and urinalysis), vital signs, or ECG in patients receiving vaborbactam (**1**) *vs.* placebo. In the Phase III clinical trial TANGO I, administration of meropenem/vaborbactam (2000 mg each) by *i.v.* infusion every 8 h for 10 days was compared to dosing with piperacillin/tazobactam (4000 mg/500 mg) in the same manner to a total of 550 patients that were diagnosed with cUTI (41%) or acute pyelonephritis (59%). The primary outcome was clinical cure and microbial eradication composite in the microbiologic modified intention-to-treat population. The primary endpoint was achieved for 98.4% of patients receiving meropenem/vaborbactam and 94% of patients receiving piperacillin/tazobactam.

A second randomized, open-label Phase III clinical trial, TANGO II, was investigated in patients with cUTI, acute pyelonephritis, hospital-acquired or ventilator-associated bacterial pneumonia, bacteremia, or complicated intra-abdominal infection due to known or suspected carbapenem-resistant Enterobacteriaceae. This trial was terminated early due to the superior performance of meropenem/vaborbactam compared to the best available treatment (BAT). For the microbiological carbapenem-resistant Enterobacteriaceae modified intention-to-treat population, it had higher cure rates (64% *vs.* 33% BAT), fewer treatment-related adverse events (24% *vs.* 44% BAT), and lower all-cause mortality (18% *vs.* 33% BAT). Based on findings of safety and efficacy, meropenem/vaborbactam was approved by the FDA in 2017 under the trade name Vabomere. It is indicated for the treatment of complicated urinary tract infections including pyelonephritis caused by designated susceptible bacteria.

4 Vaborbactam Medicinal Chemistry Synthesis

The medicinal chemistry synthesis of vaborbactam (**1**) was accomplished in seven steps and 31% overall yield, see Scheme 1. The synthesis begins with the known chiral building block **29**.[24,25] After alcohol protection with TBS, $[Ir(COD)Cl]_2$[26]-catalyzed regioselective hydroboration with pinacolborane provided intermediate **39** which was converted to the more stable pinanediol boronate to provide **32**. This group helped provide stereocontrol during the chloromethyl-group installation to **33** using the Matteson reaction.[27,28] Though the Matteson reaction required –95 °C, and was conducted at scales up to 30 kg, those cryogenic conditions were not suitable for commercial quantities of material and would be addressed later with flow or continuous processing (CP) chemistry during the process chemistry campaign. This compound contained a diastereomeric impurity at the chloro-center as 15% of the product mixture. The impurity was removed at the final step. Compound **33** was aminated with LiHMDS[29] and the resulting amine acylated with 2-thienylacetic acid **34** to provide amide **35**. This compound was conveniently deprotected and converted to the cyclic boronate **1** on treatment with 3 N HCl in dioxane. Recrystallization from EtOAc/water provided isomerically pure product.

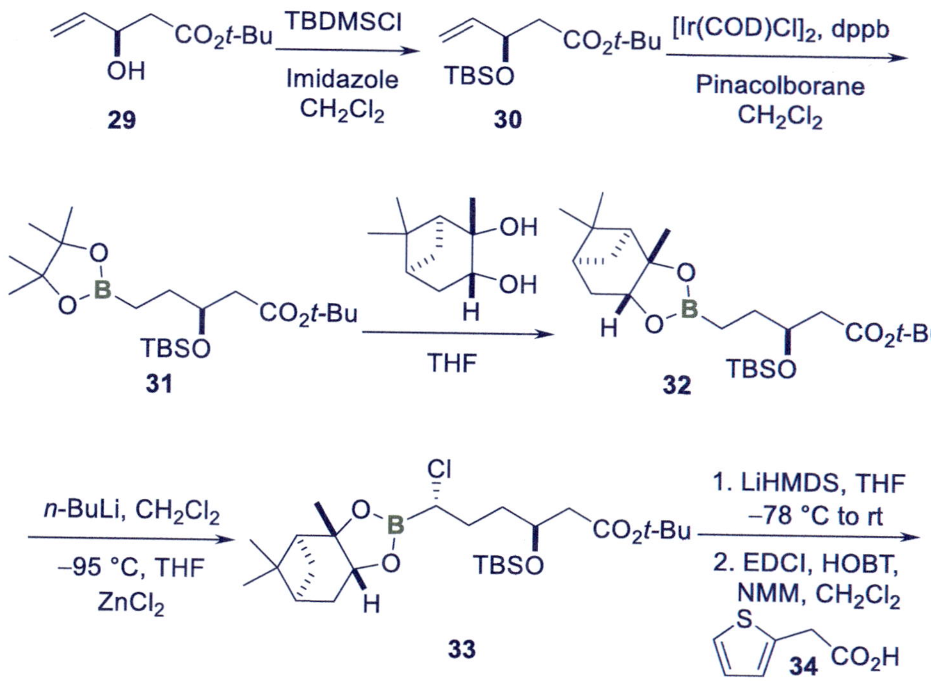

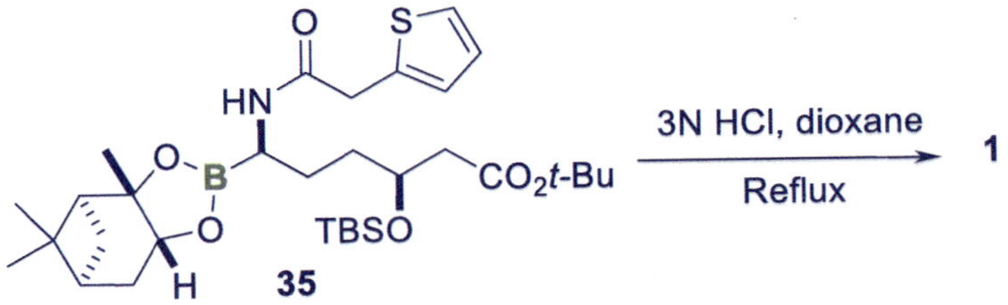

Scheme 1. Medicinal Chemistry Synthesis of Vaborbactam (**1**).

5 Vaborbactam Process Chemistry Synthesis

The discovery of a boron-containing small molecule drug presented unique challenges and opportunities for synthesis optimization on a process chemistry scale. In this case, unique boron chemistry was used to facilitate cGMP synthesis with an early-stage solid $B(IV)^-$ salt to be used as the registered starting material. Flow chemistry was applied to solve the cryogenic requirements of the Matteson chloromethylene insertion reaction. Moreover, the first part of the synthesis, preparation of the chiral building block **29**, involved more traditional process chemistry problem solving.

Scheme 2 shows the initial preparative method for chiral intermediate **29**. Aldol condensation between the lithium enolate of *t*-BuOAc and acrolein (**36**) provided racemic alcohol **37**, which was resolved by treatment with bulk lipase (non-immobilized PS Amano lipase from *Burkholderia cepacia*) and vinyl acetate in pentane.[24,25] Compound **29** prepared in this way could be isolated in yields of about 65% for the combined 2 steps after SiO_2 chromatography. Both steps needed modification for kg-level operations because of the cryogenic conditions of the aldol reaction, the pentane use, and the chromatographic purification required in the second step.

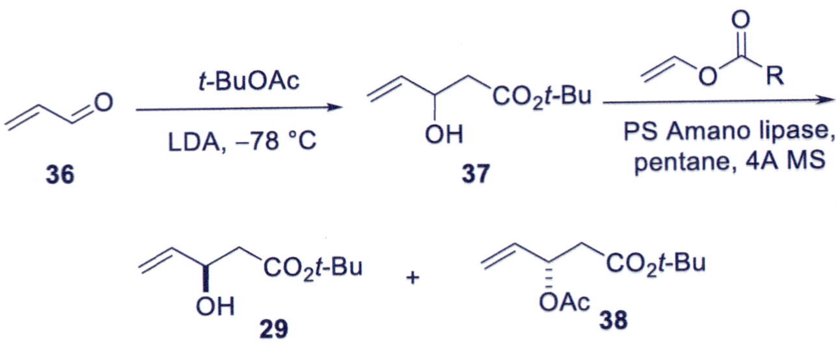

Scheme 2. Initial preparation of intermediate **29**.

Scheme 3 shows the multi-kg process scale synthesis of intermediate **29**.[30] Aldol condensation worked fine at –48 °C (79% yield) which resolved this cryogenic reaction concern. Resolution by a method without chromatography required significant investigation. After survey of several lipases, PS Amano IM was found to be the most effective and the acylation worked in the absence of pentane with vinyl acetate functioning as the solvent. To test if distillation could be used to purify **29**, lipase esterification was investigated with higher molecular weight vinyl propionate to provide **38b**. On scales below 1 kg, the double distillation of **29** (bp 87–90 °C/6 torr) worked to effect separation from **38b** (81% yield). However, at larger scales, separation by distillation was unsatisfactory. Purification of **29** from the previously optimized reaction mixture, which included **38a**, was accomplished by first derivatizing **29** as a succinate ester, and performing extractive isolation of the resulting carboxylic acid **39**. Hydrolysis was studied with K_2CO_3 and NH_4OH but worked best with 4 equiv of hydrazine at pH 9 to provide **29** pure at multi-kg scale with no chromatography and only partition extractions for purification. This hydrolysis was sensitive to pH since reactions run at pH 10 resulted in the elimination product **40** as a 40% impurity. Moreover, maintaining pH 9 eliminated production of this side product.

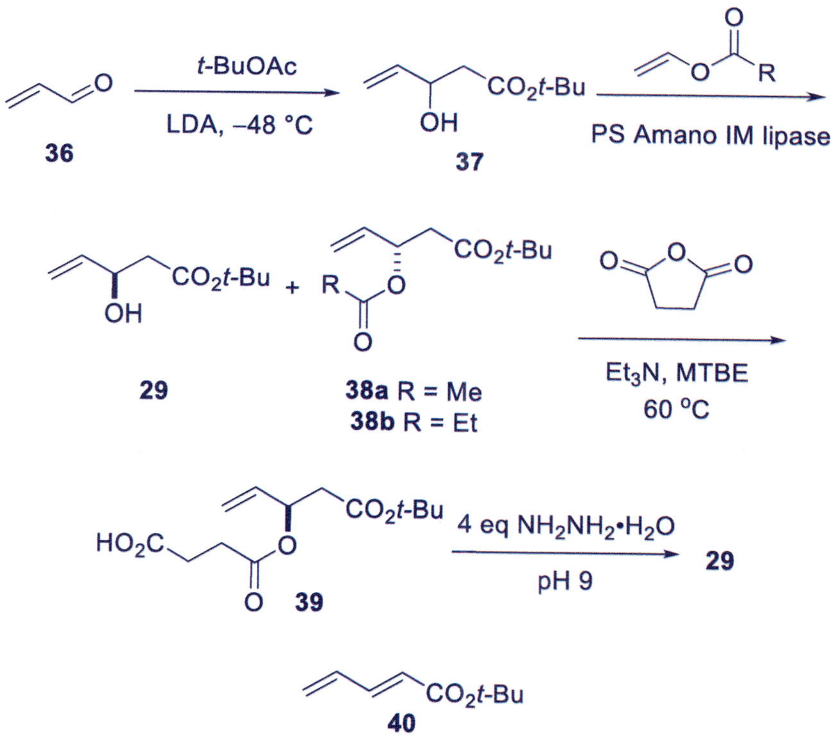

Scheme 3. Kg-scale preparation of intermediate **29**

The next stage of the vaborbactam (**1**) synthesis targeted a solid intermediate to be the registered starting material for cGMP manufacture. In the existing 7-step sequence (Scheme 1) only the penultimate intermediate **35** and the final compound **1** are crystalline solids. It was envisioned that some earlier boron-containing intermediate would be ideal because of the ready facility of boron to form $B(IV)^{-}$ salts. This would allow the number of steps conducted under GMP to be reduced which in turn would lower the overall cost of goods. Hence, a boron-containing intermediate represented by **41** was sought that could readily be converted to **32** (Scheme 4). In turn, **41** would be derived from TMS-ether protected intermediate **42** for ease of synthesis. Additionally, the resulting sequence to the $B(IV)^{-}$ salt allowed for complete removal of the toxic impurities of acrolein, hydrazine and iridium which were used in preceding reactions (*vide infra*).

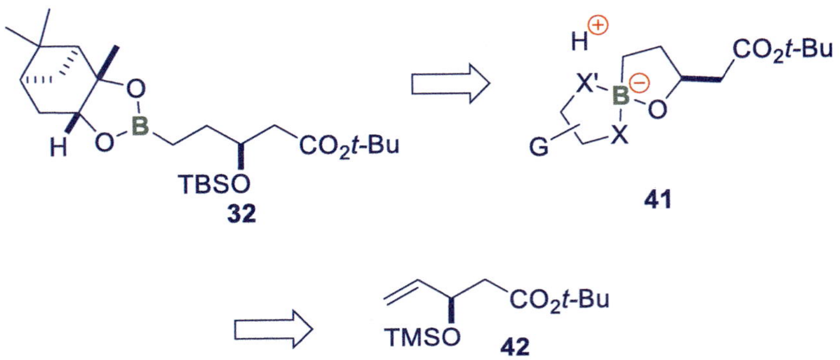

Scheme 4. The quest for solid intermediate **41** as the cGMP registered starting material.

The route to **41** and its conversion to **32** are shown in Scheme 5.[31] To simplify the preparation of **41**, the TBS-ether **30** was substituted with the more labile TMS-ether **42**. This ether was synthesized using NEt_3 as the base since imidazole use poisoned the subsequent iridium-catalyzed hydroboration.[32] After this small modification, pinacolborane hydroboration could be done with only 0.1 mol% $[Ir(COD)Cl]_2$ as catalyst under refluxing conditions to provide **43**. Exchanging the solvent to heptane and filtration through silica gel reduced residual iridium levels to < 30 ppm. Dilute acid cleaved the TMS ether to **44**. The pinacol ester was oxidatively cleaved with $NaIO_4$[33] and the free boronic acid cyclized with the proximal alcohol to form **45**. Various *N*-substituted ethanolamines were tested and only unsubstituted ethanolamine was effective at forming a solid boron centered spirocyclic salt, *i.e.* **41a**, from the MTBE/ACN reaction solvent. Functionally, the HCl desilylation and $NaIO_4$ reactions were conducted sequentially in one pot, product **45** isolated by MTBE extraction, the extract concentrated to four volumes and then treated with ethanolamine to provide the product **41a** as a filterable solid in 70–75% yield for the three steps. This compound was a mixture of two diastereomers (the boron atom is a stereocenter) and powder X-ray diffraction revealed it to be non-crystalline. Altogether in this sequence, the SiO_2 filtration, the oxidative treatment with $NaIO_4$ and the many aqueous treatments proved to remove the toxic impurities acrolein, hydrazine and iridium to acceptable low or undetectable ppm levels in the resulting solid compound **41a**. This compound was converted to the previously prepared intermediate **32** to intersect the established synthesis by sequential treatment with pinanediol in CH_2Cl_2/water and then silylation with TBSCl in the presence of 2 equivalents of imidazole where the first equivalent sequestered the boron (as **47**) to allow formation of the desired TBS-ether **32** after extractive isolation. Thus, the cGMP registered solid starting material **41a** was established early in the synthesis in high purity which greatly facilitated the manufacture of vaborbactam (**1**) free of the toxic impurities utilized in these early synthesis steps.

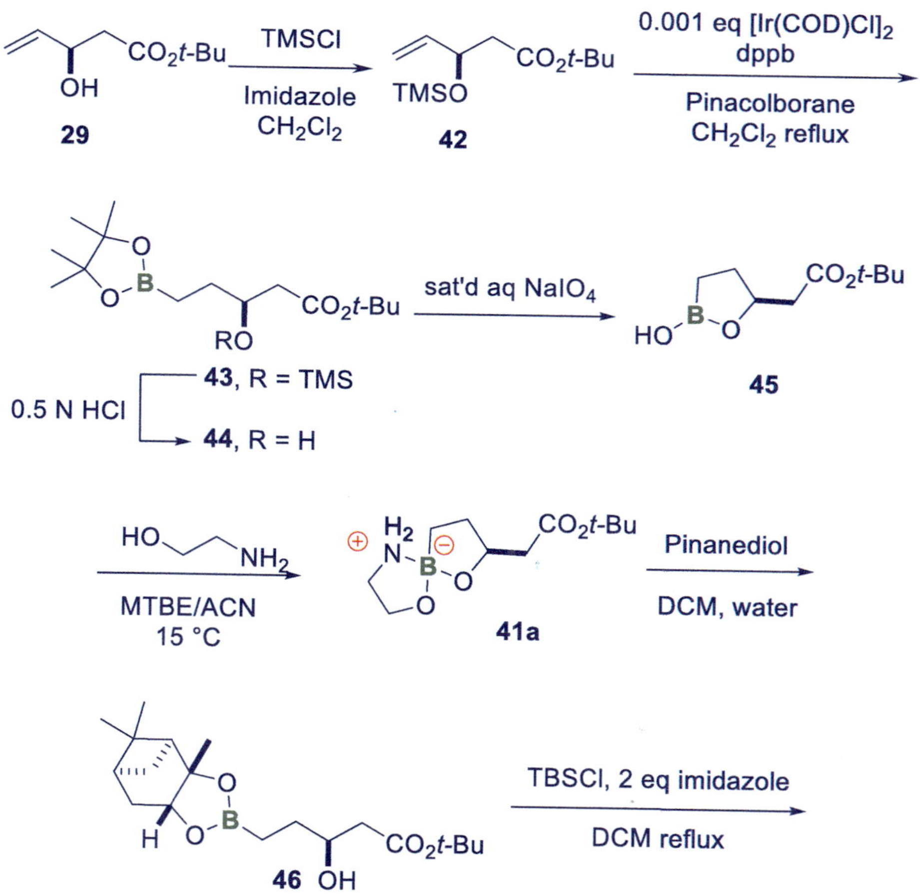

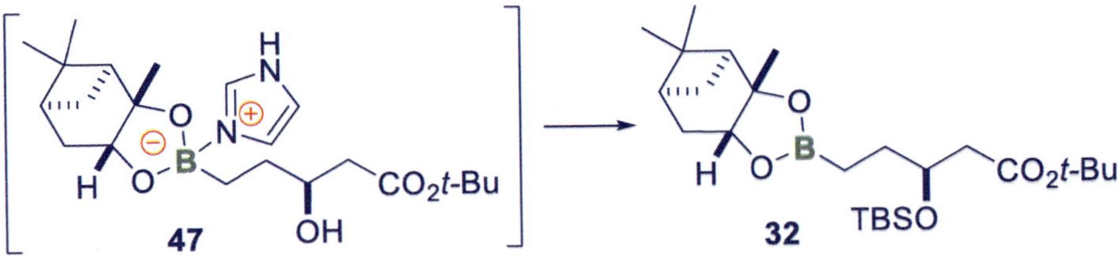

Scheme 5. Synthesis of intermediate **32** *via* the new starting material **41a**.

The next consideration for kg-level scale-up was the low temperature (–95 °C) Matteson chloromethylene-insertion reaction, see Scheme 6 for mechanism.[27,28] Low temperature is necessary since the incipient dichloromethyl-anion ($^{-}CHCl_2$) is unstable above –90 °C and decomposes to chloromethyl-carbene (:CHCl). The mechanism requires the anion to add to the boron to form a B(IV)$^{-}$ species, which rearranges by a 1,2-shift after coordination with $ZnCl_2$. The chelate with $ZnCl_2$ imparts diastereoselectivity *via* a proposed 5-member ring as shown with intermediate **48** that preferentially rearranges to the desired isomer **33**. To address the requirements for low temperature control, a flow chemistry or continuous processing (CP)[34,35] approach was investigated. The advantage of CP with a cryogenic reaction at kg scale is better temperature control of the small volumes passing through the reaction tubing compared to the variable temperatures caused by localized reaction exotherms in a large volume reaction vessel. The application of this technology to the Matteson reaction in the synthesis of vaborbactam (**1**) has been described in detail[36] and will be reviewed here.

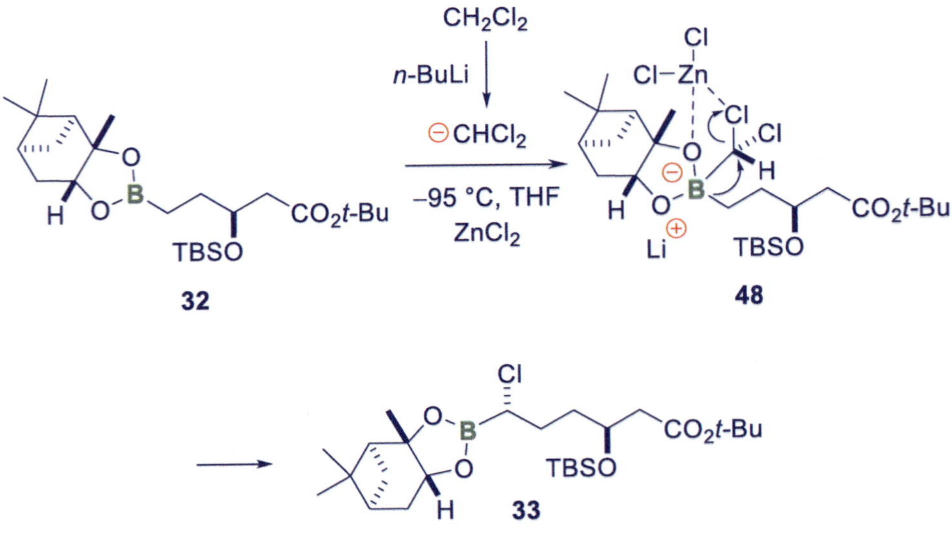

Scheme 6. Mechanism of the Matteson chloromethylene-insertion reaction.

The CP reaction setup evolved into the arrangement shown in Scheme 7. Stainless steel mixing T-joints combined two precooled solutions of CH_2Cl_2 and *n*-BuLi which then exited into a reactor loop (R2) which then combined with a solution of **32** in the second T-joint before progressing into the next reactor loop R3. Final reaction rearrangement was accomplished by transfer of this mixture into a stirred batch vessel containing a precooled solution of $ZnCl_2$. The optimal temperature was –80 °C for all vessels and residency times were optimized. The final mixing of this reaction process stream with an excess of $ZnCl_2$ in a separate batch reaction vessel created an enhanced dilution effect, avoided precipitation in the flow lines, and produced >98% conversion to target product **33** with excellent diastereomeric excess (*d.r.* = 97.7 : 2.3).

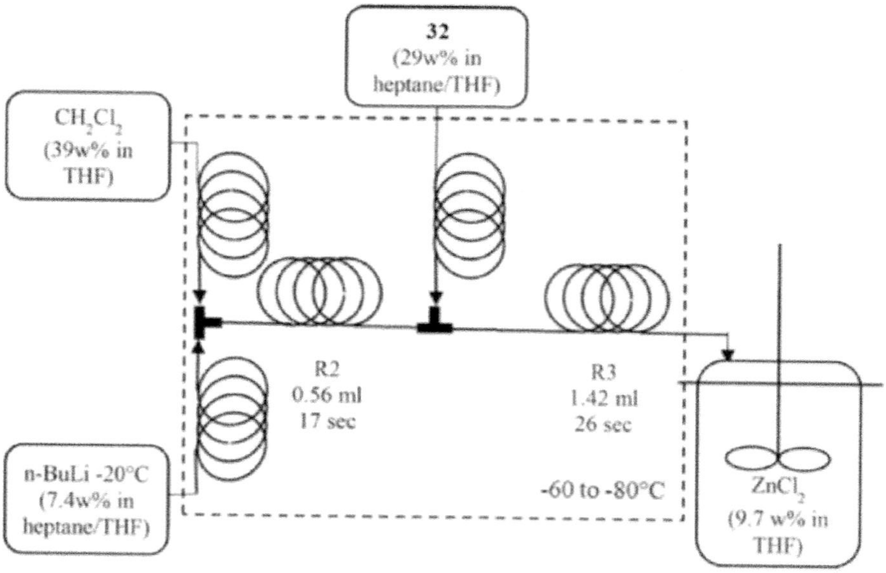

Scheme 7. Initial optimized flow chemistry protocol for the Matteson reaction. (Adapted with permission from Reference 36. Copyright 2019 American Chemical Society)

In the design of 40 kg pilot runs, tubing was arranged in a vertical reaction channel with coolant flowing around reaction tubing in a second outer tubing that acted as a cooling jacket. The length and diameter of mixing tubes was increased and radial mixing components were added inside reaction tubing. Optimization of *n*-BuLi equivalents, reaction temperatures and residency times resulted in > 90% yields and 95:5 diastereoselectivity control. Further construction and operation of refined flow reactors at larger scale produced several metric tons of intermediate **33** of high purity in a cGMP production campaign.

To achieve a fully continuous process, improvements were made to convert the batch $ZnCl_2$ quench method to a fully continuous process for commercial production. To achieve this, three operations were addressed: (1) the quench with $ZnCl_2$ at < –20 °C; (2) the batch sequence of aqueous washes and (3) the distillation procedure to concentrate the final product. The latter two were solved by established technology but the first required special study. The solution for the $ZnCl_2$ quench was studied with two strategies, a continuous loop quench plus continuous stirred tank reactor (CSTR, Scheme 8) and a sequential cascade double CSTR quench (Scheme 9). Both methods involved a smaller vigorous premixing zone of $ZnCl_2$/THF with reaction mixture from R3 at controlled temperature followed by transfer to a larger secondary mixing chamber or CSTR at controlled temperature before exiting to continuous work-up processing.

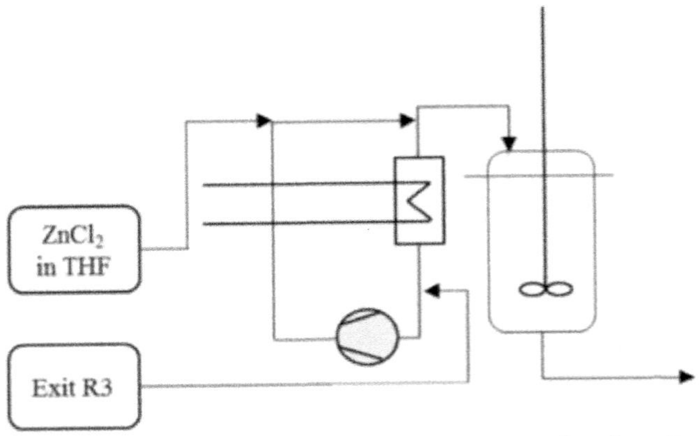

Scheme 8. Continuous loop quench. (Adapted with permission from Reference 36. Copyright 2019 American Chemical Society)

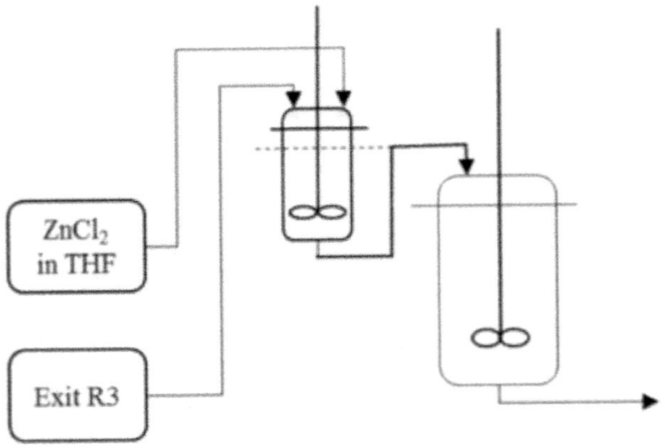

Scheme 9. Continuous stirred-tank reactor (CSTR) cascade quench. (Adapted with permission from Reference 36. Copyright 2019 American Chemical Society)

For the continuous loop method (Scheme 8), a small volume is circulated at a high flow rate and temperature is controlled by a heat exchanger. For the cascade CSTR method (Scheme 9), the temperature is controlled by a cooling jacket around the first smaller CSTR. For both, the stoichiometry of mixing is controlled by the flow rate of the reaction mixture exiting R3 at –60 °C and the $ZnCl_2$/THF stock being added as a 25 °C solution. After transfer to the larger CSTR for final mixing, the stable product solution is directed to a buffer tank for further continuous downstream processing. Better temperature control is realized because of the smaller initial mixing volumes. Also, elimination of the $ZnCl_2$ solution precooling step provides significant time and energy cost savings.

During optimization, the concentration of $ZnCl_2$/THF solution was varied (10–20%) as well as flow rates for that solution. Like before, higher $ZnCl_2$ equivalents (> 2 mol equiv) resulted in improved product quality. Studies of CSTR temperature revealed optimal results even at ≥–10 °C, which is the highest temperature reported for the Matteson reaction quenching step. This temperature tolerance was realized as a result of the fast mixing occurring at the entry point of the boronate adduct reaction stream. Comparison to the previous batch mode quenching run at –25 °C showed lower **33** diasteromer byproduct, 0.9% (CSTR) and 1.3% (loop) *vs.* 2.6% (batch mode) at 0.3 mol equiv. less $ZnCl_2$ (2.1 mol equiv.). For translation to production scale, the loop method was chosen for safety offered by the reaction chamber's pressure resistance, simplicity, lower cost and excellent scalability aspects.

For production, the modification of equipment only involved replacing the batch $ZnCl_2$ quenching vessel with a quench loop. A demonstration cGMP campaign, which included the loop reactor and subsequent continuous mode downstream operations, produced several hundreds of kilograms of intermediate **33** of high chemical purity (98% HPLC) and yield (97%) with only 0.8% undesired diastereomer. This was a true demonstration that flow reaction technology can open commercial production to traditionally temperature sensitive reactions and offer superior reaction selectivity controls.

The remainder of the synthesis scheme, the two-step conversion of **33** to vaborbactam (**1**), (the amine displacement and amide bond formation to **35** are performed without isolation of the intermediate), followed the original methodology (Scheme 1).

6 Conclusions

Discovery of this new boron-containing FDA approved drug vaborbactam (**1**) involved a design exploration which removed the aromatic ring utilized in earlier studies. Boron as the central element in a cyclic structure enabled enzyme inhibition to be covalent-reversible as a result of formation of a reversible bond with an active site serine residue. It also extended it to be a broad-spectrum β-lactamase inhibitor because of conformational flexibility observed in the boronate ring. Special synthesis challenges were solved, including boron-chemistry specific, to enable the commercial synthesis of this important molecule with methods which included (1) the traditional with enzymatic and partition extractive purification to provide chiral building block **29**; (2) the unique with preparation of the boron centered spirocyclic salt **41a** as the solid cGMP starting material; and (3) the modern with commercial scale continuous processing flow chemistry for the previously prohibitive cryogenic Matteson reaction to **33**.

7 References

1. FDA News release August 29, 2017. https://www.fda.gov/news-events/press-announcements/fda-approves-new-antibacterial-drug (accessed May 14, 2021).
2. Hecker, S. J.; Reddy, K. R.; Totrov, M.; Hirst, G. C.; Lomovskaya, O.; Griffith, D. C.; King, P.; Tsivkovski, R.; Sun, D. Sabet, M.; Tarazi, Z.; Clifton, M. C.; Atkins, K.; Raymond, A.; Potts, K. T.; Abendroth, J.; Boyer, S. H.; Loutit, J. S.; Morgan, E. E.; Durso, S.; Dudley, M. N. Discovery of a Cyclic Boronic Acid β-Lactamase Inhibitor (**RPX7009**) with Utility vs Class A Serine Carbapenemases. *J. Med. Chem.* **2015**, *58*, 3682–3692.
3. Fleming, A. Streptococcal Meningitis Treated with Penicillin. *Lancet* **1943**, *242*, 434–438.
4. For a review on recent and historical applications of co-administration of β-lactamase inhibitors with β-lactam antibiotics see Papp-Wallace, K. M. The latest advances in β-lactam/β-lactamase inhibitor combinations for the treatment of Gram-negative bacterial infections. *Expert Opin. Pharmacother.* **2019**, *20*, 2169–2184. and references therein.
5. Nordmann, P.; Dortet, L.; Poirel, L. Carbapenem resistance in Enterobacteriaceae: here is the storm! *Trends Mol. Med.* **2012**, *18*, 263–272.
6. Bortezomib: Einsele H. Bortezomib. In: Martens U. (eds) Small Molecules in Oncology. Recent Results in Cancer Research, vol 201. Springer, Berlin, Heidelberg, (2014).
7. Ixazomib: Shirley, M. Ixazomib: first global approval. *Drugs* **2016**, *76*, 405–411.
8. Tavaborole: Sharma N, Sharma D. An upcoming drug for onychomycosis: Tavaborole. *J. Pharmacol. Pharmacother.* **2015**, *6*, 236–239.
9. Crisaborole: Perry, M. A. Case Study: Eucrisa (Crisaborole Oinment, 2%), A Boron-based PDE4 Inhibitor for the Topical Treatment of Atopic Dermatitis. *Med. Chem. Rev.* **2020**, *55*, 589–605.
10. Taniborbactam: Liu, B.; Trout, R. E. L.; Chu, G.-H.; McGarry,D.; Jackson,R. W.; Hamrick, J. C.; Daigle, D. M.; Cusick, S. M.; Pozzi, C.; De Luca, F.; Benvenuti, M.; Mangani, S; Docquier, J.-D.; Weiss, W. J.; Pevear, D. C.; Xerri, L.; Burns, C. J. Discovery of Taniborbactam (**VNRX-5133**): A Broad-Spectrum Serine- and Metallo-β-lactamase Inhibitor for Carbapenem-Resistant Bacterial Infections *J. Med. Chem.* **2020**, *63*, 2789–2801.
11. Xeruborbactam (**QPX7728**): Hecker, S. J.; Reddy, K. R.; Lomovskaya, O.; Griffith, D. C.; Rubio-Aparicio, D.; Nelson, K.; Tsivkovski, R.; Sun, D.; Sabet,

M.; Tarazi, Z.; Parkinson, J.; Totrov, M.; Boyer, S. H.; Glinka, T. W.; Pemberton, O. A.; Chen, Y.; Dudley, M. N. Discovery of Cyclic Boronic Acid **QPX7728**, an Ultrabroad-Spectrum Inhibitor of Serine and Metallo-β-lactamases. *J. Med. Chem.* **2020**, *63*, 7491–7507.

12. Ledaborbactam (**VNRX-7145/VNRX-5236**) etzadroxil: Trout, R. E.; Zulli, A.; Mesaros, E.; Jackson, R. W.; Boyd, S.; Liu, B.; Hamrick, J.; Daigle, D.; Chatwin, C. L.; John, K.; McLaughlin, L.; Cusick, S. M.; Weiss, W. J.; Pulse, M. E.; Pevear, D. C.; Moeck, G.; Luigi Xerri, L; Burns, C. J. Discovery of **VNRX-7145 (VNRX-5236** Etzadroxil): An Orally Bioavailable β-Lactamase Inhibitor for Enterobacterales Expressing Ambler Class A, C, and D Enzymes. *J. Med. Chem.* **2021**, *64*, 10155–10166.
13. Smoum, R.; Rubinstein, A.; Dembitsky, V. M.; Srebnik, M. Boron-containing Compounds as Protease Inhibitors. *Chem. Rev.* **2012**, *112*, 4156–4220.
14. Hernandez, V.; Akama, X. L.; Perry, M.; Xia, Y.; Zhang, Y.-K. Boron in Medicinal Chemistry. *Med. Chem. Rev.* **2016**, *51*, 327–344.
15. Kiener, P. A.; Waley, S. G. Reversible inhibitors of penicillinases. *Biochem. J.* **1978**, *169*, 197–204.
16. Martin, R.; Jones, J. B. Rational design and synthesis of a highly effective transition state analog inhibitor of the RTEM-1 β-lactamase. *Tetrahedron Lett.* **1995**, *36*, 8399–8402.
17. Strynadka, N. C.; Martin, R.; Jensen, S. E.; Gold, M.; Jones, J. B. Structure-based design of a potent transition state analogue for TEM-1 beta-lactamase. *Nat. Struct. Biol.* **1996**, *3*, 688–695.
18. Ness, S.; Martin, R.; Kindler, A. M.; Paetzel, M.; Gold, M.; Jensen, S. E.; Jones, J. B.; Strynadka, N. C. Structure-based design guides the improved efficacy of deacylation transition state analogue inhibitors of TEM-1 beta-Lactamase. *Biochemistry* **2000**, *39*, 5312–5321.
19. Morandi, F.; Caselli, E.; Morandi, S.; Focia, P. J.; Blazquez, J.; Shoichet, B. K.; Prati, F. Nanomolar Inhibitors of AmpC β-Lactamase. *J. Am. Chem. Soc.* **2003**, *125*, 685–695.
20. Dobozy, O.; Mile, I.; Ferencz, I.; Csanyi, V. Effect of electrolytes on the activity and iodine sensitivity of penicillinase from B. cereus. *Acta Biochim. Biophys. Acad. Sci. Hung.* **1971**, *6*, 97–105.
21. Strynadka, N. C.; Adachi, H.; Jensen, S. E.; Johns, K.; Sielecki, A.; Betzel, C.; Sutoh, K.; James, M. N. Molecular structure of the acyl-enzyme intermediate in beta-lactam hydrolysis at 1.7 A resolution. *Nature* **1992**, *359*, 700–705.
22. For a detailed review on metallo-β-lactamases and the search for their pharmacological control, see Bahr, G.; González, L. J.; Vila, A. J. Metallo-β-lactamases in the Age of Multidrug Resistance: From Structure and Mechanism to Evolution, Dissemination, and Inhibitor Design. *Chem. Rev.* **2021**, *121*, 7957–8094.
23. Bolger, C. A.; Carpenter, J. E.; Dhar, T. G. M.; Pashine, A.; Dragovich, P. S.; Cook, J. H.; Gillis, E. P.; Peese, K. M.; Merritt, J. R. To Market, To Market—2017. *Med. Chem. Rev.* **2018**, *53*, 587–693.

24. Vrielynck, S.; Vanderwalle, M. A Chemoenzymatic Synthesis of A-ring Key-intermediates for 1α,25-dihydroxyvitamin D_3 and Analogues. *Tetrahedron Lett.* **1995**, *36*, 9023–9026.
25. Tan, C.-H.; Holmes, A. B. The Synthesis of (+)-Allopumiliotoxin 323B′. *Chem. Eur. J.* **2001**, *7*, 1845-1854.
26. Yamamoto, Y.; Fujikawa, R.; Umemoto, T.; Miyaura, N. Iridium-catalyzed hydroboration of alkenes with pinacolborane. *Tetrahedron* **2004**, *60*, 10695–10700.
27. Matteson, D. S. α-Halo boronic esters in asymmetric synthesis. *Tetrahedron* **1998**, *54*, 10555–10607.
28. Matteson, D. S. Boronic Esters in Asymmetric Synthesis. *J. Org. Chem.* **2013**, *78*, 10009−10023.
29. Matteson, D. S. α-Amido boronic acids: A synthetic challenge and their properties as serine protease inhibitors. *Med. Res. Rev.* **2008**, *28*, 233–246.
30. Boyer, S. H.; Hecker, S. J. unpublished observations.
31. Hecker, S.; Boyer, S. Synthesis of Boronate Salts and Uses Thereof. U.S. Patent 10,385,074 B2, Aug. 20, 2019.
32. Imidazole is a known iridium ligand: Mura, P.; Casini, A.; Marcon, G.; Messori, L. Synthesis, molecular structure and solution chemistry of the iridium(III) complex imidazolium [trans(bisimidazole)tetrachloro iridate(III)] (IRIM). *Inorg. Chim. Acta* **2001**, *312*, 74–80.
33. Coutts, S. J.; Adams, J.; Krolikowski, D.; Snow, R. J. Two efficient methods for the cleavage of pinanediol boronate esters yielding the free boronic acids. *Tetrahedron Lett.* **1994**, *35*, 5109–5112.
34. Poechlauer, P.; Skranc, W. Scale-Up of Flow Processes in the Pharmaceutical Industry, in Vaccaro, L., Ed.; *Sustainable Flow Chemistry*; Wiley-VCH Verlag GmbH & Co. KG: Weinheim, Germany, 2017; pp 73−102.
35. Hughes, D. L. Applications of Flow Chemistry in Drug Development: Highlights of Recent Patent Literature. *Org. Process Res. Dev.* **2018**, *22*, 13−20.
36. Stueckler, C.; Hermsen, P.; Ritzen, B.; Vasiloiu, M.; Poechlauer, P.; Steinhofer, S.; Pelz, A.; Zinganell, C.; Felfer, U.; Boyer, S.; Goldbach, M.; de Vries, A.; Pabst, T.; Winkler, G.; LaVopa, V.; Hecker, S.; Schuster, C. Development of a Continuous Flow Process for a Matteson Reaction: From Lab Scale to Full-Scale Production of a Pharmaceutical Intermediate *Org. Process Res. Dev.* **2019**, *23*, 1069–1077.

3

Baloxavir Marboxil (Xofluza), A Cap-Dependent Endonuclease Inhibitor for Treating Influenza

Yong-Jin Wu

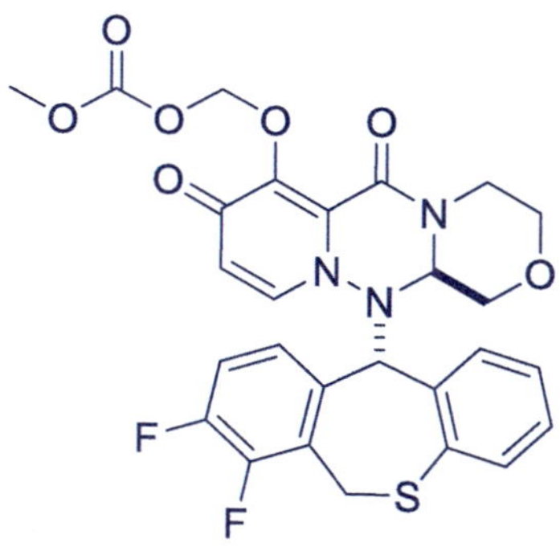

USAN: Baloxavir Marboxil
Trade name: Xofluza (**1**)
Shionoggi/Roche, 2018
Endonuclease inhibitor

1 Background

Influenza viruses are responsible for both seasonal influenza and pandemic influenza, the latter of which has occurred four times since the turn of the 20th century in the years 1918 (H1N1), 1957 (H2N2), 1968 (H3N2), and 2009 (H1N1), respectively. These instances resulted in the combined deaths of more than 52 million people (Table 1).[1] Influenza, which is an acute respiratory disease, affects approximately 13% of the world's population each year, leading to 3–5 million severe cases and 290,000–650,000 deaths worldwide.[2] The two-pronged approach to control influenza is vaccination and antiviral therapy. However, a sizable portion of the general population worldwide remains unvaccinated against seasonal flu (only 45% in the United States in 2020), and furthermore, influenza vaccines achieve approximately 60% protection rate due to

Current Drug Synthesis, First Edition. Edited by Jie Jack Li.

mismatch to one or more circulating influenza viruses.[3] Thus, anti-influenza drugs are essential for the treatment and prevention of influenza.[4]

Table 1. Timeline of influenza pandemics.

Year	Virus strain	Deaths (world)	Deaths (United States)
1918	HIN1	>50 million	675,000
1957	H2N2	1.1 million	116,000
1968	H3N2	1 million	100,000
2009	H1N1	151,700–575,400[a]	12,469

Source: data from Reference 1.

Over the past few decades, three classes of inhibitors have been developed to treat influenza, as shown in Table 2.[4] The M2 ion channel blockers, amantadine (**2**) and rimantadine (**3**), are no longer recommended because influenza A viruses have developed resistance to treatment from these drugs, and these drugs lack efficacy against influenza B viruses (influenza A and B viruses generally cause seasonal flu).[5] Three neuraminidase (NA) inhibitors are recommended by the CDC for the treatment of uncomplicated influenza: two doses per day of oral oseltamivir (**4**) or inhaled zanamivir (**5**) for 5 days, or one dose of intravenous peramivir (**6**). However, their efficacy has been limited by emergent resistance due to the mutations of the NA protein of the viruses. In October 2018, a single oral dose baloxavir marboxil (Xofluza, **1**), a cap-dependent endonuclease (CEN) inhibitor designed to block the initiation of viral mRNA synthesis, was approved by the FDA for the treatment of flu in patients 12 years and older who were symptomatic for 48 hours or less. Baloxavir marboxil (**1**) represents the first flu medication in nearly 20 years after the FDA approved oseltamivir (**4**) and zanamivir (**5**) in 1999 (Figure 1).[6]

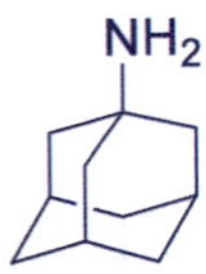

2 (amantadine)

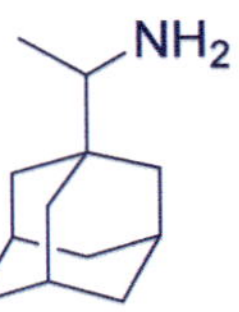

3 (ramantadine)

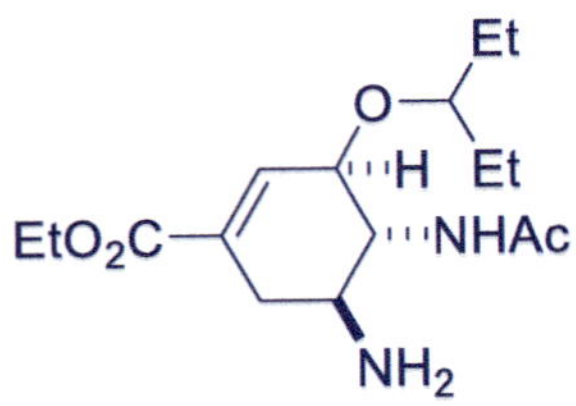

4 (osetamivir)

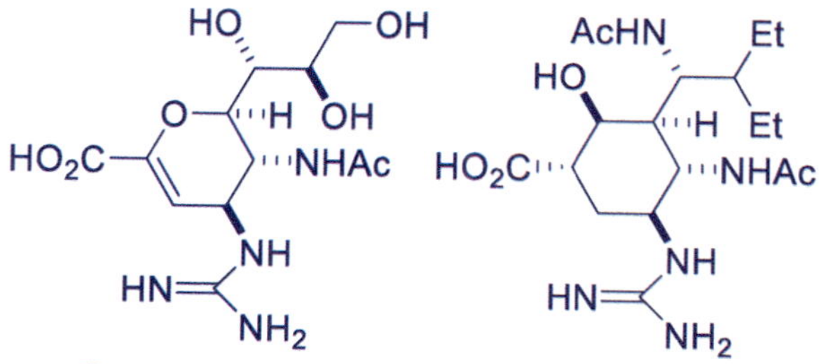

5 (zanamivir) 6 (peramivir)

Figure 1. Structures of anti-flu drugs.

Table 2. Timeline of FDA-approved antiviral drugs for influenza.

Year	Drug	Mechanism	Virus type	Route	Age
1973	Admantadine (**2**)	M2 inhibitor	Influenza A	Oral	NR
1994	Rimantadine (**3**)	M2 inhibitor	Influenza A	Oral	NR
1999	Oseltamivir (**4**)	NA inhibitor	Influenza A/B	Oral	Any age
1999	Zanamivir (**5**)	NA inhibitor	Influenza A/B	Inhaled	>7 years
2014	Peramivir (**6**)	NA inhibitor	Influenza A/B	IV	>2 years
2018	Baloxavir marboxil (**1**)	CEN inhibitor	Influenza A/B	Oral	>12 years

Source: Based on Reference 4. NR: Not Recommended by CDC.

2 Mechanism of Action

Influenza virus life cycle starts with attachment to host cell surface sialic acid (SA) receptor using the viral surface glycoprotein hemagglutinin (HA).[7,8] Upon entry, the virus undergoes endosomal acidification to activate the M2 proton channel, which causes HA conformational change. This change leads to HA-mediated membrane fusion, resulting in release of the viral ribonucleoproteins (vRNPs) containing the viral genome into the nucleus for transcription of viral mRNAs, which is dependent upon the activity of the heterotrimeric influenza RNA polymerase consisting of PA, PB1, and PB2 subunits.[9] The 7-methylguanylate cap binding domain of the PB2 subunit binds to the cap structure at the 5′ ends of pre-mRNAs (the nuclear RNA precursors to cellular mRNAs), which are cleaved by a CEN of the PA subunit at a position 10–14 bases downstream from the cap to generate the capped RNA fragments, which are used as primers for viral mRNA synthesis by RNA-dependent RNA polymerase (RdRp) of the PB1 subunit, a process known as "cap-snatching."[10,11] The mRNAs are transported to cytoplasm for translation to produce viral proteins, which are combined with the newly synthesized viral genome segments to form vRNPs in the nucleus. These new vRNPs are exported from the nucleus *via* viral nucleoprotein to the cell membrane where new virions are assembled in a process called budding. After cleaving the terminal sialic acid residues from HA using

NA, the new budding viruses are released from the cell to start a new cycle of infection and replication (Figure 2).

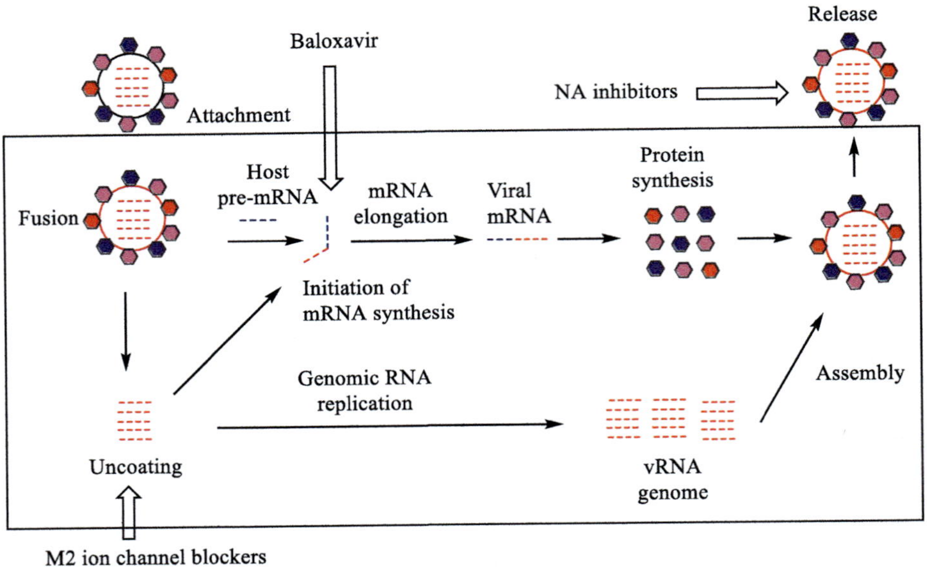

Figure 2. Mechanism of action of anti-flu drugs. (Based on References 7 and 8).

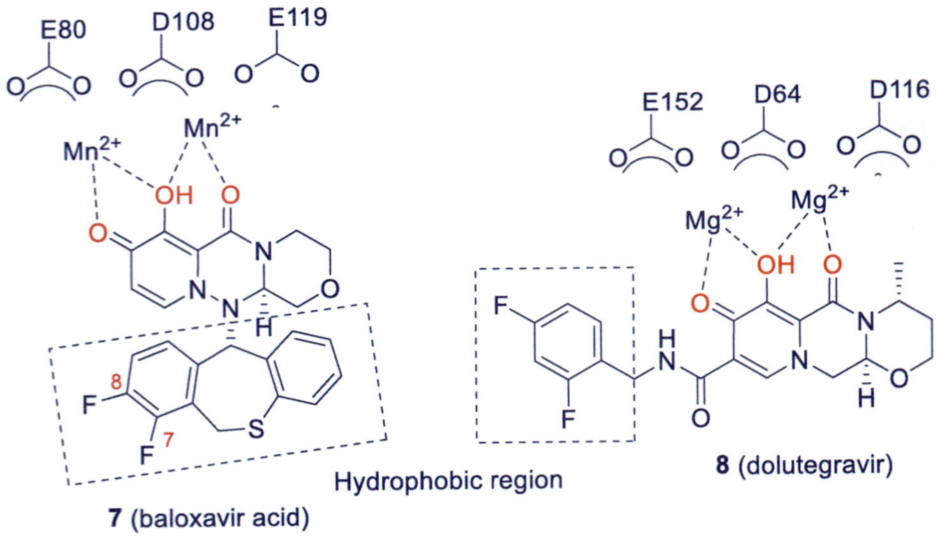

Figure 3. Binding modes of baloxavir acid (7) and dolutegravir (**1**).

The life cycle of influenza virus depends on all steps for its virulence, replication, and transmission; therefore, small molecule inhibitors designed to block any of these steps can be effective for the treatment or prevention of influenza infections.[12,13] Adamantanes (amantadine and rimantadine) inhibit viral replication by blocking the M2 proton channel which modulates acidification of the virion interior essential for its uncoating. The NA inhibitors (oseltamivir and zanamivir) block the function of viral neuraminidases, preventing viral release from the host cell.

The recently approved baloxavir marboxil (**1**) inhibits the PA subunit of the influenza virus polymerase complex by binding to the two catalytic divalent metal ions of the CEN active site,[14–18] a binding mode similar to dolutegravir (**8**, Figure 3), an inhibitor of the human immunodeficiency virus type 1 (HIV-1) integrase for the treatment of HIV infections in combination with other HIV medications.[18–20] The crystal structures of the truncated PA subunit with and without baloxavir acid (**7**) have shown that the inhibitor binds to the two metal ions with three oxygen atoms with a hydrophobic region known as "specificity domain," which provides selectivity for the target.[15]

3 Structure–Activity Relationship[14,15]

High throughput screening of the Shionogi compound library provided hit compound **9** with a bicyclic carbamoyl pyridone core as the two metal chelators.[14] This compound was a weak CEN inhibitor with IC_{50} of 69 μM due to lack of a hydrophobic region. Introduction of a C1 benzyl group to the hydrophobic domain increased potency to submicromolar range (**10**, IC_{50} = 0.24 μM), but the benzyl analog **10** still lacked activity in a cytopathic effect (CPE)-based antiviral assay (CPE EC_{50} > 25 μM). The benzhydryl substitution further enhanced the hydrophobic interaction, resulting in significant improvement in both enzyme inhibitory activity (**11**, IC_{50} = 48 nM) and antiviral activity (CPE EC_{50} = 0.29 μM). The C7 carboxylic acid was not involved in the metal chelation, but it was engaged in hydrogen bonding interactions with the enzyme. However, it contributed to poor membrane permeability; therefore, the free carboxylic acid was removed to give **12**, which showed more potent antiviral activity despite reduction in enzyme activity. Nevertheless, compound **12** exhibited very fast rat IV clearance (42 mL/min/kg), which was attributed to a high free fraction in rat plasma (f_u = 30%) rather than metabolic instability (79% remaining after 30 minutes in rat liver microsomes (RLM). Consistent with this hypothesis, replacement of the methoxymethyl side chain with an isopropyl group (i.e., **13**) increased clogP from 1.3 to 1.8, which led to a substantial drop in free fraction (f_u = 18%) and rat IV clearance (25 mL/min/kg). Further increase in protein binding would require additional hydrophobic groups, which could adversely impact both biological activity and metabolic stability. To increase lipophilicity without incorporation of new hydrophobic moieties, the C-1 methine was replaced with a non-basic nitrogen atom (due to the α effect) to form a cyclic aminal **14**, a judicious bioisosteric substitution that led to 1.4 unit increase in clogP. The lower lipophilicity transferred to a substantial decline in the free fraction (from 18% to 3.8%) and rat IV clearance (from 25 to 11 mL/min/kg), while maintaining potent antiviral activity and good metabolic stability.

Compound **14** demonstrated protection in a mouse influenza B infection model upon IV administration, thus confirming that CEN inhibition is an effective approach to fight influenza.[15] Even so, the *in vivo* activity of **14** was not robust enough even with IV dosing, let alone oral efficacy due to poor oral bioavailability and moderate antiviral activity.

9 (HTS hit)
CEN IC_{50} = 69 μM

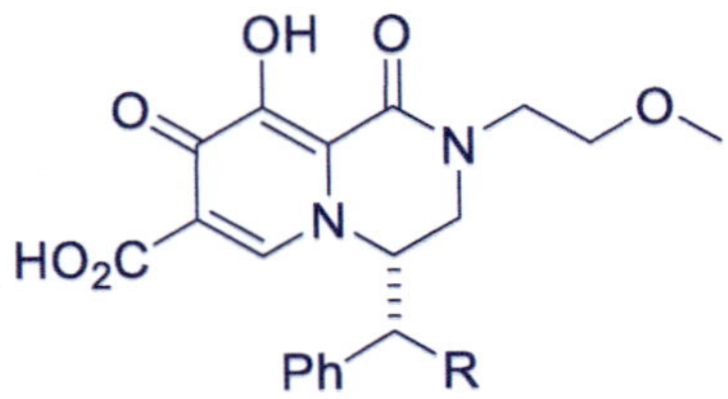

10: R = H
CEN IC_{50} = 0.24 μM
CPE EC_{50} > 25 μM

11: R = Ph
CEN IC_{50} = 48 nM
CPE EC_{50} = 0.29 μM

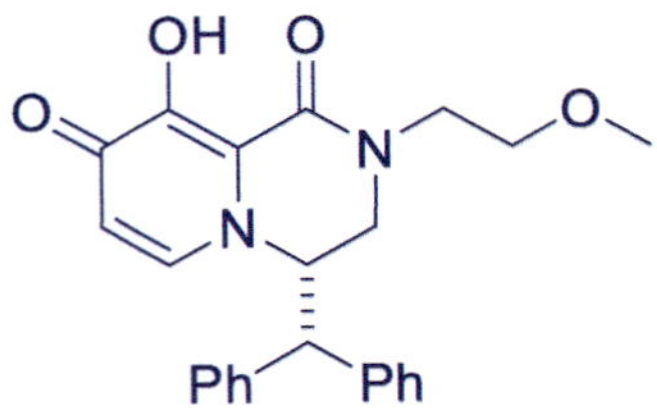

12
CEN IC_{50} = 115 nM
CPE EC_{50} = 134 nM
Met. stab. = 79% (RLM)
Rat IV Cl = 42 mL/min/kg
f_u = 30% (clogP = 1.3)

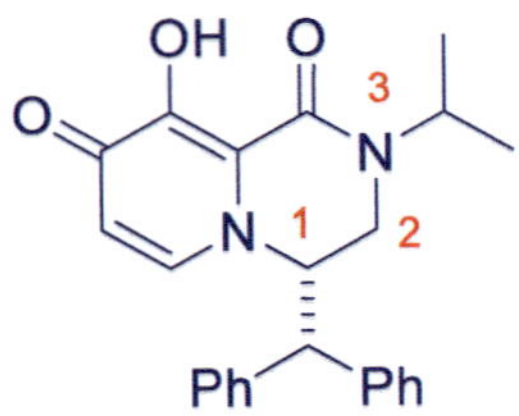

13
CEN IC_{50} = 239 nM
CPE EC_{50} = 58 nM
Met. stab. = 79% (RLM)
Rat IV Cl = 25 mL/min/kg
f_u = 18% (clogP = 1.8)

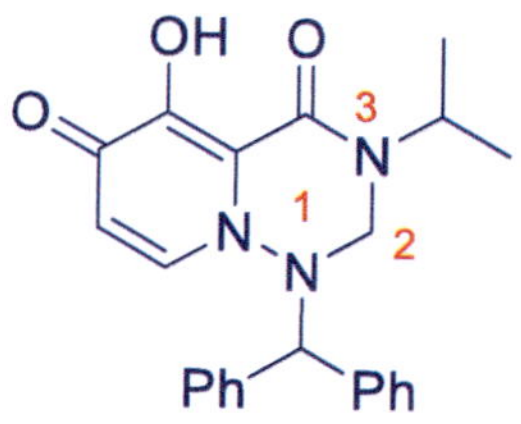

14
CEN IC_{50} = 286 nM
CPE EC_{50} = 81 nM
Met. stab. = 86% (RLM)
Rat IV Cl = 11 mL/min/kg
f_u = 3.8% (clogP = 3.2)

Figure 4. SAR leading to **14**.

The moderate antiviral activity of **14** (CPE EC_{50} = 81 nM) was addressed by optimizing the lipophilic benzhydryl group at the N1 position.[15] Conformational restriction by linking two phenyl rings together to form dihydrodibenzoannulene **15** resulted in 4-fold enhancement in both enzyme inhibition and antiviral potency (IC_{50} = 60 nM; CPE EC_{50} = 20 nM). Further incorporation of an oxygen atom led to dihydrodibenzooxepin **16** with much reduced activity, whereas racemic dihydrodibenzo-thiepine **17** was comparable in activity to its carbon counterpart **15**. As the more active enantiomer of **17** was more potent than **15**, dihydrodibenzothiepine was selected as the N1 substituent to investigate the impact of the solvent exposed N3 substituent. This substituent was expected to have marginal impact on intrinsic enzyme inhibitory activity but may improve antiviral activity due to better membrane permeability and also alter pharmacokinetic properties. Indeed, simply switching one of the methyl groups with trifluoromethyl (i.e., **18**) increased antiviral activity to subnanomolar EC_{50} range. The trifluoromethyl analog **18** also showed lower free fraction and better metabolic stability, which led to further drop in rat IV clearance.

With **18** identified as a potent antiviral agent, optimization of both phenyl rings in the hydrophobic pocket was carried out with fluorine and/or chlorine substitution, a thorough SAR exercise that identified the difluoro analog **19** with excellent antiviral activity as well as low rat IV clearance.[14] In a mouse model of influenza A virus infection, a single dose of **19** at 0.4 mg/kg (PO, BID) was shown to be more effective than that of oseltamivir phosphate at 5 mg/kg (PO, BID) in reducing viral titers in the lung.

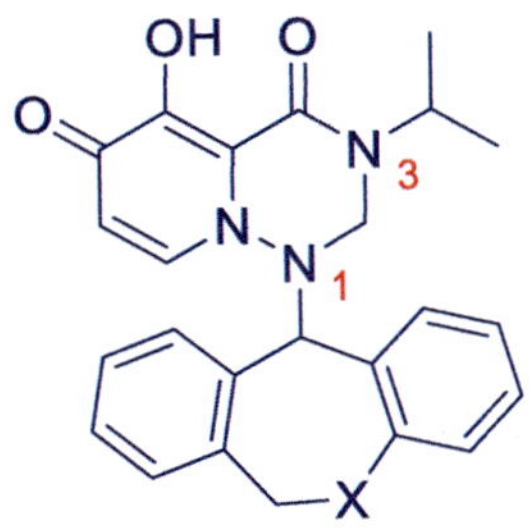

15: X = CH_2
CEN IC_{50} = 60 nM
CPE EC_{50} = 20 nM
16: X = O
CEN IC_{50} = 264 nM
CPE EC_{50} = 71 nM

17: X = S
CEN IC_{50} = 62 nM
CPE EC_{50} = 17 nM
Met. stab. = 73% (RLM)
Rat IV Cl = 33 mL/min/kg
f_u = 7.7% (clogP = 3.5)

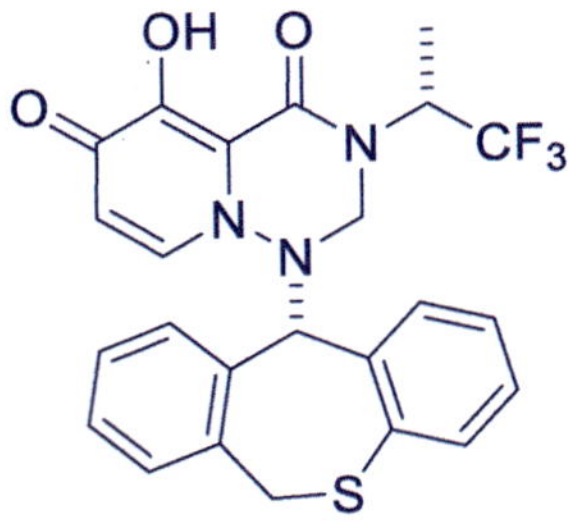

18
CEN IC_{50} = 50 nM
CPE EC_{50} = 5 nM
Met. stab. = 98% (RLM)
Rat IV CL = 16 mL/min/kg
f_u = 1% (clogP = 3.7)

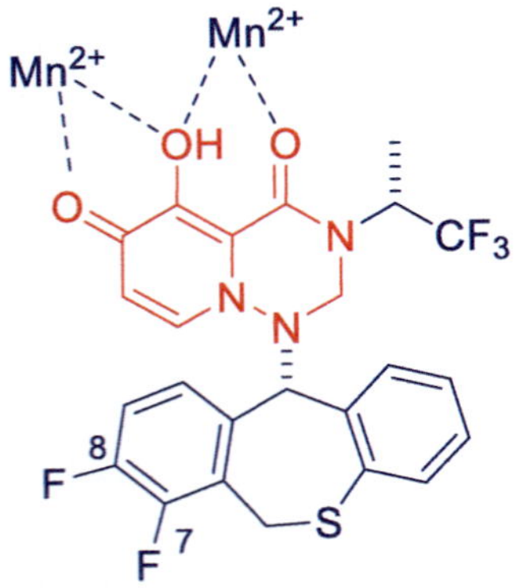

19
CEN IC_{50} = 6 nM
CPE EC_{50} = 4 nM
Met. stab. = 81% (RLM)
Rat IV CL = 12 mL/min/kg
f_u = 0.4% (clogP = 3.9)

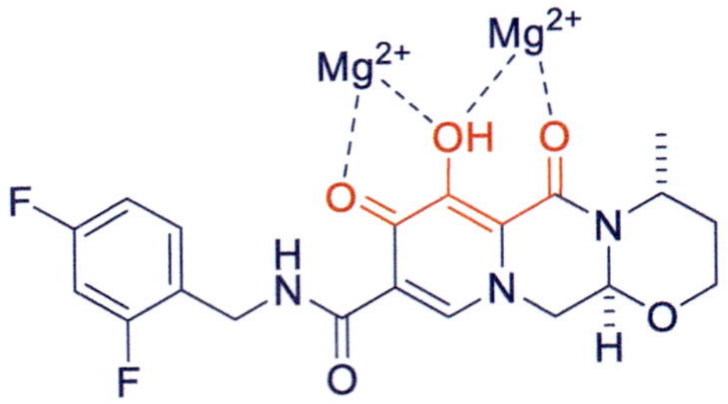

8 (dolutegravir)

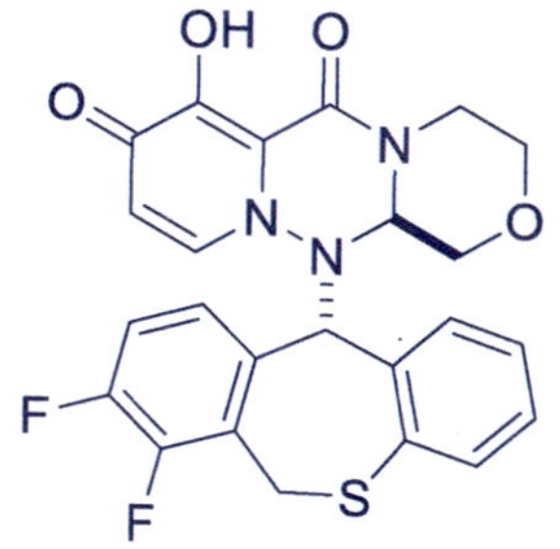

7 (baloxavir acid)
CEN IC_{50} = 1.9 nM
CPE EC_{50} = 1.1 nM
Human CL/F = 10.3 L/h
Human V_d/F = 1180 L
Human $t_{1/2}$ = 79 h
Human serum f_u = 6.1–7.1%
clogP = 2.6
$t_{1/2}$ = 79 h in humans

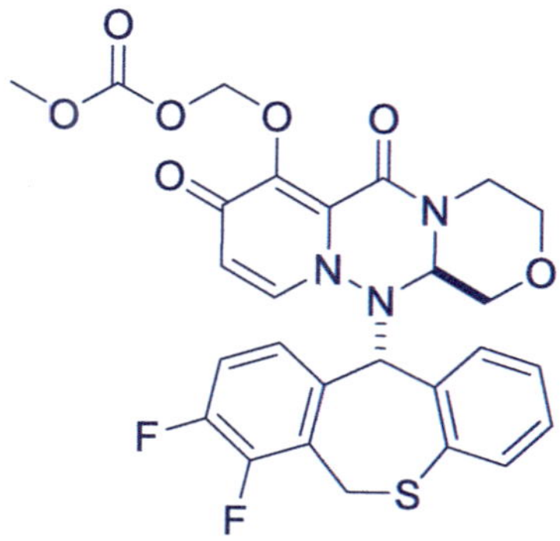

1 (baloxavir marboxil)
CEN IC_{50} = 0.53 μM

Figure 5. SAR leading to the discovery of baloxavir marboxil (**1**).

Further optimization of **19** was inspired by dolutegravir (**8**), an HIV-1 integrase inhibitor, which also engages two divalent metal ions for binding to the HIV-1 integrase active site.[19–21] To imitate the tricyclic fused metal binding scaffold of dolutegravir, a morpholine ring was fused onto the bicyclic system of **19** at the C2 and N3 positions, a transformation that culminated in the discovery of baloxavir acid (**7**), which was developed as a prodrug, baloxavir marboxil (**1**) to improve its oral absorption.

Baloxavir acid (**7**) selectively inhibits the CEN of the PA subunit with IC_{50} of 1.9 nM, with much lower activity on RdRp of the PB1 ($IC_{50} > 40$ nM) and the cap binding activity of PB2.[8] The CEN IC_{50} values against both influenza A and B viruses are within 1.4–3.1 nM and 4.5–8.9 nM, respectively, and it also demonstrated potent antiviral activity in infected cells against both influenza A and B viruses with CPE EC_{90} of 0.46–0.98 nM and 2.2–3.4 nM, respectively. Additionally, baloxavir acid shows a broad spectrum of activity against various subtypes of influenza A viruses (H1N2, H5N1, H5N2, H5N6, H7N9 and H9N2), thus supporting its clinical application to treat various influenza infections.

4 Pharmacokinetics and Drug Metabolism[22,23]

Baloxavir marboxil (**1**) is readily hydrolyzed to the active baloxavir acid (**7**) primarily by arylacetamide deacetylase (AADAC), a serine hydrolase found in the intestine, blood, and liver, with maximum plasma concentration reached at 4 hours. Baloxavir acid (**7**) exhibits modest unbound fraction (6.1–7.1%) in human serum proteins, which contributes to low clearance (CL/F = 10.3 L/h, approximately 10% of human hepatic blood flow = 109 L/h for 70 kg body weight). The low CL/F along with a huge volume of distribution (V_d/F = 1,180 L) results in a prolonged terminal elimination half-life of 80 hours ($t_{1/2}$ = 0.693•V_d/CL), supporting a QD dosing regimen.

Baloxavir acid (**7**) undergoes phase II metabolism predominantly in the liver through glucuronidation of the hydroxyl group by uridine diphosphateglucuronosyl-transferase 1A3 (UGT1A3). Oral administration of [^{14}C]-baloxavir marboxil showed predominant excretion through feces (80.1% total radioactivity) followed by urine (14.7% total radioactivity and 3.3% as baloxavir acid).

5 Efficacy and Safety[23–27]

In a phase III clinical trial (CAPSTONE-1), a single pill of baloxavir marboxil (**1**, 40 mg or 80 mg, depending on the body weight) was compared to oseltamivir (**4**, 75 mg, BID, 5 days) and to placebo for the alleviation of flu-like symptoms. Baloxavir marboxil (**1**) reduced the median time of alleviation from an average of 80.2 hours to 53.7 hours *vs*. placebo, a clinical outcome comparable to oseltamivir (53.8 hours); however, it reduced viral loads more rapidly in respiratory secretions, with a median duration of infectious virus detection of 24 hours versus 72 hours for oseltamivir and 96 hours for placebo. Baloxavir marboxil (**1**) was generally well tolerated with no serious drug drug interactions.

Baloxavir marboxil (**1**) has been approved as an influenza antiviral drug for patients 12 and older exhibiting flu-like symptoms for less than 48 hours. Despite similar clinical efficacy to oseltamivir (**4**), baloxavir marboxil (**1**) offers a significant advantage of one single pill as an effective regimen over oseltamivir that requires twice daily treatment for five days. Nevertheless, the generic oseltamivir (**4**) is less expensive than brand name Xofluza, and therefore, it remains a good option if the 5-day course of medication poses no serious issue to the patient.

6. Synthesis[28]

Baloxavir acid (**7**) was prepared *via* S_N2 type reaction from tricyclic alcohol **21** and tricyclic hydrazine **20**, which was derived from hydrazine ester **22** and morpholine ether **23** through condensation (Scheme 1).

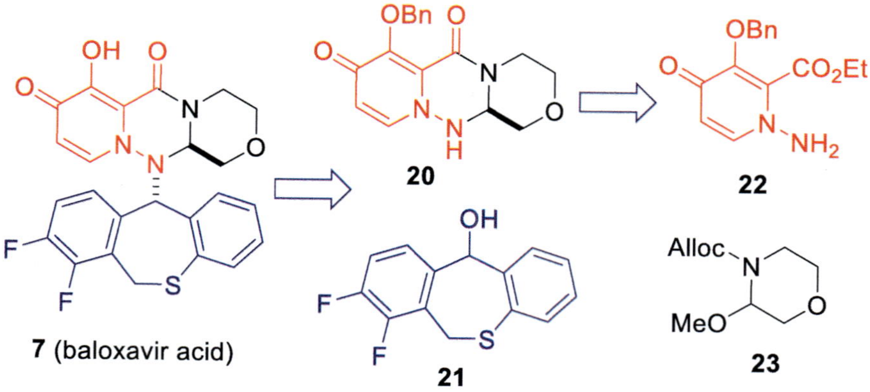

Scheme 1. Retrosynthetic analysis of baloxavir acid (**7**).

Hydrazine ester **22** was readily available from pyridone **24** by hydrazination with *tert*-butyl carbazate followed by Boc deprotection, and a three-step sequence was developed for the synthesis of morpholine ether **23** from morpholin-3-one (**26**, Scheme 2). Thus, the lactam NH was protected with allyl chloroformate to give **27**, and the lactam carbonyl was reduced to the hemiaminal **28**, which was methylated to furnish methyl ether **23**.

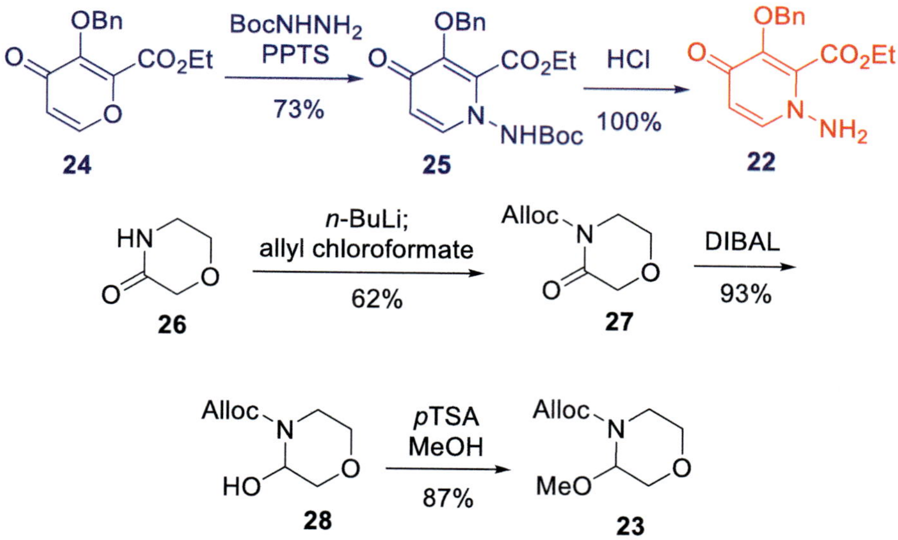

Scheme 2. Synthesis of **22** and **23**.

Sn(IV)-promoted condensation of hydrazone ester **22** with morpholine ether **23** furnished substituted hydrazine **29**, which underwent deprotection and spontaneous lactamization to give the reacemic tricyclic hydrazine **20**. Amide coupling of **20** with (*R*)-tetrahydrofuran-2-carboxylic acid gave a 1 : 1 mixture of two diastereomers, from which the desired (*R,R*)-diastereomer **31** was obtained in 45% yield *via* selective crystallization. The chiral auxiliary tetrahydrofuroyl group of **31** was removed under basic conditions to deliver the tricyclic hydrazine **20** (Scheme 3).

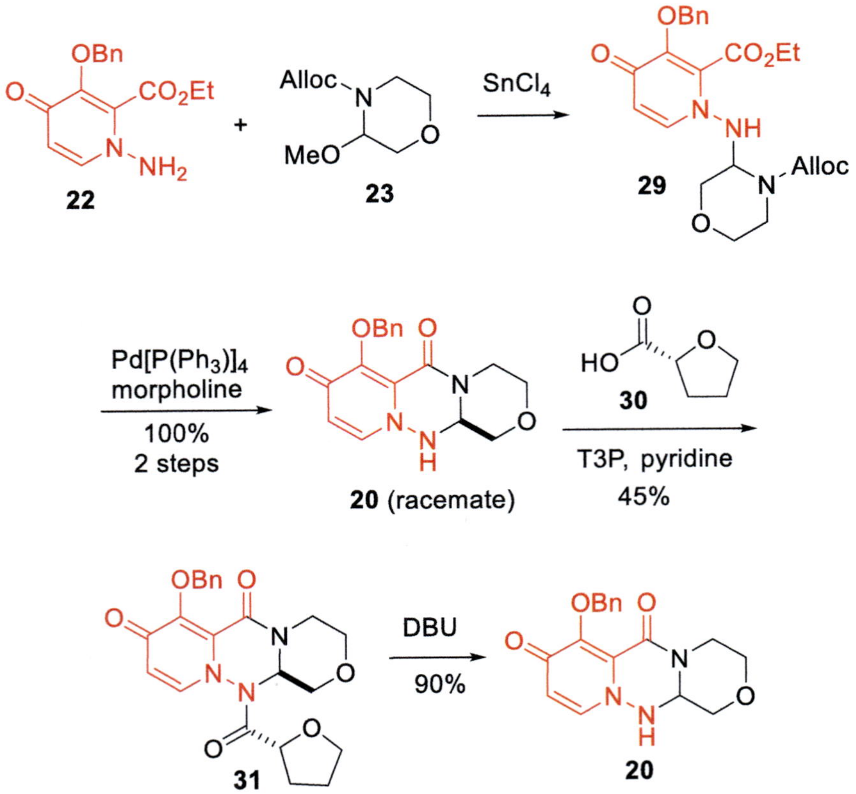

Scheme 3. Synthesis of **20**.

The tricyclic alcohol was made from 3,4-difluorobenzoic acid (**31**) in 5 steps (Scheme 4). Regioselective *ortho*-metallation followed by quenching with DMF gave the bicyclic lactol **32**, which was converted to thioacetal **33**. Reduction of **33** with 1,1,3,3,-tetramethylsiloxane and aluminum trichloride afforded acid **34**, which was cyclized to tricyclic ketone **35** *via* the Friedel–Crafts reaction. Reduction of **35** furnished the racemic tricyclic alcohol **21** (Scheme 4).

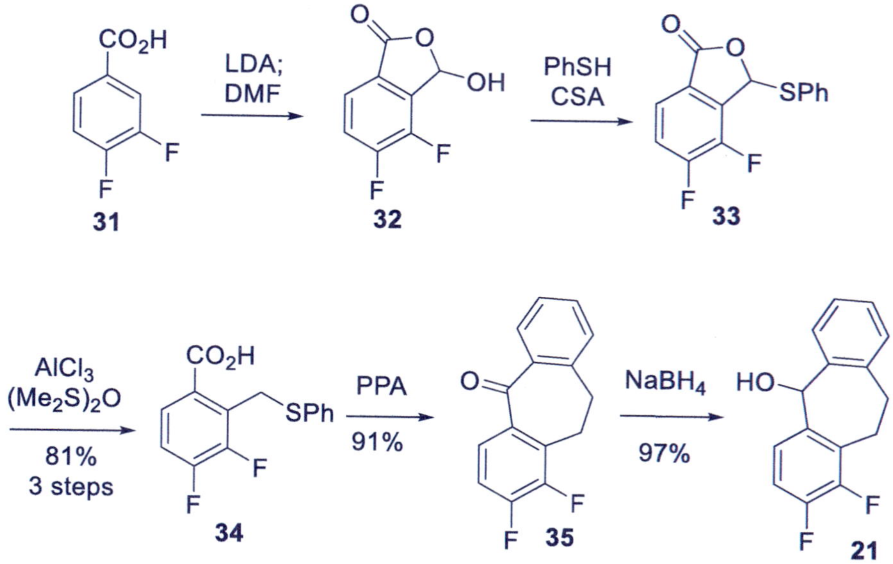

Scheme 4. Synthesis of **21**.

Coupling of **20** with **21** was carried out using a combination of propanephosphonic acid anhydride (T3P) with methanesulfonic acid to give a mixture of **36** and **7** due to partial debenzylation. This mixture was subjected to benzylation conditions to give enantiomerically pure **36** in 53% yield after crystallization. Debenzylation of **36** to **7** was carried out with lithium chloride in dimethylacetamide at 80 °C, an operationally simple condition that avoided the use of a palladium catalyst at the final steps of the synthesis. Baloxavir acid (**7**) was converted to its prodrug, baloxavir marboxil (**1**), by treatment with chloromethyl methyl carbonate.

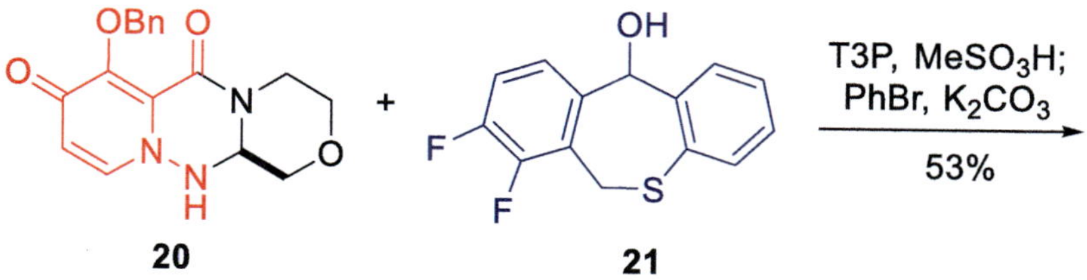

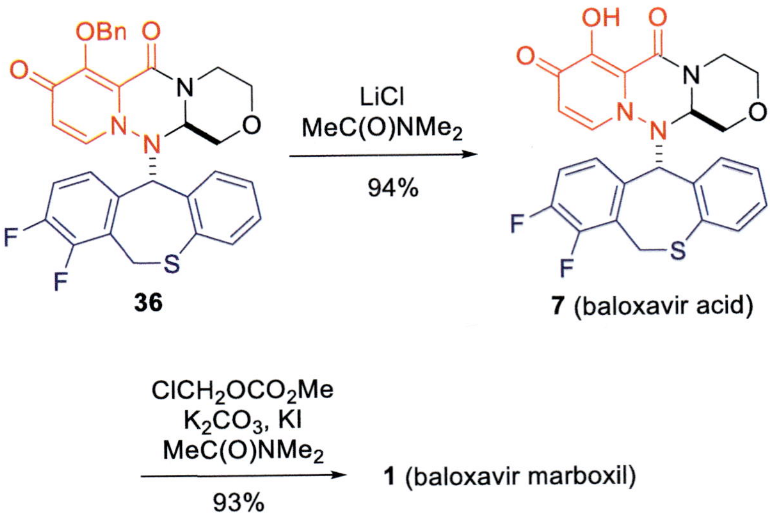

Scheme 5. Synthesis of baloxavir marboxil (**1**).

7. Summary

The long journey of baloxavir marboxil started more than 40 years ago[29] when Robert Krug and his flu team at Memorial Sloan–Kettering Cancer Center studied flu virus in the 1970s, which led to the surprising discovery in 1979 that virus produced mRNAs by "stealing a piece of a host cell messenger RNA to use as a primer to initiate the chains of viral messenger RNA", a pirating process called cap-snatching.[10,11] However, the cap-snatching PA subunit of a virus polymerase remained a mystery until 2014 when Stefan Reich et al. at the European Molecular Biology Laboratory in France revealed the detailed X-ray structure of the entire influenza virus polymerase composed of PA, PB1 and PB2 subunits.[9] The CEN of the PA subunit is essential for the cap-snatching process, therefore, a CEN inhibitor would disrupt this early step in viral replication. To this end, Shionogi discovered baloxavir acid as a potent CEN inhibitor of the viral polymerase to block its cap-snatching function. Baloxavir acid (**7**) was developed as its orally bioavailable prodrug, baloxavir marboxil (**1**), a new weapon in the arsenal for fighting potentially lethal influenza viruses for years to come.

8. References

1. History of 1918 Flu Pandemic | Pandemic Influenza (Flu). https://www.cdc.gov/flu/pandemic-resources/1918-commemoration/1918-pandemic-history.htm.
2. What is the global incidence of influenza? – Medscape.

https://www.medscape.com/answers/219557-3459/what-is-the-global-incidence-of-influenza.
3. Vaccine Effectiveness: How Well Do Flu Vaccines Work? https://www.cdc.gov/flu/vaccines-work/vaccineeffect.htm.
4. What You Should Know About Flu Antiviral Drugs. https://www.cdc.gov/flu/treatment/whatyoushould.htm.
5. Influenza Antiviral Drug Resistance. Questions & Answers. https://www.cdc.gov/flu/treatment/antiviralresistance.htm.
6. US Food and Drug Administration. FDA approves new drug to treat influenza (FDA 2018).
7. Fang, Q.; Wang, D. Advanced researches on the inhibition of influenza virus by Favipiravir and Baloxavir. *Biosafe. and Health.* **2020**, *2*, 64–70.
8. Noshi, T.; Kitano, M.; Taniguchi, K.; Yamamoto, A.; Omoto, S.; Baba, K.; Hashimoto, T.; Ishida, K.; Kushima, Y.; Hattori, K.; Kawai, M.; Yoshida, R.; Kobayashi, M.; Yoshinaga, T.; Sato, A.; Okamatsu, M.; Sakoda, Y.; Kida, H.; Shishido, T. A. Naito. In vitro characterization of baloxavir acid, a first-in-class cap-dependent endonuclease inhibitor of the influenza virus polymerase PA subunit. *Antiviral Res.* **2018**, *160*, 109–117.
9. Reich, S.; Guilligay, D.; Pflug, A.; Malet, H.; Berger, I.; Crépin, T.; Hart, D.; Lunardi, T.; Nanao, M.; Ruigrok, R. W.; Cusack, S. Structural insight into cap-snatching and RNA synthesis by influenza polymerase. *Nature.* **2014**, *516*, 361–366.
10. Plotch, S. J.; Bouloy, M.; Krug, R. M. Transfer of 5′-terminal cap of globin mRNA to influenza viral complementary RNA during transcription in vitro. *Proc. Natl. Acad. Sci. U.S.A.* **1979**, *76*, 1618–1622.
11. Plotch, S. J.; Bouloy, M.; Ulmanen, I.; Krug, R. M. A unique cap(m7GpppXm)-dependent influenza virion endonuclease cleaves capped RNAs to generate the primers that initiate viral RNA transcription. *Cell.* **1981**, *23*, 847–858.
12. Shie, J.; Fang, J. M. Development of effective anti-influenza drugs: congeners and conjugates – a review. *J. Biomed. Sci.* **2019**, *26*, 84.
13. Shen, Z.; Lou, K.; Wang, W. New small-molecule drug design strategies for fighting resistant influenza A. *Acta. Pharm. Sin. B.* **2015**, *5*, 419–430.
14. Taoda, Y.; Miyagawa, M.; Akiyama, T.; Tomita, K.; Hasegawa, Y.; Yoshida, R.; Noshi, T.; Shishido, T.; Kawai, M. Dihydrodibenzothiepine: promising hydrophobic pharmacophore in the influenza cap-dependent endonuclease inhibitor. *Bioorg. Med. Chem. Lett.* **2020**, *30*, 127547.
15. Miyagawa, M.; Akiyama, T.; Taoda, Y.; Takaya, K.; Takahashi-Kageyama, C.; Tomita, K.; Yasuo, K.; Hattori, K.; Shano, S.; Yoshida, R.; Shishido, T.; Yoshinaga, T.; Sato, A.; Kawai, M. Synthesis and SAR study of carbamoyl pyridone bicycle derivatives as potent inhibitors of influenza cap-dependent endonuclease. *J. Med. Chem.* **2019**, *62*, 8101–8114
16. Todd, B.; Tchesnokov, E. P.; Götte, M. The active form of the influenza cap-snatching endonuclease inhibitor baloxavir marboxil is a tight binding inhibitor. *J. Biol. Chem.* **2021**, *296*, 100486.
17. Tang, L.; Yan, H.; Wu, W.; Chen, D.; Gao, Z.; Hou, J.; Zhang, C.; Jiang, Y. Synthesis and anti-influenza virus effects of novel substituted polycyclic

pyridone derivatives modified from baloxavir. *J. Med. Chem.* **2021**, *64*, 19, 14465–14476.
18. Ivashchenko, A. A.; Mitkin, O. D.; Jones, J. C.; Nikitin, A. V.; Koryakova, A. G.; Ryakhovskiy, A.; Karapetian, R. N.; Kravchenko, D. V.; Aladinskiy, V.; Leneva, I. A.; Falynskova, I. N.; Glubokova, E. A.; Govorkova, E. A.; Ivachtchenko, A. V. Non-rigid diarylmethyl analogs of baloxavir as cap-dependent endonuclease inhibitors of influenza viruses. *J. Med. Chem.* **2020**, *63*, 9403–9420.
19. Johns, B. A.; Kawasuji, T.; Weatherhead, J. G.; Taishi, T.; Temelkoff, D. P.; Yoshida, H.; Akiyama, T.; Taoda, Y.; Murai, H.; Kiyama, R.; Fuji, M.; Tanimoto, N.; Jeffrey, J.; Foster, S. A.; Yoshinaga, T.; Seki, T.; Kobayashi, M.; Sato, A.; Johnson, M. N.; Garvey, E. P.; Fujiwara, T. Carbamoyl pyridone HIV-1 integrase inhibitors 3. A diastereomeric approach to chiral nonracemic tricyclic ring systems and the discovery of dolutegravir (S/GSK1349572) and (S/GSK1265744). *J. Med. Chem.* **2013**, *56*, 5901−5916.
20. Kawasuji, T.; Johns, B. A.; Yoshida, H.; Weatherhead, J. G.; Akiyama, T.; Taishi, T.; Taoda, Y.; Mikamiyama-Iwata, M.; Murai, H.; Kiyama, R.; Fuji, M.; Tanimoto, N.; Yoshinaga, T.; Seki, T.; Kobayashi, M.; Sato, A.; Garvey, E. P.; Fujiwara, T. Carbamoyl pyridone HIV-1 integrase inhibitors. 2. Bi- and tricyclic derivatives result in superior antiviral and pharmacokinetic profiles. *J. Med. Chem.* **2013**, *56*, 1124−1135.
21. Kawasuji, T.; Johns, B. A.; Yoshida, H.; Taishi, T.; Taoda, Y.; Murai, H.; Kiyama, R.; Fuji, M.; Yoshinaga, T.; Seki, T.; Kobayashi, M.; Sato, A.; Fujiwara, T. Carbamoyl pyridone HIV-1 integrase inhibitors. 1. Molecular design and establishment of an advanced two-metal binding pharmacophore. *J. Med. Chem.* **2012**, *55*, 8735−8744.
22. Heo, Y.A. Baloxavir: first global approval. *Drugs*. **2018**, *78*, 693–697.
23. https://www.accessdata.fda.gov/drugsatfda_docs/label/2018/210854s000lbl.pdf
24. Hayden, N. Sugaya, N. Hirotsu, N. Lee, M. D. de Jong, A. C. Hurt, T. Ishida, H. Sekino, K. Yamada, S. Portsmouth, K. Kawaguchi, T. Shishido, M. Arai, K. Tsuchiya, T. Uehara, A. Watanabe. Baloxavir marboxil for uncomplicated influenza in adults and adolescents. *N. Engl. J. Med.* **2018**, *379*, 913–923.
25. Koshimichi, H.; Ishibashi, T.; Kawaguchi, N.; Sato, C.; Kawasaki, A.; Wajima T. Safety, tolerability, and pharmacokinetics of the novel anti-influenza agent baloxavir marboxil in healthy adults: phase I study findings. *Clin. Drug Investig.* **2018**, *38*, 1189–1196.
26. R. O'Hanlon, M. L. Shaw ML. Baloxavir marboxil: the new influenza drug on the market. *Curr. Opin. Virol.* **2019**, *35*, 14–18.
27. Ng, K. E. Xofluza (Baloxavir Marboxil) for the treatment of acute uncomplicated influenza. *P & T.* **2019**, *44*, 9–11.
28. Hughes, D. L. Review of the patent literature: synthesis and final forms of antiviral drugs tecovirimat and baloxavir marboxil. *Org. Process Res. Dev.* **2019**, *23*, 1298–1307.
29. Airhart, M. G. The 40 year-old discovery behind a promising new flu drug. https://cns.utexas.edu/news/the-40-year-old-discovery-behind-a-promising-new-flu-drug.

4

Process Chemistry Development of the HIV Protease Inhibitor Drug Kaletra: A Mixture of Ritonavir and Lopinavir

Daniel A. Dickman

USAN: Ritonavir (**1**)
Trade name: Norvir
Abbott Laboratories, 1996
HIV protease inhibitor

USAN: Lopinavir (**44**)
Trade name: Kaletra
(Lopinavir + Ritonavir *ca.* 4:1)
Abbott Laboratories, 2000
HIV protease inhibitor

Current Drug Synthesis, First Edition. Edited by Jie Jack Li.

1 Background:

The treatment of viral diseases is an important area of medicinal chemistry with a plethora of anti-viral targets that vary in size and complexity. The advent of acquired immunodeficiency syndrome (AIDS) in the 1980s, which is a degenerative disease of the immune system, became one of the most vexing problems in medicine to curtail, and a substantial amount of work was undertaken to design and market a class of drugs that inhibited the HIV-1 aspartic protease enzyme. Norvir[1] (ritonavir, **1**) and Kaletra[2] (co-formulated lopinavir + ritonavir *ca.* 4:1, respectively) became important therapies to treat AIDS patients. Kaletra, the trade name for the lopinavir/ritonavir mixture, was approved for use in 2000 and is on the World Health Organization's list of Essential Medicines as is Norvir, which is ritonavir alone. Both ritonavir and lopinavir share the same central core structure and knowledge gained from the synthesis of ritonavir greatly shortened the development time to produce lopinavir and hence, the drug cocktail of lopinavir plus ritonavir. It is noteworthy that ritonavir can super-charge several other HIV protease inhibitors where it was shown that lopinavir is among the most efficacious of a broad class of other HIV protease inhibitors.[2,3] The scope of this chapter is how the early medicinal chemistry discovery syntheses of both ritonavir and lopinavir evolved to large-scale processes to manufacture both APIs at Abbott Laboratories (now AbbVie). This chapter is divided into sections starting from the ritonavir synthesis, which first describes the synthesis of its core or central portion and is then followed by the synthesis of the ritonavir side or wing pieces that are appended to the core, and lastly, the methods of coupling these wing pieces to the core structure. The same concept will apply to the discussion for the lopinavir synthesis. The work discussed herein for the large-scale process syntheses of ritonavir and lopinavir is limited in scope to work performed at Abbott Laboratories as is referenced from the chemical and patent literature.

2 Ritonavir Portion of Kaletra Syntheses

Ritonavir (**1**) was Abbott Laboratories' first marketed drug for treatment of the HIV virus. It was the result of a substantial research effort[4] where several novel inhibitors were designed and synthesized based on the C_2-symmetric structure of HIV protease.[5] Its retrosynthetic scheme is shown in Scheme 1.

The original class of HIV protease inhibitors studied at Abbott Laboratories involved compounds whose core structures were C_2-symmetric as depicted by diamine **2**[6,7] in Figure 1. They evolved to molecules with central core structures such as diamine **3** with equivalent potency because compounds containing the core structure of **2** tended to have lower bioavailability than those containing the core structure of **3**. It is noteworthy that early efforts to synthesize compounds containing the core structure of **3** involved prior formation of compounds like **2**[8] followed by selective deoxygenation to compounds such as **3**.[9]

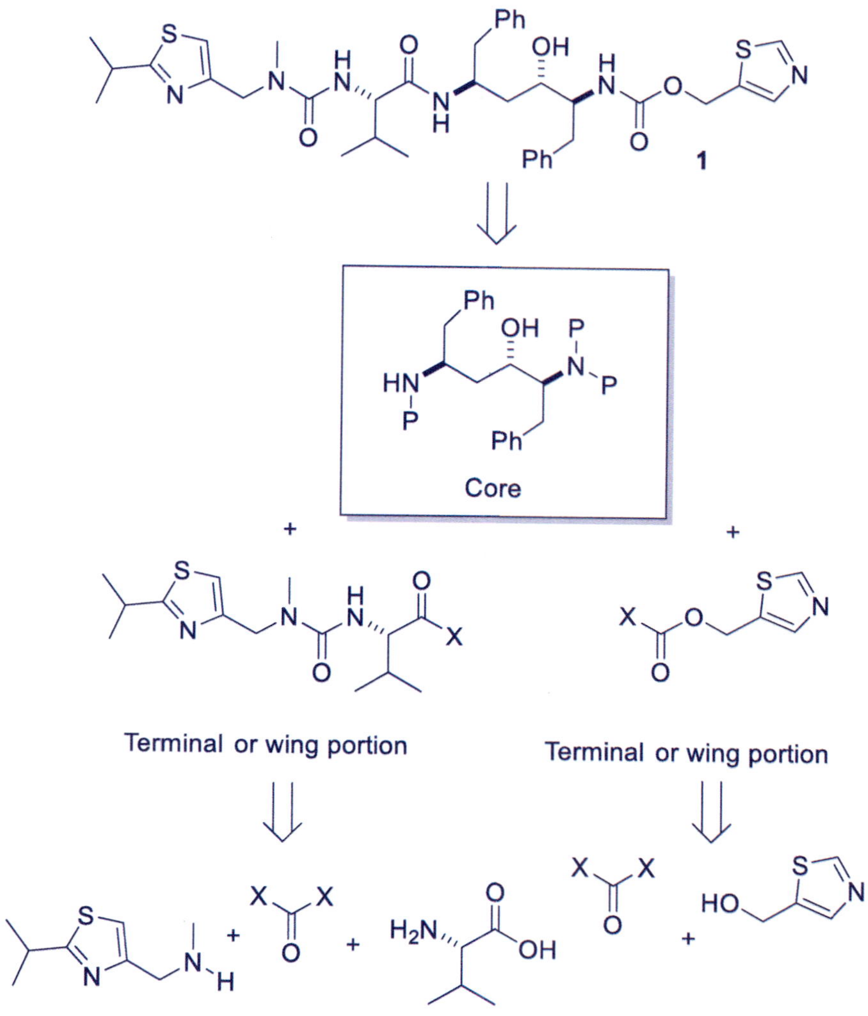

X = activating group, P = protecting group

Scheme 1. Retrosynthesis of Ritonavir (**1**)

2 **3**

Figure 1. C_2 (**2**) and non-C_2 symmetric (**3**) core structures

A retrosynthetic analysis of ritonavir (ABT-538) is depicted in Scheme 1. Displayed are the building blocks, which are the core, (*S*)-valine, two carbonyl groups and the substituted amino and hydroxy thiazoles. Of these, the core **3** is the most complex and it contains three chiral centers, where all are the (*S*)-configuration. Therefore, stereocontrol is a critical element in its synthesis. The information presented in the ritonavir section of this chapter proceeds by first discussing its discovery syntheses[4,8,9] followed by a discussion of the process syntheses that are amenable to large scale production. The information regarding the lopinavir synthesis is presented in similar fashion.

3 Discovery Synthesis of the Ritonavir Core

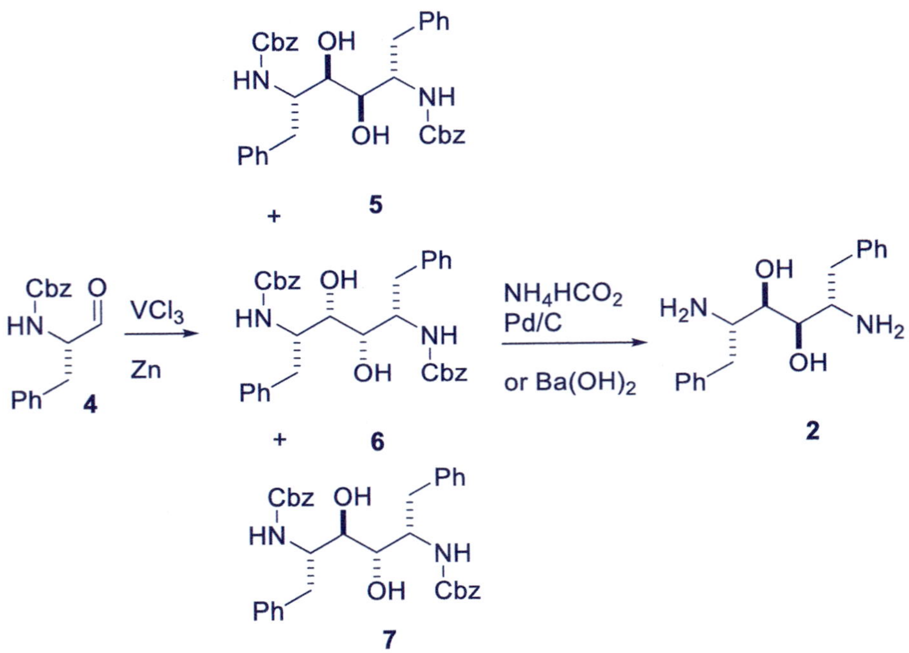

Scheme 2. Synthesis of diaminodiol **2** from *N*-Cbz *l*-phenylalaninal **4**

The discovery synthesis of the ritonavir core derives its chirality from commercially available *N*-Cbz (*S*)-phenylalaninol that is oxidized to form compound **4** by treatment with DMSO and oxalyl chloride[10] in greater than 99.5% *e.e.* It can then undergo a variety of coupling protocols to form compounds **5-7** as shown in Scheme 2. Initially in the discovery phase of the project, it was not critical for there to be a selective formation of **5** and reagents such as $TiCl_3$, Zn–Cu in DMF[11] could be employed to form compounds **5-7**, which could be purified and converted to more analogs. It was later found by using reagents such as VCl_3 $(THF)_3$ and zinc dust in dichloromethane[12,13] that the desired compound **5** was selectively formed over **6** and **7** by a ratio of (8:1:1, respectively).

Removal of the *N*-Cbz group (or *N*-Boc group in certain cases) was done with either ammonium formate and palladium (transfer hydrogenation) or by barium hydroxide hydrolysis to form **2** in approximately 50% overall yield from **4**.

Several analogs of **2**, and its isomers from deprotection of **6** or **7**, could also be prepared by sequential reaction with the appropriate activated carboxylic acids to prepare symmetrical compounds or unsymmetrical compounds after chromatography. To further the study of different analogs, it was determined that compounds containing the non-C_2 symmetric core structure of **3** depicted in Figure 1 should be explored. The discovery synthesis of **3** is shown in Scheme 3 starting from **5**. Diol **5** was treated with α-acetoxyl isobutyl bromide to provide bromide **8** with inversion of configuration of the bromine atom. Bromide **8** was de-halogenated to give the des-bromo compound **9** with tri-*n*-butyl tin hydride in THF, which was then globally deprotected with barium hydroxide in water and dioxane. The overall yields of **3** from **5** were consistently around 70%.

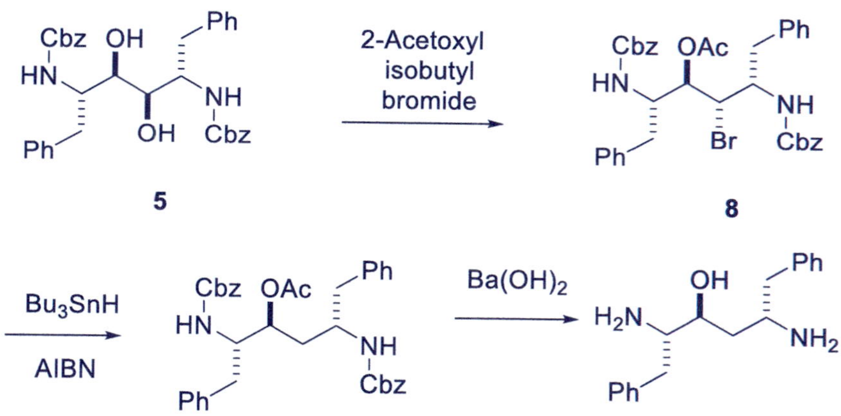

5 8 9 3

Scheme 3: Conversion of diol **5** to alcohol **3**

The next steps in the discovery synthesis of ritonavir involved coupling of (**3**) to various activated carbonyl groups, attached to a thiazole (aryl or other heterocycle) ring, where the activated carbonyl groups were usually *para*-nitrophenyl (PNP) carbonates of the general structures **X** or **Y** and R^1 and R^2 were substituted thiazoles (Scheme 4). These compounds could then be allowed to react with **3** in a non-selective fashion where each isomer (monomer or dimer) could then be separated from another by column chromatography. Subsequently, each different regio-isomer could undergo coupling with another activated carbonyl group containing a second heterocyclic moiety.

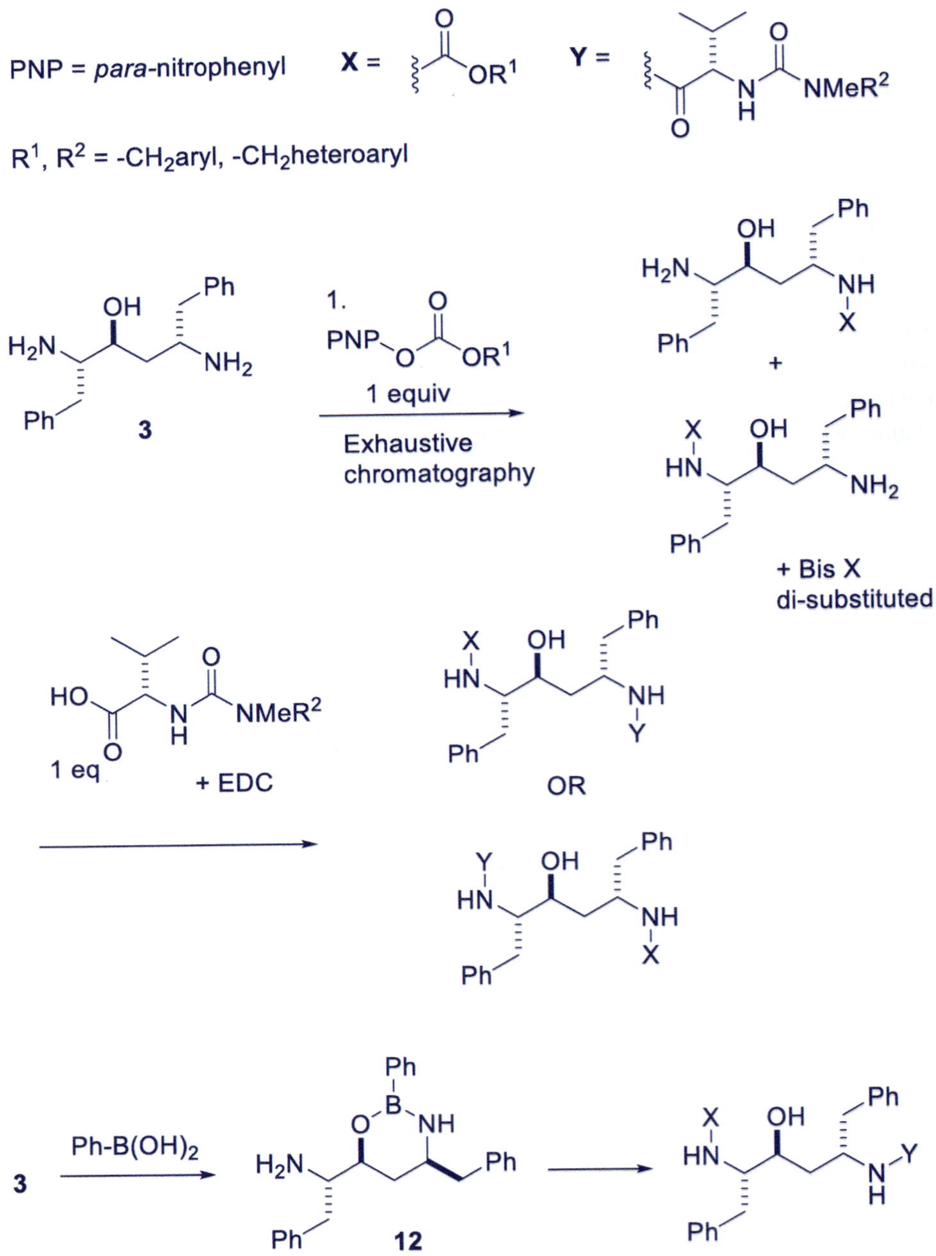

Scheme 4: Synthetic methods to prepare ritonavir and its analogs

This approach was practical for small amounts of lead compounds. However, when larger amounts of ritonavir or other analogs were desired for further evaluation, **3** was allowed to reflux in toluene with phenyl boronic acid to preferentially from the cyclic aza-oxaboronate **12**,[4,14] which protects the β-hydroxyl nitrogen atom. Compound **12** can then be allowed to react with the appropriate activated carbonyl groups to provide more ritonavir and its analogs without the need for purification by chromatography. A diagram of the types of coupling steps to prepare ritonavir and its analogs is shown in Scheme 4 where prior formation of **12**, by condensation of **3** with phenyl boronic acid is quite useful to avoid difficult column chromatographic operations.[14]

4 Discovery Synthesis of Ritonavir Wing Pieces

In the following Schemes (5–7) the thiazole wing pieces were synthesized, and they are coupled to the protected core portion **12** with the appropriate amino-acid linkers [(*S*)-valine for the case of ritonavir] are described. In Scheme 5, the synthesis of 5-(hydroxymethyl) thiazole (**16**) is shown where step one is the reaction of thio-formamide (**13**) with ethyl 2-chloro-2-formyl-acetate (**14**) in acetone in 60% yield[4,15] to form 5-carboethoxy thiazole (**15**), which is readily reduced with LAH in THF to provide alcohol **16** in 75% yield. Compound **16** is allowed to react with *p*-nitrophenyl chloroformate and *N*-methyl morpholine in dichloromethane to provide the O-PNP activated carbonate compound **17**.

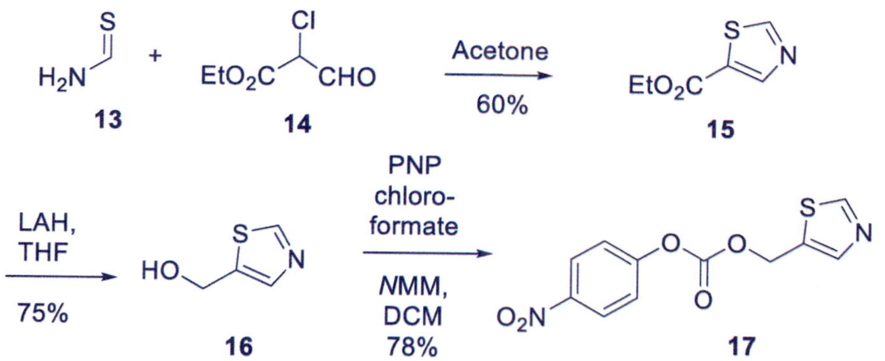

Scheme 5: The synthesis of the O-PNP carbonate ester of 5-(hydroxymethyl) thiazole **17**

The next steps of the synthesis are to couple **17** and **19** to the 2-phenyl 1,3,2-oxazaborinane protected core **12** to form ritonavir (**1**), which is shown in Scheme 6. Compounds **12** and **17** are allowed to react to provide **18**, in about 50% yield, selectively after silica-gel plug filtration, by conditions that cleave the phenyl boronic acid protecting group. Compound **18** is then allowed to react with {[(2-isopropylthiazol-4-yl)methyl]-(methyl)carbamoyl}-*l*-valine (**19**), whose synthesis is descried in Scheme 6, in the presence of HOBt and EDC to provide ritonavir (**1**) in 75% yield.

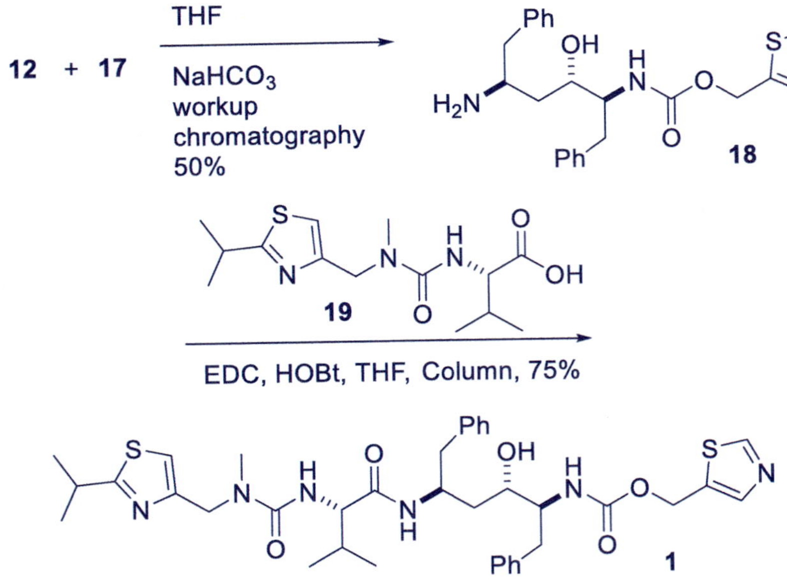

Scheme 6: The coupling of the terminal wing pieces **17** and **19** to the protected core to provide ritonavir (**1**)

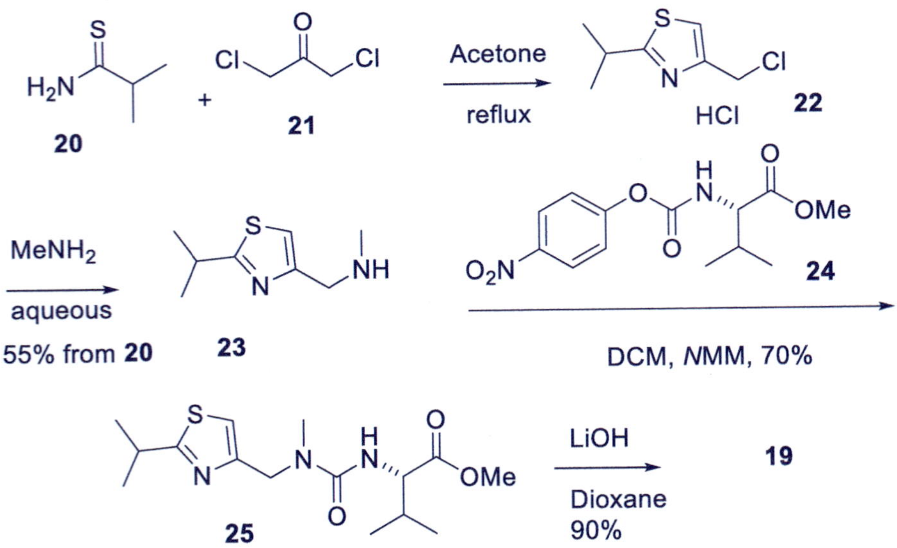

Scheme 7: The preparation of [((2-isopropylthiazol-4-yl) methyl) (methyl)carbamoyl]-*l*-valine (**19**)

The synthesis of compound **19** is described in Scheme 7 where thioamide **20** is treated with 1,3-dichloroacetone (**21**) in refluxing acetone to provide 4-(chloromethyl)-2-

isopropylthiazole hydrochloride (**22**), which is then dissolved in water and is added to a solution of 40% aqueous methyl amine to provide 1-(2-isopropylthiazol-4-yl)-*N*-methyl-methanamine **23** in 55% overall yield over two steps. Compound **23** is then treated with methyl [(4-nitrophenoxy) carbonyl]-*l*-valinate (**24**) to provide the methyl ester **25**, which is then hydrolyzed, without racemization, to **19** with lithium hydroxide and acid **19** was coupled to the mono substituted core as previously described.

In conclusion, the discovery synthesis of ritonavir was very amenable to the preparation of several analogs in short order with the caveat that extensive column chromatography operations could be run with relative ease. The early ritonavir synthesis was not amenable to large scale due to several factors, which are as follows: (1) The synthesis of the core structure was not entirely selective, and (2) it involved several cumbersome reduction and oxidation steps from phenylalanine, where the (3) tri-*n*-butyl tin hydride reduction step would most certainly need to be eliminated on larger scales due to its toxicity and (4), it would be advantageous to have different, less-expensive coupling conditions of the terminal wing pieces to the core as well as (5) improved syntheses of the wing pieces themselves. The next sections will describe how an improved core synthesis was discovered, the improvements performed in both route convergency and in the synthesis of the wing pieces individually, and the substitution of more expensive reagents with less-costly ones to provide a robust synthesis that enabled production of ritonavir at the multi-kilogram level.

5 Large-Scale Process Chemistry Synthesis of the Ritonavir Core

A much-improved ritonavir core synthesis employs commercially available (*S*)-phenylalanine (Scheme 8) as its source for chirality. (*S*)-Phenylalanine (**26**) is treated with benzyl chloride (3.3 equiv) in refluxing water with excess potassium carbonate to provide **27** in 94% yield with negligible racemization. The next step involves treatment of the *N,N*-dibenzyl-benzyl ester product **27** with excess acetonitrile anion, which is generated by addition of sodium amide in acetonitrile in THF at –40 °C, at low temperatures, less than –40 °C, to afford **28** in high yields (78–85%) after crystallization. It is noteworthy that at temperatures above –40 °C, **28** will racemize. Conversion of nitrile **28** to enaminone **30** proceeded in 94% yield after crystallization where the %*e.e.* of the product was enriched slightly on crystallization to > 99%.[16,17] Conversion of enaminone **30** to **32a** with the appropriate acid catalyzed borohydride reagents was first accomplished[16] by a two-stage process with sodium borohydride and methane sulfonic acid in isopropanol/THF to provide a mixture of (*S*):(*R*) diamines **31a**:**31b** in a 94:6 ratio, respectively, which were then subsequently reduced with a preformed solution of $NaBH_3(OTFA)$[18] to give a 93:4:2:1 mixture of (*S,S,S*):(*S,S,R*):(*S,R,S*):(*S,R,R*) amines, **32a**:**32b**:**32c**:**32d**, respectively, where **32a** was the major (93%) product.

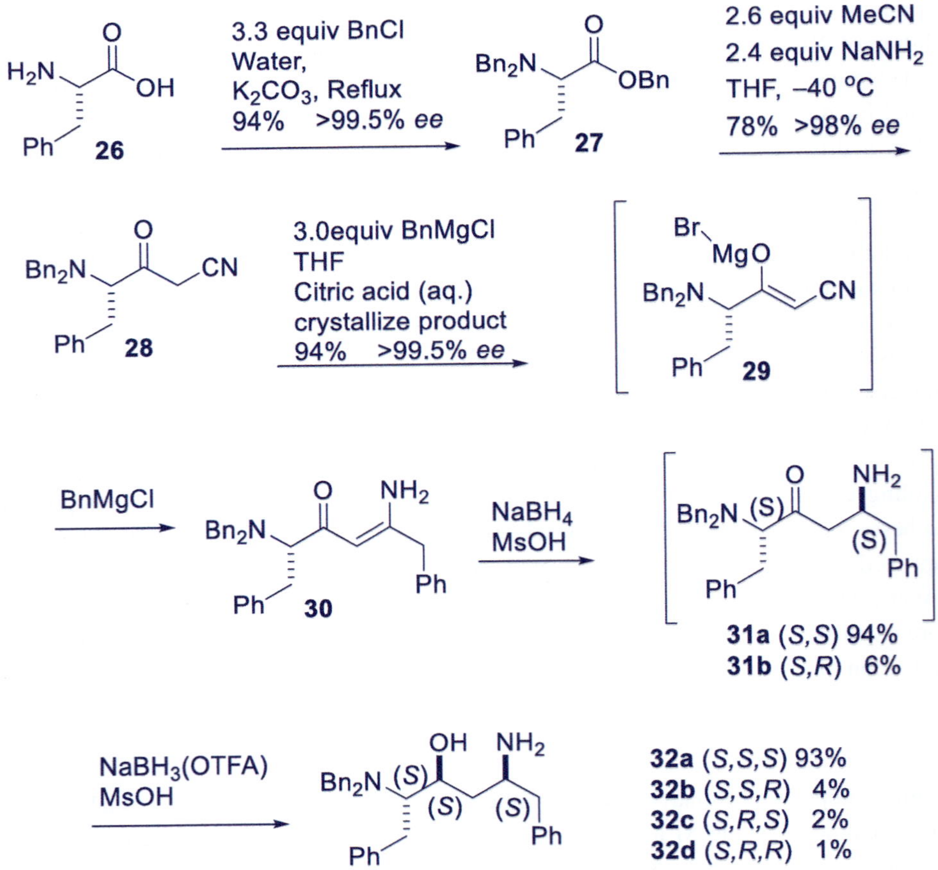

Scheme 8: The process to the core **32a** *via* (*S*)-phenylalanine **26**

Removal of the *N,N*-dibenzyl groups of **32a,-d** by transfer hydrogenation followed by formation of the dihydrochloride salt of **3** and crystallization could purge the unwanted isomers with a significant loss of material (Scheme 9).[16] Although these results were acceptable in the short-term, more work was required to determine the scalability of this process to determine if either a more selective synthesis of **32a** could be performed or an improved crystallization protocol to remove the unwanted isomers **32b-d** could be established. It was also necessary to have one of the two primary amines on the alkyl chain protected to assure selective coupling of the thiazole wing pieces during the subsequent synthetic steps to eliminate the use of phenyl boronic acid.

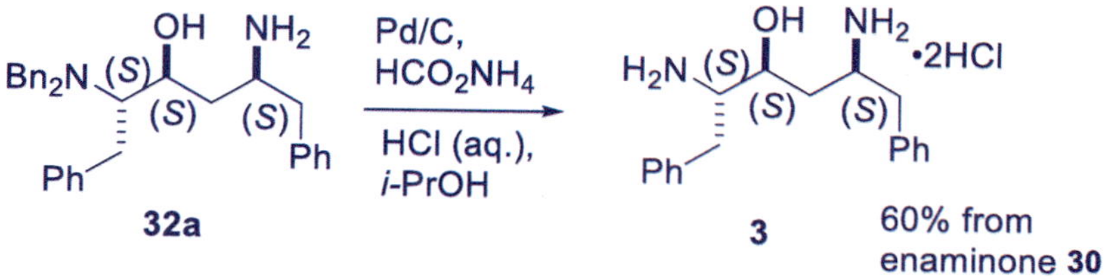

Scheme 9: Salt formation/crystallization/purification of compound **3**

As stated previously, excellent stereochemical control for the two-stage reduction of enaminone **30** to **32a** was accomplished by treatment with sodium borohydride in THF/isopropanol mixtures with methane sulfonic acid followed by treatment of the (94:6) mixture of **31a**:**31b**, respectively, with a sodium borohydride trifluoroacetic acid complex. However, upon scaling this mixture to larger volumes, it was noted that THF, the co-solvent, started to decompose due to the extended reaction times and therefore, a new protocol became necessary. An elegant solution was determined by careful screening of several different mineral and organic acids, borane reagents, and solvents.[19] It involved taking a suspension of sodium borohydride, 3.0 equiv, in dimethoxy ethane and treating it with methane sulfonic acid, 7.5 equiv at –5 °C to 5 °C followed by addition of enaminone **31**, 123 kg, in isopropanol and allowing the solution to stir at 0-10°C for 12 hours. The reaction mixture was then quenched by addition of triethanolamine, 3.0 equiv, and it was allowed to stir at less than 5 °C for 30 minutes. Then a solution of sodium borohydride, 2.5 equiv., in DMA, 7.9 equiv., was slowly added and the mixture was allowed to stir at 10-20 °C for 2 hours at which time it was quenched with water and extracted with MTBE to provide by weight % assay yield **32a**:**32b**:**32c**:**32d** of 83:4:6:2, respectively. Although the selectivity could not be improved and did diminish on larger scales, there remained several methods to remove the undesired isomers by a robust crystallization protocol as shown in Scheme 10. Crude **32a-d** are treated[20] with Boc_2O in MTBE and 10% aqueous potassium carbonate to provide the N-Boc derivatives **33a-d**, which are then subjected to transfer hydrogenation conditions where the desired major (*S,S,S*) isomer **34** is crystallized as its hemi-succinate salt. This process was performed to provide between 55–85 kg of **34**.

If necessary, enaminone **30** can be converted to the corresponding *N*-Boc enaminone **35**, which can also undergo stereoselective reduction to **36** and further to **33a** as shown in Scheme 11.[20] By formation of the *N*-Boc compound **35**, the order of reduction is now reversed in that the keto functionality is reduced first and is followed by reduction of the *N*-Boc enamine. This transformation can be selectively performed to form **33a** in several ways where N-Boc enamide **35** is selectively reduced to the (*S,S*)-*N,N*-dibenzyl amino alcohol **36** by treatment with lithium aluminum hydride or lithium triethyl borohydride, which can be further reduced to **33a** under a hydrogen atmosphere with platinum oxide or palladium hydroxide in 70–75% yields. The transformation of **35** to **33a** can also be performed in one pot, by sequential treatment with borane tetrahydrofuran complex, then lithium aluminum hydride or potassium borohydride. The other minor isomers **33b-d** are removed by following the crystallization protocol shown in Scheme 10.

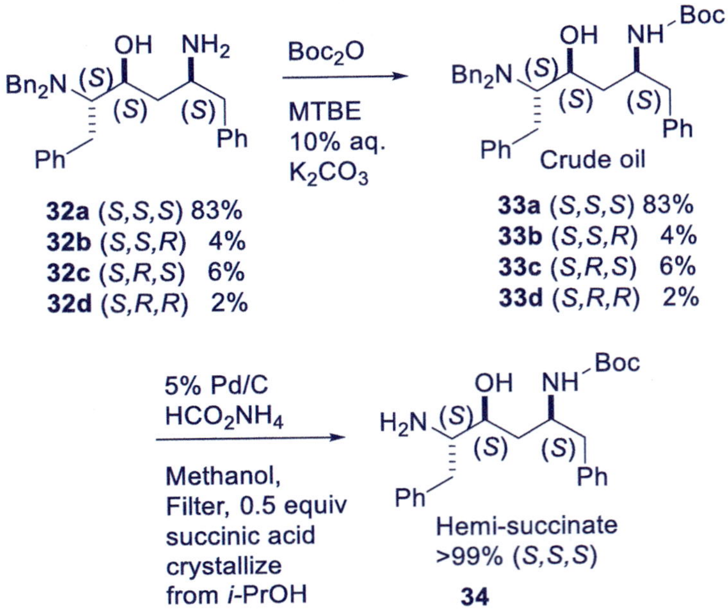

Scheme 10: The conversion of crude **32a–d** to pure *N*-Boc core hemi-succinate **34**

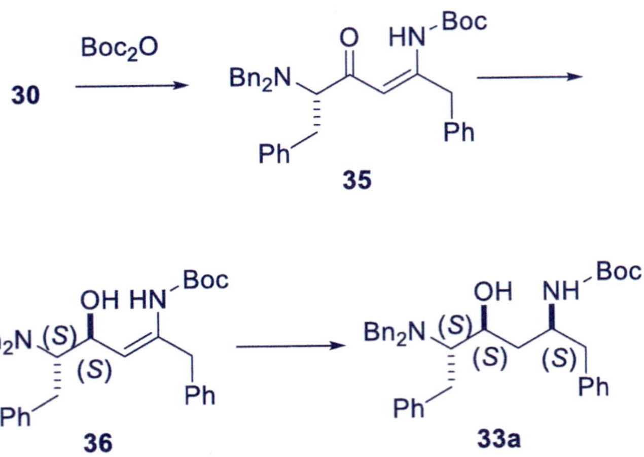

Scheme 11: Formation and reduction of *N*-Boc enaminone **35** to **33a**

In conclusion, a very scalable and robust method was discovered to produce the *N*-Boc core compound **34** as its highly pure hemi-succinate salt. The ritonavir core section is protected as its (*S*)-*N*-Boc amino derivative, which is now suitable to attach each thiazole wing piece selectively. Both methods, listed in Schemes 8 and 11, to convert enone **30** stereo-selectively to a protected version of compound **3** are suitable to couple the carbonyl-*N* or *O* thiazoles on the much larger scales. The next sections will discuss the work performed towards a robust synthesis of both 5-hydroxymethyl thiazole (**16**) and 1-(2-isopropylthiazol-4-yl)-*N*-methyl-methanamine (**23**) and how they were coupled to the protected core as their activated carbonyl derivatives **17** and **19**, respectively.

6 Large-Scale Syntheses of the 5-Hydroxymethyl Thiazole Wing Portion (16)

The initial route to 5-hydroxymethyl thiazole (**16**) had a liability for large scale preparation because the starting material, thio-formamide (**13**), (Scheme 5) is very difficult to prepare and is unstable. Therefore, an alternative route that did not use it was necessary. Two methods, which are scalable to prepare (**16**) and do not use thio-formamide (**13**) were developed. They are depicted in Schemes 12 and 13. The first[21] starts from inexpensive thiazolidine-2,4-dione (**37**), which is treated with phosphorus oxy-bromide in the presence of DMF and the mixture is heated to form 2,4-dibromo-5-thiazolecarboxaldehyde (**38**) in 72% yield after crystallization. The next two steps, reduction with sodium borohydride in methanol to alcohol **39**, and dehalogenation with hydrogen and palladium on carbon provided 5-hydroxymethyl thiazole (**16**) in 95% yield over those two steps (Scheme 12).

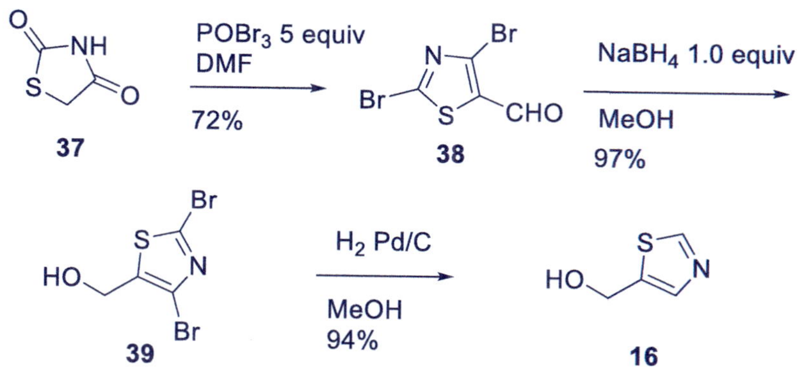

Scheme 12: The synthesis of 5-hydroxymethyl thiazole (**16**) from thiazolidine 2,4-dione (**37**)

The second method begins by treatment of *N*-allyl isothiocyanate (**40**) with excess chlorine gas or sulfuryl chloride to provide 2-chloro-5-chloromethyl-thiazole (**41**) in approximately 50% yield after distillation.[22] It is then converted to its formate ester with sodium formate in water and MTBE co-solvents followed by hydrolysis with sodium hydroxide to provide 2-chloro-5-hydroxymethyl thiazole (**42**) in 95% overall yield. The next step, reductive de-chlorination, is accomplished with hydrogen over palladium and carbon with a slight excess of sodium acetate to provide 5-hydroxymethyl thiazole in 94% yield[23] as shown below in Scheme 13.

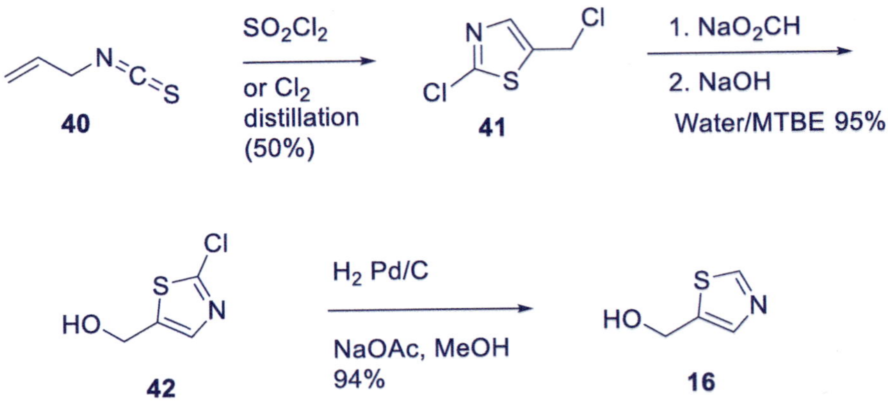

Scheme 13: Conversion of allyl isothiocyanate (**40**) to 5-hydroxylmethyl thiazole (**16**)

Although both methods are very robust, 2-chloro-5-chloromethyl-thiazole is cheaper to purchase and hence, the latter (Scheme 13) of the two methods was preferable. However, the former method (Scheme 12) may be employed if an inexpensive source of thiazolidine-2,4-dione (**37**) is available. For the preparation of 1-(2-isopropylthiazol-4-yl)-N-methyl-methanamine (**23**), the starting materials from the discovery route of thio-iso-butyramide (**20**), 1,3-dichloroacetone (**21**), and methyl amine as depicted in Scheme 7 do represent a preferred method to prepare it from readily accessible starting materials. Thus, no short-term process research was necessary. However, there were further improvements to be made in the ritonavir process from wing pieces **17** and **19**, and the core **34**, which are described in section 7.

7 The Large-Scale Coupling of the Thiazole Wing Pieces (17) and (19) to the Core (34)

The next steps in the process to supply multi-kilogram amounts of ritonavir were to devise efficient, short, and robust methods for the conversion sequence of **34** plus **17** to **43**, remove the *N*-Boc protecting group to form **18**, and then append **19** to provide ritonavir (**1**). This is shown in Scheme 14 where 65 kg of **17** was allowed to react with 75 kg of **34** in the presence of aqueous sodium bicarbonate and ethyl acetate to provide **43**, which was not isolated and was treated with HCl to remove the *N*-Boc protecting group where compound **18**, which is des-Boc **43**, was isolated as its di-hydrochloride salt.[24] The final step of the process was an improvement in the peptide coupling protocol where {[(2-isopropylthiazol-4-yl) methyl) (methyl)carbamoyl)-*l*-valine (**19**)]}, prepared by the sequence in Scheme 7, was activated with isobutyl chloroformate and *N*-hydroxy succinimide to form its activated *N*-hydroxy succinimide ester *in situ*, which was then allowed to react with **18** to provide ritonavir (**1**) after solvent swap and filtration[24].

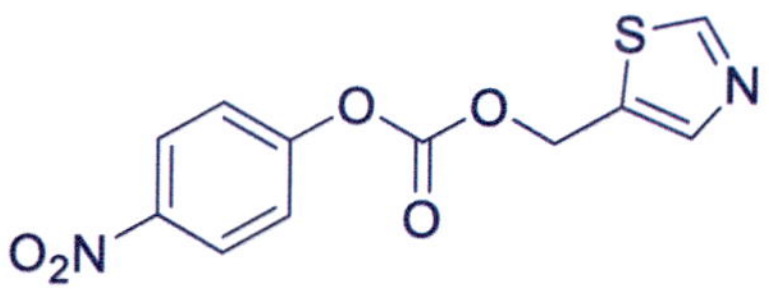

17

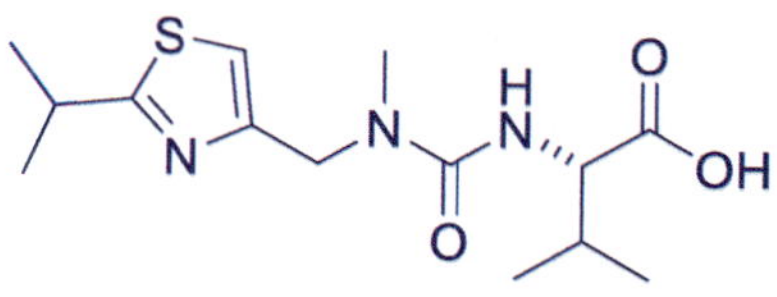

19

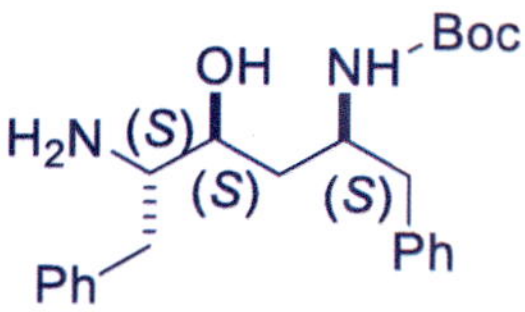

Hemi-succinate

34

Figure 2. Ritonavir advanced intermediates

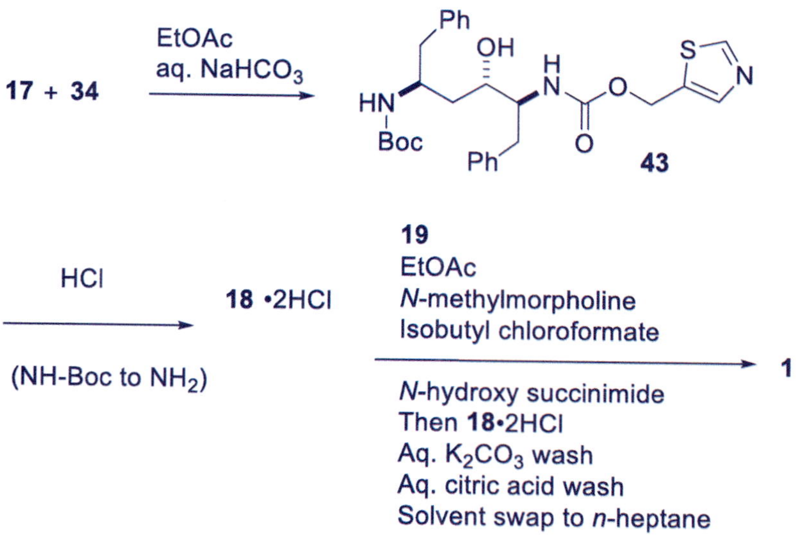

Scheme 14: The scalable conversion of *N*-Boc Core **34** to ritonavir (**1**)

In conclusion, the process to prepare ritonavir (**1**) evolved from a non-specific discovery route with several column chromatographic operations to purify all known isomers of ritonavir for biological evaluation, to several highly selective, robust, and scalable syntheses with no chromatography that was able to eventually provide the drug substance in > 100 kg lots. This was accomplished by an alternative hydroxy diamino core synthesis that was able to selectively introduce two additional chiral centers in a diastereoselective manner from *N,N*-dibenzyl-*O*-benzyl (*S*)-phenylalanine (**27**). Acetonitrile serves as the source of the other two carbons on the alkyl chain and the other benzyl group arises from benzyl magnesium chloride. Several methods are available to transform the key enamidinone intermediate **30** to the hemi-succinate salt of **34** with many reagents. The synthesis of the thiazole end or wing pieces became robust and scalable. Their coupling to the core structure (*N,N*-dibenzyl or *N*-Boc) also became robust. From a larger-scale standpoint, it would make sense to employ another phosgene equivalent instead of *p*-nitrophenol chloroformate due to lower cost as has been published in the patent literature,[25] although that work is beyond the scope of this discussion. The synthesis of intermediate **34** became a useful starting point for the synthesis of other analogs in rapid fashion. Ritonavir was shown to be an inhibitor of cytochrome P450 3A4 (CYP3A4), the major human hepatic drug-metabolizing enzyme and this was a slight drawback in its early clinical development. However, it was found that ritonavir, by administration in sub-therapeutic doses, could increase plasma concentrations of other, more potent HIV drugs metabolized by CYP3A4 to extend their clinical efficacy.[26] Therefore, it became a primary goal to find other compounds that were more potent than ritonavir (**1**), which could potentially benefit from this interaction, or be administered alone, without the undesired side-effects of ritonavir (**1**)-induced CYP3A4 inhibition. Lopinavir (**4**) was one of those types of compounds.

8 Lopinavir (44) Portion of Kaletra—Discovery Synthesis and Process Development

It was believed that it was the thiazole wing pieces that imparted the CYP3A4 inhibitory side effects to ritonavir that needed either modification or replacement. Thus, several more analogs were prepared containing the ritonavir core structure. Lopinavir (ABT-378) was found to be a highly potent inhibitor of HIV protease that had a greatly improved pharmacokinetic profile when co-administered with ritonavir.[27] From a drug development standpoint, it then became an imperative to produce it in large-scale quantities. Since a great deal of work had been placed into defining an efficient, robust, and inexpensive synthesis of the core portion of the molecule, what became necessary was the implementation of synthetic methods to prepare both the wing portions of lopinavir and to couple them to the core structure in an efficient and relatively lower-cost manner. The next sections of this chapter will discuss how this was done starting from the initial lopinavir synthesis and will conclude with the definition of a multi-kilogram processes.

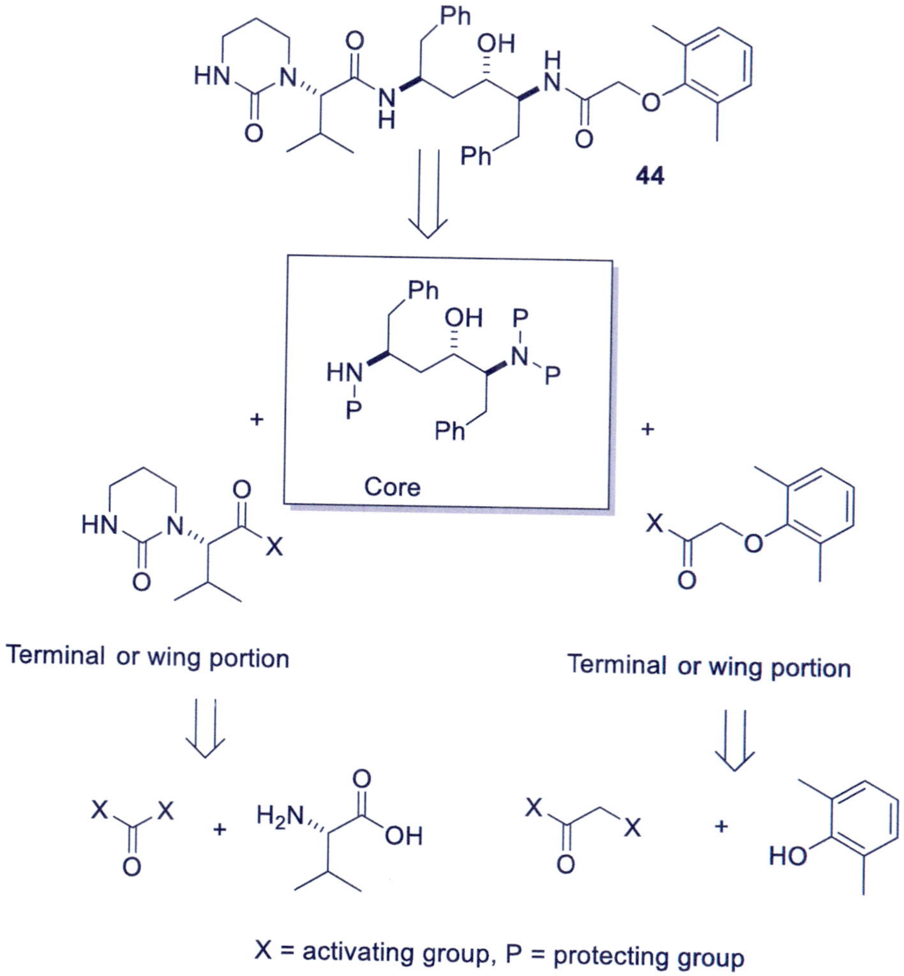

Scheme 15: Retrosynthesis of lopinavir (**44**)

9 Discovery Synthesis of Lopinavir

The discovery synthesis of lopinavir (Scheme 16) originates from *tert*-butyl [(2*S*,4*S*,5*S*)-5-amino-4-hydroxy-1,6-diphenylhexan-2-yl] carbamate hemi-succinate (**34**) or the core, from the ritonavir campaigns, which was allowed to react with 2-(2,6-dimethylphenoxy) acetic acid (**45**) in the presence of EDC and HOBt to provide the monosubstituted *N*-Boc core molecule **46** in 90% yield after crystallization from *n*-heptane. Removal of the *N*-Boc protecting group from **46** with trifluoroacetic acid in DCM followed by treatment of the des-*N*-Boc intermediate with EDC and HOBt and **47** gave lopinavir (**44**) in 86% yield over the last two steps as an amorphous solid.[28,29]

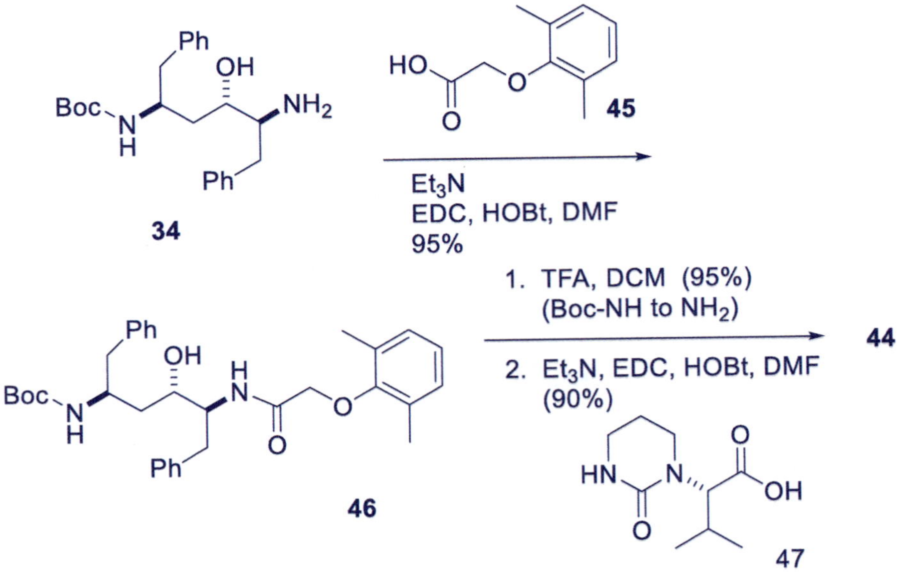

Scheme 16: Discovery lopinavir synthesis as an amorphous solid

10 Discovery Synthesis of Wing Pieces (45) and (47)

2-(2,6-Dimethylphenoxy)-acetic acid (**45**) was prepared from ethyl 2-bromoacetate (**48**) in DMF and 2,6-dimethyl phenol (**49**) with sodium hydride as the base followed by lithium hydroxide hydrolysis of ester **50** to provide **45** in 75% overall yield[28,30] as shown in Scheme 17.

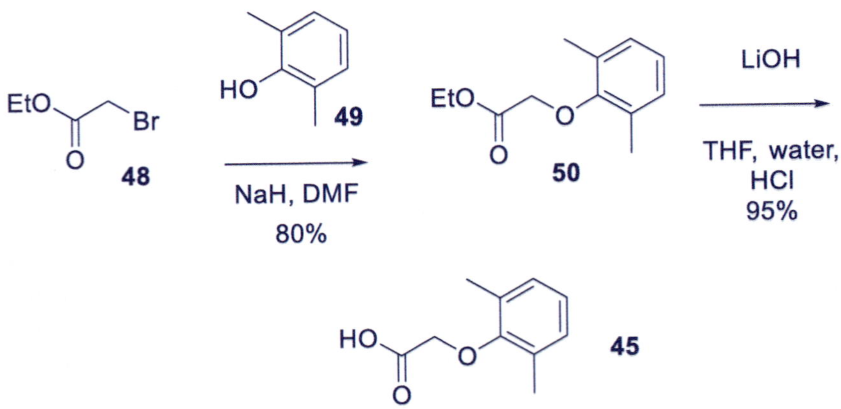

Scheme 17: Discovery preparation of 2-(2,6-dimethylphenoxy)-acetic acid (**45**)

For the synthesis of the valine substituted wing piece **47**, the discovery synthesis[28,30] (Scheme 18) was general enough to prepare 5 through 7-membered ring cyclic ureas. It started from 1-amino-3-propanol (**51**), which was converted to aldehyde **52** by a two-step process where the nitrogen atom was first *N*-protected as its *N*-Cbz derivative followed by Swern[10] oxidation to **52**. Aldehyde **52** was then condensed with (*S*)-valine methyl ester hydrochloride (**53**), sodium acetate, in acetic acid/water as solvent and the iminium intermediate was reduced with sodium cyanoborohydride to form methyl ester **54**. The next three steps involved removal of the *N*-Cbz protecting group by treatment with hydrogen, insertion of a carbonyl group to form the cyclic urea with CDI, and ester hydrolysis without racemization, by lithium hydroxide hydrolysis, to provide (*S*)-3-methyl-2-[2-oxotetrahydropyrimidin-1(2*H*)-yl]-butanoic acid (**47**) in 70% yield over the last three steps and in 45% overall yield from **51** over six total steps.

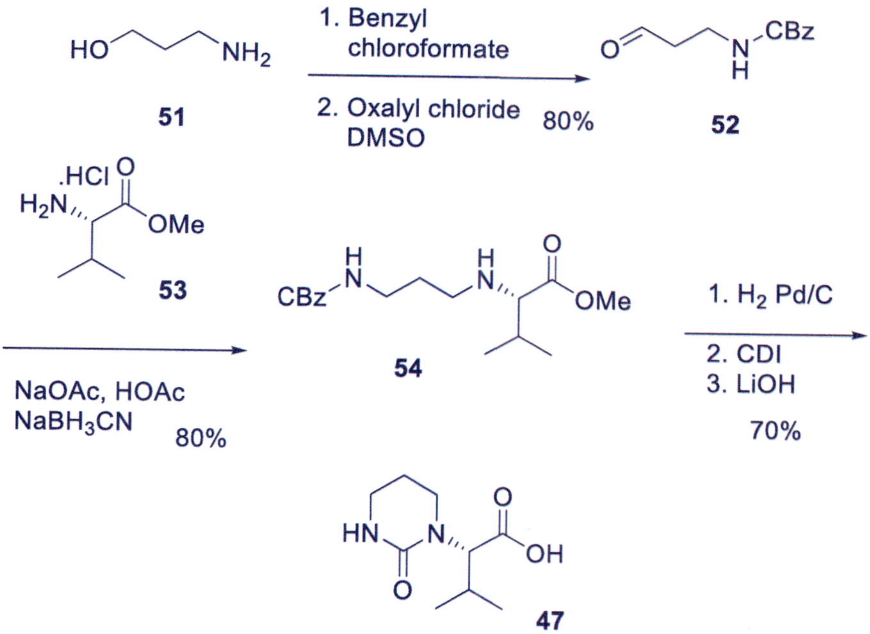

Scheme 18: Discovery preparation of (*S*)-3-methyl-2-(2-oxotetrahydropyrimidin-1(2*H*)-yl) butanoic acid (**48**) from 1-amino-3-propanol (**52**)

What became necessary were improved syntheses of both **45** and **47** and less costly methods to couple them to the core. Furthermore, in the beginning of the work, the API was an amorphous solid that needed to become crystalline to facilitate its purification, isolation, and formulation. It was also imperative from an efficiency standpoint to determine if fewer numbers of chemical steps with less-isolations could be performed with less costly reagents.

11 Process Improvements to the Wing Pieces (45) and (47)

From a cost standpoint, the preparation of **45** did not employ the least expensive starting materials because ethyl bromoacetate and sodium hydride/DMF are relatively expensive, and there is a safety concern that DMF/sodium hydride mixtures will explode on large scales at elevated temperatures. It would also be advantageous to eliminate the ester hydrolysis step for the conversion of **50** to 2-(2,6-dimethylphenoxy) acetic acid (**45**) since chloroacetic acid is known to undergo alkylation (chlorine atom displacement) in the presence of base. The solution[30,31] involved treatment of inexpensive chloroacetic acid sodium salt (**55**), with 2,6-dimethyl phenol (**49**) in aqueous sodium hydroxide to form **56**, which upon acidification with hydrochloric acid, provided the product **45** after filtration from water. With this improvement established, **45** was prepared as a crystalline solid in one step using water as the reaction solvent with no extractive workup as shown in Scheme 19.

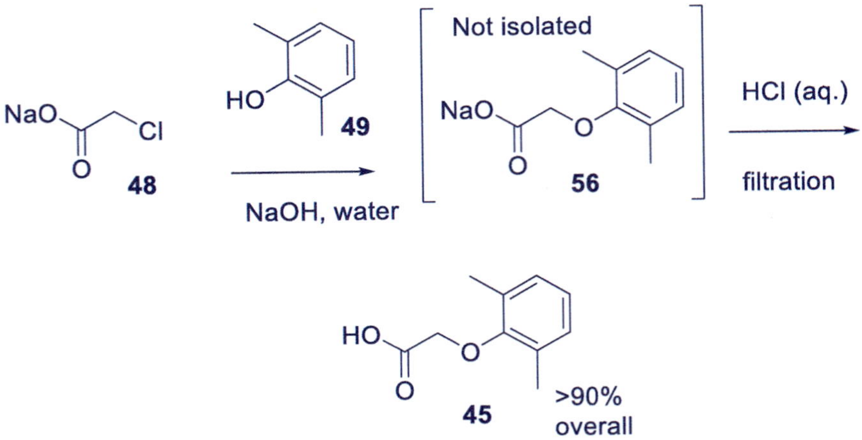

Scheme 19: Process synthesis of 2-(2,6-dimethylphenoxy) acetic acid (**45**)

Also, a cost-effective synthesis of (*S*)-3-methyl-2-(2-oxotetrahydropyrimidin-1(2*H*)-yl) butanoic acid (**47**) was necessary from (*S*)-valine (**57**) due to its lengthy (six steps) discovery synthesis shown in Scheme 18. One improvement was performed by treatment of (*S*)-valine[31] with phenyl chloroformate under buffered conditions to form the *O*-phenyl carbamate (**58**) in 92% yield. The next step was performed in THF where **58** was treated with commercially available 3-chloro-1-amino propane hydrochloride (**59**) in the presence of both sodium hydroxide and potassium *tert*-butoxide to provide intermediate **60**, which would then react further to **47** in 75% yield and in >99% *e.e.* as shown in Scheme 20.

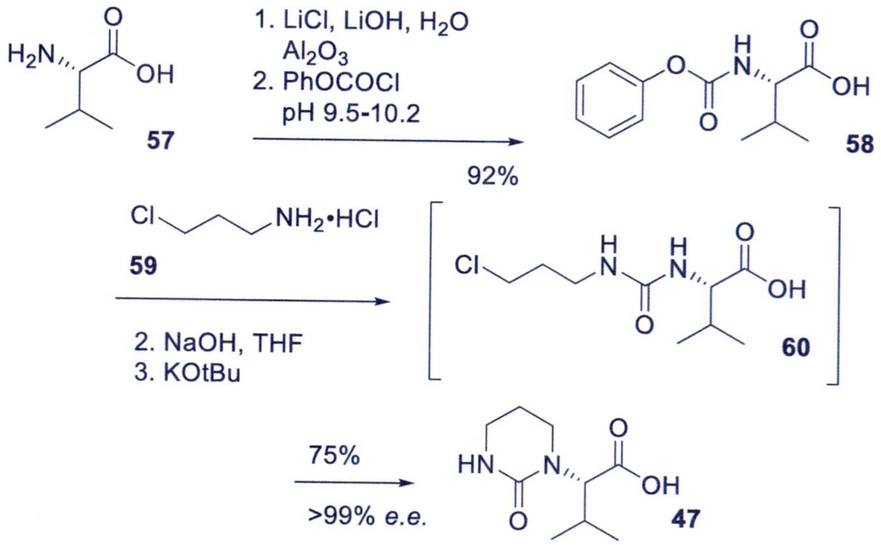

Scheme 20: An improved synthesis of **47**

The route in Scheme 20 employed phenyl chloroformate as its source for the urea carbonyl group in **47** and used 3-amino-1-chloropropane hydrochloride (**59**) to incorporate the carbon chain. These reagents are somewhat costly. The source for amine **59** is acrylonitrile. This begs the question of whether acrylonitrile can be used directly, since it is inexpensive as compared to **59**. This concept is shown in Scheme 21 where treatment of (*S*)-valine (**57**) with 1.0 equiv of sodium hydroxide and acrylonitrile in water gave the cyanoethylated adduct[30,32] **61** in quantitative yield. Subsequent addition of methyl chloroformate to **61** provided the methyl carbamate **62** in 65% overall yield in one pot. Reduction of **62** as its sodium salt at high temperature under a hydrogen atmosphere, also in water, under basic conditions with nickel aluminum alloy (Raney-type) provided the amine, which cyclized immediately to provide **47** in high yield[28,30] after acidification and filtration. Thus, the wing piece **47** was routinely prepared in moderate overall yields and in high optical purity from (*S*)-valine, acrylonitrile, methyl chloroformate, and Raney nickel reduction/cyclization in water. The process depicted in Scheme 21 to prepare (**47**) is preferred from a cost of starting materials (reagents) standpoint as methyl chloroformate is less expensive than phenyl chloroformate, acrylonitrile is cheaper than 3-amino-1-chloropropane hydrochloride, Raney nickel is not expensive, and the bases and solvents are less costly.

Now that relatively inexpensive syntheses of both **45** and **47** were established, the next phase of lopinavir process development involved research to determine ways to couple them to the core without the need of expensive peptide coupling reagents or the use of Boc_2O as a reagent.

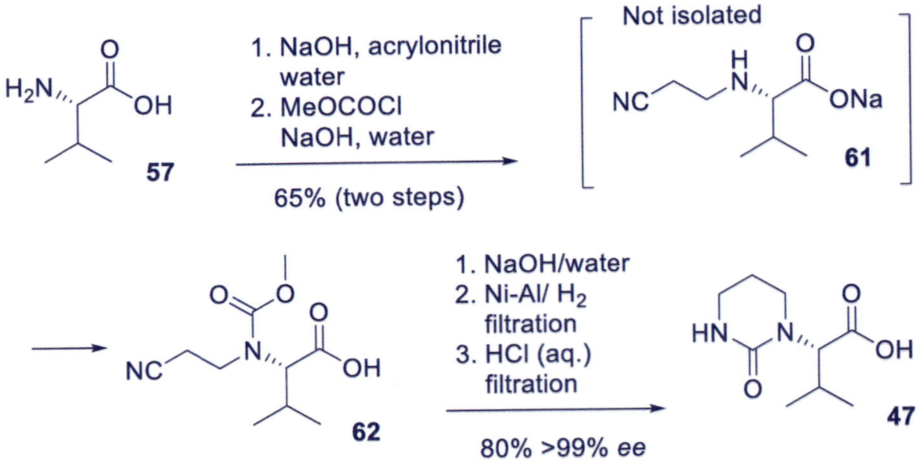

Scheme 21: Alternate process synthesis of compound **47**

12 Optimization of Lopinavir Synthesis with Intermediates

As noted, the isomerically pure (*S,S,S*) derivative **34** as its hemi-succinate salt served as an ideal starting point for large-scale lopinavir preparation. The other items on the agenda toward a lower-cost robust lopinavir synthesis were methods to couple wing pieces **45** and **47** to **34**. This was performed by taking **45** and treating it with thionyl chloride[31] to form acid chloride **63**, which was then allowed to react with **34** in the presence of excess base to form the mono *N*-substituted core structure **64** as a crystalline solid. Compound **64** is then converted to its des-*N*-Boc derivative **65** with trifluoroacetic acid to form crystalline **66**. The final step of the process by the early route in Scheme 22 is to convert **66** to lopinavir utilizing an inexpensive method to activate **47** to its acid chloride **67** and perform the coupling step. This operation was not trivial due to the lack of stability of acid chloride **67**. A solution was found by noting that acid chloride **67** was prone to polymerize[33] to form structures such as **68** or **69** under acidic media and that this polymerization could be controlled with strong base (>3.3 equiv of imidazole) to allow the final step to proceed in 95% yield (Scheme 22). After much work, it was found that lopinavir could exist as a crystalline solid with several polymorphic forms.[34] Fortunately, there were no issues with new polymorphs appearing during commercial manufacture of the API as were seen with ritonavir.[35]

The peptide coupling steps were less costly because both acid chlorides **63** and **67** could be coupled to **34**. What remained to be demonstrated was whether it was necessary to *N*-Boc protect the epimeric mixture of (**32a–d**) and form the hemi-succinate salts to purge the unwanted isomers by crystallization to provide pure **34**. In other words, could the two-step protection-deprotection sequence be eliminated from the process. Therefore, an intensive investigation involving the preparation and crystallization of several different types of acid-amine salts of **68** and **69** was conducted with the goal of purging the unwanted diastereomers as shown in Scheme 23.[31]

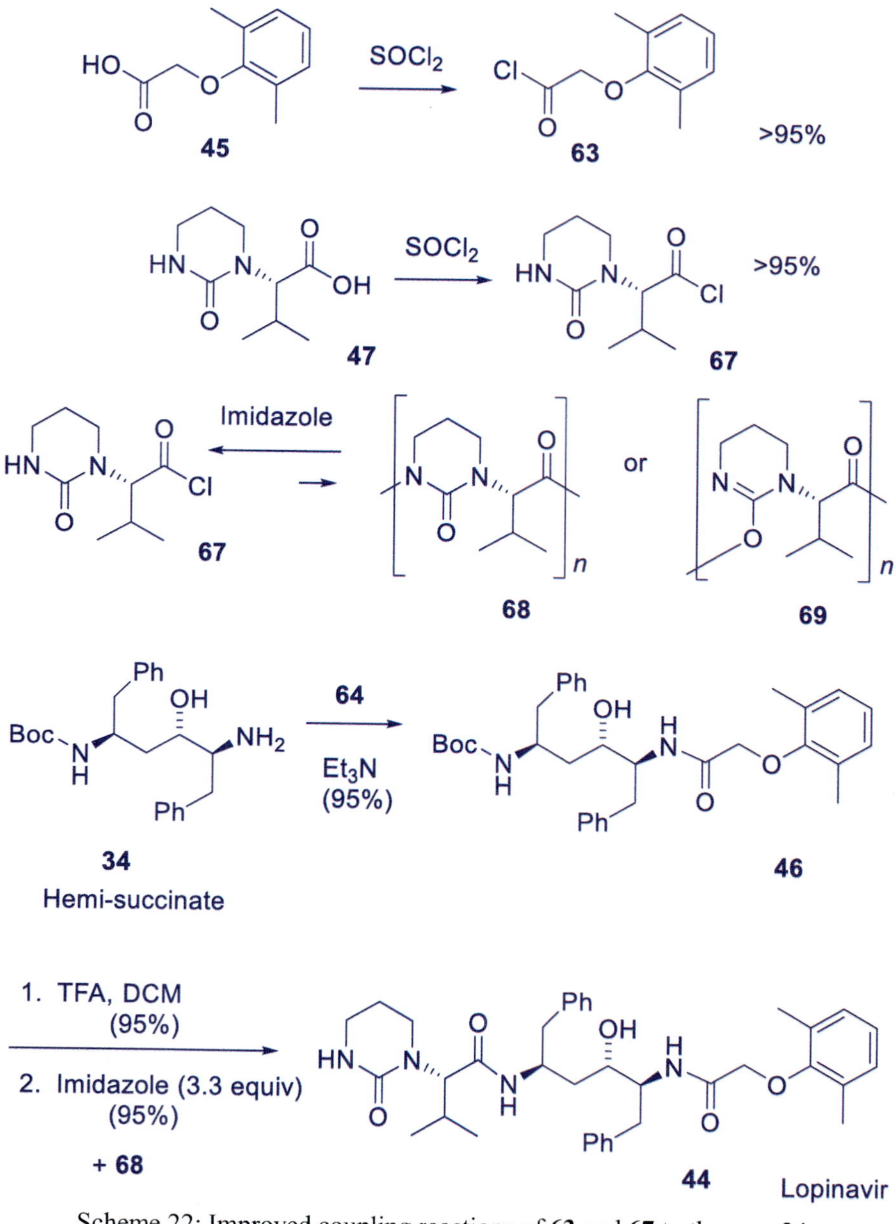

Scheme 22: Improved coupling reactions of **63** and **67** to the core **34**

After an investigation of 30 different types of acids, it was found that the (*l*) pyroglutamic acid salt of (**69**) could purge the other ~12% of the unwanted (*S,S,R*), (*S,R,S*), and (*S,R,R*) isomers by crystallization.[31] The conversion of **32a–d** to pure **69** as its (*l*)-pyroglutamate salt was accomplished in 74% overall yield after crystallization and the final step of the process, the coupling of acid chloride **64** to **69**, devoid of the unwanted epimers, was accomplished in 89% to yield large kilogram quantities of lopinavir per batch size.

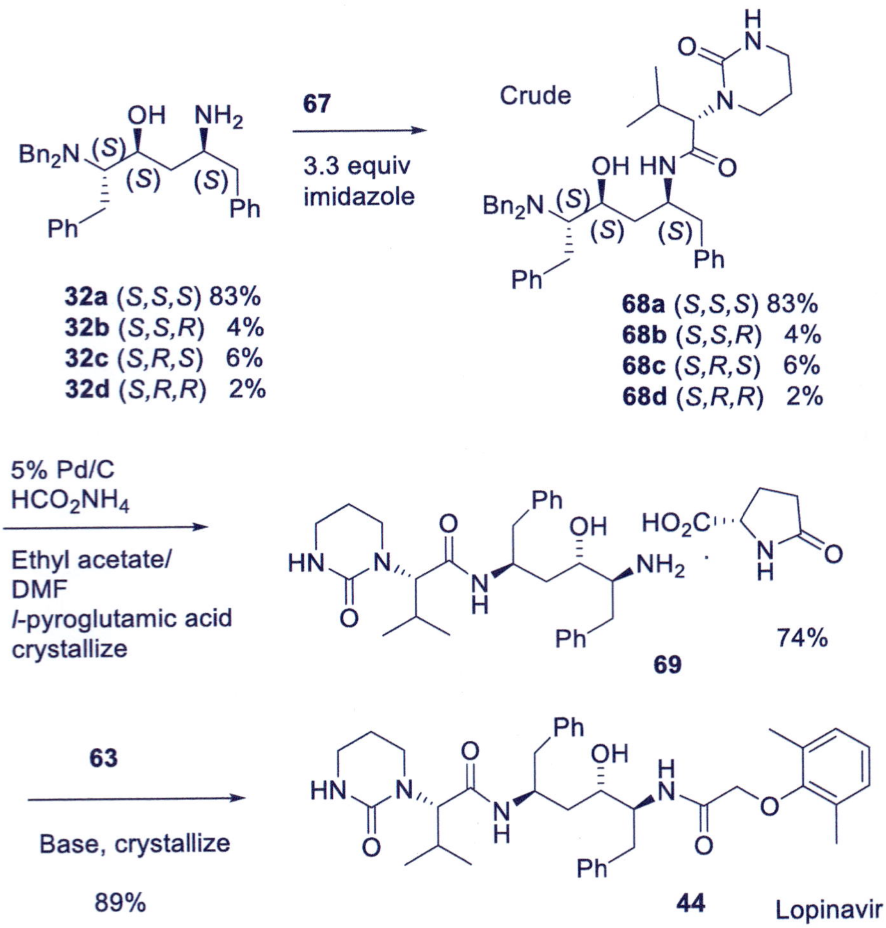

Scheme 23: The shortened synthesis of lopinavir

The superior synthesis of lopinavir is shown in Scheme 23 where there are no N-Boc protection/deprotection steps in the process. The lopinavir–ritonavir mixture (4:1), as the drug Kaletra, was brought to market in about four years after the discovery of lopinavir and was fast-tracked due to its importance. The lopinavir synthesis involved taking advantage of the novel process synthesis of the ritonavir core and coupling it to the wing pieces **45** and **47** in an optimal fashion utilizing standard *N*-acylation type chemistry with acid chlorides **63** and **67**. It is interesting to note that the lopinavir/ritonavir combination therapy (Kaletra) was more efficacious than the ritonavir combination therapy with indinavir (Crixivan) and nelfinavir (Viracept). This is postulated to arise from the fact that ritonavir and lopinavir are structurally more similar compounds to each other than are indinavir/ritonavir and nelfinavir/ritonavir combinations from a metabolic standpoint.

13 Conclusions

In conclusion the process research and plant scale up efforts to bring both ritonavir and lopinavir and thus Kaletra to the market in rapid fashion did save lives and hence these compounds are justifiably on the World Health Organization's list of Essential Medicines. What is listed in the chemical schemes represents chemical reactions that have all been disclosed in both the patent and the chemical literature and constitute a strong effort on behalf of all the medicinal and process chemists at Abbott Laboratories. The efforts of the many scientists involved at Abbott Laboratories (now AbbVie) from the biologists to the medicinal chemists through process chemistry, formulations, and production were indeed heroic. The names of those involved are listed in the references section.

14 References

1. Lea, A. P.; Faulds, D. Ritonavir. *Drugs* **1996**, *52*, 541–546.
2. Oldfield, V.; Plosker, G.L. Lopinavir/Ritonavir: A review of its use in the management of HIV infection. *Drugs* **2006**, *66*, 1275–1299.
3. Young, T. P.; Parkin, N. T.; Stawiski, E.; Pilot-Matias, T.; Trinh, R.; Kempf, D. J.; Norton, M. Prevalence, mutation patterns, and effects on protease inhibitor susceptibility of the L76V mutation in HIV-1 protease. *Antimicrob. Agents and Chemother.* **2010**, *54*, 4903–4906.
4. Kempf, D. J.; Sham, H. L.; Marsh, K. C.; Flentge, C. A.; Betebenner, D.; Green, B. E.; McDonald, E.; Vasavanonda, S.; Saldivar, A.; Wideburg, N. E.; Warren, M. K.; Ruiz, L.; Zhao, C.; Fino, L.-M.; Patterson, J.; Molla, A.; Plattner, J. J.; Norbeck, D. W. Discovery of ritonavir, a potent inhibitor of HIV protease with high oral bioavailability and clinical efficacy. *J. Med. Chem.* **1998**, *41*, 602–617.
5. Wlodawer, A.; Erickson, J. W. Structure-based inhibitors of HIV-1 protease. *Annu. Rev. Biochem.* **1993**, *62*, 543–585.
6. Kempf, D. J.; Norbeck. D. W.; Codacovi, L.; Wang, X. C.; Kohlbrenner, W. E.; Wideburg, N. E.; Paul, D. A.; Knigge, M. F.; Vasavanoda, S.; Craig-Kennard, A.; Saldivar, A.; Rosenbrook, W., Jr.; Clement, J. J.; Plattner, J. J.; Erickson, J. Structure-based C_2 symmetric inhibitors of HIV protease. *J. Med. Chem.* **1990**, *33*, 2687–2689.

7. Kempf, D. J.; Codacovi, L.; Wang, X. C.; Kohlbrenner, W. E.; Wideburg, N. E.; Saldivar, A.; Vasavanoda, S.; Marsh, K. C.; Bryant, P.; Sham, H. L.; Green, B. E.; Betebenner, D. A.; Erickson, J.; Norbeck, D. W. Symmetry-based inhibitors of HIV protease. Structure-activity studies of acylated 2,4-diamino-1,5-diphenyl-3-hydroxypentane and 2,5-diamino-1,6-diphenylhexane-3,4-diol. *J. Med. Chem.* **1993**, *36*, 320–330.
8. Kempf, D. J.; Sowin, T. J.; Doherty, E. M.; Hannick, S. M.; Codavoci, L.-M.; Henry, R. F.; Green, B. E.; Spanton, S. G.; Norbeck, D. W. Stereocontrolled synthesis of C_2-symmetric and pseudo-C_2-symmetric diamino alcohols and diols for use in HIV protease inhibitors. *J. Org. Chem.* **1992**, *57*, 5692–5700.
9. Kempf, D. J.; Marsh, K. C.; Fino, L.-M. C.; Bryant, P.; Craig-Kennard, A.; Sham, H. L.; Zhao, C.; Vasavanonda, S.; Kohlbrenner, W.E.; Wideburg, N. E.; Saldivar, A.; Green, B. E.; Herrin, T.; Norbeck, D. W. Design of orally bioavailable, symmetry-based inhibitors of HIV protease. *Bioorg. Med. Chem.* **1994**, *2*, 847–858.
10. Mancuso, A. J.; Huang, S. L.; Swern, D. Oxidation of long-chain and related alcohols to carbonyls by dimethyl sulfoxide "activated" by oxalyl chloride. *J. Org. Chem.* **1978**, *43*, 2480–2482.
11. McMurry, J. E. Carbonyl-coupling reactions using low-valent titanium. *Chem. Rev.* **1989**, *89*, 1513–1524.
12. Fruedenberger, J. H.; Konradi, A. W.; Pedersen, S. F. Intermolecular pinacol cross coupling of electronically similar aldehydes. An efficient and stereoselective synthesis of 1,2-diols employing a practical vanadium (II) reagent. *J. Am. Chem. Soc.* **1989**, *111*, 8014–8016.
13. Konradi, A. W.; Pedersen, S. F. Pinacol homocoupling of (*S*)-2-[N-benzyloxycarbonyl)amino] aldehydes by $[V_2Cl_3(THF)_6]_2[Zn_2Cl_6]$. Synthesis of C2-symmetric (1*S*,2*R*,3*R*,4*S*)-1,4-diamino 2,3-diols. *J. Org. Chem.* **1992**, *57*, 28–32.
14. Reference 4, page 603, Sowin, T. J. (unpublished results).
15. Mashraqui, S. H.; Keehn, P. M. Cyclophanes. 14. Synthesis, structure assignment, and conformational properties of [2.2] (2,5) oxazolo-and thiazolophanes. *J. Am. Chem. Soc.* **1982**, *104*, 4461–4465.
16. Stuk, T. L.; Haight, A. R.; Scarpetti, D.; Allen, M. S.; Menzia, J. A.; Robbins, T. A.; Parekh, S. I.; Langridge, D. C.; Tien, J.-H. J.; Pariza, R. J.; Kerdesky, F. A. J. An efficient stereocontrolled strategy for the synthesis of hydroxyethylene dipeptide isosteres. *J. Org. Chem.* **1994**, *59*, 4040–4041.
17. Haight, A. R.; Stuk, T. L.; Menzia, J. A.; Robbins, T. A. A convenient synthesis of enaminones using tandem acetonitrile condensation, Grignard addition. *Tetrahedron Lett.* **1997**, *38*, 4191–4194.
18. Gribble, G. W.; Nutaitis, C. F. Sodium borohydride in carboxylic acid media. A review of the synthetic utility of acyloxyborohydrides. *Org. Prep. Proc. Int.* **1985**, *17*, 317–384.

19. Haight, A. R.; Stuk, T. L.; Allen, M. S.; Bhagavatula, L.; Fitzgerald, M.; Hannick, S. M.; Kerdesky, F. A. J.; Menzia, J.; Parekh, S. I.; Robbins, T. A.; Scarpetti, D.; Tien, J.-H. J. Reduction of an enaminone: Synthesis of the diamino alcohol core of ritonavir. *Org. Process Res. Dev.* **1999**, *3*, 94–100.
20. Stuk, T. L.; Allen, M. S.; Haight, A. R.; Kerdesky, F. A.; Langridge, D. C.; Leanna, M. R.; Lijewski, L. M.; Melcher, L.; Morton, H. E.; Norbeck, D. W.; Reno, D. S.; Robbins, T. A.; Scarpetti, D.; Sham, H. L.; Sowin, T. J.; Tien, J.-H. J.; Zhao, C. Process for the preparation of a substituted 2,5-diamino-3-hydroxyhexane. US 5,491,253 (1996).
21. Kerdesky, F. A. J.; Seif, L. S. A novel and efficient synthesis of 5-(hydroxymethyl) thiazole: An important synthon for preparing biologically active compounds. *Synth. Commun.* **1995**, *25*, 2639–2645.
22. Beck, G.; Heitzer, H. Preparation of 2-chloro-5-chloromethylthiazole. US 4,748,243 (1977).
23. Leanna, M. R.; Morton, H. E. Process for the preparation of 5-hydroxymethylthiazoles. WO 96/16050 (1996).
24. Tien, J.-H. J.; Mezia, J. A.; Cooper, A.J. Process for the preparation of an HIV protease inhibiting compound. US 5,567,823 (1996).
25. Bellani, P.; Frigerio, M.; Castoldi, P. A process for the synthesis of ritonavir. EP 1 133 485 B1 (2006).
26. Rock, B. M.; Hengel, S. M.; Rock, D. A.; Wienkers, L. C.; Kunze, K. L. Characterization of ritonavir-mediated inactivation of cytochrome P450 3A4. *Mol. Pharmacol.* **2014**, *86*, 665–674. and references cited therein.
27. Sham, H. L.; Kempf, D. J.; Molla, A.; Marsh, K. C.; Kumar, G. N.; Chen, C. M. ABT-378, a highly potent inhibitor of the human immunodeficiency virus protease. *Antimicrob. Agents Chemother.* **1998**, *42*, 3218–3224.
28. Sham, H. L.; Norbeck, D. W.; Chen, X.; Betebenner, D. A. Retroviral protease inhibiting compounds. US 5,914,332 (1999).
29. Sham, H. L.; Betebenner, D. A.; Chen, X.; Saldivar, A.; Vasavanonda, S.; Kempf, D. J.; Plattner, J. J.; Norbeck, D. W. Synthesis and structure-activity relationships of a novel series of HIV-1 protease inhibitors encompassing ABT-378 (lopinavir). *Bioorg. Med. Chem. Lett.* **2002**, *12*, 1185–1187.
30. Sham, H. L.; Norbeck, D. L.; Chen, X.; Betebenner, D. A.; Kempf, D. J.; Herrin, T.; Kumar, G. N.; Condon, S. L.; Cooper, A. J.; Dickman, D.A.; Hannick, S. M.; Kolaczkowski, L.; Oliver, P. A.; Plata, D. J.; Stengel, P. J.; Stoner, E. J.; Tien, J.-H. J.; Liu, J.-H.; Patel, K. M. Retroviral protease inhibiting compounds. US 6,472,529 B2 (2002).
31. Stoner, E. J.; Cooper, A. J.; Dickman, D. A.: Kolaczkowski, L.; Lallaman, J. E.; Liu, J.-H. Oliver-Shaffer, P. A.; Patel, K. M.; Paterson, J. B.; Plata, D. J.; Riley, D. A.; Sham, H. L.; Stengel, P. J.; Tien, J.-H. J. Synthesis of HIV protease inhibitor ABT-378 (Lopinavir). *Org. Process Res. Dev.* **2000**, *4*, 264–269.

32. McKinney, L. L.; Uhing, E. H.; Setzkorn, E.; Cowan, J. C. Cyanoethylation of alpha amino acids. I. Monocyanoethyl derivatives. *J. Am. Chem. Soc.* **1950**, *72*, 2599–2603.
33. Stoner, E. J.; Stengel, P. J.; Cooper, A. J. Synthesis of ABT-378, an HIV protease inhibitor candidate: Avoiding the use of carbodiimides in a difficult peptide coupling. *Org. Process Res. Dev.* **1999**, *3*, 145–148.
34. Dickman, D. A.; Chemburkar, S.; Fort, J. J.; Henry, R. F.; Lechuga-Ballesteros, D.; Niu, Y.; Porter, W. Crystalline pharmaceutical. US 8,796,451 (2014). (And patents referenced therein).
35. Bauer, J.; Spanton, S.; Henry, R.; Quick, J.; Dziki, W.; Porter, W.; Morris, J. Ritonavir: An extraordinary example of conformational polymorphism. *Pharm. Res.* **2001**, *18*, 859–866.

5

Eravacycline (Xerava), A Novel and Completely Synthetic Fluorocycline Antibiotic

Jie Jack Li

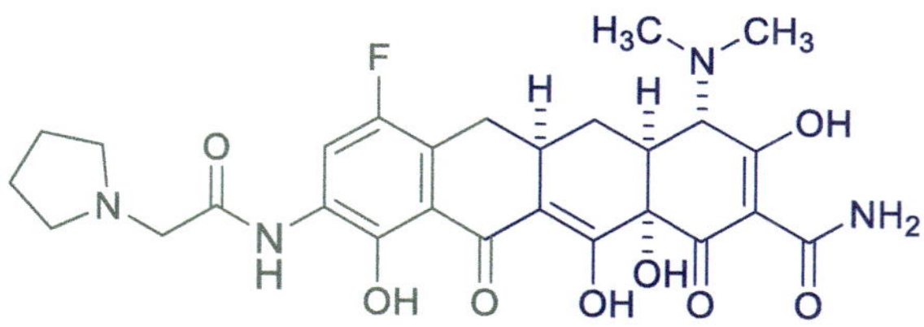

Eravacycline
(Xerava, **1**)
Tetraphase, 2018
Tetracycline antibiotic

5.1 Background

Nine tetracycline antibiotics are currently on the market. Three of them are isolated from nature. Five additional tetracycline antibiotics are prepared by semi-synthesis. Only one fully synthetic tetracycline has been approved by the FDA. That is Tetraphase Pharmaceuticals' hyper-potent eravacycline (Xerava, **1**).

In the 1940s, American Cyanamid's Lederle Laboratories at Pearl River, New York started to screen soil samples to look for antibiotics that would possess a better safety profile than streptomycin for treating tuberculosis. In 1945, 73-year old botanist Benjamin M. Duggar was hired as a consultant for Lederle and led their screening efforts in the hunt for new antibiotics. Coincidently, a sample from the University of Missouri, where Duggar taught botany forty years earlier, yielded an antibiotic later named chlortetracycline (**2**). Lederle sold it under the brand name Aureomycin in 1948. Thanks to its increased spectrum against a wide panel of bacteria compared with other antibiotics and its great oral bioavailability, Aureomycin won a good share of the antibiotics market. Nowadays, Duggar is considered the pioneer of tetracycline antibiotics.

Current Drug Synthesis, First Edition. Edited by Jie Jack Li.

The discovery of the second tetracycline antibiotic was also serendipitous. In the late 1940s Charles Pfizer and Company became deeply worried about competitors' resurgence in the antibiotic arena, especially when the price of penicillin plummeted. Like many pharmaceutical companies at that time, Pfizer plunged into research for newer antibiotics. In 1949, a yellow powder with strong antibiotic properties was isolated from a soil sample. This yellow powder was oxytetracycline (**3**) that proved to be both safe and effective against a range of bacteria that caused more than a hundred infectious diseases. Backtracking revealed that the soil sample was collected at the Terre Haute factory in Indiana owned by Pfizer. Pfizer sold oxytetracycline (**3**) under the brand name Terramycin.

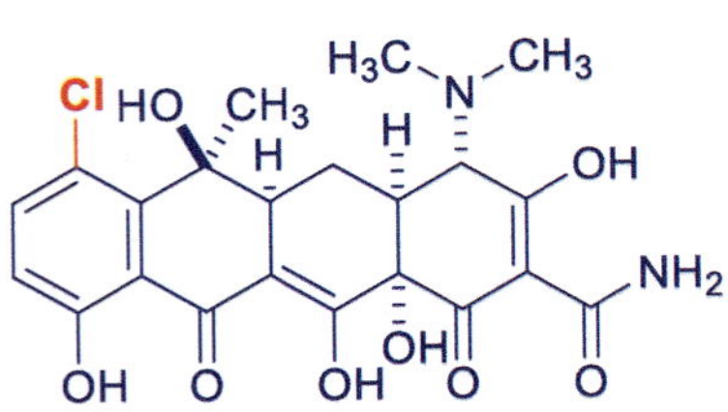

Chlortetracycline (Aureomycin, **2**)

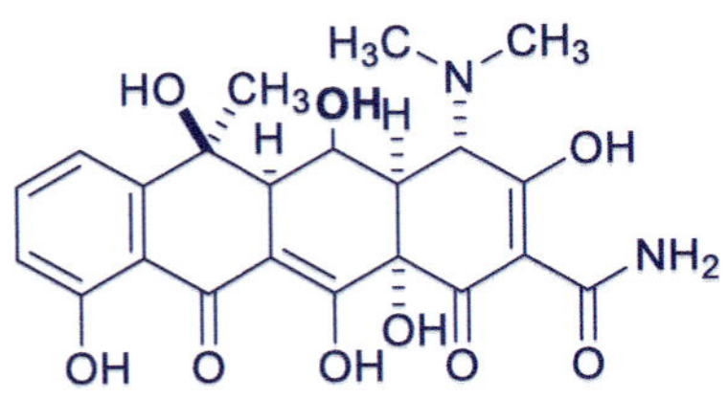

Oxytetracycline (Terramycin, **3**)

In addition, Pfizer formed a team to elucidate the chemical structure of oxytetracycline. They enlisted the help of R. B. Woodward at Harvard. In 1952, Pfizer and Woodward jointly published their elegant work on the structure of oxytetracycline. Meanwhile, a member of the team, Lloyd Conover, shocked his colleagues by preparing another powerful antibiotic chemically from chlortetracycline (**2**). Under carefully controlled catalytic hydrogenation conditions using hydrogen gas and palladium on charcoal, Conover converted Lederle's chlortetracycline (**2**) to tetracycline by stripping the chlorine atom and replacing it with a hydrogen atom to make tetracycline (Tetracyn, **4**). Chlortetracycline (**2**) and tetracycline (**4**) are unstable at the extreme of pH. In humans, they tend to undergo acid-catalyzed dehydration of the C-6 hydroxyl group on the C ring to form reportedly toxic anhydro-tetracycline derivatives.

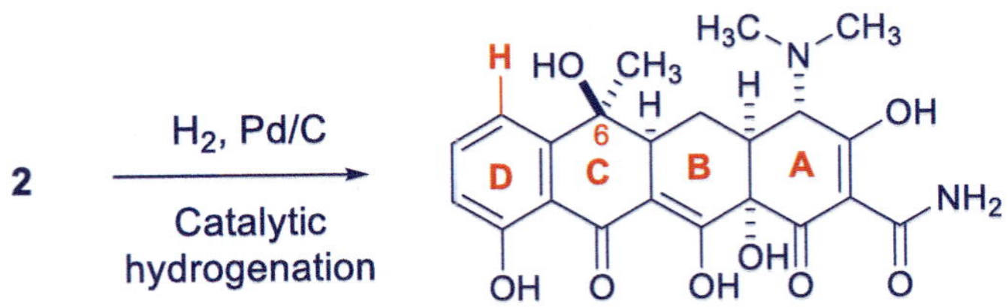

Tetracycline (**4**)

Conover's exploit was truly revolutionary. Before that, it was generally believed by all practitioners that "natural" antibiotics produced by microbial metabolism were the only ones possessing desirable biological properties. Conover demonstrated that chemical manipulations could afford active antibiotics as well. Tetracyclines became the most prescribed broad-spectrum antibiotics in the United States within three years. Conover's discovery created a brand-new field of medical research—semi-synthetic antibiotics. It also sparked a wide-scale search for superior structurally modified antibiotics, which has provided most of the important antibiotic discoveries since then. Both doxycycline (Doxylin, **5**, 1967) by Pfizer and minocycline (Minocin, **6**, 1972) by Lederle were discovered using semi-synthetic approach. Reductive removal of the C-6 hydroxyl group of the natural tetracycline gave rise to the much more stable 6-deoxytetracyclines such as doxycycline (**5**). It was prepared from oxytetracycline (**3**) in four steps and remains one of the most prescribed generic antibiotics in use today. Installation of the dimethylamino group at the C-7 position of minocycline (**6**) has been so beneficial that almost all future tetracycline antibiotics have retained it in one form or the other. Remarkably, both doxycycline (**5**) and minocycline (**6**) have nearly 100% oral bioavailability!

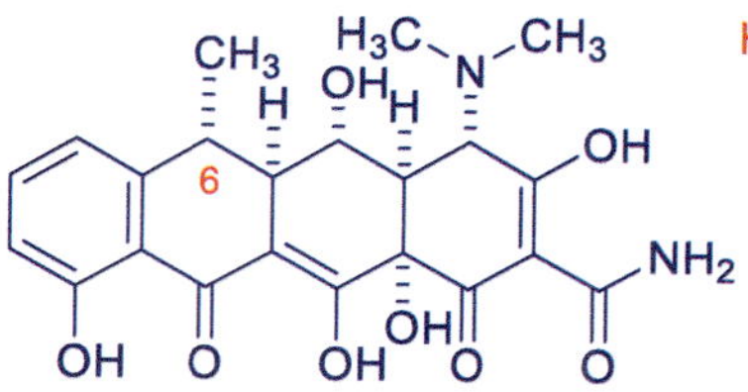

Doxycycline (Doxylin, **5**)

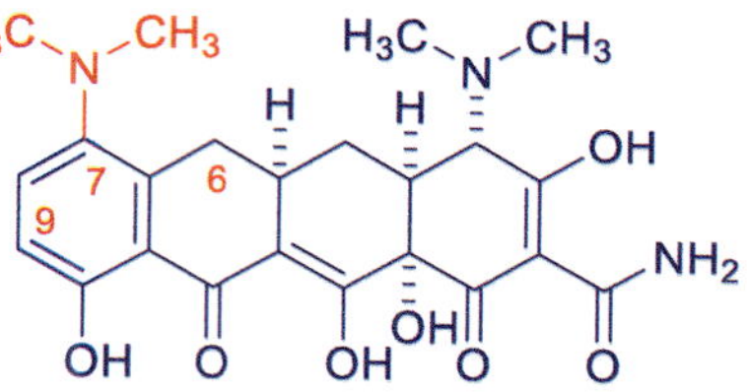

Minocycline (Minocin, **6**)

Bacteria are capable little creatures. In order to survive they develop drug resistance toward antibiotics sooner or later. There are many tetracycline resistance genes reported in many bacteria and based on their sequence similarities can be classified into many types that are named as *tet* A, B, C, D, E, K, L, M, O, P, Q, and X. Based on their mechanism of action these tetracycline resistance genes can be of three types: antibiotic efflux (*tet* A–E, K, L, P), target protection (*tet* M, O, Q), and antibiotic inactivation (*tet* X).

Overcoming drug resistance requires constant replenish our arsenal of antibiotics. There are two mechanisms for bacterial resistance to tetracyclines. One is efflux pumps of antibiotics by membrane-spanning proteins [genotypes of *tet*(A–E)] and [*tet*(K–L)]. And the other is ribosomal protection [genotypes of *tet*(M–O)]. Ribosomal protection proteins are cytoplasmic proteins that bind to the ribosome, causing an alteration in ribosomal conformation that prevents tetracycline from binding. To overcome minocycline (**6**)'s resistance to certain bacterial strains, tigecycline (Tygacil, **7**) was discovered by a team led by Phaik-Eng Sum from a program initiated by Francis

Tally at the American Cyanamid Company. It was approved by the FDA in 2005. The major difference between minocycline (**6**) and tigecycline (**7**) is that the latter has an additional 9-*t*-butylglycylamido-substituent at the C-9 position, which is believed to contribute to the greater affinity of tigecycline (**7**) for the bacterial ribosome and, as a consequence, its increased efficacy against many pathogenic bacteria that had acquired resistance to all other tetracycline antibiotics. Tigecycline (**7**) was one of the most potent members of this group and is characterized by a long half-life in patients and by broad tissue distribution.[1]

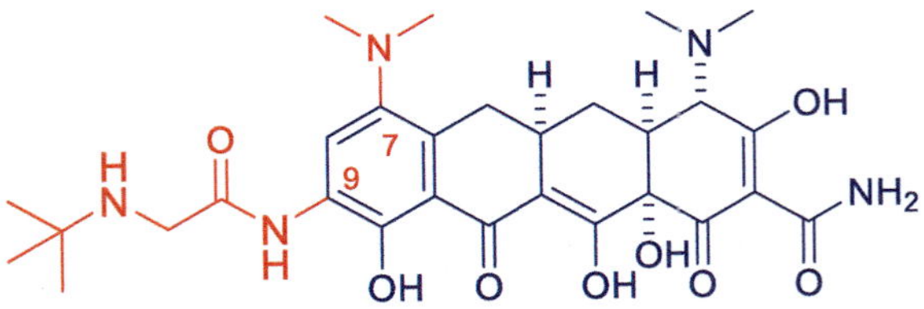

Tigecycline (Tygacil, **7**)
Wyeth, 2005
Tetracycline antibiotic

Regrettably, tigecycline (**7**) is plagued by low bioavailability and can only be given intravenously (IV). Its gastrointestinal (GI) side effects and increased mortality in patients was severe enough to warrant the black box warning by the FDA. Paratek Pharmaceuticals in Boston strived to improve upon tigecycline (**7**) by boosting its bioavailability and minimizing its adverse effects. The fruit of their labor was omadacycline (Nuzyra, **8**) and sarecycline (Seysara, **9**). Omadacycline (**8**) is a broad-spectrum antibacterial with activity against Gram-positive and Gram-negative aerobic, anaerobic and atypical bacteria and may be administered either orally or intravenously.[2] On the other hand, sarecycline (**9**) is a narrow-spectrum tetracycline approved for the treatment of moderate to severe acne vulgaris. It works by interfering with transfer ribonucleic acids (tRNAs) accommodation and tethers messenger ribonucleic acid (mRNA) to the 70 Svedberg units (70S, S is the sedimentation coefficient and S values are not additive) ribosome.[3]

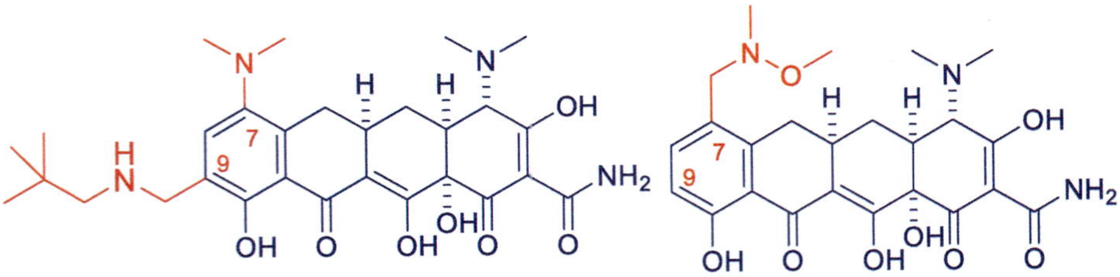

Omadacycline (Nuzyra, **8**)
Tufts/Parateck, 2018

Sarecycline (Seysara, **9**)
Parateck, 2018

5.2 Pharmacology

All tetracycline antibiotics, including eravacycline (**1**), share the same MoA. They selectively inhibit bacterial protein synthesis. In particular, tetracyclines bind to a subunit, such as 30S, of the bacterial ribosome (the site of protein synthesis) and prevent binding of amino-acyl tRNA to the acceptor site (A site) on the mRNA:ribosome complex. As a result, protein synthesis is halted since no new amino acid (AA) can be delivered to the growing peptide chain.

Protein synthesis is a complex process with the ribosome as a central player. Its task is to decode the mRNA into the corresponding sequence of amino acids with the aid of amino-acylated tRNAs.

What is a ribosome?

A ribosome is a complex nucleoprotein particle. It is a complex of two units. One is a ribosomal RNA (rRNA) and the other is a protein. Structurally, a ribosome is an mRNA sandwiched between a large subunit of 50S and a small subunit of 30S. Meanwhile, tRNAs carry amino acids to the ribosome. They act as "bridges", matching a codon in an mRNA with the amino acid it codes for (Fig. 1).

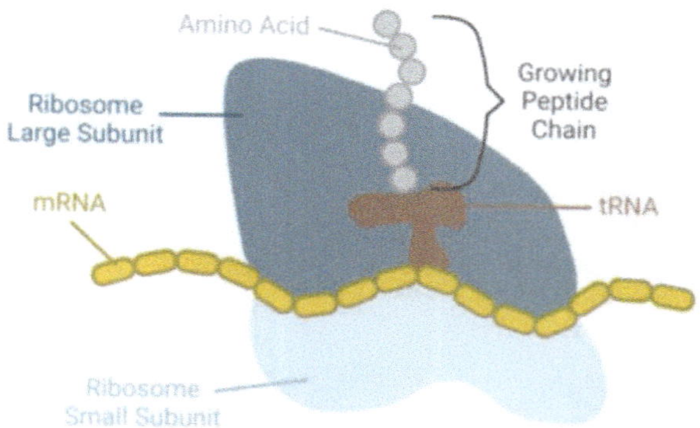

Fig. 1. Structure of ribosome, diagram by Alexandra H. Li.

Incorporation of an amino acid to peptide normally involves four steps:

a. The ribosome focuses two codons on mRNA. One has bound its tRNA–AA complex; the other is ready to receive the next tRNA–AA complex.
b. The amino acids of two adjacent complexes are linked by peptidyltransferase.
c. The tRNA dissociates from mRNA.
d. The right tRNA–AA complex shifts to the left, allowing the next complex to be bound at the right.

Tetracyclines, including eravacycline (**1**), interrupts step a, inhibiting the binding of tRNA–AA complexes. Their action is bacteriostatic and affect a broad spectrum of pathogens. Tetracycline inhibits the conversion between the first and second binding steps, i.e., prevents full accommodation of the A-site tRNA, which results in the loss of the tRNA from the A-site.

In 2000, Bodersen published the solved tetracycline (**4**)–30S ribosome cocrystal structure. It becomes abundantly clear that the "northwest" region of tetracycline (**4**) is not directly involved in interactions with the ribosome and can be modified without substantial loss of activity. On the other hand, the "southeast" region is directly involved in an extensive H-bond network in the A site of the bacterial ribosome. Therefore, conservation of this H-bond network is essential to retain tight ribosome binding, which explains why the "southeast" region is constant in all tetracycline antibiotics. As shown in Fig. 2, the picture on the left is the overview of primary binding site of tetracycline (**4**), indicating the RNA components close to the site and the interaction with A site tRNA, H34 (blue), H31 (green), H18 (orange), and H44 (cyan). The model of A site tRNA (red) and mRNA (yellow) is shown. The right picture shows chemical structure diagram of tetracycline (**4**) and possible interactions with 16S RNA at the primary site (blue). The shaded area represents positions on the molecule that can be modified without affecting its inhibitory action.[4]

Not surprisingly, this has left only the C7 and C9 as the most promising modification sites for improved antibacterial activity, which led to the discovery of minocycline (**6**) and tigecycline (Tygacil, **7**).

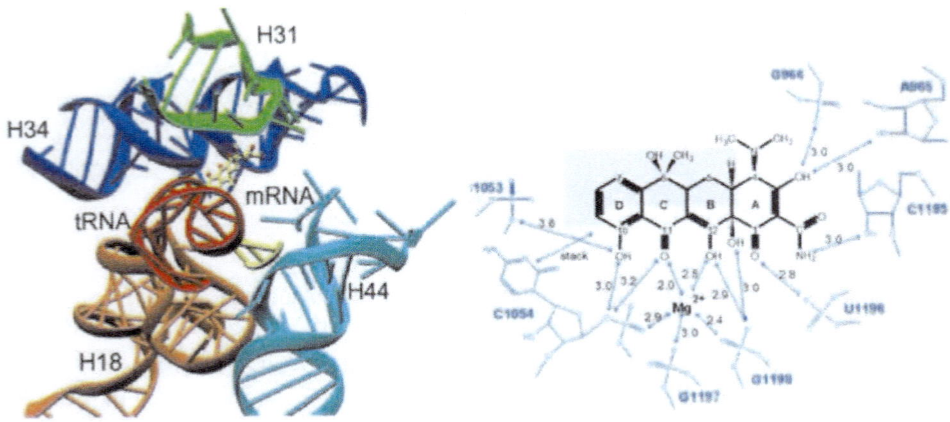

Fig. 2. Left, primary binding site of tetracycline (**4**); Right, tetracycline (**4**) and possible interactions with 16S RNA at the primary site
(Adapted from Reference 4 with permission from Elsevier)

In 2015, Zhou and co-workers described structural characterization of an alternative mode of tigecycline binding to the bacterial ribosome. It indicated extensive interactions of the southeast portion of the D-ring of the tetracycline with the A-site of the ribosome through hydrogen bonding, mutual coordination of magnesium ions, and π-stacking.[5]

The small, or 30S, ribosomal subunit has three binding sites for tRNA molecules, designated the A (aminoacyl), P (peptidyl), and E (exit) sites. Tetracyclines inhibit protein synthesis by binding to the 30S subunit of the ribosome. Similar to the aminoglycosides, tetracyclines inhibit the binding of amino-acyl tRNA to the A site of the ribosome. The 7S ribosomal protein is part of the binding site. A highly conserved region of 16S rRNA may also be part of the binding site, which explains the broad spectrum of tetracycline. Tigecycline (**7**) also binds at the A site of the 30S subunit of the ribosome and inhibits bacterial protein synthesis with potency 3- and 20-fold greater than that of minocycline (**6**) and tetracycline (**4**), respectively. The increased potency of tigecycline (**7**) stems from additional interaction made between the *tert*-butyl-glycylamindo side chain and the ribosome, in particular, a stacking interaction between the glycylamine moiety and C1054 (H34, see Fig. 1).

Tetracyclines work by their A-site occupation of the ribosome.

Tetracycline is an inhibitor of the elongation cycle, where it specifically prevents binding of tRNA to the A-site. Though non-enzymatic binding (without EF-Tu) of tRNA to the A-site is totally inhibited, the first step of tRNA binding in the form of the ternary complex aa-tRNA•EF-Tu•GTP is possible. However, upon hydrolysis of GTP to GDP by EF-Tu the incoming tRNA is lost from the ribosome.

The translation machinery ensures accurate conversion of the genetic information of the messenger RNA (mRNA) into the corresponding polypeptide sequence. The ribosome provides the platform on which the mRNA can be decoded by transfer RNAs (tRNAs). Each tRNA carries a specific amino acid, which is faithfully incorporated into the growing polypeptide chain. Three tRNA-binding sites exist on the ribosome: The A-site is the site at which decoding occurs; here the correct aminoacyl tRNA (aa-tRNA) is selected on the basis of the mRNA codon displayed at this site. Before peptide-bond formation, the P-site carries the peptidyl-tRNA, the tRNA bearing the elongating polypeptide chain. The E-site binds exclusively deacylated tRNAs, i.e., those tRNAs that having undergone peptide-bond formation are ready to exit from the ribosome

5.3 Structure–Activity Relationship (SAR)

Tetracycline antibiotics are quite fragile. They decompose readily at pH10 or higher, which explains why limited semi-synthetic tetracyclines are available on the market.

Around 1995, Myers at Cal Tech began an ambitious project to synthesize tetracycline derivatives *via* total synthesis, thus gaining access to tetracyclines not possible from semi-synthesis. The efforts continued after Myers moved to Harvard in 1998. Myers's convergent synthesis of tetracyclines created revolutionary discovery engine! Thus far, more than 3,000 tetracycline derivatives have been prepared using his practical and scalable synthesis, which led to access of many tetracycline analogs that were not accessible via semi-synthesis. The best in the completely synthetic tetracycline is probably eravacycline (**1**). The structure-activity relationship of marketed tetracycline antibiotics is shown in Table 1, where MIC stands for mean inhibition concentration. In order to test the antibiotics' activities against resistant strains, their *in vitro* activities were also tested in ribosomal protection protein M, *tet*(M).[6]

Table 1. The structure–activity relationship of marketed tetracycline antibiotics

	In vitro MIC	(mg/mL)
Antibiotics	Control	*tet*(M)
Eravacycline (**1**)	0.063	0.063
Tigecycline (**7**)	0.063	0.13
Minocycline (**6**)	0.5	64
Doxycycline (**5**)	2	64
Tetracycline (**4**)	2	128

As Fig. 3 has made it abundantly clear, eravacycline (**1**) is significantly more potent in both naïve and resistant bacterial strains than tetracycline (**4**) and tigecycline (**7**). Source: Based on reference 6.

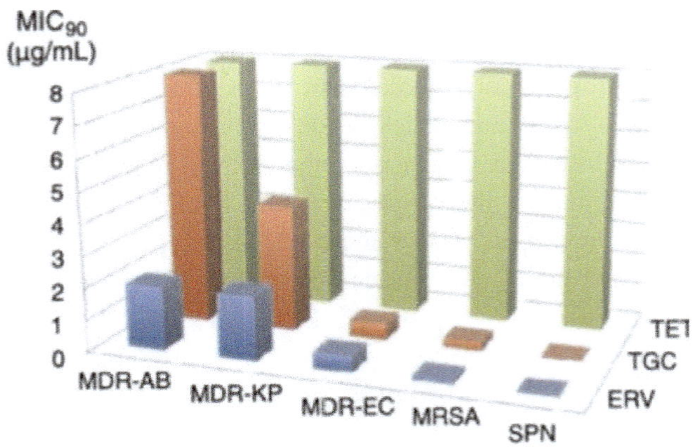

Fig. 3. Relative potencies of eravacycline (**1**) vs. tetracycline (**4**) and tigecycline (**7**) (Adapted from Reference 6 with permission from Elsevier)

5.4 Pharmacokinetics and Drug Metabolism

Eravacycline (**1**) is given orally at 300 mg or 400 mg once daily, or 100 mg every 12 hours. However, the oral administration was discontinued after poor outcome in a phase III clinical study in complicated urinary tract infections (cUTIs).

For intravenous (IV) administration of eravacycline (**1**), it is given at 1 mg/kg concentration over 12 h. Its maximal concentration C_{max} is 1.82 μg/mL and its half-life $t_{1/2}$ is 20 h, which is considered very high. Its volume of distribution V_d is 321 L, indicating that eravacycline (**1**) is highly distributed into body tissues. The area under the curve (AUC) for eravacycline (**1**) is 6.31 μg•h/mL, whereas its plasma protein binding (PPB) has a range of 79–90%, which is considered moderate for human subjects.[7]

The major enzymes that metabolize eravacycline (**1**) are CYP3A4 and flavin-containing mono-oxygenase (FMO).[8]

5.5 Efficacy and Safety

Eravacycline (**1**) exhibited greatly improved broad-spectrum antibiotic activity relative to other tetracyclines, particularly against multidrug-resistant bacteria, with MIC_{90} values ranging from < 0.008 to 2 μg/mL. It has been shown to evade major tetracycline-specific resistance mechanisms such as ribosomal protection and efflux.

Eravacycline (**1**) potently inhibits growth of a collection of clinically important Gram-negative bacterial species, a number of which are resistant to many or all third-generation cephalosporins, aminoglycosides, and fluoroquinolones.

The most common treatment-emergent adverse events included vomiting (5.7%), blood amylase increase (5.7%), lipase increase (5.7%), abdominal pain (3.8%), ileus (3.8%), nausea (1.9%), and thrombophlebitis (1.9%).[9]

A couple of papers have been published on meta-analysis of efficacy and safety of eravacycline (**1**) for treating complicated intra-abdominal infections (cIAIs). Eravacycline (**1**) maintains stability against the efflux pumps and ribosome protection protein that typically confer resistance to other members of tetracyclines. While the efficacy of eravacycline (**1**) is commensurate to reference carbapenems such as ertapenem and meropenem, it seems to be associated with higher risk of treatment-emergent adverse effects (TEAEs), especially for nausea and vomiting.[10,11]

5.6 Synthesis

The first- and the second-generation syntheses of eravacycline (**1**) were developed by Myers' group at Harvard. After Tetraphase Pharmaceuticals was founded in 2006, the company carried out extensive process chemistry development work to optimize the large-scale synthesis of eravacycline (**1**). Here in this chapter, only the process/manufacture synthetic route is discussed to save space.

The retro-synthetic scheme is shown. In essence, eravacycline (**1**) is assembled *via* a crucial Michael–Dieckmann cyclization between A-ring **10** and A–B ring **11**. Tricycle 11, on the other hand, is prepared by the coupling between furfural **12** and isoxazole **13**.

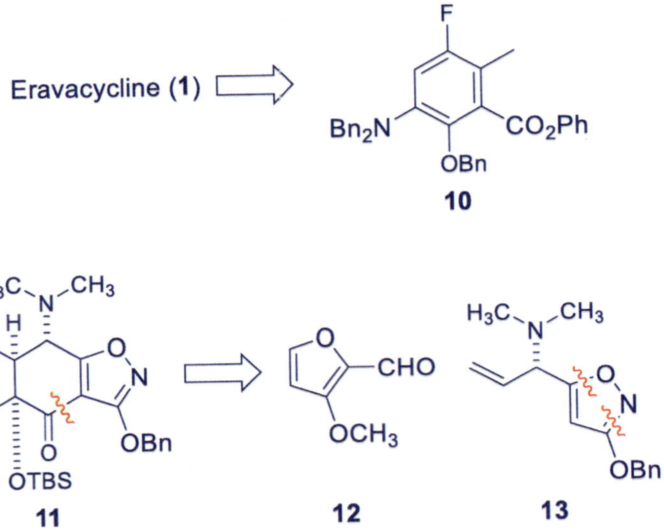

In practice, the process synthesis of enantiomerically enriched isoxazole allylic amine **13** began with a radical-initiated dibromination of inexpensive dimethyl maleate (**14**). The bromination was done at a 200 kg batch. Condensation of the resulting dibromide **15** and *N*-hydroxyurea (**16**) was facilitated by *t*-BuOK to assemble hydroxy-isoxazole **17**. Subsequent benzylation of its hydroxyl group was followed by a DIBAL-H reduction to furnish aldehyde **18**. Condensation of aldehyde **18** with the (*S*)-variant of the Ellman auxiliary [(*S*)-*tert*-butylsulfinylamide **19**] was aided by $CuSO_4$ to prepare chiral sulfinimine **20** in 85% yield. The key vinylation of sulfinimine **20** was carried out using vinylmagnesium chloride in the presence of methyllithium and zinc chloride to afford adduct **21** in 98% *de*. The *tert*-butylsulfinyl chiral auxiliary was readily removed under acidic condition and the exposed primary amine was treated with formaldehyde in the presence of sodium acetate, followed by reduction with a picoline–borane complex to give dimethylamine **13** in 95% *ee*. The *ee* value was greatly enhanced by forming its corresponding tartrate salt **13′**, which had a 99% *ee* after recrystallization.[12,13]

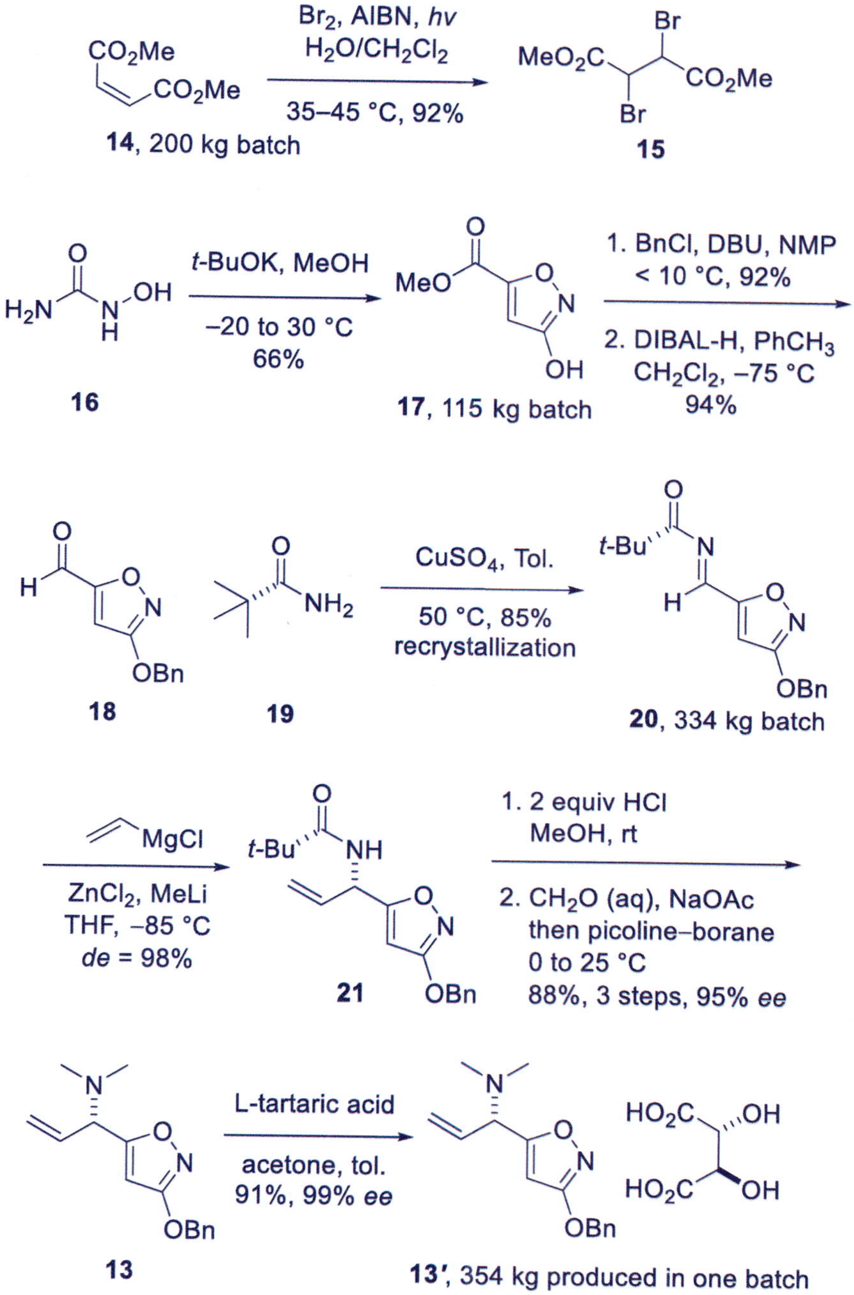

Tetraphase disclosed their process preparation of A–B ring intermediate TP-808 (**11**) in 2017. At first, the tartrate salt **13′** was treated with sodium hydroxide to provide free base **13**. Deprotection of the 4-position of isoxazole **13** was achieved using Knochel's Turbo Grignard reagent. To the deprotonated intermediate was added aldehyde **12** to produce adduct **22** as a mixture of diastereomers with a 3.57:1 *de* with the major diastereomer as shown. Subsequently, alcohol **22** was refluxed in DMSO with Hünig's base to promote the crucial intramolecular Diels–Alder reaction and the resulting alcohol intermediate was oxidized with sulfur trioxide–pyridine complex to give ketone **23**. Demethylation of methyl enol ether **23** was accomplished using boron trichloride at cold temperature, which concurrently promoted a Lewis acid-promoted ring opening of the oxacycle to produce enone **24** in 90% yield. Protection of the alcohol on **24** as its *tert*-butyldimethylsilyl ether was followed by recrystallization in isopropanol to furnish the A–B ring intermediate **11**.[14,15]

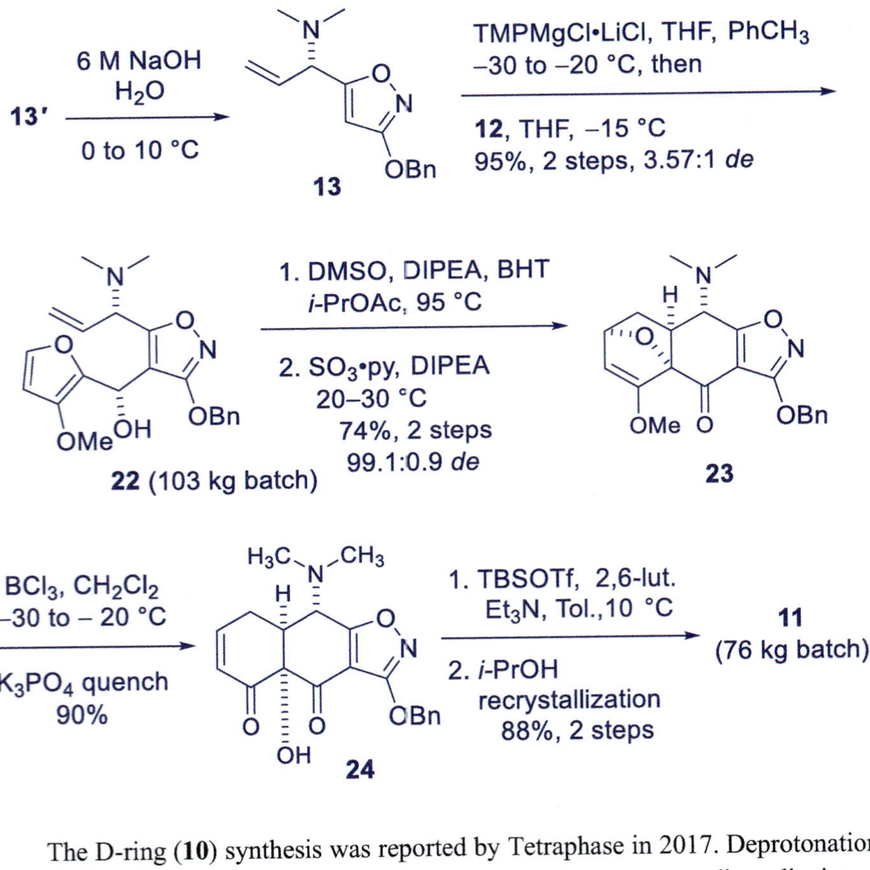

The D-ring (**10**) synthesis was reported by Tetraphase in 2017. Deprotonation of benzoic acid **25** using 2 equiv of LDA provided the corresponding dianion and methylation selectively took place at the 2 position to give **26**. The carboxylate anion was not esterized to give the methyl ester. After converting carboxylic acid **26** to its corresponding benzyl ester, de-methylation using BBr_3 reveal phenol **27**. To make the synthesis more convergent, the 9-aniline was installed sooner rather than later. To that end, nitration of **27** was followed by benzylation of the phenol functionality to produce nitroarene **28**. The nitro group was reduced with sodium bisulfite to afford aniline **29**. Since transformations **25** to **29** were conveniently "telescoped", Tetraphase finally tallied the overall yield of the last six steps to be a remarkable 83%! Finally, bis-benzylation of aniline **29** furnished the D-ring as **10** in 80% yield.[16]

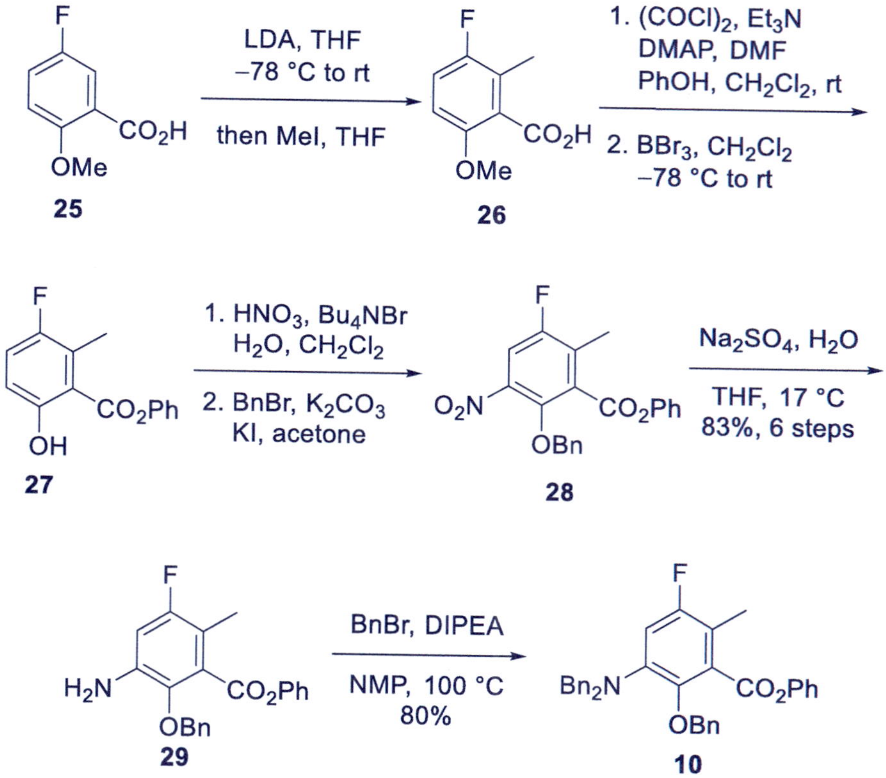

The grand finale was to assemble all four rings of eravacycline (**1**) as the endgame. It was achieved by coupling between fragments **10** and **12** using the Michael–Dieckmann cyclization by the treating **10** with LDA, followed by addition of **12** and LiHMDS to assemble adduct **30** in 94% yield. Subsequently, aqueous HF proved to be the best reagent for the removal of the TBS protective group despite of its highly corrosive nature to give alcohol **31**. The three benzyl protective groups were summarily removed using a palladium-catalyzed hydrogenation to reveal aniline **32**. Amidation of aniline **32** with acid chloride **33** delivered eravacycline (**1**) as a free base. Finally, the free base was treated with excess HCl to give its corresponding bis-HCl salt as the active pharmaceutical ingredient (API) for the drug.[17]

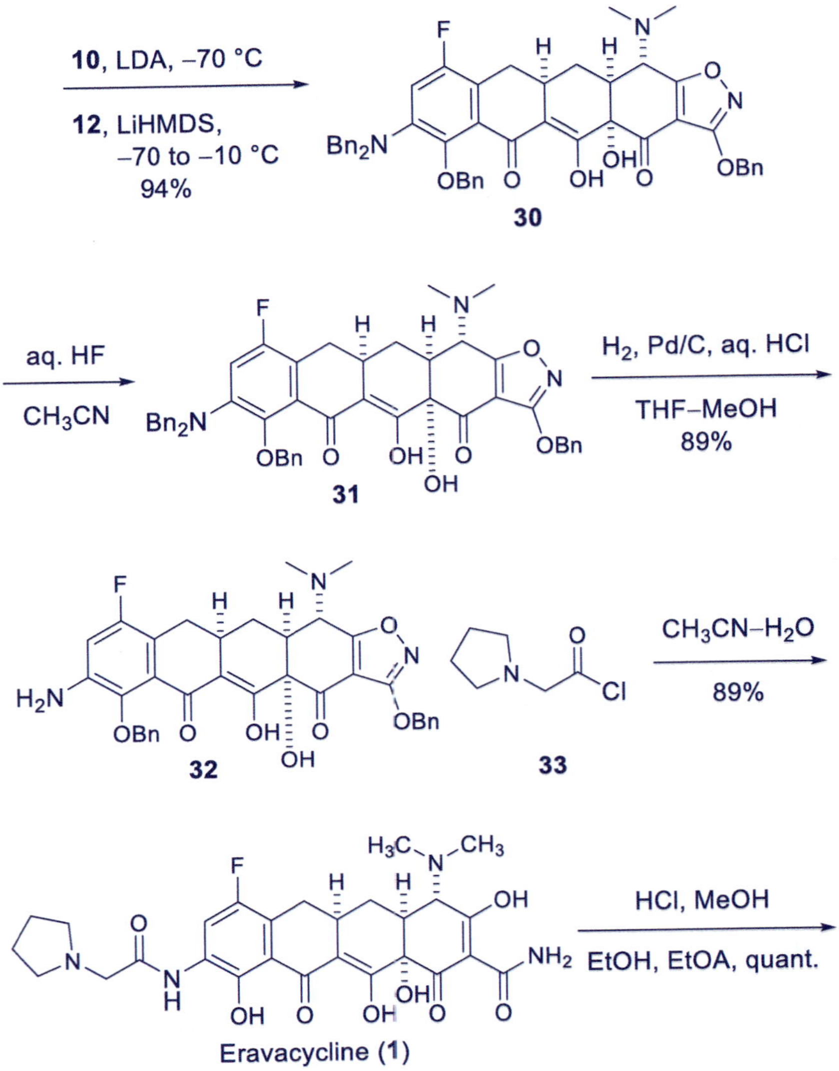

Eravacycline•2HCl

5.7. Summary

Myers' convergent and practical synthesis of eravacycline (**1**) demonstrated the power of total synthesis, which led to access to many tetracyclines that were not possible from semi-synthesis. Drug design to enhance potency and minimize drug resistance led to the discovery and development of a hyper-potent tetracycline antibiotics that proved to be efficacious to treat bacterial pathogens that developed resistance to older tetracyclines.

Discovery of a successful antibiotic is a great scientific achievement and greatly benefits humanity. Yet, scientific success does not always translate to financial success. Most of biotech companies specializing on antibiotics have not done well financially. For instance, Achaogen Inc., a company in South San Francisco, won the FDA approval of its novel aminoglycoside antibiotic plazomicin (Zemdri) in 2018. Since sales of the drug were so abysmal that the company went bankrupt in 2019.

Meanwhile, despite a great scientific and medical triumph in discovery and development of eravacycline (Xerava, **1**), Tetraphase Pharmaceuticals was not a profitable company. Its stocks were worth only a couple of dollars in 2020 when La Jolla Pharmaceutical Company acquired Tetraphase for a poultry $59 million.

5.8. References

1. (a) Jones, C. H.; Petersen, P. J. Tigecycline: a review of preclinical and clinical studies of the first-in-class glycylcycline antibiotic. *Drugs Today* **2005**, *41*, 637–659. (b) Sum, P.-E.; Petersen, P. Synthesis and structure-activity relationship of novel glycylcycline derivatives leading to the discovery of GAR-936. *Bioorg. Med. Chem. Lett.* **1999**, *9*, 1459–1462.
2. Tanaka, S. K.; Steenbergen, J.; Villano, S. Discovery, pharmacology, and clinical profile of omadacycline, a novel aminomethylcycline antibiotic. *Bioorg. Med. Chem.* **2016**, *24*, 6409–6419.
3. (a) Deeks, E. D. Sarecycline: First Global Approval. *Drugs* **2019**, *79*, 325–329. (b) Batool, Z.; Lomakin, I. B.; Polikanov, Y. S.; Bunick, C. G. Sarecycline interferes with tRNA accommodation and tethers mRNA to the 70S ribosome. *Proceed. Natl. Acad. Sci.* **2020**, *117*, 20530–20537.
4. Bodersen, D. E.; Clemons, W. M.; Carter, A. P.; Morgan–Warren, R. J.; Wimberly, B. T.; Ramakrishan, V. The Structure Basis for the Action of the Antibiotics Tetracycline, Pactamycin, and Hygromycin B on the 30S Ribosomal Subunit. *Cell* **2000**, *103*, 1143–1154.
5. Schedlbauer, A.; Kaminishi, T.; Ochoa-Lizarralde, B.; Dhimole, N.; Zhou, S.; Lo'pez-Alnoso, J. P.; Connell, S. R.; Fucini, P. Structural characterization of an alternative mode of tigecycline binding to the bacterial ribosome. *Antimicrob Agents Chemother* **2015**, *59*, 2849–2854.
6. Liu, F.; Myers, A. G. Development of a platform for the discovery and practical synthesis of new tetracycline antibiotics. *Cur. Opin. Chem. Biol.* **2016**, *32*, 48–57.
7. (a) Heaney, M.; Mahoney, M. V.; Gallagher, J. C. Eravacycline: The Tetracyclines Strike Back. *Ann. Pharmacother.* **2019**, *53*, 1124–1135. (b) McCarthy, M. W. Clinical Pharmacokinetics and Pharmacodynamics of Eravacycline. *Clin. Pharmacokinet.* **2013**, *58,* 1149–1153.

8. Newman, J. V.; Zhou, J.; Izmailyan, S.; Tsai, L. Mass balance and drug interaction potential of intravenous eravacycline administered to healthy subjects. *Antimicrob. Agents Chemother.* **2019**, *63*, e01810-18/1-e01810-18/11.
9. Thakare, R.; Dasgupta, A.; Chopra, S. Eravacycline for the treatment of patients with bacterial infections. *Drugs Today* **1998**, *54*, 245–254.
10. (a) Eljaaly, K.; Ortwine, J. K.; Shaikhomer, M.; Almangour, T. A.; Bassetti, M. Efficacy and safety of eravacycline: A meta-analysis. *J. Global Antimicrobial Resis.* **2021**, *24*, 424–428. (b) Tang, H.-J.; Lai, C.-C. The safety of eravacycline in the treatment of acute bacterial infection. *Clin. Infect. Dis.* **2020**, *70*, 2750–2751.
11. Scott, L. J. Eravacycline: A Review in Complicated Intra-Abdominal Infection. *Drugs* **2019**, *79*, 315–324.
12. Xiao, X.-Y.; Hunt, D. K.; Zhou, J.; Clark, R. B.; Dunwoody, N.; Fyfe, C.; Grossman, T. H.; O'Brien, W. J.; Plamondon, L.; Ronn, M.; et al. Fluorocyclines. 1. 7-Fluoro-9-pyrrolidinoacetamido-6-demethyl-6-deoxytetracycline: A Potent, Broad Spectrum Antibacterial Agent. *J. Med. Chem.* **2012**, *55*, 597–605.
13. (a) Zhang, W.-Y.; Hogan, P. C.; Chen, C.-L.; Niu, J.; Wang, Z.; Lafrance, D.; Gilicky, O.; Dunwoody, N.; Ronn, M. Process Research and Development of an Enantiomerically Enriched Allylic Amine, One of the Key Intermediates for the Manufacture of Synthetic Tetracyclines. *Org. Process Res. Dev.* **2015**, *19*, 1784–1795. (b) Brubaker, J. D.; Myers, A. G. A Practical, Enantioselective Synthetic Route to a Key Precursor to the Tetracycline Antibiotics. *Org. Lett.* **2007**, *9*, 3523–3525.
14. Zhang, W.-Y.; Chen, C.-L.; He, M.; Zhu, Z.; Hogan, P.; Gilicky, O.; Dunwoody, N.; Ronn, M. Process Research and Development of TP-808: A Key Intermediate for the Manufacture of Synthetic Tetracyclines. *Org. Process Res. Dev.* **2017**, *21*, 377–386.
15. Clark, R. B.; Hunt, D. K.; He, M.; Achorn, C.; Chen, C.-L.; Deng, Y.; Fyfe, C.; Grossman, T. H.; Hogan, P. C.; O'Brien, W. J.; et al. Fluorocyclines. 2. Optimization of the C-9 Side-Chain for Antibacterial Activity and Oral Efficacy *J. Med. Chem.* **2012**, *55*, 606–622.
16. (a) Myers, A. G.; Kummer, D. A.; Li, D.; Hecker, E.; Dion, A.; Wright, P. M. Synthesis of Tetracyclines and Intermediates for the Treatment of Infection. WO 2010126607A2, 2010. (b) Zhang, W.-Y.; Che, Q.; Crawford, S.; Ronn, M.; Dunwoody, N. A. Divergent Route to Eravacycline. *J. Org. Chem.* **2017**, *82*, 936–943.
17. Ronn, M.; Zhu, Z.; Hogan, P. C.; Zhang, W.-Y.; Niu, J.; Katz, C. E.; Dunwoody, N.; Gilicky, O.; Deng, Y.; Hunt, D. K.; et al. Process R&D of Eravacycline: The First Fully Synthetic Fluorocycline in Clinical Development. *Org. Process Res. Dev.* **2013**, *17*, 838–845.

6

Albuvirtide (Aikening), A gp41 Analog as an HIV-1 Fusion Inhibitor

Yvonne M. Angell, Wendy J. Hartsock, Timothy M. Reichart

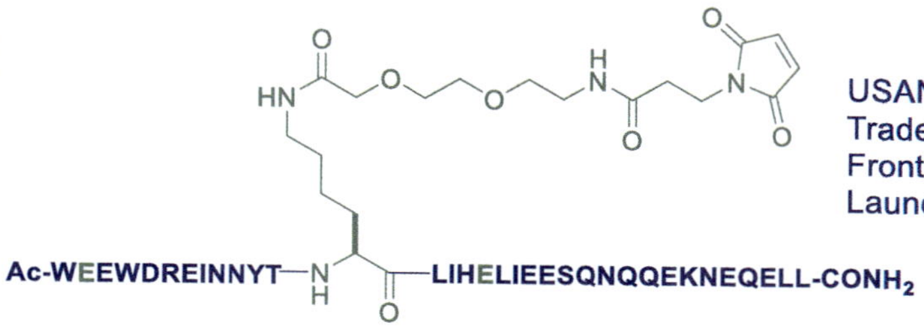

USAN: Albuvirtide
Trade name: Aikening
Frontier Biotechnologies
Launched: 2018

1

1 Background

On June 6, 2018, without any reports from the US FDA, a press release from Frontier Biotech announced the approval of a new HIV drug in China that is given by once-weekly injection.[1] This is a rare example of the first approval of an HIV treatment not occurring in either the US or Europe. The approval of albuvirtide (Aikening, **1**), as an HIV-1 Fusion inhibitor and gp41 analog, is the first approval of a second-generation, long-acting peptide HIV fusion inhibitor. Albuvirtide (**1**) is a maleimide-modified peptide derived from the HIV-1 gp41 sequence and demonstrates an impressive plasma half-life of 11 days in antiretroviral-naive adults treated intravenously.[2]

The first peptide HIV fusion inhibitor, enfuvirtide (T20, Fuzeon), was approved by the US FDA in 2003 as a rescue therapy for HIV/AIDS patients who failed to respond to antiretroviral therapeutics.[3,4] Enfuvirtide was the first of a new class of HIV/AIDS drugs (HIV fusion inhibitors) and was approved for use with other anti-HIV medicines to treat adults and children ages 6–16 years with HIV infection.[5] Enfuvirtide binds to a region of the HIV-1 envelope glycoprotein gp41 and prevents subsequent viral fusion with the target cell membrane. However, as a first-generation compound in the class, enfuvirtide's clinical

Current Drug Synthesis, First Edition. Edited by Jie Jack Li.

application has been limited primarily by its relatively low potency, short *in vivo* half-life leading to inconvenient dosing by twice daily subcutaneous injections,[6] and the rapid development of drug-resistant mutations.[7–9] Significant research has been undertaken since then to develop next generation peptide HIV fusion inhibitors with enhanced antiviral potency, mutation resistance, and improved pharmacokinetic profiles.

To improve enfuvirtide's anti-HIV efficacy, drug-resistance profile, half-life and pharmaceutical properties, some of the most successful approaches under development include the addition of the pocket-binding domain (PBD) to the N-terminus of T20, linking of the M-T hook to the N-terminus of the PBD, as well as conjugation of cholesterol, a serum albumin-binding motif or gp120-binding fragment with a PBD-containing C-terminal heptad repeat-peptide. Therefore, sifuvirtide from Tianjin FusoGen Pharmaceuticals,[10] albuvirtide (**1**) from Frontier Biotechnologies,[11] cholesterol-conjugated HIV fusion inhibitor from the Institute of Pathogen Biology,[12] Chinese Academy of Medical Science, 2DLT, a bivalent HIV fusion inhibitor/inactivator,[13] and a enfuvirtide/sifuvirtide combination regimen from the New York Blood Center[14] are all being pursued as second-generation versions of enfuvirtide. Compared with T20, these newly designed PBD-containing HIV fusion inhibitors exhibit improved anti-HIV efficacy, drug resistance, pharmacokinetic/pharmacodynamic profile and pharmaceutical properties, and may improve patient's quality of life as the next generation HIV fusion inhibitors targeting gp41 for clinical use. The path to a successful outcome, the approval of albuvirtide (**1**), stemming from these novel and exciting peptide therapeutic designs will be discussed in this chapter.

2 Pharmacology

The human immunodeficiency virus (HIV) is one of the most challenging to medically manage and continues to be a major global public health issue. The World Health Organization estimates that HIV has claimed 36.3 million lives so far, and there were an estimated 37.7 million people living with HIV at the end of 2020, over two thirds of whom (25.4 million) are in the WHO African Region.[15]

HIV is speculated to have evolved from the simian immunodeficiency virus.[16] HIV infects and destroys CD4 cells (T cells), and undermines the human immune system in AIDS. In general, HIV cannot be cured, and therefore requires life-long treatment. When not treated, a compromised immune system surrenders to co-infections, and eventually leads to significant patient mortality. HIV is highly heterogeneous, mutates quickly, and can exist in a latent state for over 10 years, which makes prevention and treatment challenging.[17]

Antiretroviral therapy (ART) has been proven effective in reducing viral loads in patients, viral infectivity, and the morbidity and mortality of acquired immunodeficiency syndrome (AIDS)-related diseases. The International Antiviral Society recommends all persons with HIV who have detectable viremia, regardless of their CD4 cell count, should begin ART as soon as possible after diagnosis.[18] Multidrug regimes are used to reduce the progression of disease, occurrence of opportunistic infections, hospitalizations, and death. Marketed therapeutics aim to suppress viral replication below the level of detection (< 50 RNA copies/mL) and to restore immunity by increasing CD4+ T cells.[19] There are currently more than 25 antiretroviral medications available in five drug classes. Aside from fusion inhibitors, other classes used in initial treatment are nucleoside reverse transcriptase inhibitors (NRTIs), non-nucleoside reverse transcriptase inhibitors (NNRTIs), protease inhibitors, and integrase strand transfer inhibitors.

For decades, a combination of two NRTIs plus a third agent has been the mainstay of HIV therapy.[20] Recommended initial regimens consist of three drugs: two NRTIs and an integrase strand transfer inhibitor (INSTI) or a two-drug regimen of dolutegravir (**2**)/lamivudine (**3**). In recent years, combinations that contain only a single NRTI lopinavir (**4**)/ritonavir (**5**) plus lamivudine or no NRTI darunavir (**6**)/ritonavir plus raltegravir (**7**) have shown comparable virological results as three-drug regimens, but both regimens have disadvantages. Lopinavir/ritonavir plus lamivudine has a larger pill burden and greater toxicity than other regimens and darunavir/ritonavir plus raltegravir (**7**) is not as effective as a three-drug regimen in people with a CD4 cell count below 200/µL or an HIV RNA level above 100,000 copies/mL.[18] New data demonstrate the efficacy of two-drug dolutegravir/lamivudine therapy as initial and maintenance treatment.[18] In addition, recommended for the first time, a long-acting antiretroviral regimen injected once every 4 weeks for treatment or every 8 weeks pending approval by regulatory bodies and availability.[18]

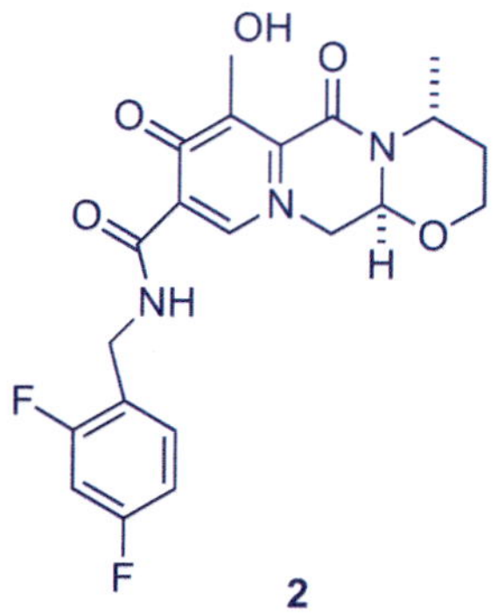

2

Dolutegravir (Dovato)
GSK, 2019
Integrase strand transfer inhibitor (INSTI)

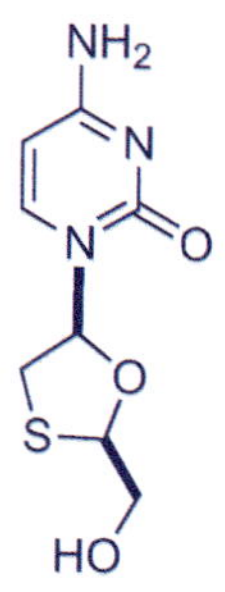

3

Lamivudiine (Epicir)
GSK, 1995
Nucleoside reverse transcriptase inhibitor (NRTI)

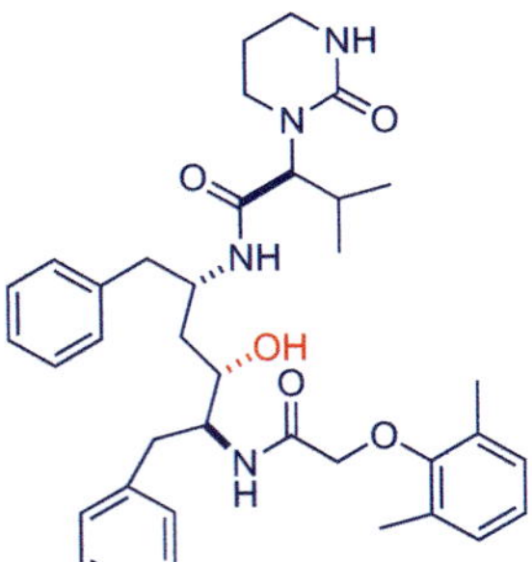

4

Lopinavir (ABT-378)
Kaletra (lopinavir/ritonavir coformulation)
Abbott, 2000
Protease inhibitor (PI)

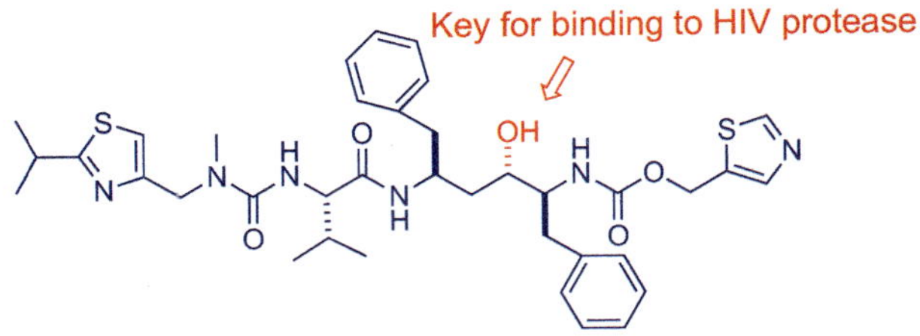

5

Ritonavir (Norvir)
Abbott, 1996
HIV protease inhibitor

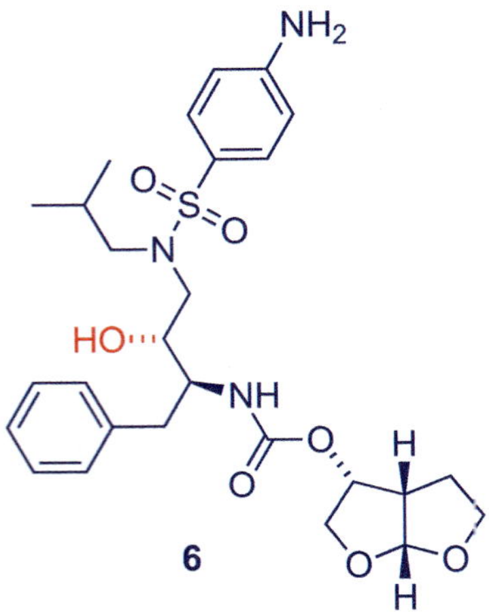

6

Darunavir (Prezista)
Tibotec, 2006
Nonpeptidic protease inhibitor (PI)

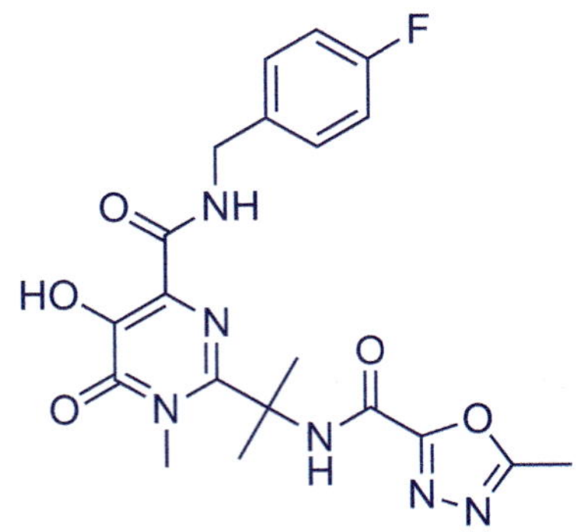

7

Raltegravir (Isentress)
Merck, 2007
Integrase inhibitor

Current treatment of HIV with a combination of antiretroviral drugs has markedly decreased the mortality and morbidity from HIV-1 infections; nevertheless, a report from WHO showed that a significant proportion (13.5%) of patients failed in their current treatment because of drug resistance, with NRTI and NNRTI resistant mutations as the most common mutations.[21] In addition, current treatment regimens are limited by the emergence of side effects and long-term toxicity over the course of lifelong daily oral treatment. Therefore, agents with new mechanisms of action are still needed.

HIV-1 fusion inhibitors block a novel target, virus-cell membrane fusion, in the viral life cycle compared with the HIV standard of care therapeutics. Albuvirtide (**1**) is a new peptide-based and first long-acting HIV fusion inhibitor. It was designed to rapidly form conjugates by covalently linking an anti-HIV fusion peptide to serum albumin at a 1:1 molar ratio to prevent proteolytic degradation and extend its *in vivo* half-life ~10-fold longer than enfurvitide.[22]

In addition, previous *in vitro* studies showed that albuvirtide (**1**) has a potent and broad anti-HIV activity, with an even higher inhibitory effect than T20. In studies with the HIV-1 subtypes that predominate the worldwide AIDS epidemics (subtypes A, B and C) and China (subtype B′, CRF07_BC and CRF01_AE recombinants), the IC_{50} of albuvirtide (**1**) ranged from 2.92 to 27.41 nM; whereas, the IC_{50} of enfurvitide ranged from 14.47 to 214.04 nM. Moreover, albuvirtide (**1**) is also effective for the induced variants of the virus that are resistant to enfurvitide.[22]

The strategy that enveloped viruses, such as HIV-1 (human immunodeficiency virus type 1), use to enter their host cells is membrane fusion.[23] Virus-encoded fusion proteins undergo structural rearrangements from a metastable prefusion conformation to a stable post fusion conformation.[24–26] In the case of HIV-1, its envelope glycoprotein (Env) functions as the fusion protein. The Env protein is produced as a precursor, gp160, which forms a trimer $(gp160)_3$ and is then cleaved by a cellular furin-like protease into two noncovalently associated subunits: the receptor-binding subunit gp120 and the fusion subunit gp41.[27] Three copies of each subunit constitute the mature envelope spike protein $(gp120/gp41)_3$. It is thought that sequential binding of gp120 to primary CD4 receptor and co-receptor (e.g., chemokine receptor CCR5 or CXCR4) triggers a cascade of refolding events in gp41 that cause the membrane fusion process.[28,29] Recent cryo-EM structures of two full-length HIV-1 Env constructs purified in detergent have been reported.[30,31] The prefusion gp41 adopts a radically different conformation from the post-fusion six-helix bundle structure. The fusion-promoting conformational changes of Env expose some highly conserved, functionally critical regions of Env and are promising targets for developing broad and effective fusion inhibitors. Additionally, fusion inhibitors do not need to cross the cell membrane to reach their target.

The mature Env spikes are antigens on the surface of the virion and induce strong immune responses in infected individuals;[32,33] thus, HIV-1 Env is an important target for the development of both vaccines and therapeutics against the virus. In fact, the first successful peptide fusion inhibitor, enfurvitide (T20), was serendipitously discovered from the failed attempts to produce AIDS vaccines.

The formation of the six-helix bundle structure of gp41 is essential for viral infectivity,[26] and there have been many peptide-based inhibitors designed and taken into development for blocking this step. Peptides derived from the N-terminal heptad repeat (NHR) or C-terminal heptad repeat (CHR) domains were designed as alpha-helix mimics that bind the fusion intermediates of gp41 and prevent the formation of the six-helix bundle.[34,35] The first FDA approved peptide fusion inhibitor, enfuvirtide (Fuzeon, T-20), codeveloped by Roche and Trimeris, is a 36-residue peptide derived from the C-terminal heptad repeat region (CHR) of gp41.[36,37]

Although enfuvirtide was initially successful, it has several shortcomings such as the requirement to be stored at low temperature, freshly reconstituted, and injected subcutaneously twice a day. Furthermore, injection site reactions, the rapid emergence of resistant virus strains, and the high cost of production limited its long-term use.[38] In addition, enfuvirtide has a very limited window of action when gp41 is exposed after a conformational change in HIV envelope structure during the entry/fusion process. Thus, a high concentration of enfuvirtide must be maintained in the patient, which requires dosing at 90 mg subcutaneously twice daily, which is very difficult to achieve in resource-limited settings or without good patient compliance.

3 Structure–Activity Relationship (SAR)

A number of clinical candidates that have shown anti-fusogenic activity *in vitro* have been based on the sequence of the C-heptad repeat of the HIV envelope glycoprotein gp41 (Figure 1). T20 is also called enfuvirtide and was approved by the FDA in 2003. C34 is a peptide corresponding to a region of the C-heptad repeat that partially overlaps with T20 but extends farther to the N terminus (Figure 2). Peptides have also been investigated that have sequences either C-terminal[39], or N-terminal[40], to that of C34.

C34 has been the subject of significant modification and SAR studies. In one study, attaching cholesterol by means of a short linker on the C-terminus of C34 improved antiviral potency by a factor of 20–50 depending on the HIV isolate.[41] This increased potency was not observed with an N-terminal cholesterol, and the cholesterol variant was more efficacious than a palmitoylated version. Interestingly, the increased potency was not due to tighter binding but was instead due to affinity to a cellular membrane (increasing its local concentration) and extending the half-life of the peptide *in vivo*.

C34-chol:	Ac-WMEWDREINNYTSLIHSLIEESQNQQEKNEQELLGSGC(Chol)-NH_2
C34:	Ac-WMEWDREINNYTSLIHSLIEESQNQQEKNEQELL-NH_2
Albuvirtide:	Ac-WEEWDREINNYTKLIHELIEESQNQQEKNEQELL-NH_2
Sifurvitide:	SWETWEREIENYTRQIYRILEESQEQQDRNERDLLE-NH_2
Enfuvirtide:	Ac-YTSLIHSLIEESQNQQEKNEQELLEDKWASLWNWF-NH_2

Figure 1. Sequence comparison.

Another set of modifications to C34 resulted in sifurvitide, a peptide currently in clinical trials. Sifurvitide has a series of modifications designed to enhance the peptide helicity and thereby reduce the ability of gp41 to form functional six-helix bundles. Chief among these modifications are the introduction of a series of *i, i+4* salt bridges, indicated in blue in Figure 1. These modifications resulted in roughly 10× more effective inhibition of HIV-1 *in vitro* than T20 while maintaining efficacy against T20-resistant strains of HIV.[10]

Albuvirtide (**1**) is also based very closely on the sequence of C34, with three key modifications: modification of M2 and S17 to glutamic acid residues, and modification of S13 to a lysine as a handle for further chemical modification. Chemical modifications to the lysine 13 side chain include attaching a short polyethylene glycol linker and then a maleimide moiety. The maleimide moiety is designed to increase the half-length of drug *in vivo* by chemical conjugation to serum albumin. Serum albumin has one free thiol group that readily and covalently attaches to the maleimide functionality of albuvirtide (**1**). This covalent attachment means that albuvirtide (**1**) is less likely to be inactivated by proteases and also less likely to be rapidly excreted through the kidneys as is common with other peptide drugs. Much like C34-chol, albuvirtide (**1**) trades slightly reduced binding affinity for significantly increased half-life *in vivo*, resulting in significantly increased efficacy against HIV.

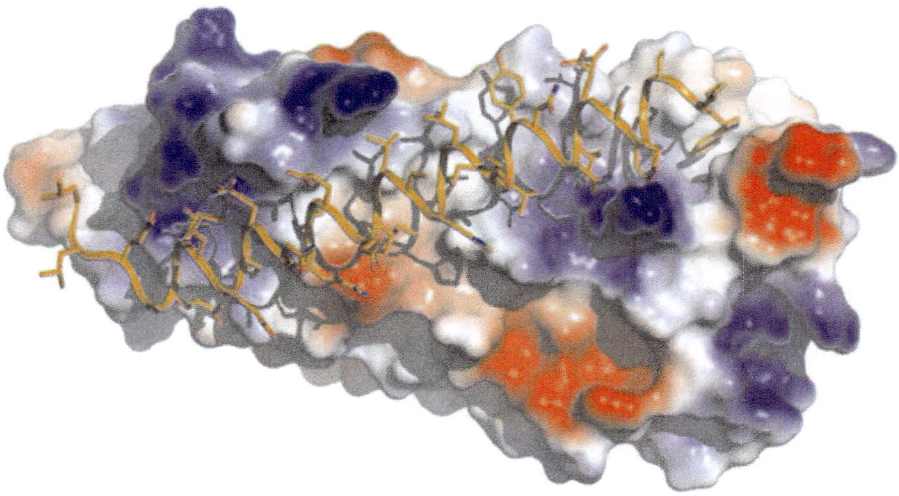

Figure 2. Crystal structure of C34. Data from PDB:3o3x.[42]
(Image design by Dr. Marc Adler)

4 Pharmacokinetics and Drug Metabolism

Preclinical evaluation of albuvirtide (ABT) demonstrated distribution in all tissues and organs in rat. The highest amounts of ABT were found in whole blood, kidneys, and ovaries and the lowest amounts were found in brain, body fat, and testes. Plasma protein binding was >96%.[43]

The elimination of ABT *in vivo* is primarily through the kidney and the expected route of metabolism, as with other peptide therapeutics, is expected to be by catabolism. Albuvirtide (**1**) did not exhibit any demonstrable inhibition of six major cytochrome P450 enzymes (Cyp1A2, 2C8, 2C9, 2C19, 2D6, and 3A4) in human microsome assays.[43] The half-life of the unmodified peptide in rat is 1.7 hours compared with 25.8 hours for the 3-maleimidopropionic acid (MPA)-modified peptide.[11] Similarly, in monkeys, the half-life is extended from 10.9 hours to 102.4 hours by incorporation of the MPA moiety and subsequent *in vivo* conjugation to albumin.

The pharmacokinetics of ABT have been assessed in single and multidose clinical studies. The 3-maleimidopropionic acid-modified peptide is delivered intravenously following reconstitution in 5% sodium bicarbonate and dilution in 0.9% sodium chloride. In a phase I study, a single dose of ABT in adults with HIV-1 gave a half-life of 10–12 days and led to viral suppression for 6–10 days.[2]

Combination therapy is employed in phase two and phase three trials of ABT (Table 1) to avoid potential emergence of early resistance that can occur during prolonged monotherapy. This is a result of the extended half-life of ABT, which requires prolonged time to achieve steady state concentrations.[43]

Table 1. Clinical trials summary

Clinical trial (location)	Experimental group[*1]	Control or active comparator[*¥]	Primary efficacy endpoint	Estimated completion date
Phase two ChiCTR-TRC-13003140 (China)	160 mg ABT qw + LPV/r or 320 mg ABT qw + LPV/r		Viral suppression at week 7	Complete
Phase three TALENT (China)	320 mg ABT qw + LPV/r	LPV/r + TDF + 3TC	Proportion of subjects with HIV-1 RNA < 50 copies/mL at week 48	Complete
Phase two Albuvirtide and 3BNC117 as Long-Acting Maintenance Therapy in Virologically Suppressed	Part 1 320 mg ABT q2w + 3BNC117 2g q2w Or 320 mg ABT q4w + 3BNC117 2g q4w	Subjects continuing baseline antiretroviral therapy	Proportion of participants with HIV-1 RNA < 50 copies/mL at the end of treatment phase in part 3 of the study	December 2022

Subjects, 3 part study (US)	Part 2 160 or 320 mg ABT q4w + 3BNC117 q4w Part 3 Optimal Dose		
Phase two Albuvirtide (1) in combination With 3BNC117 in virologically suppressed subjects with HIV-1 Infection (China)	320 mg ABT q2w + 3BNC117 2 g q2w	Proportion of participants with HIV-1 RNA < 50 copies/mL at week 26	December 2022
Phase two albuvirtide in combination With 3BNC117 in patients with multi-drug resistant (MDR) HIV-1 infection (US)	320 mg ABT qw + 3BNC117 2 g q2w or 320 mg ABT q2w + 3BNC117 2 g q2w	Proportion of participants (%) achieving a viral load reduction of at least 0.5 log from baseline	December 2022

[*]LPV/r is administered as a 400 mg/100 mg bid dose of lopinavir/ritonavir.
[1]3BNC117 Recombinant, fully human mAb of the IgG1κ isotype that specifically binds to HIV-1 gp120[¥]. TDF = tenofovir 300 mg qd; 3TC = lamivudine 300 mg qd.

Pharmacokinetics and drug–drug interactions of ABT with the protease inhibitors lopinavir and ritonavir were assessed in a phase two study utilizing a two-drug regimen (160 mg ABT + LPV/r or 320 mg ABT + LPV/r) in antiretroviral naïve, HIV-1 infected subjects (18–50 years of age). Subjects (N = 10 per group) were dosed twice daily (days 1–46) with LPV/r (400/100 mg). On days 5–7 ABT was given for 3 days then weekly until day 40 followed by a 47-day observation period.

As noted, the pharmacokinetic analysis of ABT required an alternative to conventional approaches because of ethical considerations in generating resistance during monotherapy.[43] Thus, a population structural pharmacokinetic model was employed to simulate the steady state concentration-time curves of ABT alone. The model was built from prior clinical trial data and used in conjunction with clinical study data to evaluate

PK DDIs. There were 10 patients enrolled in the 320 mg ABT + LPV/r (400/100 mg) group and nine subjects completed the trial, with one subject discontinuing.

Virtual studies (100 virtual trials with nine subjects, equivalent to the number in the physical study) were conducted in conjunction with the physical trial. Measured drug concentrations from co-therapy were used to determine interaction ratios by comparison of actual values with predicted values of $AUC_{(0-t)}$ (area under the plasma concentration–time curve in one dosing interval at steady state), C_{end} (plasma concentration of ABT at the end of the infusion), and C_{trough} (pre-dose plasma concentration at steady state) from each individual virtual trial.

Compared to LPV/r alone, co-administration of ABT and LPV/r led to lower values in AUC, C_{max}, and C_{trough} of lopinavir and ritonavir. This decrease may not impact effectiveness of the combination therapy. In contrast, co-administration of LPV/r with ABT did not significantly influence ABT exposure except for an increase in C_{end} (concentration of ABT at the end of infusion) of ABT. The cause for an increase in the observed C_{end} was not definitively correlated with co-administration of LPV/r and the clinical significance is not established. In summary, the pharmacokinetic outcomes of the phase one and phase two trials demonstrated a half-life of ABT to be 10–12 days leading to viral suppression for 6–10 days. Co-administration of ABT and LPV/r did not decrease exposure to ABT but did decrease exposure to LPV/r.

5 Efficacy and Safety

Short-term efficacy of ABT in combination with LPV/r was assessed in a phase two study in which HIV-1 infected subjects (18–50 years of age) were dosed with LPV/r BID and once weekly with ABT at 160 mg or 320 mg.[44] At week 7, both groups of subjects experienced viral suppression (HIV-1 RNA < 50 copies/mL) with a greater percentage achieving suppression in the group treated with 320 mg ABT qw (55.6%) than the group treated with 160 mg ABT qw (11.1%).

Interim results from the phase three, Test Albuvirtide in Experienced Patients (TALENT) study, a randomized, open-label, multicenter, parallel group, non-inferiority trial have been reported.[45] Subjects in TALENT are treatment experienced adults (16–60 years of age) infected with HIV-1 that have a plasma viral RNA load of at least 1000 copies/mL. Subjects were required to have been treated with two classes of antiretroviral agents, NRTIs + NNRTIs, for at least 6 months.

The objective of TALENT (Table 1) is to demonstrate non-inferiority of ABT in combination with LPV/r compared to standard second line combinations of LPV/r with two NRTIs (tenofovir and lamivudine). The non-inferiority margin was specified as 12%. Non-inferiority was defined as achieving a difference of viral suppression (<50 copies/mL of viral RNA at week 48) in the treatment arm vs active comparator such that the lower limit of the two tailed 95% confidence interval had to be equal to or greater than –12%.

The study was designed to test for superiority at a significance level of 5% in the event of establishing non-inferiority. At the time of the interim report, 347 patients had been randomized and 208 subjects were included in the interim analysis.

The primary efficacy outcome is the proportion of subjects with HIV-1 RNA < 50 copies/mL at week 48. Secondary outcomes include the percentage of subjects with HIV-1 RNA < 400 copies/mL at week 48, changes of HIV-1 RNA relative to baseline at week 48, and changes of CD4 cell count relative to baseline at week 48.

A modified intent to treat (mITT) population (N = 83 treatment arm, N = 92 active comparator) was comprised of subjects that received either the experimental treatment (ABT + LPV/r) or active comparator (LPV/r + tenofovir + lamivudine). Subjects that had serious protocol violations or did not have efficacy data were removed from the mITT population. The per protocol (PP) population (N = 76 experimental arm, N = 85 active comparator) had additional removal of subjects that were lost to follow up, withdrew consent, removed at investigators discretion, or discontinued due to an adverse event.

In the interim analysis of subjects at week 48 (primary endpoint), 80.4% (N = 46) of subjects in the treatment arm (ABT + LPV/r) and 66% (N = 50) of the active comparator arm (LPV/r + TDF + 3TC) had HIV-1 RNA levels below 50 copies/mL. Thus, non-inferiority was met with a difference of 14.4%, 95% CI (–3 ~ 31.9). established between the treatment arm and active comparator.

Table 2. Clinical trials secondary outcomes

	Week 48		Week 24	
Study arm	ABT + LPV/r (N = 46)	LPV/r + NRTIs (N = 50)	ABT + LPV/r (N = 83)	LPV/r + NRTIs (N = 92)
HIV-1 RNA <400 copies/mL	84.8%	74.0%	89.2%	82.6%
HIV-1 RNA log10 copies/mL change	-2.27 ± 0.96	-1.77 ± 1.33	-2.00 ± 1.01	-1.85 ± 1.16
CD4 T-cell count change (cells/μL)	120.5	150.3	79.0	86.0

Week 24 data from the mITT population also established non-inferiority with 79.5% (N = 83) of subjects in the treatment arm and 78.3% (N = 92) of subjects in the active comparator arm achieving viral RNA levels below 50 copies/mL with a difference of 1.2%, 95% CI (–10.8 ~ 13.4). The proportion of subjects in the PP population that achieved < 50 copies/mL of viral RNA at week 48 was 94.9% (N = 39) for the treatment arm and 74.4% (N = 43) for the active comparator. At week 24 the proportions were 84.2% (N = 76) for the treatment arm and 83.5% (N = 83.5) for the active comparator. Secondary outcomes were also achieved (Table 2). This interim analysis demonstrated non-inferiority and statistical superiority ($p = 0.01$) of ABT + LPV/r over NRTI + LPV/r based predominantly on virologic outcomes.

5.1 Safety (Adverse Effects, AEs)

Adverse event profiles were similar between the two study arms and most AEs were mild to moderate. The most common AEs were diarrhea, upper respiratory tract infection, and increases in triglycerides (Grade 3–4). No injection site reactions were reported.

5.2 Mutations

Genotypic resistance against at least one antiretroviral therapy was observed in 81.7% of subjects at baseline. The emergence of resistant mutations was assessed in subjects with viral load >400 copies/mL from the phase three TALENT study (N = 5 from ABT group, N = 13 from NRTI group). While NRTI and NNRTI resistant mutations were observed in at least one subject in each arm, none of the subjects in the ABT arm showed resistance mutations in the gp41 gene.

6 Synthesis

Albuvirtide (**1**) is unusual among peptide drugs of its size in that its synthesis is accomplished entirely on the solid phase. The initial report,[10] subsequent studies,[22] and patents[46,47] all report that albuvirtide (**1**) is synthesized using standard solid-phase peptide synthesis techniques. Specifically, Fmoc-protected amino acids are coupled stepwise to a resin depending on the C-terminus functionality. C-terminal acids are produced using 4-hydroxymethylphenoxyacetyl-4′-methylbenzhydrylamine (HMP) resin, while C-terminal amides are produced using Fmoc-protected Ramage resin (Figure 3A).

Fmoc-protected amino acids are activated using hexafluorophosphate benzotriazole tetramethyl uranium (HBTU, Figure 3B) in dimethylformamide in the presence of diisoproylethylamine (DIEA). The side chains of amino acids were protected with acid-labile protecting groups.

Albuvirtide (**1**) is synthesized first by using an automated peptide synthesizer to build the peptide chain, then the Alloc group of Lys13 is manually removed using $Pd(PPh_3)_4$, allowing sequential manual coupling of the linker and maleimide (Scheme 1).

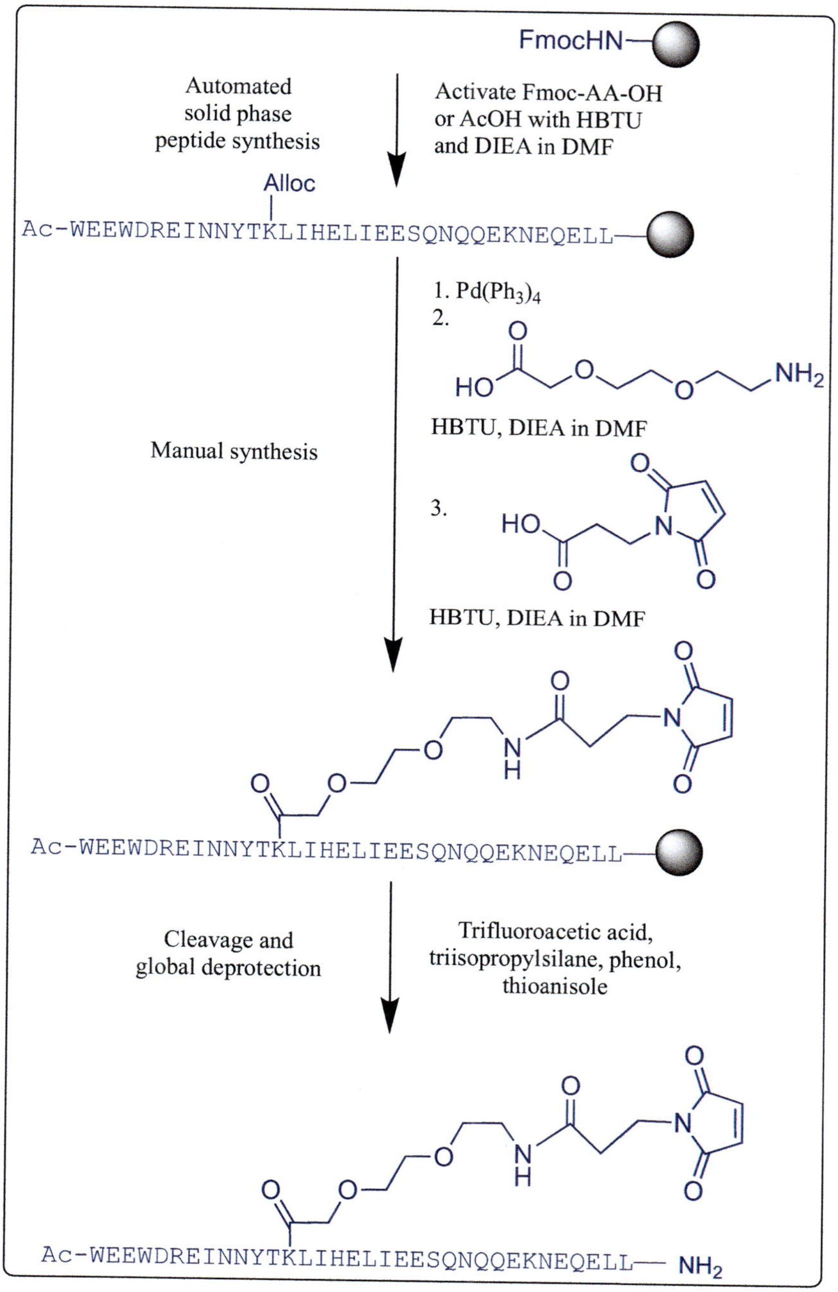

Scheme 1. Synthesis of albuvirtide (**1**).

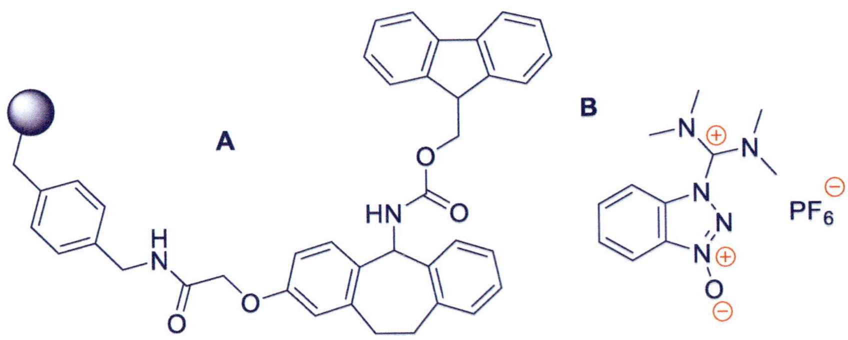

Figure 3., A) The structure of the Fmoc-protected Ramage resin., B) The structure of the coupling agent HBTU.

7 Summary

Albuvirtide (**1**) was first approved by the Chinese State Drug Administration in 2018.[1] This approval was based on positive top-line results achieved by Frontier Biotechnologies in its phase three TALENT study, and in a recent press release interim analysis demonstrated that an ABT-based two-drug treatment arm was non-inferior to a lopinavir (LPV)-based three-drug arm (75.7% vs 77.3%).[48] The results of the randomized controlled, open-labelled, multicenter, non-inferiority study showed successful achievement of the primary endpoint, with a good proportion of patients on the ABT regimen with HIV RNA less than 50 copies/mL at 48 weeks. The drug demonstrated efficacy against major strains of HIV, including resistant strains. Additionally, the subjects who experienced virological failure at 48 weeks did not present with treatment emergent resistant associated mutations with gp41. The high genetic barrier to resistance of ABT+LPV/r meant no further resistance developed against these agents, which is important to avoid compromising future drug options for these treatment-experienced HIV patients.[43]

The Republic of Ecuador has become the first country outside of China to authorize the sale and use of a HIV drug developed in China by Frontier Biotech.[49] Aikening, albuvirtide (**1**), 320 mg, is designed to be administered once a week. The drug will be marketed under the brand name Alfusid in Ecuador and other South American countries. Albuvirtide (**1**) is used in combination with other antiretrovirals to treat patients who have not achieved viral suppression with other regimens. The drug is indicated to be effective against most known strains of HIV, including those with drug resistance. Frontier Biotech is actively working with international partners to obtain marketing authorizations for more than 20 countries across South East Asia, Africa, South America and Europe.

In clinical trials, and practical applications, no drug interactions have been found with the antibacterial, antifungal, anti-tuberculosis and anticancer drugs commonly used in the treatment of patients with advanced HIV. Aikening is listed in the Liverpool Drug Interactions Database (https://www.hiv-druginteractions.org/).[44]

The approval of albuvirtide (**1**), resulting from 15 years of peptide fusion inhibitor research since the first approval of Fuzeon, is a significant advancement in the treatment of HIV-1 patients. There is an urgent need to develop new drugs or prevention strategies to address the issues of poor patient compliance with current anti-HIV-1 treatments, emerging drug resistance, drug intolerance and adverse effects, and a lack of alternative regimens. Albuvirtide (**1**) is improving the lives of HIV-1 patients who have developed resistance to current treatment regimens due to a favorable side effect profile, limited once-weekly administration, and its effectiveness in viral suppression.

8. References

1. Frontier Biotech press release. Frontier Biotech Receives Marketing Authorization from China FDA for Aikening (Albuvirtide for Injection), China's first new drug for the treatment of HIV. (06 June 2018). http://www.frontierbiotech.com/en/news
2. Wu, H.; Yao, C.; Lu, R. J. Albuvirtide, the first long-acting HIV fusion inhibitor, suppressed viral replication in HIV-infected adults. 52nd Interscience Conference on Antimicrobials and Chemotherapy (ICAAC). September 9–12, 2012. San Francisco. Abstract H-554.
3. Kilby, J. M.; Hopkins, S.; Venetta, T. M.; DiMassimo, B.; Cloud, G. A.; Lee, J. Y.; Alldredge, L.; Hunter, E.; Lambert, D.; Bolognesi, D.; Matthews, T.; Johnson, M. R.; Nowak, M. A.; Shaw, G. M.; Saag, M. S. Potent suppression of HIV-1 replication in humans by T-20, a peptide inhibitor of gp41-mediated virus entry. *Nat. Med.* **1998**, *4,* 1302–1307.
4. Lalezari, J. P.; Henry, K.; O'Hearn, M.; Montaner, J. S. G.; Piliero, P. J.; Trottier, B.; Walmsley, S.; Cohen, C.; Kuritzkes, D. R.; Eron Jr., J. J.; Chung, J.; DeMasi, R.; Donataci, L.; Drobnes, C.; Delehanty, J.; Salgo, M.; TORO 1 Study Group. Enfuvirtide, an HIV-1 fusion inhibitor, for drug-resistant HIV infection in North and South America. *N. Engl. J. Med.* **2003**, *348,* 2175–2185.
5. FDA. FDA Approves Fuzeon. Available from: http://www.fda.gov/ForConsumers/ByAudience/ForPatientAdvocates/HIVandAIDSActivities/ucm125088.htm2003.
6. Patel, I. H.; Zhang, X.; Nieforth, K.; Salgo, M.; Buss, N. Pharmacokinetics, Pharmacodynamics and Drug interaction potential of enfuvirtude. *Clin. Pharmacokinet.* **2005**, *44*, 175–186.
7. Rimsky, L. T.; Shugars, D. C.; Matthews, T. J. Determinants of human immunodeficiency virus type 1 resistance to gp41-derived inhibitory peptides *J. Virol.* **1998**, *72*, 986–993.
8. Poveda, E.; Briz, V.; Soriano, V. Enfuvirtide, the first fusion inhibitor to treat HIV infection. *AIDS Rev.* **2005**, *7*, 139–147.
9. Lu, J.; Deeks, S. G.; Hoh R.; Beatty, G.; Kuritzkes, B. A.; Martin, J. N.; Kuritzkes, D. R. Rapid emergence of enfuvirtide resistance in HIV-1-infected patients: results of a clonal analysis. *J. Acquir. Immune Defic. Syndr.* **2006**, *43*, 60–64.

10. He, Y.; Xiao, Y.; Song, H.; Liang, Q.; Ju, D.; Chen, X.; Lu, H.; Jing, W.; Jiang, S.; Zhang, L. Design and evaluation of sifuvirtide, a novel HIV-1 fusion inhibitor. *J. Biol. Chem.* **2008**, *283*, 11126–11134.
11. Xie, D.; Yao, C.; Wang, Li.; Min, W.; Xu, J.; Xiao, J.; Huang, M.; Chen, B.; Liu, B.; Li, X.; Jiang, H. An albumin-conjugated peptide exhibits potent anti-HIV activity and long in vivo half-life. *Antimicrob. Agents Chemother.* **2010**, *54*, 191–196.
12. Zhu, Y.; Chong, H.; Yu, D.; Guo, Y.; Zhou, Y.; He, Y. Design and characterization of cholesterylated peptide HIV-1/2 fusion inhibitors with extremely potent and long-lasting antiviral activity. *J. Virol.* **2019**, *93*, e02312–e02318.
13. Lu, L.; Pan, C.; Li, Y.; Lu, H.; He, W.; Jiang, S. A bivalent recombinant protein inactivates HIV-1 by targeting the GP41 prehairpin fusion intermediate induced by CD4 D1D2 domains. *Retrovirology* **2012**, *9,* 104–117.
14. Pan, C.; Lu, H.; Qi, Z.; Jiang, S. Synergistic efficacy of combination of enfuvirtide and sifuvirtide, the first- and next generation HIV-fusion inhibitors. *AIDS* **2009**, *23*, 639–641.
15. World Health Organization. HIV/AIDS Fact Sheet, 2021.
16. Burke, D. S. Recombination in HIV: an important viral evolutionary strategy. *Emerg. Infect. Dis*. **1997**, *3*, 253.
17. Levy, J. A. Dispelling myths and focusing on notable concepts in HIV pathogenesis. *Trends Mol. Med.* **2015**, *21*, 341–353.
18. Saag, M. S.; Gandhi, R. T.; Hoy, J. F.; Landovitz, R. J.; Thompson, M. A.; Paul, Sax, P. E.; Smith, D. M.; Benson, C. A.; Buchbinder, S. P.; del Rio, C.; Eron Jr, J. J.; Fätkenheuer, G.; Günthard, H. F.; Molina, J.-M.; Jacobsen, D. M.; Volberding, P. A. Antiretroviral drugs for treatment and prevention of HIV infection in adults: 2020 recommendations of the international antiviral society-USA panel. *J. Am. Med. Assoc.* **2020**, 312.
19. Este, J. A.; Cihlar, T. Current status and challenges of antiretroviral research and therapy. *Antivir. Res.* **2010**, *85*, 25–33.
20. Saag M. S., Benson C. A., Gandhi R. T., Hoy, J. F.; Landovitz, R. J.; Mugavero, M. J.; Sax, P. E.; Smith, D. M.; Thompson, M. A.; Buchbinder, S. P.; del Rio, C.; Eron, J. J.; Fatkenheuer, G.; Gunthard, H. F.; Molina, J.-M.; Jacobsen, D. M.; Volberding, P. A. Antiretroviral drugs for treatment and prevention of HIV infection in adults: 2018 recommendations of the international antiviral society–USA panel. *J. Am. Med. Assoc.* **2018**, *320*, 379–396.
21. World Health Organization. WHO HIV Drug Resistance Report 2012, Geneva, World Health Organization.
22. Chong, H.; Yao, X.; Zhang, C.; Cai, L.; Cui, S.; Wang, Y.; He, Y. Biophysical property and broad anti-HIV activity of albuvirtide, a 3-maleimimidopropionic acid-modified peptide fusion inhibitor. *PLoS One* **2012**, *7*, e32599.
23. Rand, R. P.; Parsegian, V. A. Physical force considerations in model and biological membranes. Can. *J. Biochem. Cell Biol.* **1984**, *62*, 752–759.
24. Harrison, S. C. Viral membrane fusion. *Virology* **2015**, 479–480, 498–507.
25. Kielian, M. Mechanisms of virus membrane fusion proteins. *Ann. Rev. Virol.* **2014**, *1*, 171–189.

26. Weissenhorn, W.; Dessen, A.; Calder, L. J.; Harrison, S. C.; Skehel, J. J.; Wiley, D. C. Structural basis for membrane fusion by enveloped viruses. *Mol. Membr. Biol.* **1999**, *16*, 3–9.
27. Harrison, S. C. Viral membrane fusion. *Nat. Struct. Mol. Biol.* **2008**, *15*, 690–698.
28. Weissenhorn, W.; Dessen, A.; Harrison, S. C.; Skehel, J. J.; Wiley, D. C. Atomic structure of the ectodomain from HIV-1 gp41. *Nature* **1997**, *387*, 426–430.
29. Chan, D. D.; Fass, D.; Berger, J. M.; Kim, P. S. Core structure of gp41 from the HIV envelope glycoprotein. *Cell* **1997**, *89,* 263–273.
30. Pan, J.; Peng, H.; Chen, B.; Harrison, S. C. Cryo-EM structure of full-length HIV-1 Env bound with the Fab of antibody PG16. *J. Mol. Biol.* **2020**, *432*, 1158–1168.
31. Torrents de la Pena, A.; Rantalainen, K.; Cottrell, C. A.; Allen, J. D.; van Gils, M. J.; Torres, J. L.; Crispin, M.; Sanders, R. W.; Ward, A. B. Similarities and differences between native HIV-1 envelope glycoprotein trimers and stabilized soluble trimer mimetics. *PloS Pathog.* **2019**, *15*, e1007920.
32. Wei, X.; Decker, J. M.; Wang, S.; Hui, H.; Kappes, J. C.; Wu, X.; Salazar-Gonzalez, J. F.; Salazar, M. G.; Kilby, J. M.; Saag, M. S.; Antibody neutralization and escape by HIV-1. *Nature* **2003**, *422*, 307–312.
33. Richman, D. D.; Wrin, T.; Little, S. J.; Petropoulos, C. J. Rapid evolution of the neutralizing antibody response to HIV Type 1 infection. *Proc. Natl. Acad. Sci.* **2003**, *100,* 4144–4149.
34. Jiang, S.; Lin, K.; Strick, N.; Neurath, A. R. HIV-1 entry inhibitor. *Nature* **1993**, *365,* 113.
35. Root, M. J.; Kay, M. S.; Kim, P. S. Protein design of an HIV-1 entry inhibitor. *Science* **2001**, *291*, 884–888.
36. Kilby, J. M.; Eron, J. J. Novel therapies based on mechanisms of HIV-1 cell entry. *NEJM* **2003**, *348*, 2228–2238.
37. Robertson, D. US FDA approves new class of HIV therapeutics. *Nat. Biotechnol.* **2003**, *21*, 470–471.
38. Poveda, E.; Rodes, B.; Lebel-Binay, S.; Faudon, J. L.; Jimenez, V.; Soriano, V. Dynamics of enfuvirtide resistance in HIV-infected patients during and after long-term enfuvirtide salvage therapy. *J. Clin. Virol.* **2005,** *34*, 295–301.
39. He, Y.; Cheng, J.; Lu, H.; Li, J.; Hu, J.; Qi, Z.; Liu, Z.; Jiang, S.; Dai, Q. Potent HIV fusion inhibitors against enfuvirtide-resistant HIV-1 strains. *PNAS* **2009**, *105,* 16332–16337.
40. Crespillo, S.; Cámara-Artigas, A.; Casares, S.; Morel, B.; Cobos, E.S.; Mateo, P.L.; Mouz, N.; Martin, C.E.; Roger, M.G.; Habib, R.E.; Su, B.; Moog, C.; Conejero-Lara, F. Single-chain protein mimetics of the N-terminal heptade repeat region of gp41 with potential as Anti-HIV-1 drugs. *PNAS.* **2014**, *111*, 18207–18212.
41. Ingallinella, P.; Bianchi, E.; Ladwa, N. A.; Wang, Y. J.; Hrin R.; Veneziano, M.; Bonelli, F.; Ketas, T. J.; Moore, J. P.; Miller, M. D.; Pessi, A. Addition of a cholesterol group to an HIV-1 peptide fusion inhibitor dramatically increases its antiviral potency. *PNAS* **2009**, *106*, 5801–5806.
42. Johnson, L. M.; Horne, W. S.; Gellman, S. H. Broad distribution of energetically important contacts across an extended protein interface. *J. Am. Chem. Soc.* **2011**, *133*, 10038–10041.

43. Yang, W.; Xiao, Q.; Wang, D.; Yao, C.; Yang, J. Evaluation of pharmacokinetic interactions between long-acting HIV-1 fusion inhibitor albuvirtide and lopinavir/ritonavir, in HIV-infected subjects, combined with clinical study and simulation results, *Xenobiotics* **2017**, *47*, 133–143.
44. Zhang, H.; Jin, R.; Yao, C.; Zhang, T.; Wang, M.; Xia, W.; Peng, H.; Wang, X.; Lu, R.; Wang, C.; Xie, D.; Wu, H. Combination of long-acting HIV fusion inhibitor albuvirtide and LPV/r showed potent efficacy in HIV-1 patients. *AIDS Res. Ther.* **2016**, *13*, 8–11.
45. Su, B.; Yao, C.; Zhao, Q.-X.; Cai, W.-P.; Wang, M.; Lu, H.-Z.; Chen, Y.-Y.; Liu, L.; Wang, H.; et al. TALENT Study Team. Efficacy and safety of the long-acting fusion inhibitor albuvirtide in antiretroviral-experienced adults with human immunodeficiency virus-1: interim analysis of the randomized, controlled, phase 3, non-inferiority TALENT study. *Chin. Med. J.* **2020**, *133*, 2919–2927.
46. Xie, D.; Jiang, H. Peptide derivative fusion inhibitors of HIV infection. Patent application AU 2010200823 Al (2010).
47. Lu, R.; Min, W. Stable albuvirtide compositions. World Patent Application WO 2020/223906 A1 (2020).
48. Press Release First Word Pharma. Frontier Biotechnologies' First Long-acting Injectable (Aikening), in a Two Drug Regimen for HIV, Proves Safe and Efficacious for Patients. Berlin, July 19, 2021, PRNewswire, https://www.firstwordpharma.com/node/1846974?tsid=17
49. Haynes, T. Chinese HIV drug, Aikening, gets first approval outside China. 4Life4Me+ 24 March 2021, https://life4me.plus/en/news/china-aikening-Albuvirtide-7327/.

II

CANCER DRUGS

7

Darolutamide (Nubeqa); An Androgen Receptor Antagonist for Treating Nonmetastatic, Castration-Resistant Prostate Cancer

Dao-Qian Chen and Ji Zhang

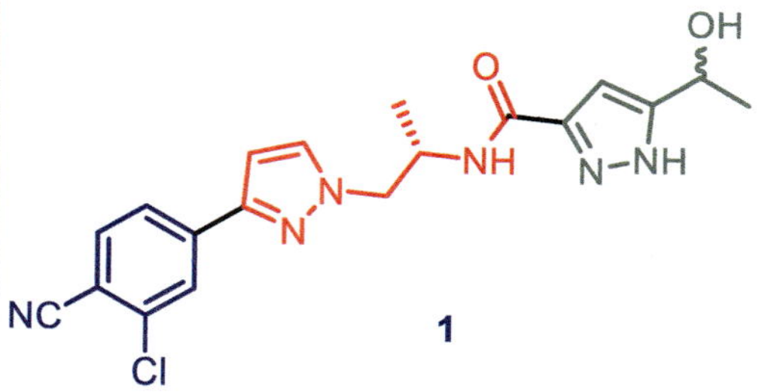

1

USAN: Darolutamide
Trade name: Nubeqa
Bayer/Orion Corporation
Launched: 2019 (United States)

7.1 Background

Based on Clinical trial-ARAMIS (NCT02200614),[1] darolutamide (formerly ODM-201 or BAY-1841788) was approved to treat nonmetastatic, castration-resistant prostate cancer (nmCRPC) in men by FDA in 2019 then successively launched in EU (2020), followed in Japan (2020) and China (2021) as of drafting time. Darolutamide, initially developed by Orion Corporation in 2010, was co-developed and commercialized by Bayer Health Care in 2014 before it entered phase III study.

According to the Global Cancer Statistics 2020,[2] prostate cancer (an estimated almost 1.4 million among 19.3 million new cancer cases, 7.3%) occupies top four cancer types for estimated cases and deaths worldwide, and is the fifth leading cause of cancer death among men in 2020. The highest incidence rate occurs in Northern and Western Europe, while the lowest in South-Central Asia. Initially, 80–90% of men with advanced

Current Drug Synthesis, First Edition. Edited by Jie Jack Li.

prostate cancer can get remission from androgen deprivation therapy (ADT) for a median progression-free survival of 12–33 months,[3] while it switches to castration-resistant prostate cancer with median overall survival (OS) of 23–37 months from the time of ADT.[3]

An increased understanding of CRPC has led to the harvest of novel drugs,[4] including agents targeting Hsp900 (IPI-504), agent targeting androgen biosynthesis (abiraterone acetate, CYP17A1 inhibitor), agents targeting androgen receptor (AR) (steroidal or non-steroidal AR antagonists), agent targeting DNA repair (olaparib, PARP inhibitor), chemotherapeutic agents (docetaxel and cabazitaxel), an immunomodulant agent sipuleucel-T, and a radiopharmaceutical agent Radium-223.

To date, extensive investigations for antiandrogens make progress to improvement for prostate cancer patient quality of life to a considerable extent. Among them, steroidal anti-androgens (cyproterone acetate, medroxyprogesterone acetate and megestrol acetate) were first introduced in the late 1950s, although no steroidal anti-androgens have been approved by FDA for use of prostate cancer since Huggins and Hodges reported the efficacy of androgen ablation therapy in patients with metastatic carcinoma of the prostate in 1941.[5] In respect of non-steroidal AR antagonists, two antagonist generations have been developed into prostate cancer patients (Figure 1). However, first-generation AR antagonists [flutamide (**2**)[6] and bicalutamide (**4**)[7]] can switch from antagonist to agonist due to the mutations in ligand-binding domain LBD of AR. Therefore, second-generation AR antagonists (Table 1) focusing on optimization to maintain antagonism and overcome resistance have been developed.

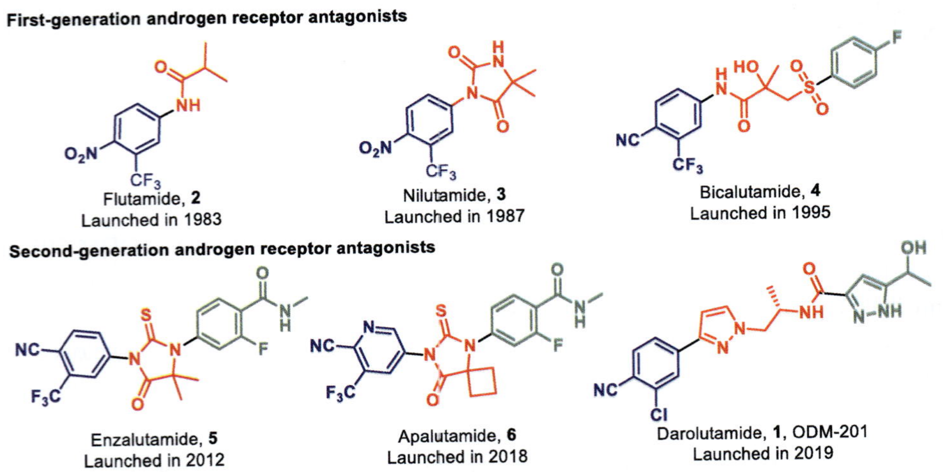

Figure 1. Non-steroidal androgen receptor antagonists.

To deserve to be mentioned, enzalutamide (**5**) has been approved to treat three forms of advanced prostate cancer (Table 1), followed by apalutamide (**6**) and darolutamide (**1**). Enzalutamide (**5**) exhibits greater affinity to AR than bicalutamide (**4**). In a phase I/II clinical trial including 140 patients with progressive, metastatic, castration-resistant prostate cancer, 78 (56%) patients showed a reduction in serum prostate-specific antigen (PSA) of 50%. 11% of adverse event was dose-dependent fatigue.[8]

Table 1. FDA-approved AR antagonists for prostate cancer

Generic name	mCRPC	nmCRPC	mCSPC	Company	Sales in 2020[a]
Enzalutamide, 5	√	√	√	University of California Medivation/Astellas/Pfizer	3.716
Apalutamide, 6	×	√	√	University of California Johnson & Johnson	0.76
Darolutamide, 1	×	√	×	Orion/Bayer	-

mCRPC: metastatic castration resistant prostate cancer
nmCRPC: nonmetastatic castration-resistant prostate cancer
mCSPC: metastatic castration-sensitive prostate cancer
a: USD billion

The phase III trial assigned 1,199 men with castration-resistant prostate cancer after chemotherapy in a 2:1 ratio, to receive oral enzalutamide at a dose of 160 mg per day (800 patients) or placebo (399 patients).[9] Patients significantly benefited from enzalutamide group (18.4 months-enzalutamide vs. 13.6 months-placebo) in the median overall survival and in serum PSA of 50% (54% vs. 2%). Seizures were reported in five patients (0.6%) receiving enzalutamide (**5**). In a phase III trial, administration to apalutamide (**6**, 240 mg/day) helps men with nmCRPC to gain longer metastasis free survival and time to symptomatic progression than placebo. A higher rate with apalutamide than with placebo showed rash (23.8% vs. 5.5%), hypothyroidism (8.1% vs. 2.0%), and fracture (11.7% vs. 6.5%) as adverse events, including two events of seizures observed. Hager and Korpal revealed that both enzalutamide (**5**) and apalutamide (**6**) exhibited an antagonist-to-agonist switch due to a missense mutation F876L (Phe 876 to Leu) in the LBD of AR.[10,11] Remarkably, darolutamide (ODM-201, **1**) is a full AR antagonist including F876L, M896V/T, I882L, V731M, H875Y, T878A, W742C/L mutations.[12] Studies based the brain/plasma ratio reveals that darolutamide exhibits minimal brain penetrance (highest blood–brain barrier, BBB) in contrast to enzalutamide (**5**) and apalutamide (**6**), illustrating low incidence of central nervous system-related adverse events in clinical development.[13]

Mechanistically, AR signaling plays a critical role in prostate cancer cell proliferation, survival, and differentiation. Blocking AR signaling can arrest nuclear translocation, recruitment of AR cofactors, and AR binding to DNA.[4] When being inactive, AR binds to heat shock proteins in the cytoplasm. The activation of AR is driven by endogenous ligand containing testosterone and dihydrotestosterone (DHT) with different affinities to AR[14] and potency of inducing downstream signaling in body, dissociating from heat shock proteins once ligand binding. When administration with AR antagonists, they can compete with endogenous ligands for binding AR, then induce a conformational change of AR that blocks AR signaling, finally accomplish the treatment with prostate cancer. The competitive relationship about AR antagonists and T (**7**) and DHT (**8**) was illustrated in Figure 2.

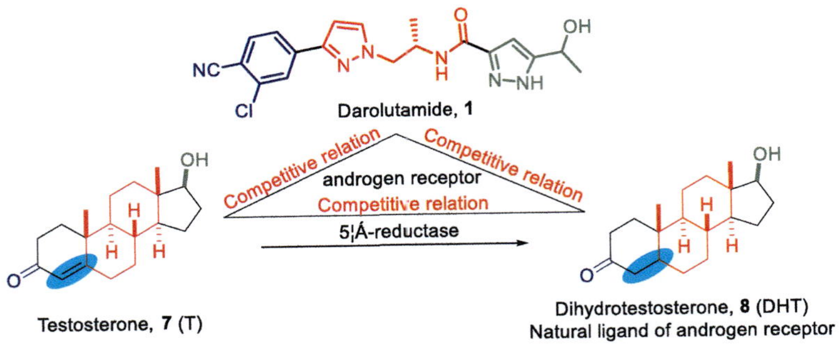

Figure 2. The relationship about darolutamide (**1**), **7** (T) and **8** (DHT).

Supplied as film-coated tablets containing 300 mg of darolutamide (**1**) for oral use, darolutamide is an selective nonsteroidal second-generation AR receptor, with a remarkably unique molecular structure distinct from bicalutamide (**4**), enzalutamide (**5**) and apalutamide **6**.[15] It is worth noting that a 1:1 mixture of diastereomers (*S*,*S*)-darolutamide (**9**) and (*S*,*R*)-darolutamide **10**, and the major metabolite, keto-darolutamide **11** (ORM-15341), have performed similar pharmacologic activity *in vitro* (Figure 3),[16] and therefore a diastereomers of darolutamide (**1**) were used as the final API which will make the process chemistry much simple in principle.

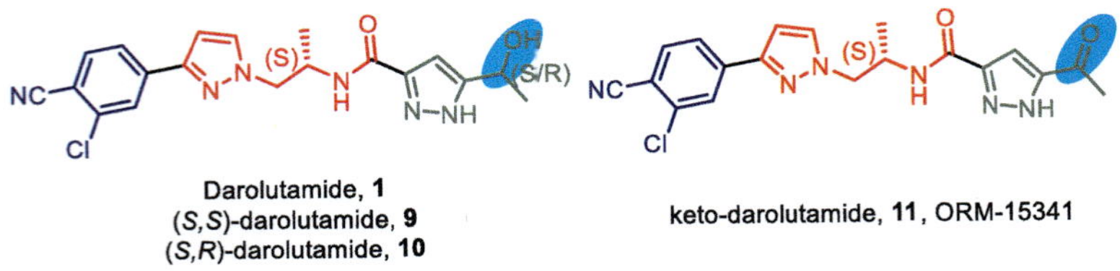

Figure 3. Diastereomers of darolutamide **1** and metabolite keto-darolutamide **11**.

7.2 Pharmacology

When it comes to darolutamide (**1**), two diastereomers (*S*-Me, *S*-OH and *S*-Me, *R*-OH) and the major metabolite, keto-darolutamide **11** naturally bring some questions to the public: Why darolutamide (**1**) was developed to be a mixture of diastereomers but not a single pure chiral diastereomer? Is the major metabolite safe? Preclinical studies and clinical studies have answered both questions from pharmacology and pharmacokinetics.

Darolutamide exposure at 600 mg twice daily results in PSA mean reduction of more than 90% from baseline. AR mutations is responsible for resistance to anti-androgen therapy, accounting for 15–20% of castration-resistant prostate cancer patients.[17] In Moilanen's study,[13] the AR mutants (F876L, T877A or W741L) encoding by cells transiently transfected with expression vectors are administrated to three second-generation AR antagonists and bicalutamide (**4**), supporting the switch antagonist-to-agonist bicalutamide (**4**) for mutant AR (F876L), as well as enzalutamide **5** and apalutamide **6** for mutant AR (W741L). Darolutamide (**1**), two diastereomers **9**, **10** and keto-darolutamide **11** functioned as strong antagonists for activity against W742C and W742L mutants. More mutations including F877L, M896V/T, I882L, V731M, H875Y, T878A, and W742C supported the antagonist of darolutamide **1** with different activities while still prior to enzalutamide **5** and apalutamide **6**. *In vitro*, good inhibition of VCaP and LAPC-4 prostate cancer cells (IC_{50} values of 0.25–0.5 and 0.44–0.84 mol/L) demonstrates that darolutamide (**1**), its diastereomers and keto-darolutamide **11** have antagonist with same efficacy in prostate cancer models. Interestingly, another vitro study reported that a mixture of darolutamide diastereomers bearing (*R*)-Me in isopropylamine linker presented same antiproliferative activity against LNCaP/AR cell with IC_{50} 1.65 μM, highly suggesting activity is independent on (*S*)-Me configuration of isopropylamine linker and it is not the key to antiproliferative activity.[18]

To demonstrated the inhibition of nuclear translocation of darolutamide (**1**) and keto-darolutamide **11**, date based in immunocytochemical labeling study in both HS-HEK293 cells and LNCaP cells showed darolutamide (**1**) and keto-darolutamide **11** blocked the translocation in the cytoplasmic, enzalutamide **5**, and apalutamide **6** as well. However, bicalutamide **4** failed.[13]

Prior to those of enzalutamide **5** (86 nM, 219 nM) and apalutamide **6** (93 nM, 200 nM), the inhibition constant (K_i) value and IC_{50} value of darolutamide (**1**) were 11 nM and 26 nM, similar with keto-darolutamide **11** (8 nM, 38 nM). *In vitro* and *in vivo*, darolutamide (**1**) showed a good specificity to inhibition of AR-dependent PC cells, and showed a remarkable antitumor activity with 50 mg/kg dose twice daily compared with control group ($p < 0.001$).[13]

When administrated to male mices containing mean serum PSA values≈5.5 g/L for 3 weeks, enzalutamide **5** (20 mg/kg, qd) increased serum testosterone concentrations ($p < 0.05$), while darolutamide **1** (50 mg/kg, bid) had no effect in the testosterone level compared to vehicle, suggesting no progress to the hypothalamic-pituitary-gonadal axis to produce androgens.[13] The study of phase I reveals the same result that darolutamide has no drug effect on serum concentrations of follicle-stimulating hormone (FSH), luteinizing hormone (LH), testosterone, or DHT.[19]

In vivo study using whole-body autoradiography, concentrations of [^{14}C]-darolutamide (**1**) ~ is 46× and 26× lower than [^{14}C]-enzalutamide **5** and [^{14}C]-apalutamide **6** brain concentrations.[20]

High protein binding of keto-darolutamide **11** means low concentration in the circulation of human.[16] The majority of pharmacological activity is dominated by darolutamide (**1**), resulting from its concentration far beyond its metabolite.

7.3 Structure–Activity Relationship (SAR)

In 2010, Orion Corporation reported their discovery of darolutamide (**1**), with a distinct structure from other non-steroidal AR antagonists (first generation and second generation). Its structure mainly referred to patent that was developed to pest control agents (nematicides).[21] As to date no detailed works about how to design darolutamide **1** by Orion Corporation were reported. However, advantages of darolutamide **1** have been verified in preclinical and clinical benefits based in this special structure. Li and coworkers concluded the structure–activity relationship (SAR) based in two generations of non-steroidal AR antagonists targeting LBD (Figure 4), in which the necessity of electron-deficient group (–CN, –CF_3) in aromatic ring and the hydrophobic ring and the side chain can be clarified.[4]

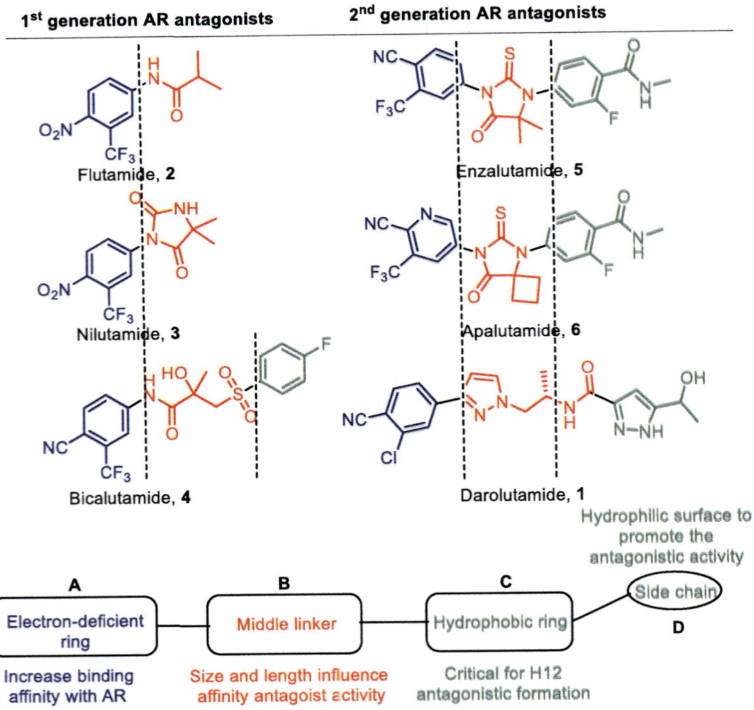

Figure 4. Structure–activity relationships (SAR) of non-steroidal AR antagonists.[4]

Both first- and second-generation AR antagonists contain the electron-withdrawing groups (–CN, $-NO_2$, $-CF_3$) (Figure 4A), because these molecules can bind N atom of CN to Gln711 and Arg752 like ketone of DHT to increase the affinity based in binding model of DHT (Figure 5), in which aryloxy tetramethylcyclobutanes demonstrated favorable efficacy after optimization of ligand efficiency.[22] In agonistic conformation of AR W742L form, a higher flexibility of isopropylamine linker of darolutamide (**1**) may maintain van der Waals interaction to the leucine side chain (Figure 4B), providing the most activity of antagonist for darolutamide (**1**) compared with enzalutamide **5** and apalutamide **6**, while agonist for bicalutamide **4**.[12] The restraint of imidazolidine ring (Figure 4B) of enzalutamide **5** and apalutamide **6** is prone to lose the van der Waals interaction because of the amino acid exchange. The hydrophobic C ring binding a side chain is responsible for eliminating the "latch" interaction that can result in agonist effect for AR signaling, and also generates the H12 antagonistic formation *via* pi stacking and hydrophobic interactions.[4] To preserve antagonistic activity, hydrophilic groups of the side chain (Figure 4D) are helpful to form a hydrophilic surface. For example, the OH group of darolutamide (**1**) maybe generates a hydrogen bond with the cysteine SH group in AR wild type.[12] Exactly, based in the equal activity, the hydrogen bond formation can be considered for darolutamide (**1**), diastereomers **9**, **10** and keto-darolutamide **11**, the ligand oxygen of which acts as hydrogen bond acceptor.

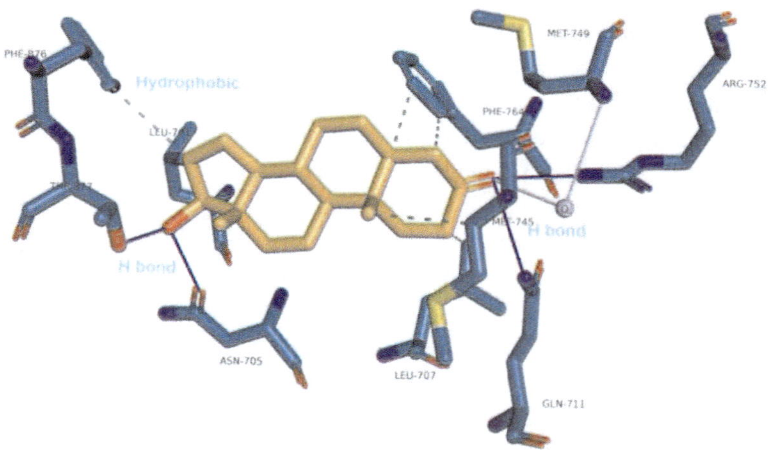

Figure 5. DHT interactions in the LBD of AR (PDB: 1T7R).[22]

More SAR about darolutamide (**1**) have been gained along with structural modification of darolutamide **1**, during which compounds **12** and **13** exhibit good antiproliferative and antitumor activities (Figure 6).[18]

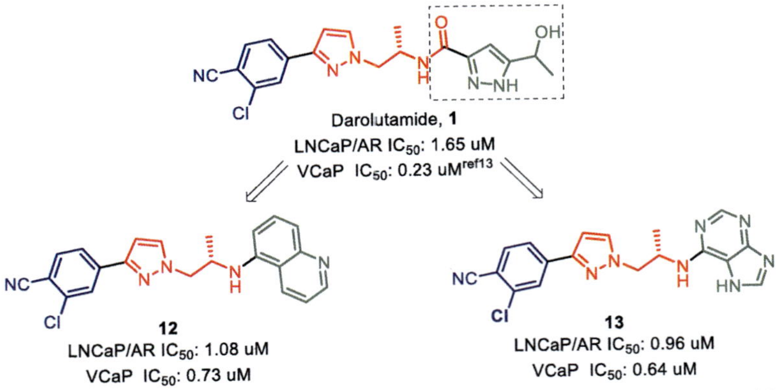

Figure 6. Structural modification AR antagonists (**12** and **13**) of darolutamide (**1**).

Remarkably, as shown in Table 2,[18] various configurations of darolutamide (**1**) including (mixture **1**; (*S,S*)-**9**; (*S,R*)-**10**; (*R,S/R*)-**14**) exhibit similar antiproliferative activity against VCaP (AR-positive human PC cell) and inhibit the proliferation of LNCaP/AR with the same antiproliferative activity. When changed from OH group to ketone at R^1 position, both compounds **11** and **15** exhibit the same inhibitory effect against LNCaP/AR with IC_{50} values 1.13 μM and 1.08 μM, respectively. Together, configurations of darolutamide (**1**) or the major metabolite, keto-darolutamide **11** have little influence on the inhibitory activity. Compound **14** without substituent of 1H-pyrazole at R^1 position even exhibits better activity than darolutamide **1**. Besides, compounds featuring 1H-pyrazole at R^1 position perform a concentration-independent manner to inhibit the proliferation of VCaP cells (note: no date for (*S,S*)-**9**; (*S,R*)-**10**). Unfortunately, the antiproliferative activity seems to decrease with compounds **17–22** which are replaced from the 1H-pyrazole to phenyl **17–19** or alkyl groups **20–22**, highlighting the effect of substituent at R^1 position.

The SAR of amide group at R^2 are shown in Table 3. The reversion of amine and carbonyl group at R^2 position (compounds **23** and **24**) result in reduced or completely vanished activity compared with compounds **16** and **18** (Table 2), lower than darolutamide (**1**), demonstrating the position of amide at R^2 position is of great influence. The replacement of amide with sulfonamide **25–27** or urea group binding aromatic rings **28** decreases the activities against LNCaP/AR and VCap slightly.[18]

Table 2. *In vitro* activity of darolutamide (**1**) derivatives

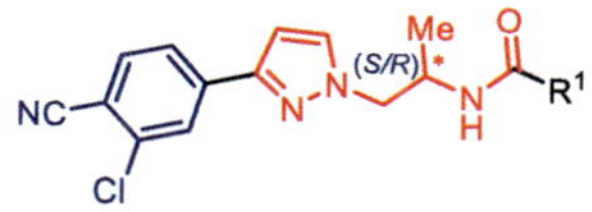

Compound	(*S/R*)-Me	R^1	IC_{50} (uM)	
			LNCaP/AR	VCaP
Enzalutamide, **5**			0.19	30
Darulutamide, **1**	*S*	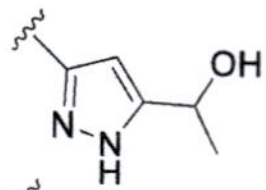	1.65	0.26-0.56[ref12]
(*S,S*)-Darulutamide, **9**	*S*	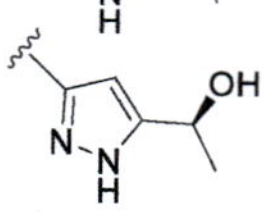	-	0.27-0.47[ref12]
(*S,R*)-Darulutamide, **10**	*S*	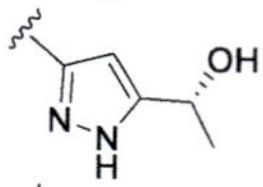	-	0.19-0.31[ref12]
14	*R*	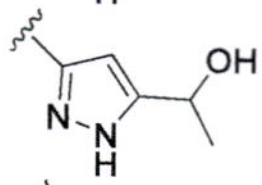	1.65	NC
Keto-darolutamide, **11**	*S*	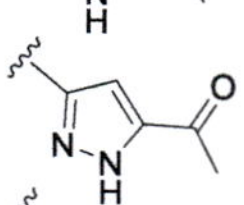	1.13	0.36-0.64[ref10]
15	*R*	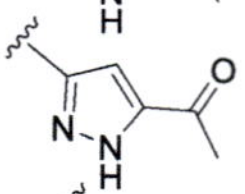	1.08	NC
16	*S*	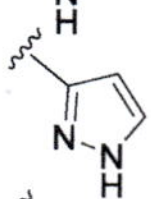	1.00	NC
17	*S*	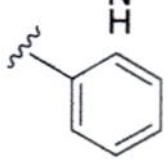	1.83	NC
18	*S*	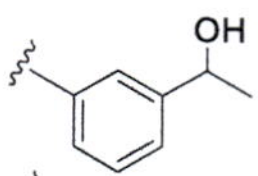	2.28	0.90
19	*S*	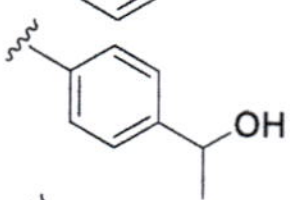	3.69	0.91
20	*S*	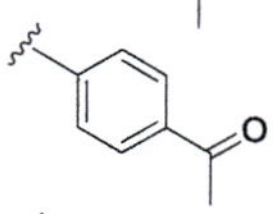	3.60	1.40
21	*S*	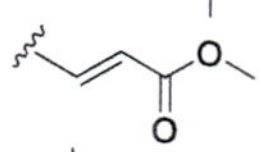	2.46	5.23
22	*S*	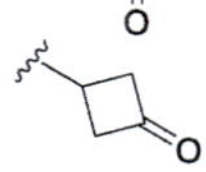	3.04	22.47

Source: Adapted from reference 18.

Table 3. *In vitro* activity of darolutamide (1) derivatives.

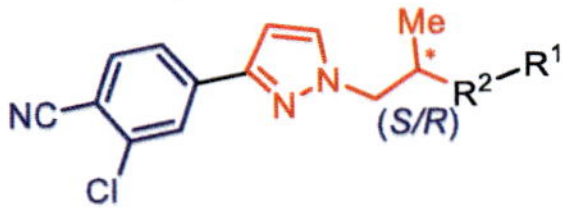

Compound	(*S/R*)-Me	R^2	R^1	IC_{50} (uM) LNCaP/AR	IC_{50} (uM) VCaP
Darulutamide, **1**	*S*	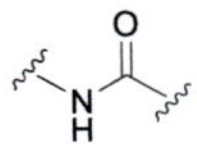	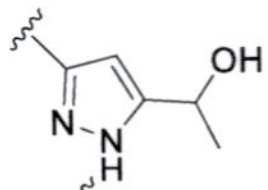	1.65	0.26-0.56[ref12]
23	*racemate*	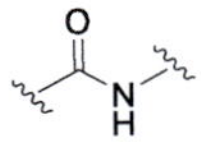	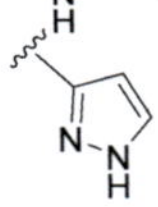	>30	>30
24	*racemate*	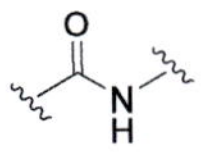	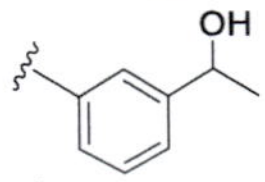	4.15	>30
25	*S*	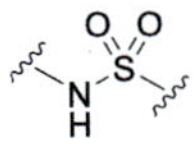	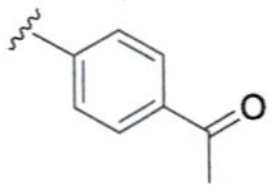	5.23	3.95
26	*S*	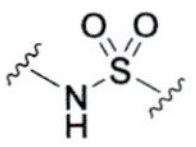	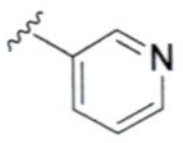	6.14	2.45
27	*S*	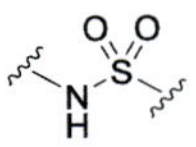	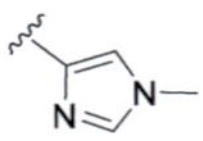	>30	13.59
28	*racemate*	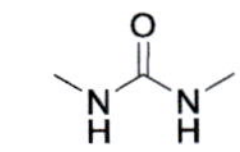	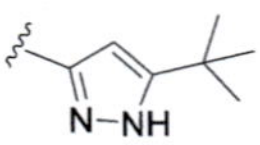	1.63	5.12

What happen to compounds without the carbonyl group in amide at the R^2 position? 22Rv1 cell line (also known as CWR22Rv1) represents one of human prostate cancer model.[23] In Table 4, an effort was taken to modify the R3 position with various aromatic rings phenyl **29**, pyridyl (**30**, **31**), pyrimidinyl **32**, indolyl **33**, isoquinolyl **34**, quinolyl (**12**), and purine group (**13**) binding with amine directly (Table 4). Even though sub-micromolar IC_{50} values were observed for LNCaP/AR (compounds **29**–**33**), higher IC_{50} values of VCaP were gained than darolutamide **1**. Besides, compounds **29**, **30**, **31** showed lower potency against 22Rv1 than compounds **32**, **33**. More compounds with bicyclic rings were evaluated. Among them, compound **34** featuring N located in the 6-R^3 shows no inhibition for VCaP, while the anti-proliferative activity regains when the nitrogen changes from 6 to 5-R^3 (Table 4, compound **12**). More importantly, compounds **12** and **13** bearing quinoline and purine obviously perform excellent inhibitory activities for LNCaP/AR, VCaP and AR-V7-positive 22Rv1 cells. Compound **12** and **13** can suppress the transcriptional activities of wild-type AR and strongly decreased the protein expression of AR in LNCaP cell.[18]

Table 4. *In vitro* activity of darolutamide (**1**) derivatives.

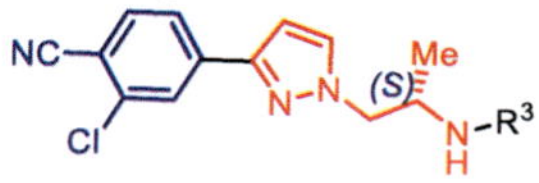

Compound	R^3	IC_{50} (uM)		
		LNCaP/AR	VCaP	22Rv1
Darulutamide, **1**	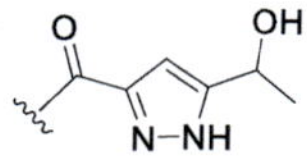	1.65	0.26-0.56[ref12]	>30
29	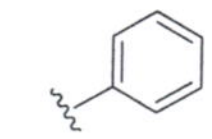	0.71	>30	>30
30	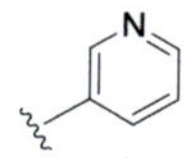	0.43	>30	>30
31	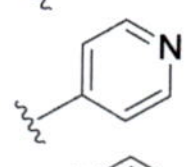	2.56	9.90	>30
32	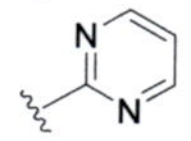	1.18	0.48	4.76
33	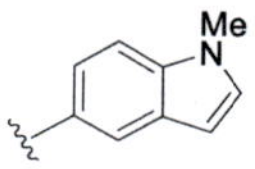	0.49	2.13	11.98
34	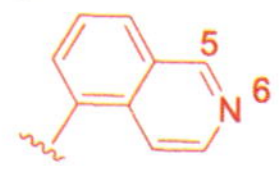	0.87	>30	21.20
12	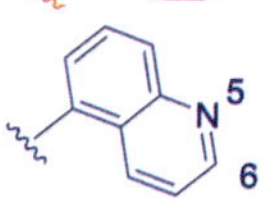	1.08	0.73	3.92
13	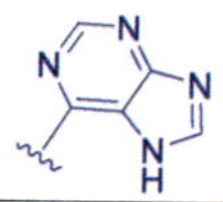	0.96	0.64	4.24

In vitro, both of compounds **12** and **13** exhibit the inhibitory potency to wild-type AR and the F876L mutant that exhibits resistance to enzalutamide based in the result of luciferase expression and nuclear translocation assay. In *vivo* studies (Table 5), compound **13** shows higher C_{max} (maximum plasma concentration = 2648.88 ng/mL) and more acceptable systemic exposure (AUC_{0-t} =7280.76 ng∗h/mL) than compound **12** (C_{max} = 521 ng/mL; AUC_{0-t} =783.55 ng*h/mL). Significantly, 77.8% and 64.7% VCaP tumor growth inhibition (TGI) were observed in male Balb/c nude mice administrated to compound **13** (60 mg/kg and 30 mg/kg), prior to enzalutamide **5** (TGI = 43%). Unfortunately, we cannot gain information about the difference between darolutamide (**1**) and compound **13**.[18]

Together, not considered to the species and dose differences, darolutamide (**1**) exhibits superiority to systemic exposure than compounds **12** and **13** according to previous report (Table 5).[24]

Table 5. Pharmacokinetic parameters of darolutamide (**1**), **12** and **13**.

	Darolutamide, 1, bid		compound 13	compound 12
dose	25mg	50 mg	30 mg, qd	30 mg, qd
$T_{1/2}$(h)	1.4	1.6	1.79	3.14
T_{max}(h)	8.5	8.5	0.65	0.33
C_{max}	27608	42712	2648.88	521
$AUC_{0\text{-}24}$(ng*h/mL)	108618	174807	7379.8	796.54

Source: Adapted from reference 18 and 24.

7.4 Pharmacokinetics and Drug Metabolism

Moilanen initially reported the mouse pharmacokinetic profile of darolutamide (**1**) and the major metabolite keto-darolutamide **11** to demonstrate the priority of brain penetrance compared with enzalutamide **5** and apalutamide **6** in 2015.[13] However, no pharmacokinetic profiles of single diastereomer [(*S,S*) or (*S,R*)] have been published before Moilanen's work.

AUC and C_{max} of darolutamide (**1**) can benefit from food for 2-fold greater along with benefit of t_{max}. In a phase I study[19] in Japanese patients with mCRPC, similar significant food effect was observed in darolutamide (**1**), single diastereomers [(*S,S*) and (*S,R*)], and metabolite keto-darolutamide **11** in both dose 300 mg and 600 mg (C_{max} ratio fed/fast = 2.11–3.29:1; AUC (0–t_{last}) ratio fed/fast = 2.02–2.83:1), similar with the result reported in ARAFOR trial.[25] In the fed state, AUC (0–t_{last}) shows high ratio (5:1 for dose 300 mg; 8:1 for dose 600 mg) of (*S,S*)-darolutamide **9** to (*S,R*)-darolutamide **10**, suggesting higher exposure and bioavailability of (*S,S*)-darolutamide. Also, metabolite keto-darolutamide **11** shows higher exposure than darolutamide (**1**).[19] In a six doses-escalation ARADES trail with western patients, a linear, dose-related performance was observed for the exposure of darolutamide (**1**), entering into plateau for 1400 mg (Figure 7). The mean keto-darolutamide **11**: parent ratio at steady state was 1.6–2.3:1. The median t_{max} are 3.0–5.1 h and 1.5–5.0 h for darolutamide (**1**) and keto-darolutamide **11**, respectively.[26]

In human, the mean half-life of keto-darolutamide **11** (10 h) is less than that of darolutamide (**1**) (15.8 h) at steady state. The clearance (%CV) of darolutamide (**1**) following intravenous administration is 116 mL/min (39.7%).[26]

The ratio of diastereomers [(*S,S*) and (*S,R*)] changes once adiministration in animals, resulting from the interconversion *via* the major metabolite keto-darolutamide **11** catalyzed by two cytosolic reductases (mainly AKR1C3 and minorly AKR1D1).[27]

Koshinen observed the conversion of (*S,S*) to (*S,R*)-darolutamide but no detection of reconversion in cryopreserved Balb/c mouse hepatocytes and a profound stereoselectivity (*S,R*)-darolutamide **10** in mouse plasma after oral dosing of darolutamide (**1**). The stereochemistry of diastereomers [(*S,S*) and (*S,R*)] was identified by single-crystal X-ray diffraction.[24] Contradictorily, previous research by Mullangi discovered only (*S,R*)-darolutamide **10** can converse the (*S,S*)-darolutamide **9** in mouse, without single-crystal X-ray diffraction.[24] Together, the different ratio is likely attributed to species differences (Figure 8).

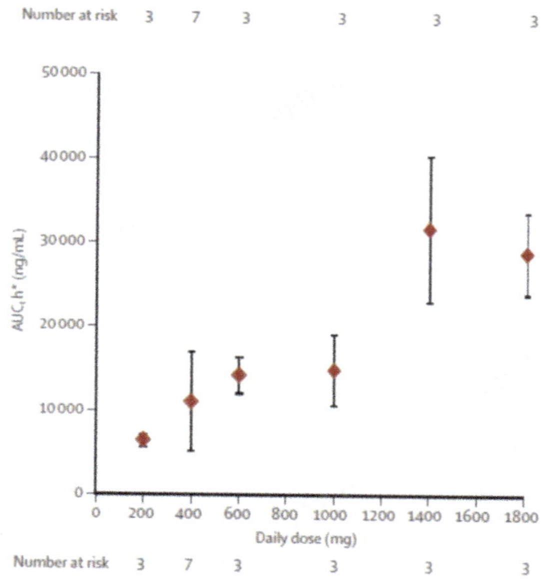

Figure 7. Darolutamide (**1**) mean area under the curve by dose in human.[26]

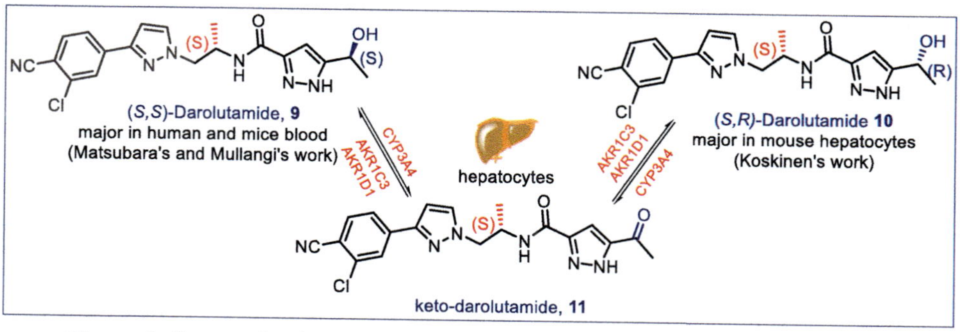

Figure 8. Conversion between diastereomers **9**, **10** and keto-darolutamide **11**.

In vitro study, darolutamide (**1**) is predominately oxidized to the major metabolite keto-darolutamide **11** by CYP3A4, which is responsible for 30% of darolutamide clearance, and is a substrate for P-gp and BCRP. A phase I clinical study in healthy volunteers evaluate the effect of co-combined drug furtherly. 1.75-fold increase in darolutamide (**1**) exposure was observed when combined with itraconazole (a CYP3A4, P-Gp and BCRP inhibitor) while a 72% decrease combined with rifampicin (a CYP3A4 and P-gp inducer). Diastereomers **9**, **10** and keto-darolutamide **11** showed similar results.[27]

Protein binding is 92% for darolutamide **1** and 99.8% for the active metabolite, keto-darolutamide **11**. Serum albumin is the main binding protein for darolutamide (**1**) and keto-darolutamide **11**.

After administration of a radiolabeled dose of darolutamide **1**, 63.4% of darolutamide-related material was recovered from urine (≈7% as unchanged drug) and 32.4% from feces (≈30% as unchanged drug).

7.5 Efficacy and Safety

Darolutamide (**1**) is a highly and fully non-steroidal AR antagonist that prevents androgen from binding to the LBD of AR in competitive AR assays with 11 nM lower than enzalutamide **5** (K_i = 86 nM) and apalutamide **6** (K_i = 93 nM), and inhibits the AR-induced nuclear translocation in *in vitro* assays. Further, darolutamide (**1**) specifically suppressed VCaP cell proliferation in *in vitro* assays, in which the IC_{50} value is 230 nM less than the values of enzalutamide **5** (IC_{50} = 410 nM) and apalutamide **6** (IC_{50} = 420 nM). *In vivo*, darolutamide (**1**) demonstrated a significant antitumor activity in mice with 50 mg/kg twice daily. During administration for 37 days, no sign of treatment-related toxicities and no decrease of body weights were observed in mice. The administration of darolutamide (**1**) (50 mg/kg, bid) did not increase serum testosterone level for three weeks and exhibited a negligible brain penetrance. It is worthy to note that darolutamide (**1**) keeps antagonization for AR mutants (F876L, W741L, and T877A), which are known to facilitate resistance to first or second-generation antiandrogens.[13] In patients, the treatment of darolutamide (**1**) actually maintains the serum testosterone at the castrate level, while an increased level of testosterone in plasma and bone marrow have been observed with enzalutamide **5**.[28]

In clinical, darolutamide (**1**) exhibits a low potential for CYP or P-gp-mediated drug–drug interactions (DDIs) except for BCRP (breast cancer resistance protein) such as rosuvastatin and organic anion-transporting polypeptide (OATP) substrates, while enzalutamide **5** and apalutamide **6** perform the potential for DDIs with anticoagulants and opioid analogues.[27]

In the ARAMIS phase III study, enrolled patients with a total of 1,509 men (2:1) with nmCRPC, a PSA no less than 2ng/mL, and a PSA doubling time no more than 10 months received darolutamide (600 mg twice daily). Metastasis-free survival (MFS) is a strong surrogate for overall survival for clinically localized prostate cancer.[29] Darolutamide (**1**) in combination with ADT significantly prolongs longer median metastasis-free survival 22 months, longer time to first use of cytotoxic chemotherapy (hazard ratio, 0.58), longer time to pain progression (40.2 months vs. 25.4 months), and longer time to symptomatic skeletal event (hazard ratio, 0.48) than placebo. Date based in different trails with enzalutamide **5** and apalutamide **6** for nmCRPC showed similar results with darolutamide (**1**) than placebo in metastasis free survival (Table 6).[30] Seizures show no difference between apalutamide and placebo, and darolutamide (**1**) shows lower risks of adverse effects than enzalutamide **5** and apalutamide **6**, which is in concert with low penetration of the BBB. As reflected in the US Prescribing Information for darolutamide (**1**), the most common side effects of darolutamide (**1**) include fatigue, pain in extremities, and rash.[15]

Table 6. Clinical benefits for nmCRPC with approved 2nd-generation AR antagonists

	PROSPER		SPARTAN		ARAMIS	
End Point	**Enzalutamide, 5**	**placebo**	**Apalutamide, 6**	**placebo**	**Darolutamide, 1**	**placebo**
Median MFS	36.6	14.7	40.4	16.2	40.4	18.4
Median time to PSAP	37.2	3.9	NR	3.7	33.2	7.3
Median time to PFS	-	-	40.5	14.7	36.8	14.8
Median time to PP	18.5	18.4	-	-	40.3	25.36
Seizures	-	-	0.2%	0	0.2%	0.2%
Fatigue	33%	14%	30.4%	21.1%	13.2%	8.3%

MFS: Metastasis Free survival; PSAP: PSA progression;
PFS: Progression-free survival; PP: Pain Progression; NR: not reached; Time: month

7.6 Synthesis

Orion Corporation developed and manufactured chemical process and product development of darolutamide (**1**) under the FDA priority review designation.[31] As to date, the strategy to darolutamide (**1**) devised by Orion Corporation remained unchanged and mainly focused on process improvements[32] *via* crystallization, elimination of undesirable reagents since the original disclosure was published in 2012.[33] Here in this chapter, we disclosed three synthetic routes with the same intermediate **40**.

Developed by the well-known reaction, the route of Orion Corp. starts with the Suzuki coupling via boronate ester **36** containing THP-protected pyrazole (Figure 9).[34] The Pd-catalyed Suzuki coupling of **36** with 4-bromo-2-chlorobenzonitrile **35** was carried out to afford **37** in 92% yield, followed by workup and crystallization. The deprotection of THP is easy to accomplish with methanolic HCl at 10 °C to isolate pyrazole **38** *via* a simple crystallization operation in 96% yield. However, to accomplish the coupling of C-

N of **40** (also is the key intermediate of other two routes) regioselectively, process chemists chose Mitsunobu reaction which is rarely used in plant due to difficulty of post-process (the removal of hydrazide and triphenylphosphine oxide). They ingeniously operated pH adjustment to protonate amine **40**, that can be attributed into aqueous phase, followed by adjustment to pH = 9, organic solvent extraction and crystallization. Keto-darolutamide **11** can be finished *via* T3P mediated condensation of amine **40** and carboxylic acid **41** in 88% yield. Finally, the reduction of keto-darolutamide **11** are operated with $NaBH_4$ in EtOH to obtain darolutamide (**1**) in 76% yield.

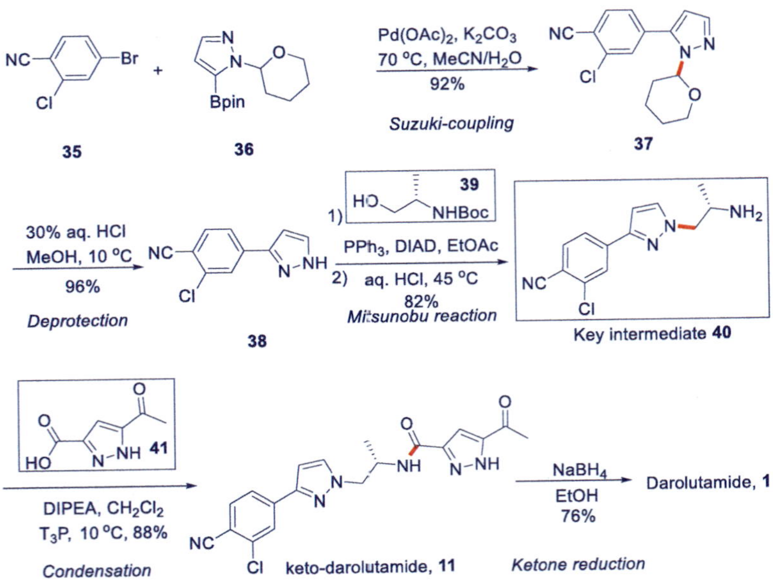

Figure 9. Orion original route to darolutamide (**1**).

The second routes used to synthesize each diastereomer of darolutamide (**1**) were still being developed by Orion in the laboratory (Figure 10).[35] Enzyme-catalytic reduction was introduced to afford enantioselective **43** with ketone **42** and catalytic loading KRED130 in 87% yield with 100% enantioselectivity (Figure 10, pathway A). Pathway B displays an alternate to **43**, in which 3-butyn-2-ol undergoes [3 + 2] cyclization with ethyl diazoacetate **44**. Next, **47** can be afforded after regular silico-protection of TBDPSCl (t-$BuPh_2SiCl$) of **45** and hydrolysis of ester **46** with NaOH. The similar condensation with intermediate **40** mediated by condensation reagent EDI furnishes silyl ether **48**, which undergoes deprotection with TBAF to give (*S,S*)-darolutamide **9**.

In 2018, Pan and coworkers developed an alternate route featuring cyclization reaction of **53** (Figure 11).[36] Aldehyde **51** can be afforded in high yield following by TBS-protection and DIBAL-H mediated reduction. LiHMDS-promoted condensation with **44** obtained **52** in 81% yield. The key intermediate **53** was prepared in 85% yield, undergoing dehydration of **52**. [1,3]-Dipolar cyclization was conducted in refluxed *n*-octane followed by ester hydrolysis with 10% aqueous NaOH to give **54** in 76% yield. Subsequently, the same strategy undergoes condensation of intermediate **40** and TBAF-mediated deprotection to give darolutamide (**1**).

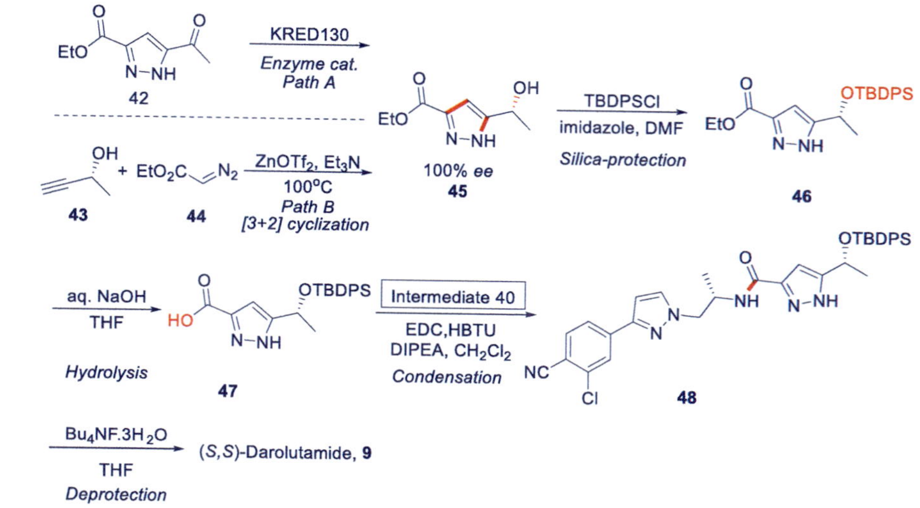

Figure 10. Orion chiral route to (*S,S*)-darolutamide **9**.

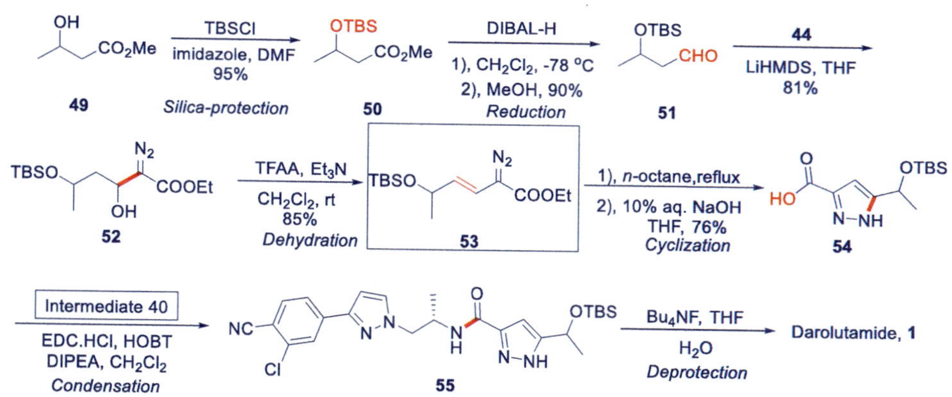

Figure 11. Alternate route to darolutamide (**1**).

7.7 The Future

When we consider something in retrospect, we can look for a progress bar about the story of drug. Clinical needs are always the foremost of new drug development. Initially, the first generation of antiandrogen drugs, such as flutamide **2**,[6] nilutamide **3**, and bicalutamide **4**, proved efficacious for treatment of early stage prostate cancer. When patients progress to hormone-refractory stage, the first generation of antiandrogen drugs switch anti-agonist to agonist. The second generation of antiandrogen drugs comes into being in 2012. The design of second generation of antiandrogen drugs focuses on avoiding agonist activity and keeping anti-agonist activity for AR mutants. As we know, darolutamide (**1**) can inhibit wild-type, AR(F876L) mutation that induces enzalutamide **5** and apalutamide **6** antagonist to agonist switch, as well as AR (W742L) mutation and AR (T876L) causing bicalutamide **4** agonist switch. Importantly, darolutamide **1** exhibits low penetrance of the BBB in mice and rats. Decreased the resistance of PC is an important advancement for this disease. As to date, darolutamide (**1**) has been approved to treat nonmetastatic castration-resistant prostate cancer nmCRPC. More new indications (mCRPC, mCSPC, mHSPC (Metastatic Hormone-sensitive Prostate Cancer)) are waiting for clinical date of darolutamide **1**. The exact mechanism of resistance is currently under active investigation as this could influence both future treatment and outcomes in CRPC patients. It is also worth noting that novel oral ARV-110 (**57**)[37] developed by Arvinas has entered into the phase I/II study (NCT03888612, ARV-110-mCRPC-101) to treat mCRPC, which can degrade AR protein to prevent AR signaling based upon proteolysis targeting chimera (PROTAC) concept and could be potentially very effective for the treatment of CRPC when the disease becomes resistant to AR antagonists or to androgen synthesis inhibitors (Figure 12).

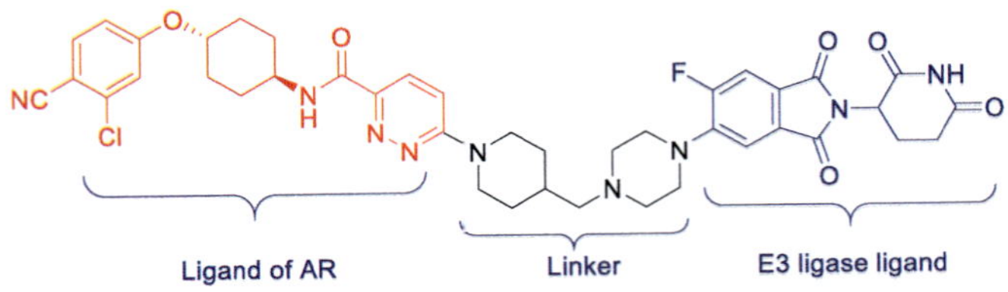

ARV-110, 57
Molecular Weight: 812.3

Figure 12. Structure of ARV-110 (**57**).

7.8 References

1. Fizazi, K.; Shore, N.; Tammela, T. L.; et al. Nonmetastatic, castration-resistant prostate cancer and survival with darolutamide. *N. Eng. J. Med.* **2020**, *383*, 1040–1049.
2. Sung, H.; Ferlay, J.; Siegel, R. L.; et al. Global cancer statistics 2020: GLOBOCAN estimates of incidence and mortality worldwide for 36 cancers in 185 countries. *CA Cancer J. Clin.* **2021**, *71*, 209–249.

3. Hellerstedt, B. A.; Pienta, K. J. The current state of hormonal therapy for prostate cancer. *CA Cancer J Clin.* **2002**, *52*, 154–179.
4. Zuo, M.; Xu, X.; Li, T.; Ge, R.; Li, Z. Progress in the mechanism and drug development of castration-resistant prostate cancer. *Future Med. Chem.* **2016**, *8*, 765–788.
5. Huggins, C.; Hodges, C. V. Studies on prostatic cancer. I. The effect of castration, of estrogen and of androgen injection on serum phosphatases in metastatic carcinoma of the prostate. *Cancer Res.* **1941**, *1*, 293.
6. Kelly, W. K.; Scher, H. I. Prostate specific antigen decline after antiandrogen withdrawal: the flutamide withdrawal syndrome. *J. Urol.* **1993**, *149*, 607–609.
7. Culig, Z.; Hoffmann, J.; Erdel, M.; Eder, I. E.; Hobisch, A.; Hittmair, A.; Bartsch, G.; Utermann, G.; Schneider, M. R.; Parczyk, K.; Klocker, H. Switch from antagonist to agonist of the androgen receptor bicalutamide is associated with prostate tumour progression in a new model system. *Br. J. Cancer* **1999**, *81*, 242–251.
8. Scher, H. I.; Beer, T. M.; Higano, C. S.; et al. Antitumour activity of MDV3100 in castration-resistant prostate cancer: a phase 1–2 study. *The Lancet* **2010**, *375*, 1437–1446.
9. Scher, H. I.; Fizazi, K.; Saad, F.; et al. Increased survival with enzalutamide in prostate cancer after chemotherapy. *N. Engl. J. Med.* **2012**, *367*, 1187–1197.
10. Joseph, J. D.; Lu, N.; Qian, J.; Sensintaffar, J.; Shao, G.; Brigham, D.; Moon, M.; Maneval, E. C.; Chen, I.; Darimont, B.; Hager, J. H. A clinically relevant androgen receptor mutation confers resistance to second-generation antiandrogens enzalutamide and ARN-509. *Cancer Discov.* **2013**, *3*, 1020–1029.
11. Korpal, M.; Korn, J. M.; Gao, X.; et al. An F876L mutation in androgen receptor confers genetic and phenotypic resistance to MDV3100 (enzalutamide). *Cancer Discov.* **2013**, *3*, 1030–1043.
12. Sugawara, T.; Baumgart, S. J.; Nevedomskaya, E.; et al. Darolutamide is a potent androgen receptor antagonist with strong efficacy in prostate cancer models. *Int. J. Cancer* **2019**, *145*, 1382–1394.
13. Moilanen, M.; Riikonen, R.; Oksala, R.; et al. Discovery of ODM-201, a new-generation androgen receptor inhibitor targeting resistance mechanisms to androgen signaling-directed prostate cancer therapies. *Sci. Rep.* **2015**, *5*, 1–11
14. Vis, A. N.; Schröder, F. H. Key targets of hormonal treatment of prostate cancer. Part 1: the androgen receptor and steroidogenic pathways. *BJU Int.* **2009**, *104*, 438–448.
15. Markham, A.; Duggan, S. Darolutamide: First Approval. *Drugs* **2019**, *79*, 1813–1818.
16. Saini, N. K.; Gabani, B. B.; Todmal, U.; Sulochana, S. P.; Kiran, V.; Zainuddin, M.; Balaji, N.; Polina, S. B.; Srinivas, N. R.; Mullangi, R. Pharmacokinetics of

Darolutamide in Mouse - Assessment of the Disposition of the Diastereomers, Key Active Metabolite and Interconversion Phenomenon: Implications to Cancer Patients. *Drug Metab. Lett.* **2021**, *14*, 9–16.

17. Robinson, D.; Van Allen, EM.; Wu, Y.; et al. Integrative clinical genomics of advanced prostate cancer. *Cell.* **2015**, *161*, 1215–1228.

18. Yu, J.; Zhou, P.; Hu, M.; Yang, L.; Yan, G.; Xu, R.; Deng, Y.; Li, X.; Chen, Y. Discovery and biological evaluation of darolutamide derivatives as inhibitors and down-regulators of wild-type AR and the mutants. *Eur. J. Med. Chem.* **2019**, *182*, 111608.

19. Matsubara, N.; Mukai, H.; Hosono, A.; Onomura, M. et al. Phase 1 study of darolutamide (ODM-201): a new-generation androgen receptor antagonist, in Japanese patients with metastatic castration-resistant prostate cancer. *Cancer Chemother. Pharmacol.* **2017**, *80*, 1063–1072.

20. Zurth, C.; Sandman S.; Trummel, D, Seidel, D.; Nubbemeyer, R.; Gieschen, H.; Higher blood-brain barrier penetration of [^{14}C] apalutamide and [^{14}C] enzalutamide compared to [^{14}C] darolutamide in rats using whole-body autoradiography. *J. Clin. Oncol.* **2019**, *37*, 156. abstract. DOI: 10.1200/JCO.2019.37.7.

21. Tanaka, K.; Motohiro, H.; Nobutaka. K.; Akiyuki, S. Novel pyrazole derivative, harmful organism control agent, and use of the control agent. WO-2008062878A1 (2008).

22. (a) Guo, C.; Linton, A.; Kephart, S.; et al. Discovery of aryloxy tetramethylcyclobutanes as novel androgen receptor antagonists. *J. Med. Chem.* **2011**, *54*, 7693–7704. (b) Hur, E.; Pfaff, S. J.; Payne, E. S.; Gron, H.; Buehrer, B. M.; Fletterick, R. J. Recognition and accommodation at the androgen receptor coactivator binding interface. *PLoS Biol.* **2004**, *2*, e274.

23. Watson, P. A.; Arora, V. K.; Sawyers, C. L. Emerging mechanisms of resistance to androgen receptor inhibitors in prostate cancer. *Nat. Rev. Cancer.* **2015**, *15*, 701–711.

24. Nykänen, P.; Korjamo, T.; Gieschen, H.; Zurth, C.; Koskinen, M. Pharmacokinetics of Darolutamide, its Diastereomers and Active Metabolite in the Mouse: Response to Saini NK et al. *Drug Metab. Lett.* **2021**, *14*, 9–16.

25. Massard, C.; Penttinen, H. M.; Vjaters, E.; Bono, P.; Lietuvietis, V.; Tammela, T. L.; Vuorela, A.; Nykanen, P.; Pohjanjousi, P.; Snapir, A.; Fizazi, K. Pharmacokinetics, antitumor activity, and safety of ODM-201 in patients with chemotherapy-naive metastatic castration-resistant prostate cancer: an open-label phase 1 study. *Eur. Urol.* **2016**, *69*, 834–840.

26. Fizazi, K.; Massard, C.; Bono, P.; Jones, R. Activity and safety of ODM-201 in patients with progressive metastatic castration-resistant prostate cancer (ARADES): an open-label phase 1 dose-escalation and randomized phase 2 dose expansion trial. *Lancet Oncol.* **2014**, *15*, 975–985.

27. Zurth, C.; Koskinen, M.; Fricke, R.; Prien, O.; Korjamo, T.; Graudenz, K.; Denner, K.; Bairlein, M.; von Bühler, C. J.; Wilkinson, G.; Gieschen, H. Drug-Drug interaction potential of darolutamide: In vitro and clinical studies. *Eur. J. Drug Metab. Pharmacokinet.* **2019**, *44*, 747–759.
28. Efstathiou, E.; Titus, MA.; Tsavachidou, A.; et al. MDV3100 effects on androgen receptor (AR) signaling and bone marrow testosterone concentration modulation: A preliminary report. *J. Clin. Oncol.* **2011**, *29*, abstract 4501.
29. Xie, W.; Regan, M. M.; Buyse, M.; et al. Metastasis-Free Survival Is a Strong Surrogate of Overall Survival in Localized Prostate Cancer. *J. Clin. Oncol.* **2017**, *35,* 3097–3104.
30. (a) Hussain, M.; Fizazi, K.; Saad, F.; et al. Enzalutamide in men with nonmetastatic, castration-resistant prostate cancer. *N. Engl. J. Med.* **2018**, *378*, 2465–2474; (b) Smith, M. R.; Saad, F.; Chowdhury, S.; et al. Apalutamide treatment and metastasis free survival in prostate cancer. *N. Engl. J. Med.* **2018**, *378*, 1408–1418.
31. U.S. FDA approves darolutamide (NCE) with Orion and Fermion supply chain. https://www.fermion.fi/about-us/news/u.s.-fda-approves-darolutamide-nce-with-orion-and-fermion-supply-chain.
32. Hughes, D. L. Review of Synthetic Routes and Crystalline Forms of the Antiandrogen Oncology Drugs Enzalutamide, Apalutamide, and Darolutamide *Org. Process Res. Dev.* **2020**, *24*, 347–362.
33. Wohlfahrt, G.; Törmäkangas, O.; Salo, H.; Höglung, L.; Karjalainen, A.; Knuuttila, P.; Holm, P. Androgen Receptor Modulating Compounds WO2011051540A1 (2011).
34. Laitinen, I.; Karjalainen, O. Process for the Preparation of Androgen Receptor Antagonists and Intermediates Thereof WO2016162604A1 (2016).
35. Törmäkangas, O.; Heikkinen, R. A Carboxamide Derivative and Its Diastereomers in Stable Crystalline Form WO20161230530A1 (2016).
36. Pan, T.; Xia, C.; Yang, Y.; Zhang, A. Process for Preparation of Novel Androgen Receptor Antagonist WO2018108130A1 (2018).
37. Crew, A. P.; Snyder, L. B.; Wang, J.; Haskell, R. J. III; Moore, M. D. Methods of treating prostate cancer US20210113557A1 (2021).

8

Venetoclax (Venclexta): A BCL-2 Antagonist for Treating Chronic Lymphocytic Leukemia

Nadia M. Ahmad and Jie Jack Li

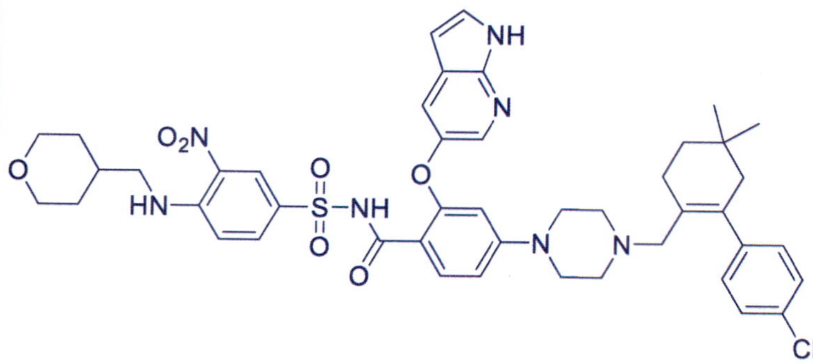

Venetoclax
(Venclexta, **1**)
Abbvie, 2016
BCL-2 inhibitor

8.1 Background

Chronic lymphocytic leukemia (CLL) is the most common form of leukemia in the Western world. Abbvie's venetoclax (Venclexta, **1**) is a treatment for CLL with the 17p deletion as detected by an FDA-sanctioned test. Approved in 2016, venetoclax (**1**) is a monotherapy for those patients who have received at least one prior therapy. It is an orally administered, selective antagonist of the B-cell lymphoma (BCL-2) proteins. BCL-2 proteins are a group of pro-survival proteins, that is, they enable malignant tumors to evade cell death by impairment of the intrinsic (mitochondrial) apoptotic pathway and hence cell death.[1]

In terms of drug design, the discovery of venetoclax (**1**) as the first-in-class antagonist of the B-cell lymphoma (BCL-2) proteins, is a great example of drug discovery using fragment-based drug design (FBDD). In fact, it was the second drug to be approved by the FDA discovered using the FBDD approach, the first being Plexxikon's B-Raf inhibitor vemurafenib (Zelboraf), approved in 2011. Instead of the co-crystallography

Current Drug Synthesis, First Edition. Edited by Jie Jack Li.

tactic used in the discovery of vemurafenib, "SAR by NMR" method was key to the generation of the fragment hits to bind to BCL-2. More impressively, venetoclax (**1**) is one of the few inhibitors of protein–protein interactions (PPIs), a notoriously challenging class of drug targets.

Venetoclax (**1**, also known as ABT-199 or GDC-0199, respectively) reinstates the apoptotic ability of malignant tumors. It was co-developed by Abbot Laboratories (now Abbvie) and Genentech. On approval in 2016, it became the latest in the new-wave arsenal of CLL treatments targeting the BCL-2 signaling pathways. These include ibtutinib (**2**), a Bruton tyrosine kinase (BTK) inhibitor and idelalisib (**3**), a selective phosphatidylinositol-3-kinase (PI3K)-δ isoform inhibitor.

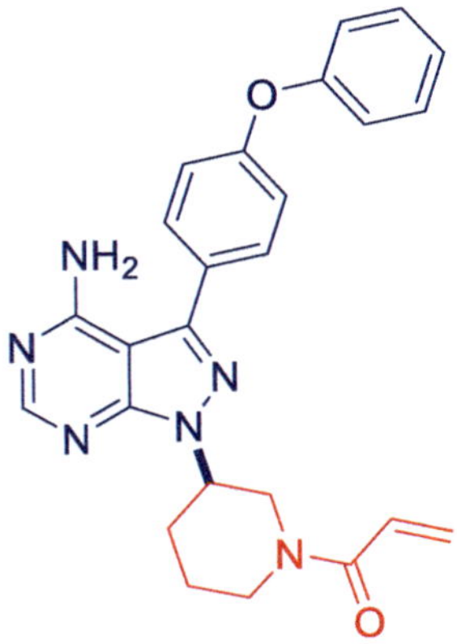

Ibrutinib (Imbruvica, **2**)
Pharmacyclics/Jansen, 2013
BTK Inhibitor

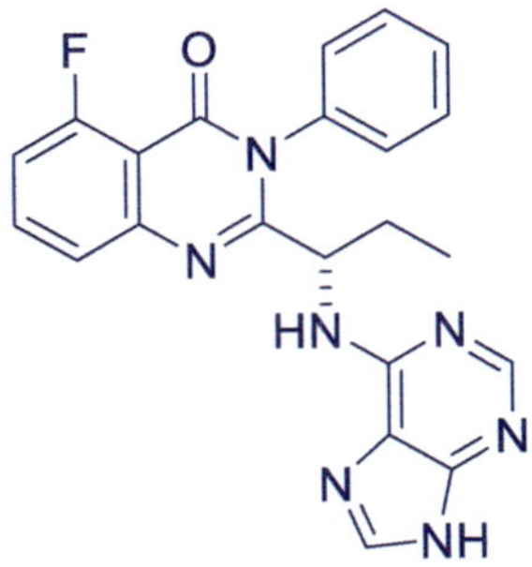

Idelalisib (Zydelig, **3**)
Gilead, 2014
PI3Kδ inhibitor

Currently, venetoclax (**1**) is on the market to treat adults with CLL or small lymphocytic lymphoma (SLL). It is also prescribed as part of a combination therapy with azacytidine or decitabine to treat adults aged 75 or over that have been newly diagnosed with acute myeloid leukemia (AML).[2]

8.2 Pharmacology

8.2.1 BCL-2 Protein Is a Driver of Cancer Development

Evasion of apoptosis is a hallmark of cancer.

Apoptosis, a term coined in 1972, means programmed cellular death. It is responsible for the removal of aged and damaged cells from healthy multicellular organisms as part of homeostasis. Indeed, apoptotic dysregulation is a characteristic feature observed in all cancers! This process of intrinsic apoptosis is mediated by certain proteins including the BCL-2 family of proteins. Inside the cytoplasm, BCL-2 proteins regulate the process of apoptosis by directly modulating the mitochondrial outer membrane permeabilization *via* PPIs. In fact, members of the BCL-2 family proteins are widely distributed in the human body, where they all operate *via* the same mitochondrial apoptotic pathway.[3]

The *pro-apoptotic proteins*, such as BCL-2-associated X protein (BAX), BCL-2-antagonist/killer 1 (BAK), and BCL-2-interacting mediator of cell death (BIM), etc., *promote cell death*. Meanwhile the *anti-apoptotic proteins*, such as BCL-2, B-cell lymphoma X long (BCL-X_L), B-cell lymphoma-like 2/W (BCL-W), myeloid cell leukemia sequence 1 (MCL-1), and A-1, etc., *keep cells alive*. BAX and BAK induce mitochondrial outer membrane (MOM) permeabilization, leading to caspase activation and programmed cell death. This is why pro-apoptotic proteins, such as BAX are BAK, are known as death proteins whereas BCL-2 protein is also known as pro-survival protein.

However, pro-survival protein BCL-X_L is also needed for platelet survival. Therefore, thrombocytopenia (i.e., an abnormal decrease in number of platelets in the blood) is a dose-limiting toxicity for drugs inhibiting BCL-X_L.

The BCL-2 homology (BH) regions have conserved residues shared by all antiapoptotic proteins, which are classified as BH1, BH2, BH3, and BH4 domains.[4] Venetoclax (**1**) is one of the small-molecule BH3 peptide mimetics.

8.2.2 Functions of the BCL-2 Family

The BCL-2 family of proteins are the master regulators of apoptosis and BCL-2 is a driver of tumor development.

Diverse functions of the BCL-2 proteins have been proposed, including anti-oxidant defense, calcium signaling, autophagy and mitochondrial dynamics; however, the *predominant* (*if not exclusive*) role of the family is to regulate cell survival by regulating the integrity of the MOM, which, if breached, sounds the death knell for a cell.[5]

In cancer cells, overproduction of BCL-2 protein leads to binding to death proteins such as BIM and shuts off the death switch, allowing cancerous cells to stay alive thus survive uncontrollably. BCL-2 family protein inhibitors, on the other hand, can bind to BCL-2 competitively, blocking death protein (e.g., BAX) from binding and allowing death protein to kill cancer cells.

As a consequence, cancer cells not only have the capacity to proliferate uncontrollably, but also can survive uncontrollably.

Several hematologic malignancies overexpress pro-survival proteins of the BCL-2 family and, as such, these proteins are attractive targets for treatment of these types of malignancies. In 1988, the BCL-2 protein was found to promote cancer by limiting cell death rather than enhancing proliferation. This discovery set the wheels in motion for an almost 30-year journey involving many international research teams that culminated in the approval in 2016 for of venetoclax (**1**) that targets this protein in the treatment of CLL.[5]

8.2.3 The Binding of Venetoclax (1) to the BCL-2 Protein

Venetoclax (**1**) *per se* was designed through reverse engineering of one of the first small molecule pro-survival protein inhibitors, navitoclax (**9**, ABT-263),[6] which was a dual inhibitor of both BCL-2 and BCL-X_L. However, this lack of selectivity led to on-target thrombocytopenia, causing dose-limitations to the efficacy obtainable with this compound. This side effect can be attributed to inhibition of BCL-X_L, which is recognized as the primary survival protein for platelets. Navitoclax (**9**) in turn resulted from targeted modifications of ABT-737 (**8**, the first reported *bona fide* BH3 mimetic),[7] a compound that inhibited BCL-2, BCL-X_L, and BCL-W, and showed efficacy in murine tumor xenograft models, but had no oral bioavailability.[7]

In contrast, venetoclax (**1**) has a 200-fold higher affinity for BCL-2 than BCL-X_L, allowing for efficacy in reduction of xenograft tumors without adversely targeting platelets. This improvement in selectivity was achieved through the discovery of a unique BCL-2-complex crystal packing dimer. It was observed that a tryptophan side-chain residue (Trp30) from the second BCL-2 intercalated into what was known as the P4 hot spot of the first BCL-2 protein that had a small molecule bound. This tryptophan-30 formed an intermolecular π-stacking interaction with the nitroaryl of venetoclax (**1**), firstly highlighting the importance of a hydrophobic interaction in the P4 region; and secondly, bringing into prominence a hydrogen bond between the nitrogen atom of the indole with Asp103 of BCL-2 (Figure 1c)— a key residue difference between BCL-2 and BCL-X_L in which it is a Glu96.[8,9]

See Figure 1 for pictorial depiction of the binding pose of venetoclax (**1**) to BCL-2 at the end of Section 3.1 (*vide infra*).

8.2.4 The Selectivity of Venetoclax (1)

One of the key challenges in the optimization of navitoclax (**9**)-type structures was to obtain a better selectivity profile. Following on from the aforementioned rational drug design, venetoclax (**1**) was determined to bind at over less than *three orders of magnitude* to BCL-X_L when compared with BCL-2 in a time-resolved fluorescence resonance energy transfer (FRET) competition binding assay (Table 1). This selectivity profile was needed to manifest in a reduction of the thrombocytopenia previously observed with navitoclax (**9**).[8]

Table 1. Binding affinity of navitoclax (**9**) and ABT-199 (**1**) for BCL-2 protein family members by TR-FRET.

	TR-FRET K_i (nM)			
Compound	BCL-2	BCL-X_L	MCL-1	BCL-W
Navitoclax (**9**)	0.044	0.055	>224	7
Venetoclax (**1**)	< 0.010	48	>444	245

8.3 Structure–Activity Relationship (SAR)

8.3.1 The Discovery Venetoclax (1) Using FBDD

A long and winding path led to the discovery of venetoclax (**1**). The basic premise that such a broad and large hydrophobic protein:protein interface could be targeted by a small molecule at all was at first questioned, with such interfaces largely perceived as "undruggable."

When Abbott began their ambitious goal to discover BCL-2 inhibitors, they looked for small molecule compounds that competed directly with BH3-motif binding, also known as BH3 mimetics ("magic bullets"). The goal was to design a compound to mimic the binding of the death protein BAK. But designing inhibitors selective for specific antiapoptotic BCL2 family members was especially challenging due to the extended hydrophobic nature of the target BH3-binding sites, the structural similarities among BCL-2 family members, and the necessity to compete for binding with high-affinity endogenous ligands. Abbott focused on taking advantage of its *hydrophobic residues* pointed out into the solvent: Arg76, Asp83, and Asp84.

Armed with the generation of an NMR structure of a BCL-X_L/BAK peptide complex, Abbott embarked on an ambitious journey of discovery small molecule antagonists of the BCL-2 PPIs. After high-throughput screening (HTS) effort yielded poor quality hits, Abbott sought to use the FBDD approach and employed Fesik's "SAR by NMR" method to generate fragment hits. From initial screening of a 10,000-compound library with MW < 215 at 1 mM concentration, *p*-fluorophenyl-benzoic acid (**4**, LE = ligand efficiency) emerged as one of the first-site (P1) ligands to the BCL-X_L protein. Later on, the second screening of a 3,500-compound library with MW ~ 150 at 5 mM concentration identified 5,6,7,8-tetrahydro-naphthalen-1-ol (**5**) as a second site (P2) ligand.[10] After lengthy efforts to enhance potency, the choice of an acyl-sulfonamide **6** as the carboxylic acid isostere enabled elongation of right-hand portion to provide a potent compound **7**, which now occupied third binding site (P3). Regrettably, **7** suffered from serum deactivation due to high levels of binding to domain III of human serum albumin (HSA-III). Careful structure-based optimization resulted in ABT-737 (**8**), which had more polar amines and had the desired, reduced protein binding.

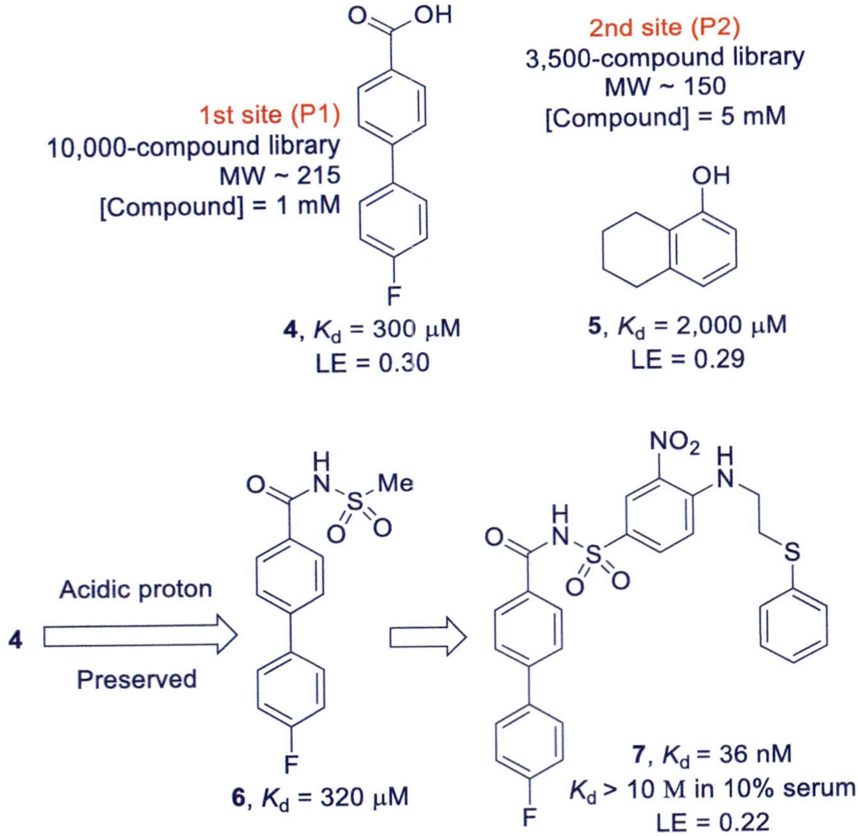

ABT-737 (**8**) was the first reported *bona fide* BH3 mimetic reported, and binds to BCL-2, BCL-X_L, and BCL-W with high affinity ($K_i \leq 1$ nM). However, poor physiochemical properties precluded its further development, ABT-737 (**8**) is not orally bioavailable and has low aqueous solubility. Its rigid and flat biphenyl moiety ("molecular flatland") probably contributed to its extensive pi-stacking and formation of a tight crystalline lattice.

To reduce the rigidity of the rigid biphenyl structure at the bottom of ABT-737 (**8**), one phenyl ring was transformed to cyclohexene to provide an opportunity to make ring modifications that were fundamentally different from those that were made to the aromatic ring in its place. Thereafter, simply bulking up the ring by adding alkyl groups produced the dimethylcyclohexene navitoclax (**9**), which had still higher plasma levels, and also appeared to improve tissue/plasma distributions in various pharmacokinetic models.[8] Addition of the dimethyl group was a great idea since it eliminated the metabolic liability associated with simple cyclohexenes, which are prone to oxidation to the corresponding aromatic phenyl analog.

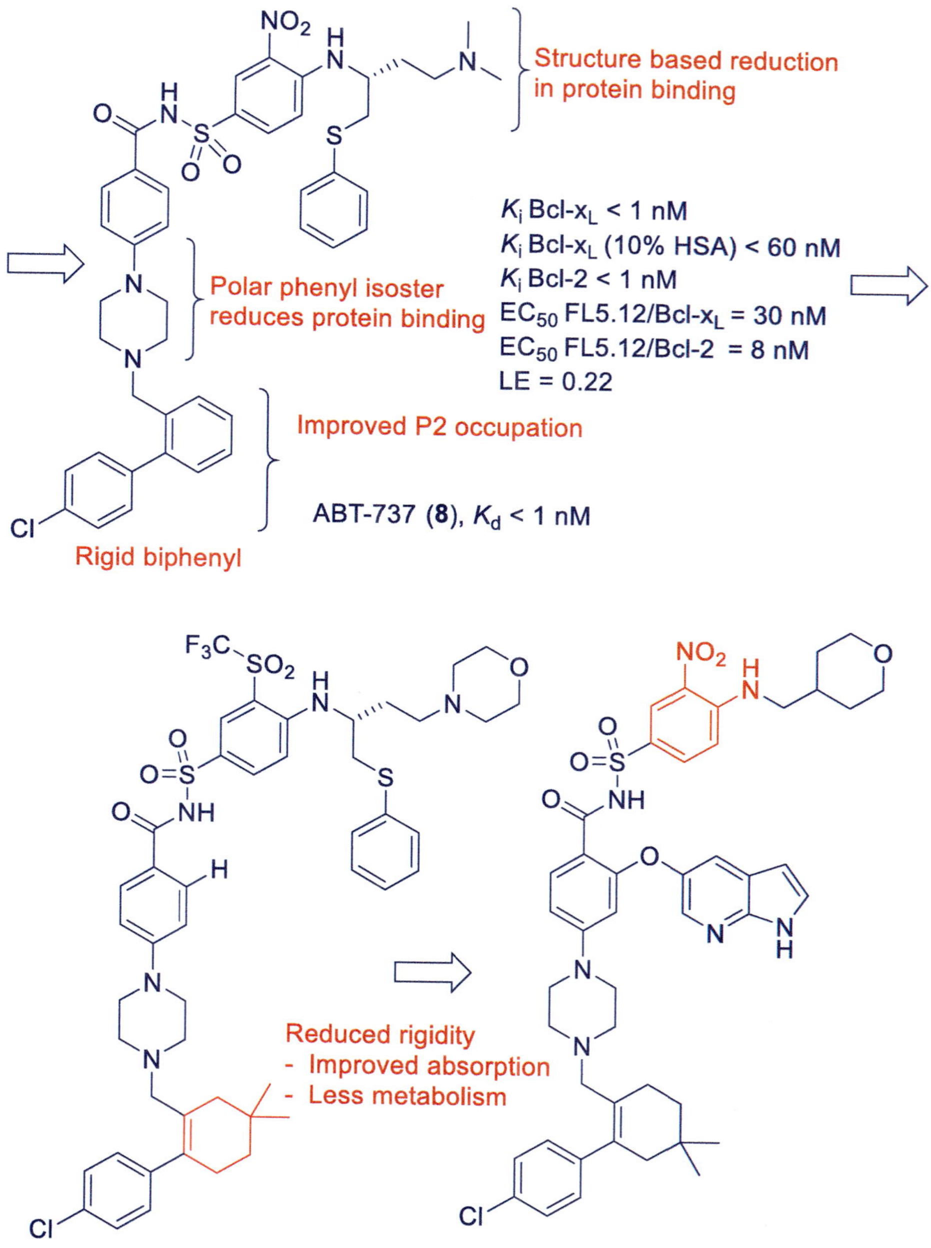

Navitoclax (**9**)
K_i = 0.04 nM, F ~ 30%, LE > 0.20

Venetoclax (Venclexta, **1**)
K_i = 0.01 nM, F ~ 29%, LE > 0.25

Navitoclax (**9**), as a potent and the first orally bioavailable BCL-2 inhibitor, was not selective against BCL-X_L (see Table 1). However, BCL-X_L was found to be essential for the survival of mature platelets, and navitoclax (**9**) induced rapid, concentration-dependent decreases in circulating platelets. This was distinct from most types of chemotherapy-induced thrombocytopenia, as navitoclax (**9**) did not kill megakaryocytes or inhibit their production of young platelets.[7]

Eventually, navitoclax (**9**)'s dose-limiting platelet depletion side effect such as thrombocytopenia was overcome by the discovery of the fourth binding site (P4). Attachment with a 7-azaindole ether at the P4 binding site enabled its N atom to capture an additional hydrogen bond with Arg104 on the target, giving rise to venetoclax (**1**) as a potent, selective (against BCL-X_L, BCL-W, and BCL-1), and orally bioavailable BCL-2 inhibitor. Venetoclax (**1**) is 10-times more potent than navitoclax (**9**) with regard to induction of apoptosis in CLL cells. Compared with navitoclax (**9**), venetoclax (**1**) requires a 200-fold higher concentration for the off-target effect of platelet apoptosis. Selective inhibition of BCL-2 while sparing BCL-X_L allows for therapeutic potentiation of apoptosis without the negative effects of thrombocytopenia.[8,11]

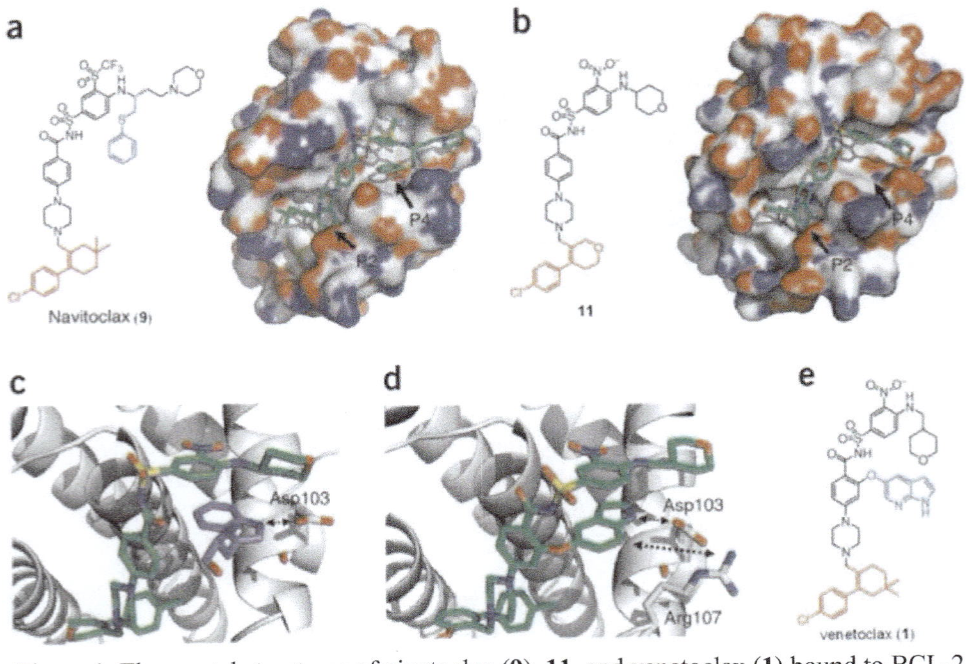

Figure 1. The crystal structures of nivatoclax (**9**), **11**, and venetoclax (**1**) bound to BCL-2 (Adapted with permission from Reference 8, Copyright 2013 Nature Publishing Group).

8.3.2 Structure–Activity Relationship of Venetoclax (**1**)

Table 2. SAR for the substitution of terminal phenyl group on **10**.

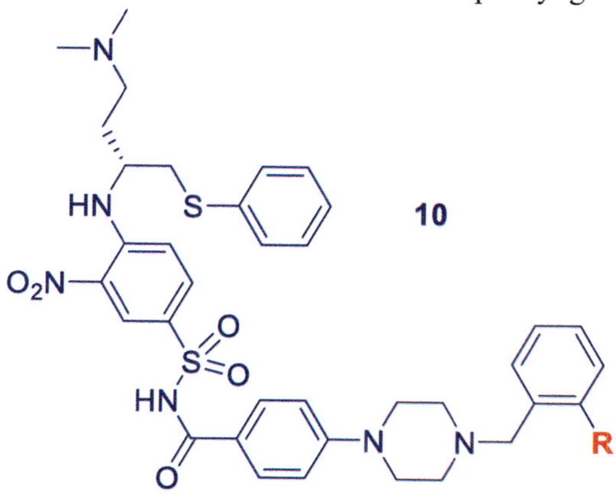

10

Compound	R	BCL-2 K_i (μM)	BCL-X_L K_i (μM)
10a	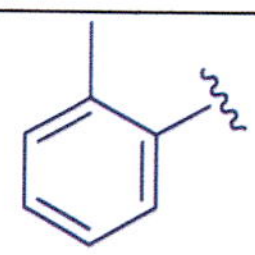	0.016	0.018
10b	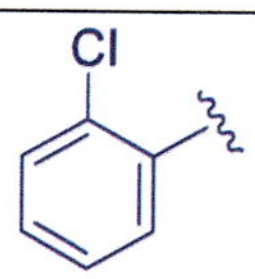	0.027	0.056
10c	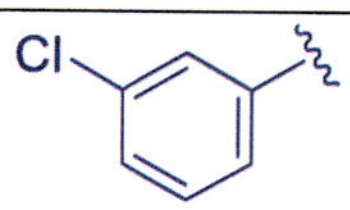	0.033	0.18
ABT-737 (**8**)	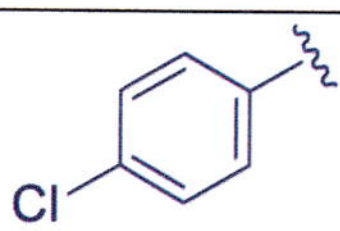	0.008	0.030
10d	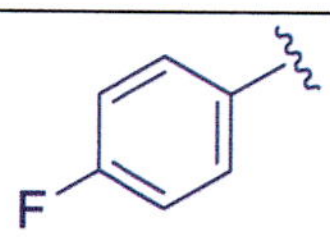	0.027	0.66
10e	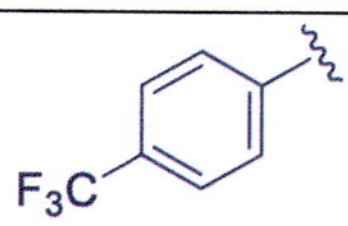	0.013	0.040

En route to the discovery of ABT-737 (**8**), the SAR of the substitution of terminal phenyl group on **10** was investigated. As shown in Table 2, the SAR investigation revealed that the *para*-chlorophenyl group gave the most potent derivative in comparison with other substituents of terminal phenyl group.[6]

Fairbrother et al. revealed certain SAR for the R substituent on acylsulfonamides **11**.[12] As shown in Table 3, R has profound impact on their binding affinities. Compounds lacking the P4-binding "bent-back" thiophenyl moiety demonstrated dramactic loss in binding affinity to both BCL-2 and BCL-X_L. For instance, **11c**–**11f** are significantly weaker compared to ABT-7373 (**8**), **11a**, and **11b**. However, the change was more deleterious to BCL-X_L binding, thus affording moderately selective inhibitors of the target protein.

Table 3. Acylsulfonamides lacking the P4-binding "bent-back" thiophenyl moiety are modestly selective for BCL-2.

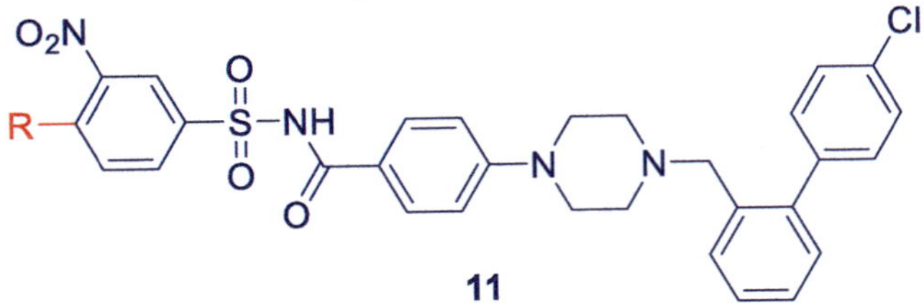

11

Compound	R	BCL-2 K_i (μM)	BCL-X_L K_i (μM)	BCL-X_L K_i / BCL-2 K_i
ABT-737 (**8**)	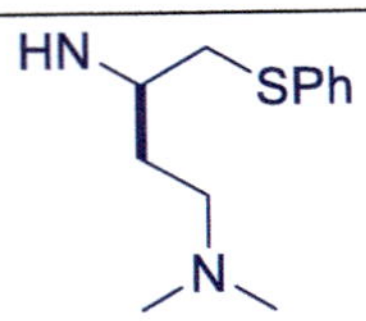	< 0.001	< 0.001	–
11a	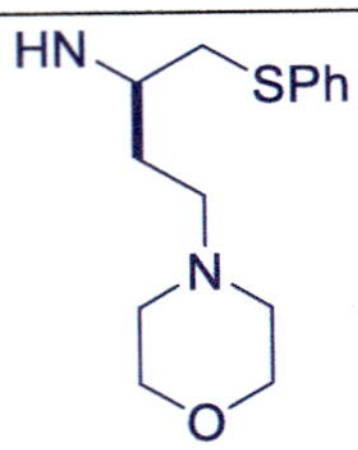	< 0.001	< 0.001	–
11b	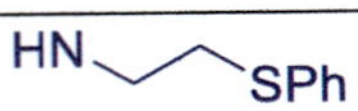	< 0.001	< 0.001	–
11c	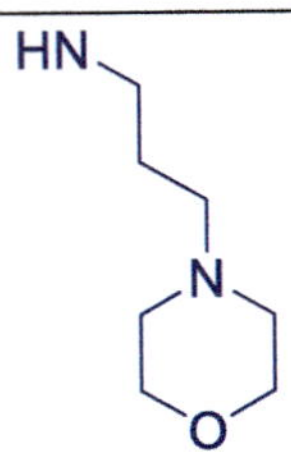	0.034	0.335	10
11d	HN–CH_3	0.088	2.12	24

11e	H	0.170	3.54	21
11f	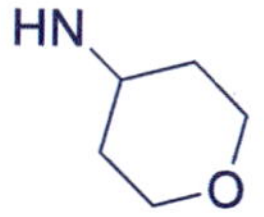	0.019	0.496	27

While the aforementioned efforts resulted in the achievement of selectivity, the insolubility of the compounds meant that progression was limited. Accordingly, the late-stage intermediate was altered to introduce analogues containing an oxygenated ring in the P2 or P4 binding regions to afford dihydropyran **12**. Although **12** exhibited 3-fold loss in affinity for BCL-2 relative to the parent compound **11f**, it was significantly more soluble. More importantly, more soluble dihydropyran **12** ultimately enabled a successful crystallization experiment to afford the first crystal structure of a moderately BCL-2 selective acylsulfonamide.[12]

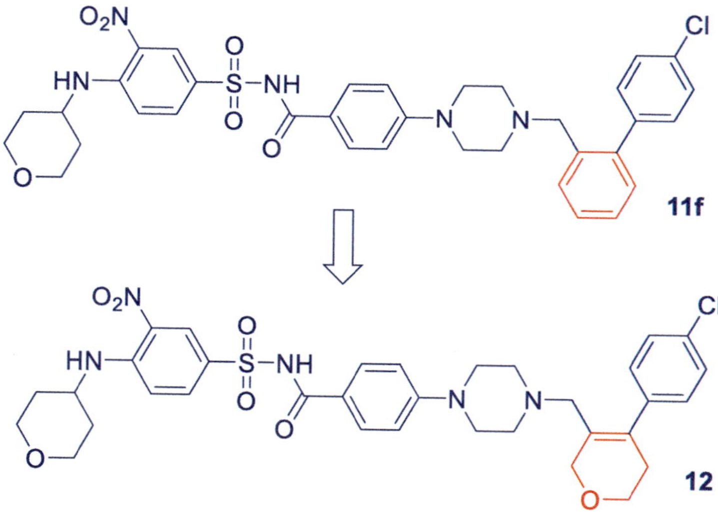

8.4 Pharmacokinetics and Drug Metabolism

It took a large team of diverse scientists across many disciplines at Abbott/Abbvie and many years of intensive focus and stamina to reach venetoclax (**1**), a selective, highly-affinity (low picomolar) BCL-2 binding molecule capable of competing with endogenous death protein as binding partner.

It was especially challenging to engineer the molecules to obtain physiochemical properties competent to afford oral bioavailability, which required ingenuity of applying many tricks in the medicinal chemistry tool box. Not surprisingly, the type of hydrophobic, high molecular-weight compounds required to inhibit BCL-2 family PPIs are not "rule of five"[13] compliant.[12]

Venetoclax (**1**) is given orally *via* a dose titration regime, or ramp-up schedule. This is 20 mg daily for 1 week followed by weekly dose increases to 50, 100, and 200 mg. After this period, 400 mg is the daily dose *ad infinitum*. Venetoclax (**1**) achieves C_{max} (maximum plasma concentration) approximately 5–8 h after oral dose following repeated administration. Exposure is increased in the fed state approximately 3- to 5-fold, depending on the fat content, when compared with the fasted state.[14]

The half-life $t_{1/2}$ of venetoclax (**1**) is 26 h, with major path of elimination *via* feces (> 99.9% of the dose, 20.8% as unchanged drug) and minor elimination *via* urine (< 0.1%). It has a large apparent volume of distribution V_d at 256–321 L, indicating that venetoclax (**1**) is highly distributed into body tissues. It exhibits high plasma protein binding (PPB) with < 0.01% of the drug unbound in plasma, not really surprising considering the molecule's extremely high hydrophobicity. Venetoclax (**1**) is principally metabolized by CYP3A4, indicating its potential for drug–drug interaction (DDI) is high. Venetoclax (**1**) is also a substrate and an inhibitor of P-gp, therefore, concomitant use of venetoclax (**1**) with P-gp inhibitors should be avoided.[15–17]

8.5 Efficacy and Safety

In preclinical studies, *in vivo* efficacy of orally administered venetoclax (**1**) was demonstrated in multiple tumor xenograft models, both as a monotherapy and in combination with anti-cancer biologics. Venetoclax (**1**) demonstrated robust efficacy in animal models primarily representing hematologic malignancies that overexpressed BCL-2 proteins. Those cancer cells are highly dependent on BCL-2 for survival or are "primed for death" based on high levels of BCL-2:BIM complexes. Combination treatments are extremely common in cancer therapy and, gratifyingly, combinations of venetoclax (**1**) with standard of care (SOC) drugs, such as bendamustine and rituximab (BR) as well as rituximab plus cyclophosphamide, hydroxydaunomycin, vincristine, and prednisone (R-CHOP), provided increased efficacy in the treatment of CLL.[18]

In clinical trials, venetoclax (**1**)'s preclinical efficacy transferred well to humans. A phase I study evaluated safety and efficacy of venetoclax (**1**) as a single agent in high risk relapsed/refractory CLL and SLL. In the dose-escalation phase, patients received venetoclax (**1**) ranging from 150 to 1200 mg/day (n = 56). Based on balance of overall response and safety data the expansion cohort was treated at a dose of 400 mg/day (n = 60). Overall, the safety profile was considered manageable.

A subsequent phase II trial enrolled 107 patients with CLL expressing the 17p deletion who had received at least one prior therapy. Venetoclax (**1**)'s efficacy and safety from this trial was sufficient to convince the FDA to approve the drug for marketing in April 2016 even before the commencement of a phase III clinical trial. Interestingly, the efficacy of venetoclax (**1**) in patients who relapsed after or became refractory to ibrutinib (**2**) or idelalisib (**3**) was also confirmed in a specific phase II study that enrolled 54 patients.[19]

Phase III trials of venetoclax (**1**) examined its efficacy and safety as both monotherapy and combination with novel anti-CD20 or chemo-immunotherapies. Due to its peculiarity, venetoclax (**1**) allows for fixed durations of therapy of 12 months in the frontline and 24 months in the relapsed/refractory setting, with a favorable impact on compliance and pharmacoeconomics. Thanks to its substantial clinical activity, some oncologists dubbed venetoclax (**1**) “a magic pill” and “a real game changer” in the treatment of CLL.[20]

Venetoclax (**1**) has a much-improved selectivity profile over its predecessors ABT-737 (**8**) and navitoclax (**9**). Accordingly, the propensity for side-effects, most notably, reduction of platelets leading to thrombocytopenia, is absent. In a trial of 116 patients presenting with either CLL or SLL, the overall response rate among all patients was 79% with a complete response achieved by 20%. All patients had undergone prior therapies and 89% also had adverse prognosis features including, in particular, the 17p deletion. The estimated overall survival rate over 2 years for all patients was 84%.[16]

Although approved initially for CLL, venetoclax (**1**) has shown great potential in the treatment of other cancers, including non-Hodgkin’s lymphoma (NHL), mantle cell lymphoma (MCL), multiple myeloma (MM), and follicular lymphoma. Many of these diseases can be treated using a combination therapy of venetoclax (**1**) and other small molecule inhibitors or monoclonal antibodies. Venetoclax (**1**) has been approved for some of these indications while others are undergoing clinical trials.[15]

Prolonged dosing of venetoclax (**1**) does not generally cause severe thrombocytopenia. The most common treatment-emergent adverse events include tumor lysis syndrome, neutropenia, thrombocytopenia, diarrhea, pyrexia, constipation, anemia, and peripheral edema.[15] The most severe toxic effect of venetoclax (**1**) was tumor lysis syndrome, essentially occurring because the drug was too effective at killing the tumor cells, which led to the tragic death of one patient who had been treated with a high dose of 1200 mg and the halting of the trial.[5]

The second most common adverse effect was neutropenia.

8.6 Synthesis

The medicinal chemistry route to venetoclax (**1**) is a convergent one, relying on the synthesis of three main fragments, which are then coupled to assemble venetoclax (**1**). Both the medicinal chemistry and subsequent manufacturing route modifications are outlined.[21]

The medicinal chemistry route began with the synthesis of diaryl ether **17**. Commercially available 5-bromoindole (**13**) was appropriately protected with a TIPS group to give **14**. Conversion of the bromo substituent to a hydroxyl group was accomplished by using palladium-catalyzed Miyaura boronation followed by oxidation to afford hydroxy-indole **15**. Diaryl ether **17** was then assembled *via* an S_NAr reaction between hydroxy-indole **15** and methyl 2,4-difluorobenzoate (**16**). Needless to say, the S_NAr reaction was *not selective* for the *ortho*-fluorine versus the *para*-position, the reaction gave a mixture of two regio-isomers.

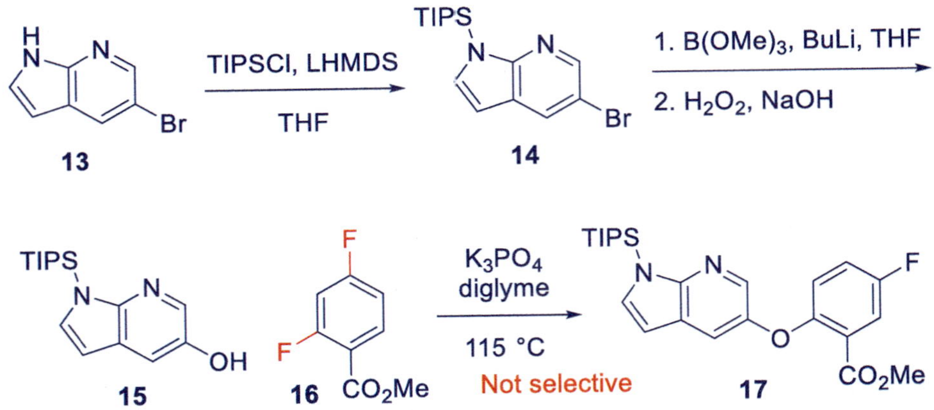

Meanwhile another fragment, sulfonamide **20,** was synthesized *via* a mild S_NAr reaction of 4-fluoro-3-nitrobenzene **18** with amine **19**. With two electron-withdrawing groups, the fluorine electrofuge in **18** was highly activated. For the corresponding 4-chloro-3-nitrobenzene analogue of **18**, the S_NAr reaction needed to be heated to 80 °C, which was the route used for the manufacturing process due to ready availability of the starting material.

With these two moieties in hand, ketoester **21** was converted to the enol triflate **22**. Palladium-mediated Suzuki cross-coupling of enol triflate **22** with *p*-chlorophenyl-boronic acid (**23**) afforded the α,β-unsaturated ester **24**.

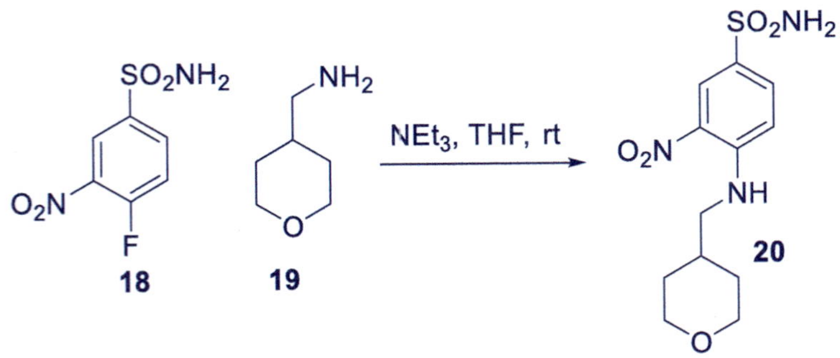

Ester **24** was reduced using $LiBH_4$ to provide allylic alcohol **25**, which was converted to the corresponding mesylate. The resulting mesylate underwent an S_N2 replacement with Boc-protected piperazine **26** to give adduct **27**. Removal of the Boc group under acidic conditions led to "naked" piperazine **28**. Subsequently, an S_NAr reaction between **28** and indole **29**, prepared from indole **17** *via* de-silylation, yielded advanced intermediate as adduct **30**. Hydrolysis of the methyl ester followed by amide coupling with sulfonamide **20** delivered then venetoclax (**1**).

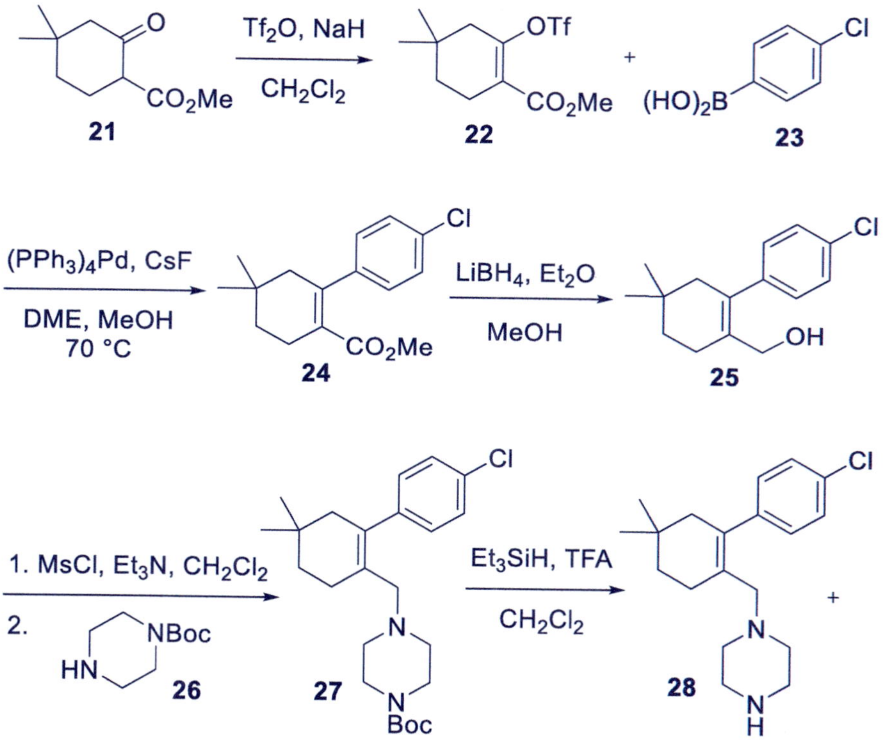

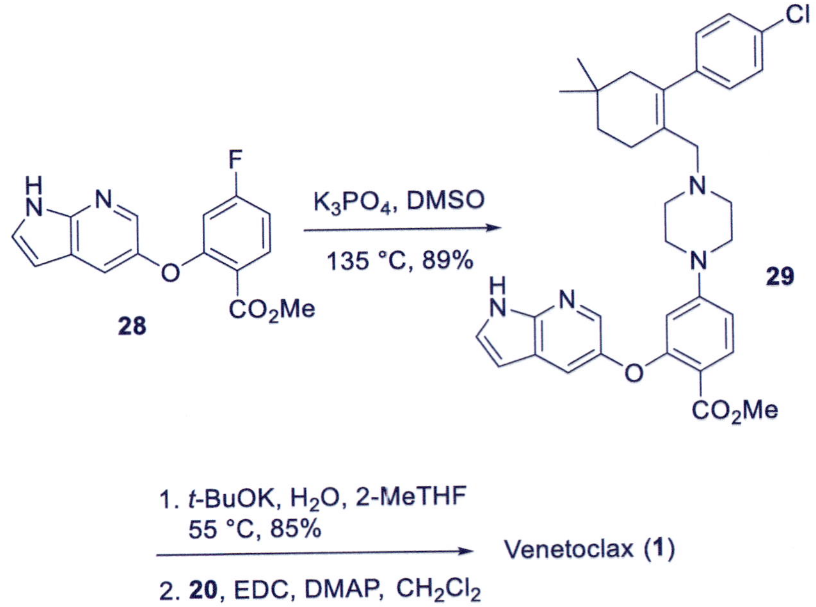

There were several issues with the discovery route that needed addressing for manufacturing purposes. These included the availability of raw materials, the length of synthesis, and the lack of regioselectivity in the SNAr reaction of hydroxy-indole **15** with methyl 2,4-difluorobenzoate (**16**).

To combat the lack of regioselectivity, 1-bromo-3-fluoro-4-iodobenzene (**30**) was instead utilized as the starting material of choice. The iodide was treated with *i*-PrMgCl and the resulting organomagnesium intermediate, from the halogen-metal exchange was treated with Boc anhydride to give ester **31**. An SNAr reaction between **31** and hydroxy indole **32** assembled adduct **33** in 86% overall yield after recrystallization using EtOAc/heptane.

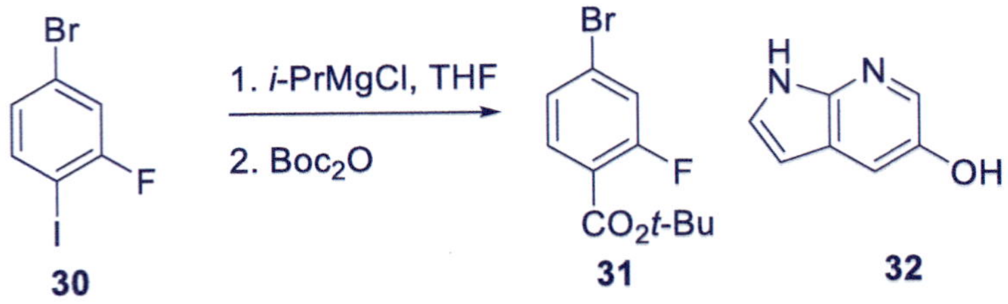

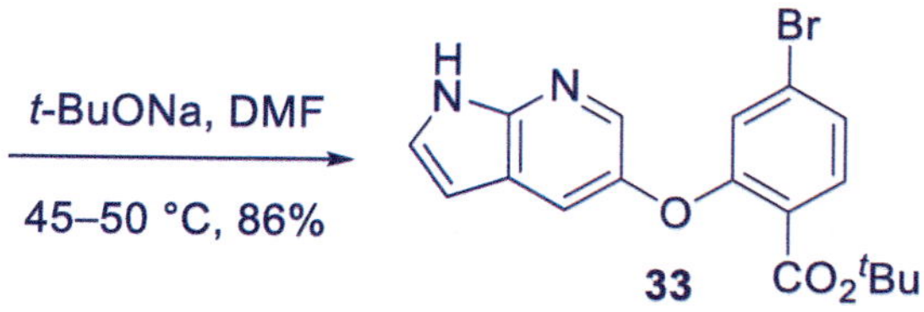

To overcome the cost and availability associated with starting materials to construct sulfonamide **20**, the 4-chloro-3-nitrobenzenesulfonamide was used for the manufacturing processes instead of 4-fluoro-3-nitrobenzenesulfonamide (**18**) used in the medicinal chemistry route.

Intermediate **28** is common to both the medicinal chemistry route and the manufacturing route. The latter began with the formation of vinyl chloride **35** by treating 3,3-dimethylcyclohexane (**34**) with $POCl_3$. Palladium-catalyzed cross-coupling of **35** with *p*-chlorophenylboronic acid (**23**) gave adduct aldehyde **36**, which immediately underwent a reductive amination with Boc-piperazine **26**, yielding adduct **37** in 86% yield following crystallization from PhMe/MeCN. Boc removal using acid produced the key intermediate **28** and subsequent Buchwald–Hartwig reaction with bromo-indole **33** afforded **38** wherein palladium was removed from the reaction by utilizing a cysteine work-up. Intermediate **38** was converted to venetoclax (**1**) using similar chemistry to that outlined above. The overall yield for the seven-step linear sequence was 52%.[21]

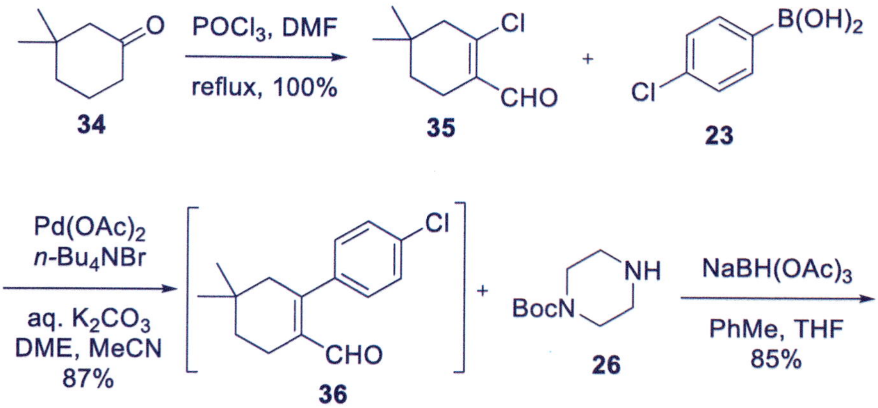

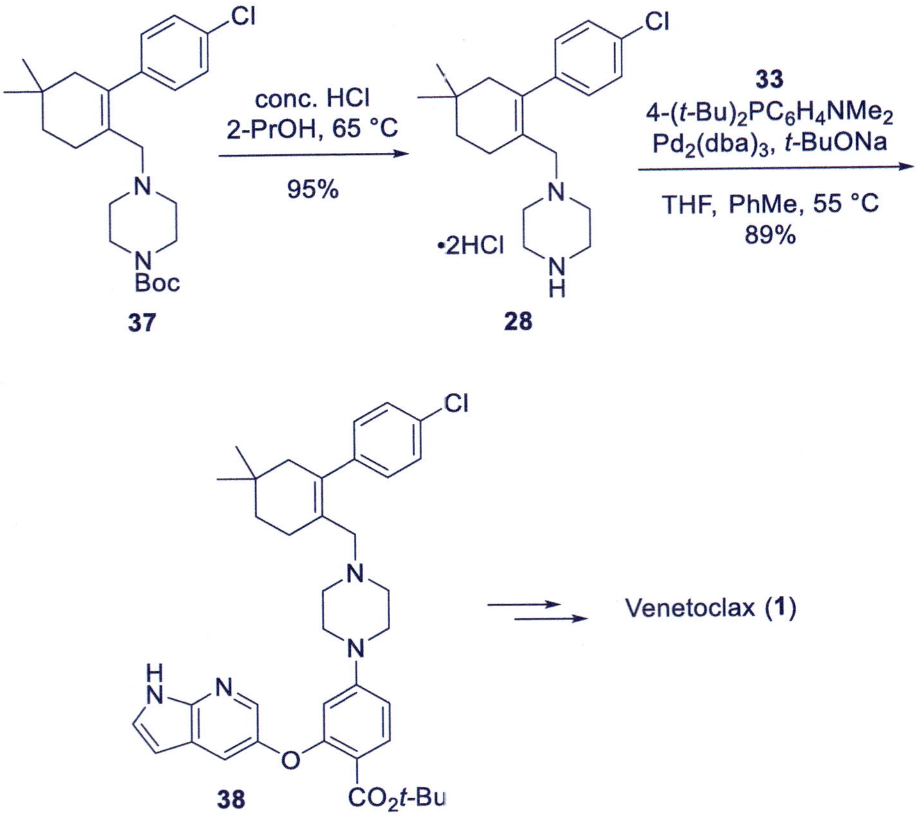

The final form of venetoclax (**1**) is claimed in US patent 8,722,657 as a crystalline anhydrous free base, form A, obtained by drying dichloromethane and ethyl acetate solvates. A further form, Form B, is described as an anhydrous free base polymorph, obtained by drying of the acetonitrile solvate.[22]

8.7. Summary

Since the discovery of apoptosis as a form of programmed cell death, targeting the apoptosis pathway to induce cancer cell death has been a high-priority goal for cancer therapy.

Beginning from the non-selective BH3 mimetics ABT-737 (**8**) and navitoclax (**9**), rational drug design to enhance potency and improve selectivity against other BCL-2 proteins led to the discovery and development of an efficacious molecule, venetoclax (**1**), to treat common leukemias for those that had developed resistance to prior therapies or did not respond at all to available medications.

Venetoclax (**1**) offers a novel strategy to combat lymphoid malignancies *via* therapeutic modulation of the BCL-2 pathway. Patients who experience severe toxicities or relapse on kinase inhibitors experience poor outcomes. With a favorable safety profile and substantial antitumor activity, venetoclax (**1**) provides a therapeutic option for patients who have historically been difficult to treat.

This highlights that the real power of these agents in the clinic is likely to be in combination with other anti-cancer drugs, with targeting of a cancer-specific "Achilles heel" (e.g., using inhibitors of oncogenic kinases) likely the safest as in such drug combinations only the BH3-mimetic would be expected to cause collateral damage to healthy tissues. Highlights also include crystallographic observations of the presence of a BCL-2 dimer which ultimately led to the exploitation of a specific hydrophobic binding pocket and residue differences. This resulted in a selective BCL-2 inhibitor with great efficacy and an improved safety profile.

As the role of BCL-2 in various malignancies becomes clear, venetoclax (**1**) may offer a therapeutic benefit to many other patient populations, including solid tumor patients.

8.8 References

1. Deeks, E. D. Venetoclax: First Global Approval. *Drugs* **2016**, *76*, 979–987.
2. King, A. C.; Peterson, T. J.; Horvat, T. Z.; Rodriguez, M.; Tang, L. A. Venetoclax: A First-in-Class Oral BCL-2 Inhibitor for the Management of Lymphoid Malignancies. *Ann. Pharmacother.* **2017**, *51*, 410–416.
3. Besbes, S.; Mirshahi, M.; Pocard, M.; Billard, C. New dimension in therapeutic targeting of BCL-2 family proteins. *Oncotarget* **2015**, *6*, 12862–12871.
4. Hafezi, S.; Rahmani, M. Targeting BCL-2 in Cancer: Advances, Challenges, and Perspectives. *Cancers* **2021**, *13*, 1292.
5. Schenk, R. L.; Strasser, A.; Dewson, G. BCL-2: Long and winding path from discovery to therapeutic target. *Biochem. Biophys. Res. Comm.* **2017**, *482*, 459–469.
6. Wendt, M. D. Discovery of ABT-263, a Bcl-family protein inhibitor: observations on targeting a large protein–protein interaction. *Expert Opin. Drug Discov.* **2008**, *3*, 1123–1143.
7. Park, C.-M.; Bruncko, M.; Adickes, J.; Bauch, J.; Ding, H.; Kunzer, A.; Marsh, K. C.; Nimmer, P.; Shoemaker, A. R.; Song, X.; et al. Discovery of an Orally Bioavailable Small Molecule Inhibitor of Prosurvival B-Cell Lymphoma 2 Proteins. *J. Med. Chem.* **2008**, *51*, 6902–6915.
8. Souers, A. J.; Leverson, J. D.; Boghaert, E. R.; et al. ABT-199, a potent and selective BCL-2 inhibitor, achieves antitumor activity while sparing platelets. *Nat. Med.* **2013**, *19*, 202–208.

9. Yap, J. L.; Chen, L.; Lanning, M. E.; Fletcher, S. Expanding the Cancer Arsenal with Targeted Therapies: Disarmament of the Antiapoptotic Bcl-2 Proteins by Small Molecules. *J. Med. Chem.* **2017**, *60*, 821–838.
10. Wendt, M. D.; Shen, W.; Kunzer, A.; McClellan, W. J.; Bruncko, M.; Oost, T. K.; Ding, H.; Joseph, M. K.; Zhang, H.; Nimmer, P. M.; et al. Discovery and Structure–Activity Relationship of Antagonists of B-Cell Lymphoma 2 Family Proteins with Chemopotentiation Activity in Vitro and in Vivo. *J. Med. Chem.* **2006**, *49*, 1165–1181.
11. Valenti, D.; Hristeva, S.; Tzalis, D.; Ottmann, C. Clinical candidates modulating protein-protein interactions: The fragment-based experience. *Eur. J. Med. Chem.* **2019**, *167*, 76–95.
12. Fairbrother, W. J.; Leverson, J. D.; Sampath, D.; Souers, A. J. Discovery and development of Venetoclax, a Selective Antagonist of BCL-2. In *Successful Drug Discovery*; Fischer, J.; Klein, C.; Childers, W. E., eds.; Wiley: Weinheim, Germany, 2019, *Vol. 4.* pp 225–245.
13. Lipinski, C. A. Lead- and drug-like compounds: the rule-of-five revolution. *Drug Discov. Today Technol.* **2004**, *1*, 337–341.
14. Package insert, available at: https://dailymed.nlm.nih.gov/dailymed/drugInfo.cfm?setid=b118a40d-6b56-cee3-10f6-ded821a97018
15. Salem, A. H.; Dunbar, M.; Agarwal, S. Pharmacokinetics of venetoclax in patients with 17p deletion chronic lymphocytic leukemia. *Anticancer Drugs* **2017**, *28*, 911–914.
16. Roberts, W. W.; Davids, M. S.; Pagel, J. M.; et al. Targeting BCL2 with venetoclax in relapsed chronic lymphocytic leukemia. *N. Engl. J. Med.* **2016**, *274*, 311–322.
17. Waldron, M.; Winter, A.; Hill, B. T. Pharmacokinetic and pharmacodynamic considerations in the treatment of chronic lymphocytic leukemia: ibrutinib, idelalisib, and venetoclax. *Clin. Pharmacokinet.* **2017**, *56*, 1255–1266.
18. Leverson, J. D., Sampath, D., Souers, A. J. et al. Found in translation: how preclinical research is guiding the clinical development of the BCL2-selective inhibitor venetoclax. *Cancer Discov.* **2017**, *7*, 1376–1393.
19. Vitale, C.; Griggio, V.; Todaro, M.; Salvetti, C.; Boccadoro, M.; Coscia, M. Magic pills: new oral drugs to treat chronic lymphocytic leukemia. *Exp. Opin. Pharmacother.* **2017**, *18*, 411–425.
20. Molica, S. Venetoclax: a real game changer in treatment of chronic lymphocytic leukemia. *Int. J. Hematol. Oncol.* **2020**, *9(4)*, IJH31.
21. Hughes, D. L. Patent Review of Manufacturing Routes to Oncology Drugs: Carfilzomib, Osimertinib, and Venetoclax. *Org. Process. Res. Dev.* **2016**, *20*, 2028–2042.

22. Catron, N.; Chen, S.; Gong, Y.; Zhang, G. G. Salts and crystalline forms of an apoptosis-inducing agent. US patent 8,722,657 B2 (2014).

9

Osimertinib (Tagrisso), A Potent and Selective Third-Generation EGFR Inhibitor for the Treatment of Both Sensitizing and T790M-Resistance Mutations

Fengtao Zhou and Ke Ding

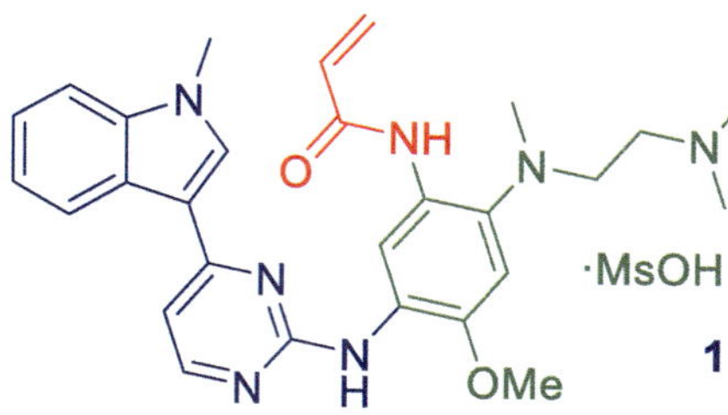

1

USAN: Osimertinib
Trade name: Tagrisso
Third-generation EGFR inhibitor
Launched: 2015

9.1 Background

Lung cancer is one of the most common cancers and is the major cause of cancer related death worldwide. Non-small-cell lung cancer (NSCLC), a heterogeneous class of tumors including squamous cell carcinoma, adenocarcinoma (gland-forming), and large-cell carcinoma histosubtypes, accounts for almost 85% of all new lung cancer cases.[1]

The epidermal growth factor receptor (EGFR) belonging to HER subfamily of receptor tyrosine kinase (RTK), plays an important role in cellular signaling related to tumor cell proliferation, survival, adhesion, migration, and decrease apoptosis. EGFR was found to be overexpressed in 40–85% of patients with NSCLC.[2] Its abnormal activity involves in the development and growth of progression of NSCLC.[3] Inhibition of the kinase catalytic activity of EGFR by small molecule inhibitors is believed to be promising treatment approach for NSCLC. The most frequently activating mutations are L858R mutation (a single point mutation in exon 21) and the exon 19 deletion (delE746-

Current Drug Synthesis, First Edition. Edited by Jie Jack Li.

A750)[4]. Patients with NSCLC suffering from these specific activating mutations display a very positive clinical response to the first-generation EGFR inhibitors such as gefitinib[5] and erlotinib[6] (Figure 1). However, the acquired resistance to the first-generation of EGFR inhibitors (gefitinib, erlotinib) commonly occurs and results in patients would have a recurrence within fewer months. In approximately 50% of these cases, the acquired resistance is caused by a mutation of the gatekeeper residue (T790M), which increased ATP affinity that decreases the inhibitory activity of ATP-competitive kinase inhibitors.[1]

Therefore, the second-generation EGFR inhibitors, including neratinib,[7] dacomitinib,[8] and afatinib[9] (Figure 1), bearing with electrophilic Michael acceptor systems such as acrylamide functional group, were designed to overcome T790M drug resistance by the formation of a covalent bond with the active thiols group (–SH) of Cys797. Despite exciting *in vitro* results, these inhibitors do not display sufficient efficacy in patients with NSCLC in clinical tests. For example, second-generation irreversible inhibitor such as Neratinib (Figure 1) was not efficacious in clinical tests because it is no selectivity against the L858R/T790M mutation over wild-type EGFR.[10]

As a consequence, their clinical efficacy in patients with T790M mutated NSCLC is limited because of the side effects such as diarrhea and skin rash caused by the dose-limiting toxicity related to the inhibition of WT EGFR. Therefore, third-generation EGFR TKIs have been designed to target EGFR T790M mutation to overcome acquired EGFR TKI resistance. Furthermore, these inhibitors display a good selectivity against EGFR mutations over wild-type EGFR and reduce the adverse effects associated with the inhibition of WT EGFR. AstraZeneca's osimertinib mesylate (Tagrisso, **1**) is one of typical third-generation EGFR inhibitor which was approved from the FDA in 2015 for the therapy of (EGFR) T790M mutation non-small cell lung carcinoma (Figure 1). Osimertinib mesylate (**1**) was designed as an irreversible covalent inhibitor to target the T790M mutation with high activity while displays low activity against wild-type EGFR.[3, 11, 12]

First Generation

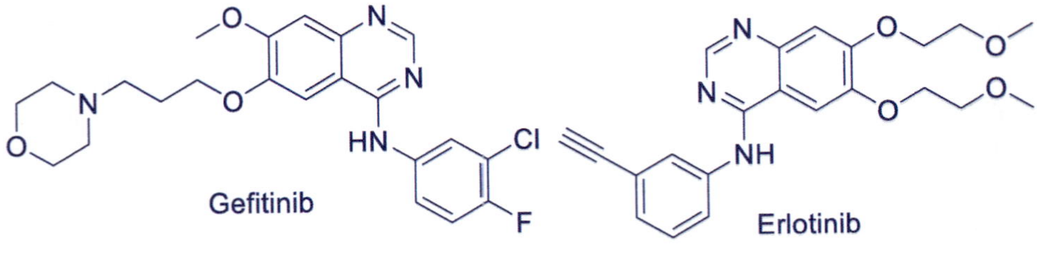

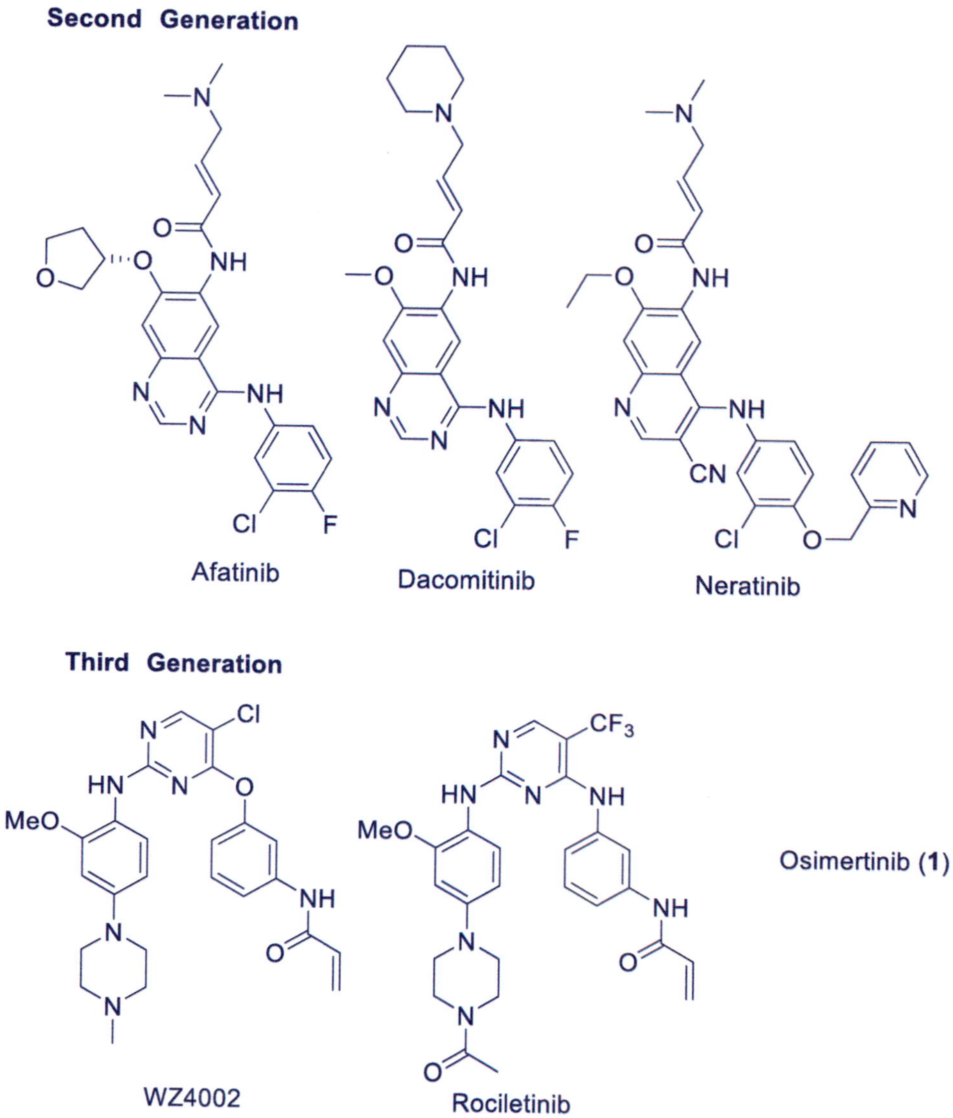

Figure 1. First-, second-, and third-generation EGFR inhibitors.

9.2 Pharmacology

EGFR is a RTK with a molecular mass of 170 kDa, encoded by a gene located on chromosome 7p11.2. It contains an α C-terminal tail harboring an autophosphorylation sites and a cysteine-rich extracellular region, an intracellular kinase domain harboring a tyrosine kinase site.[2] EGFR is vitally important in cell proliferation, survival, differentiation, migration, and adhesion. Under normal condition, binding of ligands (e.g., EGF, TGF-α) triggers the intracellular tyrosine kinase activity of EGFR *via* homo- or heterodimerization with EGFR or other HER family members, leading to recruiting downstream signaling molecules that adjust the functions of EGFR (Figure 2)[3].

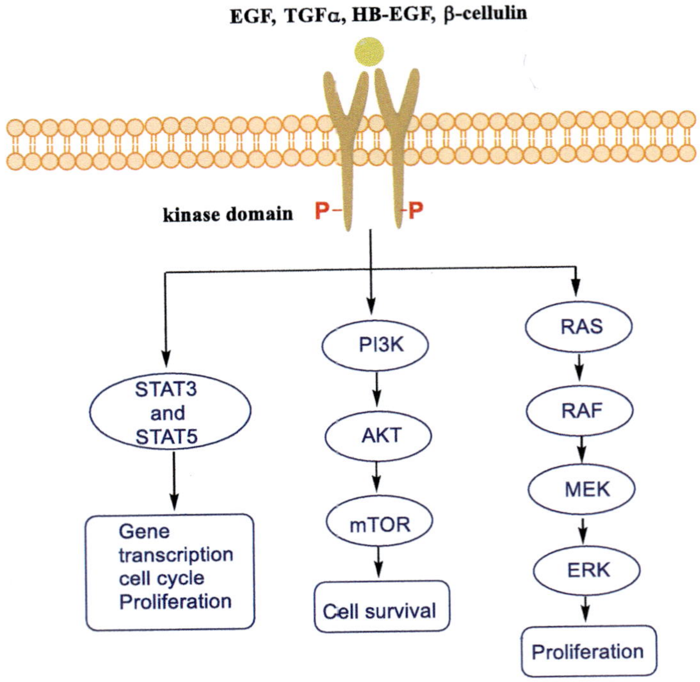

Figure 2. EGFR induces receptor homo- and heterodimerization, which activates downstream pathways.

In fact, deletions in exon 19 (44%) and the point mutation of L858R (41%) are most leading activating mutations observed in patients with NSCLC.[13] Many mutations such as G719X and exon 20 insertion have also been reported. Another major point mutation (T790M) of exon 20 of the EGFR gene has been thought to be acquired resistance during treatment, accounts for approximately 50% of all patients who are treated with EGFR tyrosine kinase inhibitors (TKI) therapy (Figure 3).[14] Patients with the common EGFR mutations such as L858R mutation in exon 21 and delE746-A750 deletions in exon 19 have obtained good benefit from treatment with the first-generation reversible EGFR TKIs, namely, erlotinib or gefitinib. However, the second site mutation usually occurred after 9 to 14 months of the treatment of erlotinib or gefitinib. The most common mutation T790M (~ 50% of cases) was observed, which leads to acquired resistance to the first-generation reversible EGFR TKIs[1]. Therefore, Osimertinib (**1**) is designed to overcome these EGFR mutations including L858R and T790M mutations, which is a third-generation EGFR TKI, with approximately 200-fold selectivity for EGFR with the L858R/T790M mutations than WT EGFR *in vitro*. In fact, osimertinib (**1**) is multi-target kinase inhibitor, HER2, HER3, HER4, ACK1 and BLK activity was also could be inhibited by the use of osimertinib (**1**).[15]

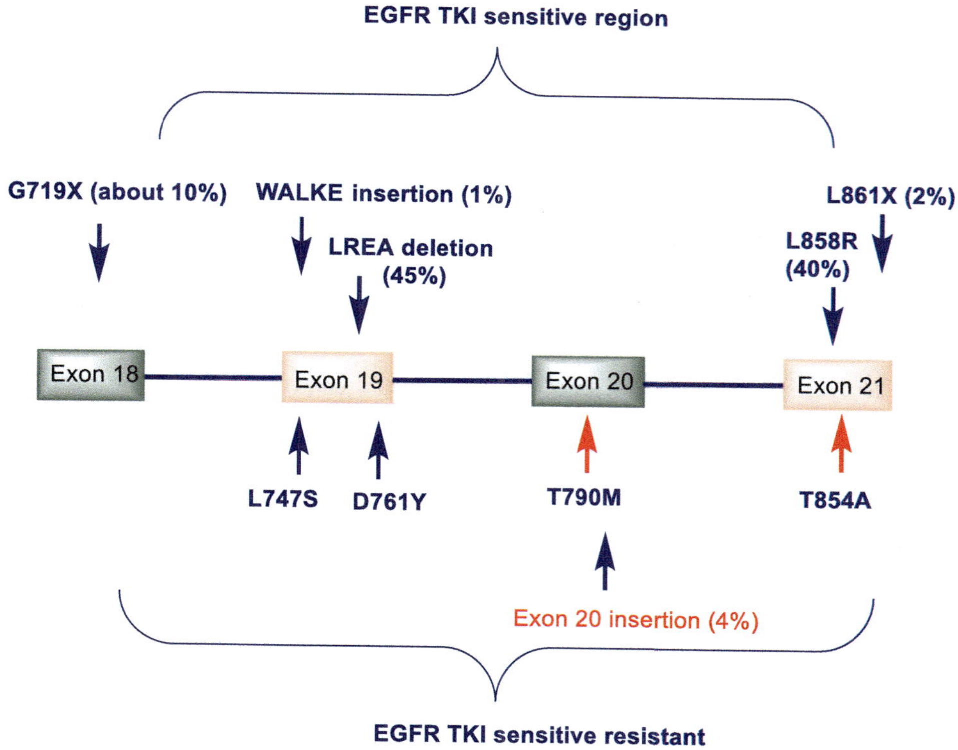

Figure 3. Distribution of EGFR mutations in lung cancer (Based on Reference 3).

3 Structure–Activity Relationship (SAR)

The development of the third generation EGFR inhibitors should meet the clinic demands: (1) could overcome acquired drug resistance resulting from the double mutation L858R/T790M; (2) display good selectivity against EGFR mutations over wild-type enzyme to reduce the side effects. In the early study of AstraZeneca, compound **2** contains an indole ring on anilino-pyrimidine (MAP) scaffold with (IC_{50} = 0.009 μM) against the L858R/T790M double-mutant (DM) enzyme and 88-fold selectivity over wild-type enzyme (Figure 4). Although compound **2** was high activity in this biochemical

DM enzyme assay, its activity (IC_{50} = 0.77 μM) was decreased approximately 90-fold in H1975 DM cellular assay.[16]

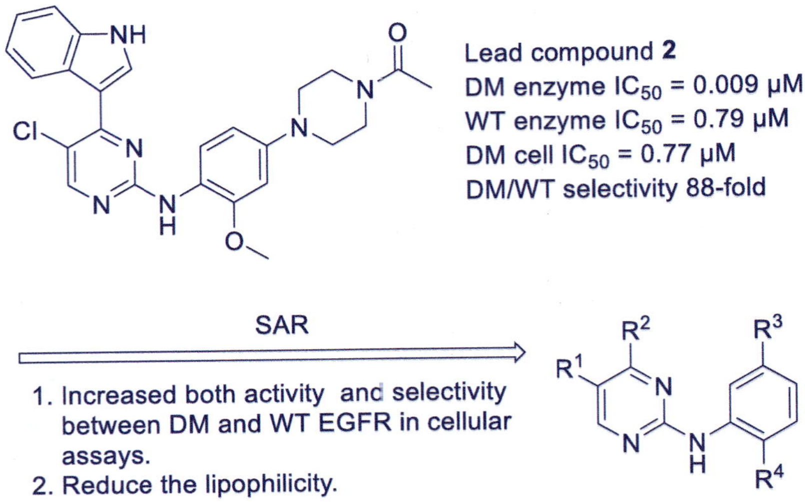

Figure 4. The conception of lead compound optimization.

The introduction of acrylamide side chain at the *meta*-position of aniline converts **2** into an irreversible covalent inhibitor. Compound **3** showed higher selectivity in double mutant (DM) cell assay (IC_{50} = 0.081 μM) than Wild type (WT) cell assay (IC_{50} = 3.5 μM), and it also displayed good activity in PC9 (exon 19-del) AM cell assay (IC_{50} = 0.27 μM) (Table 1). Similarly, compound **4** harboring an ortho-methoxy group at R^4 also exhibits increased activity and kinase selectivity. Protein co-crystal structure of compound **4** bound in wild-type EGFR, implying that a covalent bond was formed between the double bond of acrylamide with Cys-797 (PDB code 4LI5).[16] This binding mode is very useful for the further modification of lead compound to improve both the activity and selectivity. However, the high lipophilicity of these compounds (Log $D_{7.4}$ > 4.3) was the major issue because it will result in both poor pharmacokinetics and physicochemical properties, hampering the progress of drug development. Thus, the next main task is to improve the physical properties of the compounds by modifying indole groups and installing basic functional groups to the acrylamide. Compound **5** containing a pyrazolopyridine ring maintained excellent potency both in DM and AM assays with an increased selectivity over WT EGFR (390-fold). Next, compounds **6** and **7** containing the basic acrylamide-based side chains result in decreasing activity against DM cell over WT cell slightly, however, Log $D_{7.4}$ was decreased significantly. Further investigation was conducted to check whether the 5-position chlorine atom is important or not for both the activity and selectivity. In fact, the substituent at position 5 of this scaffold affected the primary potency and WT selectivity of the compound. The removal of the chlorine atom at the 5-position of the template resulted in decreasing the Log $D_{7.4}$ by an order of magnitude (3.6 compared with 2.6). The biological data of these 5-substituted compounds demonstrate that it is possible to achieve promising activities both in EGFR DM and AM assay with improved the physicochemical properties by the removal of chlorine atom in the 5-position (Table 1).[16]

Table 1. The optimization of lead compound **2**.

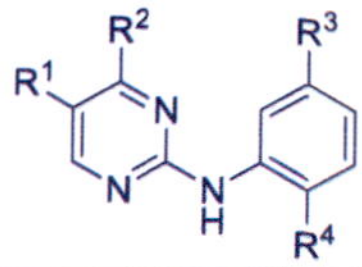

Compound	R^1	R^2	R^3	R^4	DM/WT cell (μM)	DM/WT cell (selectivity)	AM cell assay	Log $D_{7.4}$
3	Cl	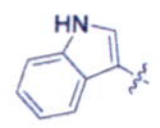	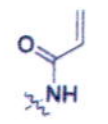	–H	0.081/3.5	43	-	>4.3
4	Cl	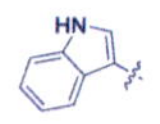	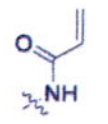	–OMe	0.022/0.55	25	0.029	>4.3
5	Cl	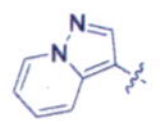	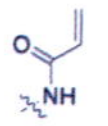	–OMe	0.033/12.8	390	0.15	>4.3
6	Cl	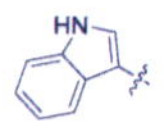	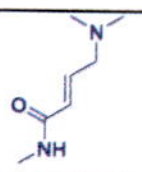	–OMe	0.053/1.55	29	0.085	4.1
7	Cl	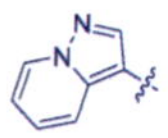	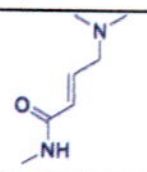	–OMe	0.096/23	240	0.40	3.6
8	H	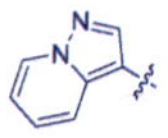	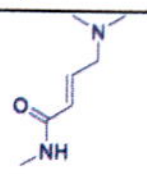	–OMe	0.25/20	80	0.39	2.6
9	F	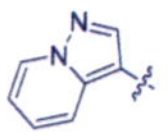	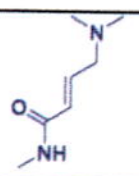	–OMe	0.22/19	85	0.38	3.3
10	Me	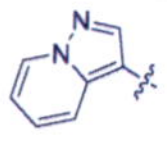	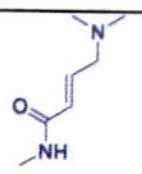	–OMe	0.29/24	82	0.25	3
11	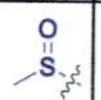	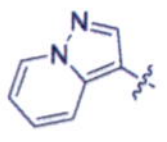	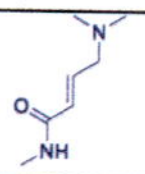	–OMe	15.2/>30	>2	1.3	1.3

Next, the installation of piperazine substituent adjacent to acrylamide could affect the conformation and geometry of the two substituents. The data show that the potency of the compound **12** is greatly improved by the introduction of piperazine substituent. Compound **12** is more potency than compound **7** both in DM and AM assay, while maintaining the good selectivity of DM over WT. When the piperazine group is replaced by other basic side chains, the cellular potency of the compound is enhanced. Those compounds (**16-18**) showed excellent cellular potency (Table 2).[15] The SAR of 5-position of the pyrimidine ring affects the potency of DM cells and selectivity of DM over WT, hERG binding and IGF1R potency. Compared compounds **21** with **23**, obviously, the chlorine substituent on 5-position is very important to maintain DM cellular potency and WT selectivity, but it has adverse effects on the activities of IGF1R and HERG. Comparing compound **1** with **23**, displayed decreased potency and low selectivity of DM over WT, however, both IGF1R and hERG activity were reduced significantly. However, it could be rescued to a certain extent by the introduction of *N*-methylation on the indole group, thus improving the selectivity of DM over WT. Finally, compound **1** was selected as a clinical candidate namely AZD9291, afterwards, named osimertinib(Table 2).[15]

Table. 2. Structures and selected properties of compounds **12–23**

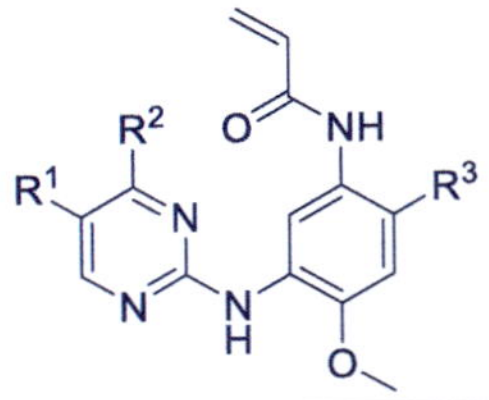

Compound	R^1	R^2	R^3	AM cell (μM)	DM/WT cell (μM)	DM/WT cell (selectivity)	Log D7.4	LLE (DM)	hERG IC_{50} (μM)	IGF1R (enzyme) IC_{50} (μM)
12	Cl	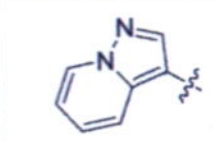	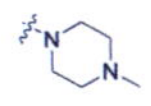	0.14	0.019 /12	620	4.1	4.1	7.1	0.002
13	H	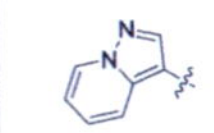	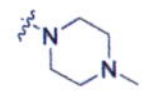	0.14	0.071 / 19	270	NT	-	-	-
14	Cl	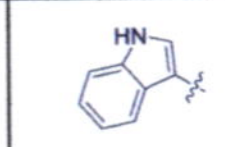	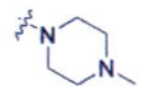	0.023	0.0056/0.77	140	3.7	4.6	-	-
15	H	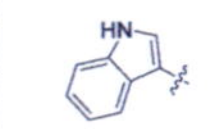	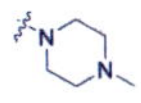	0.017	0.068 /1.3	19	3.4	3.7	9.1	0.813 (*n* = 1)
16	Cl	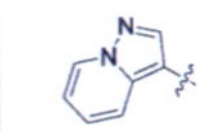	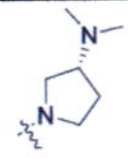	0.016	0.002 /0.357	179	3.1	5.6	5.6	0.007 (*n* = 1)
17	Cl	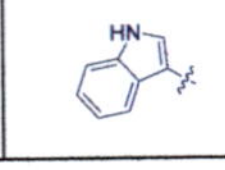	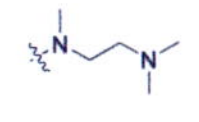	0.002	0.0006/0.145	241	2.8	6.4	6.7	0.006 (*n* = 1)

18	Cl	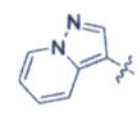	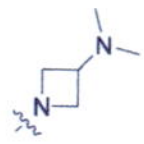	0.021	0.004/0.938	235	3.3	5.1	4.0	0.026 (*n* = 1)
19	CN	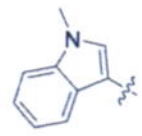	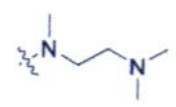	0.001	0.0009/0.046	51	2.7	-	4.3	0.038 (*n* = 1)
20	Cl	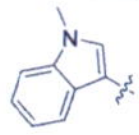	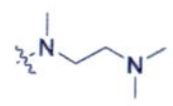	0.002	0.002/0.058	29	3.3	-	7.4	0.04 (*n* = 1)
21	Cl	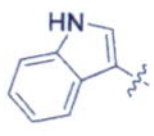	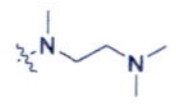	0.0006	0.0002/0.011	55	3.3	-	14.8	0.007 (*n* = 1)
22	Me	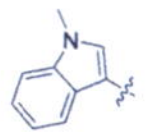	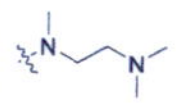	0.002	0.001/0.071	71	2.8	-	15.6	0.196 (*n* = 1)
1	H	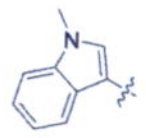	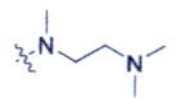	0.017	0.015/0.48	32	3.4	-	16.2	2.9
23	H	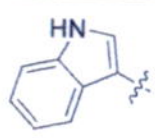	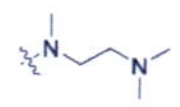	0.002	0.002/0.033	17	2.9	-	17.5	0.263

(Based on references 15 and 16)

9.4 Pharmacokinetics and Drug Metabolism

In a phase I/II AURA trial, in patients with EGFR TKI-resistant NSCLC, the median time to C_{max} was 6 hours (range 3–24 hours) when osimertinib (**1**) was absorbed after single-dose orally capsule administration of osimertinib in the range of 20–240 mg dosages.[17] The pharmacokinetics of osimertinib (**1**) were almost linear during multiple once-daily administration in patients with EGFR TKI-resistant NSCLC, with a steady-state exposures, accumulation obtained approximately 4.5-fold after 22 days of administration of osimertinib (**1**). The C_{max} to C_{min} ratio of osimertinib (**1**) was 1.6-fold with a steady state. The mean volume of distribution (V_{ss}/F) of osimertinib (**1**) was 996 L at steady-state. Osimertinib (**1**) plasma concentrations were reduced with time and an estimated mean half-life of osimertinib (**1**) was 55 hours, and oral clearance (CL/F) was 14.2 (L/h). Osimertinib is predominantly eliminated through the feces (68%), with a small amount of osimertinib eliminated *via* the urine (14%), while only 2% of unchanged osimertinib (**1**) was excreted.[18]

In vitro, the most common metabolic pathways of osimertinib were oxidation and *N*-dealkylation. In fact, there are two pharmacologically-active metabolites, namely, AZ7550 and AZ5104, found in the plasma after oral administration of osimertinib (**1**, Figure 5). AZ7550 exhibited a very similar activity with osimertinib (**1**), while AZ5104 displayed better activity against WT EGFR than osimertinib (approximately 15-fold) and its del19 deletion and T790M mutations (approximately 8-fold). As a result, this increased potency against WT EGFR and low selectivity probably contributed to the adverse effects when the patients were exposed to osimertinib (**1**).[19,20]

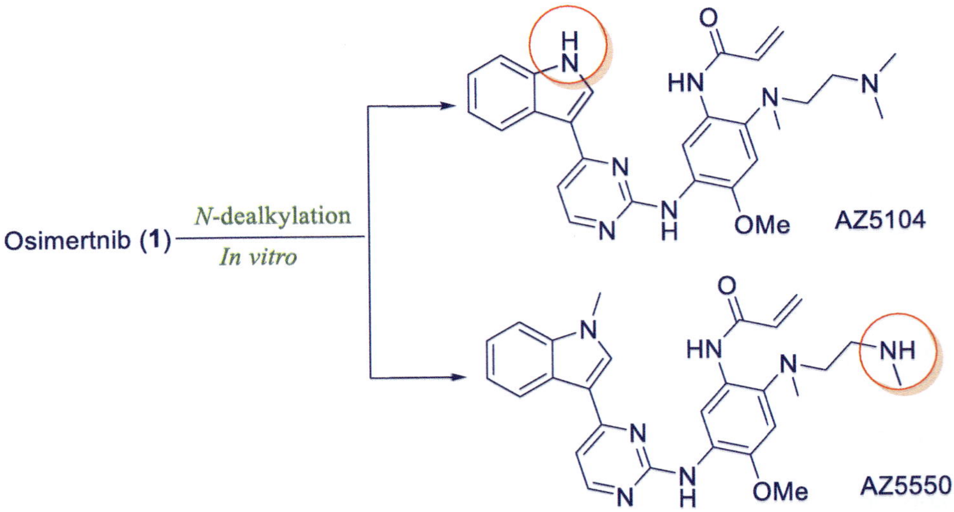

Figure 5. Two pharmacologically-active metabolites AZ7550 and AZ5104.

5 Efficacy and Safety

In murine models of lung cancer harboring EGFR T790M mutation, osimertinib (**1**) was taken in 5 mg/kg/day dosage for 1–2 weeks and tumor volume was remarkable reduced compared with afatinib and vehicle-only control. Notably, osimertinib could inhibit EGFR phosphorylation and downregulate downstream signaling after only a single dose of osimertinib 5 mg/kg once daily.[19]

Osimertinib tablets (available as 40 and 80 mg) can once daily be taken orally without regard to food. The administration of osimertinib is 80 mg once daily recommended by FDA. If patients have difficulty in swallowing solids, osimertinib was also could be taken orally after it is dispersed in 50 mL of non-carbonated water. [17]

Most of the NSCLC patients harboring advanced or metastatic EGFR TKI-resistant exhibit a positive clinical response in the randomized, multinational, open-label phase I/II AURA trial (NCT01802632). All NSCLC patients harboring EGFR TKI-resistant in the phase I part of AURA (n = 239 evaluable), showed an objective response rate (ORR) of 51% and a disease control rate (CR, PR or SD) of 84% after osimertinib administration. [17] In the all population, the ORR had any obvious difference at each of the dosages of osimertinib. Patients harboring EGFR T790M mutation (n = 127 evaluable) showed an good ORR (61%) and an excellent disease control rate (95%) after osimertinib administration, however, patients with no EGFR T790M mutation (n = 61 evaluable) display a very lower ORR (21%) and a disease control rate (61%).[17] It is indicated that osimertinib exhibits better therapeutic efficacy in patient with EGFR T790M mutation than patients without EGFR T790M mutation.

Furthermore, osimertinib (**1**) can cross the blood–brain barrier (BBB) and is distributed to the brain in many animal species such as monkey, rat, and mouse, while the AUC ratios of between brain and plasma is approximately two after oral dosing of osimertinib. These data suggest the osimertinib (**1**) as a promising drug for the treatment and prevention of CNS metastases of EGFR mutation-positive NSCLC.[19] In the phase I analysis of AURA, the most frequently adverse events of any grade were observed such as diarrhea (47%), rashes and acne (group term; 40%), nausea (22%), decreased appetite (21%), and dry skin (20%) when the patients were treated with osimertinib (**1**).[17]

Taken together, osimertinib (**1**) displays good therapeutic efficacy in patient harboring EGFR T790M mutation with an acceptable tolerability profile. Furthermore, There were no dose-limiting toxicities was observed in 20–240 mg/day dosages of osimertinib in NSCLC patients harboring EGFR TKI-resistant in the dose-escalation cohort of AURA.[17]

6 Synthesis

Based on the retro-synthetic analysis of osimertinib (**1**) is shown earlier (Scheme 1). Osimertinib (**1**) is assembled from aromatic ring A, ring B, and ring C. In fact, ring A and ring B is connected by Friedel–Crafts arylation. Fragment **33** is connected from three fragments **27**, **29**, and **34** by an amidation and an S_NAr reaction. On other hand, osimertinib (**1**) is also assembled from fragments **26** and **33** *via* an S_NAr reaction. The fragments **26** is connected by a Friedel–Crafts arylation of **24** with **25**. The fragments **33** is assembled by an amidation **35** with **34**. The fragment **35** is assembled from **27** and **29** *via* a S_NAr reaction.

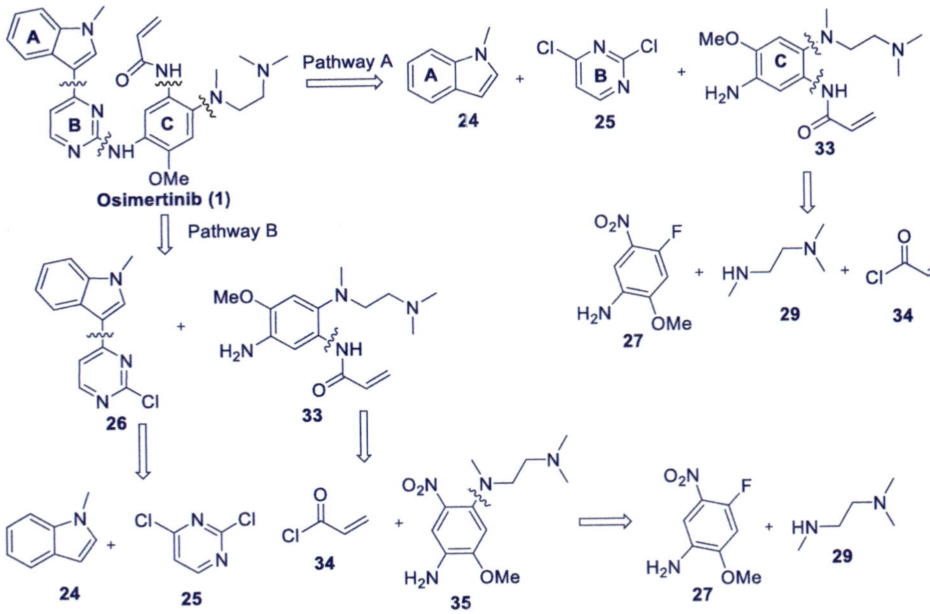

Scheme 1. The retro-synthetic route of osimertinib (**1**).

AstraZeneca disclosed the six-step (longest linear sequence) synthetic route of osimertinib (**1**), using *N*-methylindole (**24**), 2,4-dichloropyrimidine (**25**), and 2-methoxy-4-fluoro-5-nitroaniline (**27**) as starting materials (Scheme 2).[21,22] Intermediate **26** was obtained in one step from *N*-methylindole (**24**) with 2-dichloropyrimidine (**25**) in the presence of $FeCl_3$ as Lewis acid catalyst in MDE at 60 °C in yield of 82% by crystallizing from the mixture solvents MeOH/water). S_NAr reaction of intermediate (**26**) with 4-fluoro-2-methoxy-5-nitroaniline (**27**) in presence of *p*-toluenesulfonic acid at 105 °C in 2-pentanol to give intermediate **28** in 95% yields. Subsequent the conversion of intermediate **28** to **30** by employing of DIPEA in DMA at 85 °C in yield of 96%. Reduction of nitro group with iron under acidic conditions in ethanol to deliver smoothly the intermediate **31** in 85% yield. Finally, amidation **32** with 3-chloropropanoyl chloride (**32**) using K_2CO_3 as base in acetone in yield of 95%, and is followed by elimination with trimethylamine as base in MeCN in yield of 94% to deliver the osimertinib (**1**). In summary, it takes a six-step process that starts from readily available starting materials delivering the product in an overall yield of 57% (Scheme 2).[16]

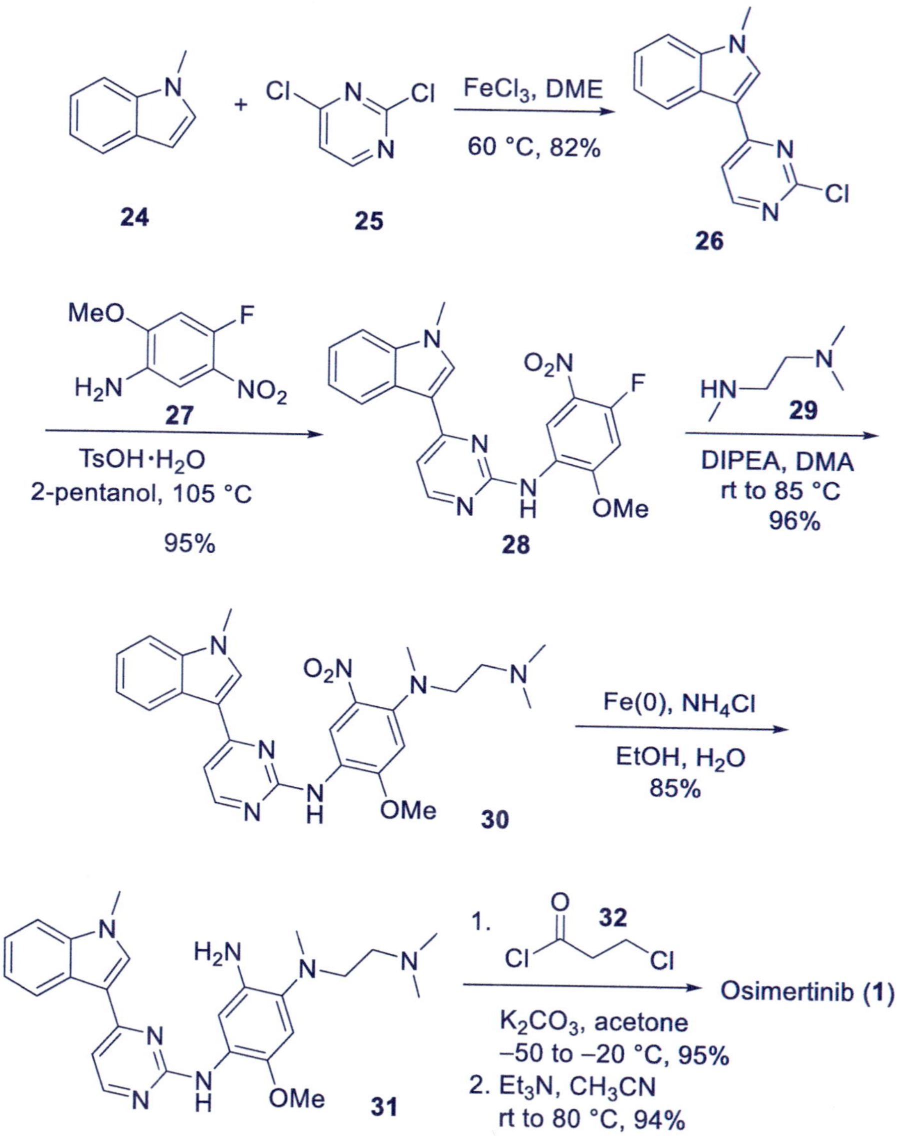

Scheme 2. AstraZeneca's synthesis route of osimertinib (**1**)

Recently, Qiu et al.[23] modified AstraZeneca's approach for the synthesis of osimertinib (**1**). Under the catalysis of $CoCl_2$, intermediate **26** was obtained in 70% yield by Friedel–Crafts arylation of 2,4-dichloropyrimidine (**25**) with *N*-methylindole (**24**). Subsequent S_NAr reaction of intermediate **26** with 4-fluoro-2-methoxy-5-nitroaniline (**27**) in presence of *p*-toluenesulfonic acid to deliver intermediate **28** in 90% yield. Subsequent

S_NAr reaction of **28** with **29** afford intermediate **30** in 91% yield using *N*-ethyl-*N*-isopropylpropan-2-amine (DIPEA) as the base. Intermediate **31** was obtained in 92% yield by reducing nitro group with hydrazine hydrate as reducing agents under catalysis of $CoSO_4 \cdot 7H_2O$. Finally, intermediate **31** reacted with acryloyl chloride (**34**) in presence of sodium carbonate as base to deliver osimertinib (**1**). In summary, this synthetic route is concise and efficient, which starts from readily available starting materials *N*-methylindole (**24**), 2,4-dichloropyrimidine (**25**), and 2-methoxy-4-fluoro-5-nitroaniline (**27**), delivering the final product in nearly 42.5% overall yield. Furthermore, the work-up operation is easily carried out. Simple filtration and recrystallization are enough to purify the crude products instead of column purification in most case (Scheme 3).

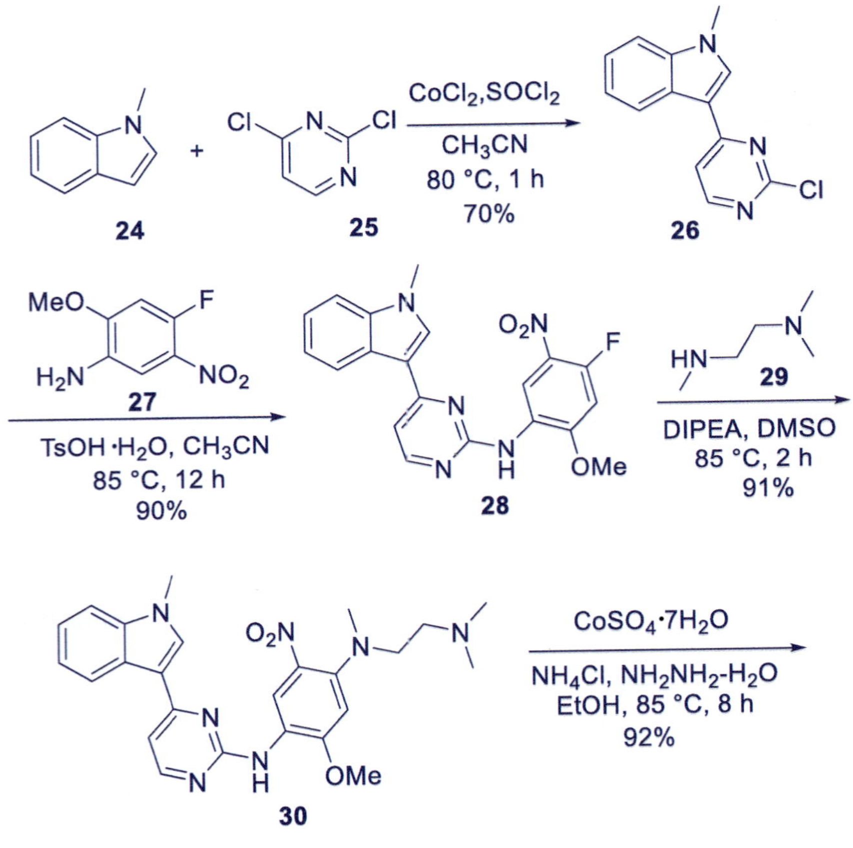

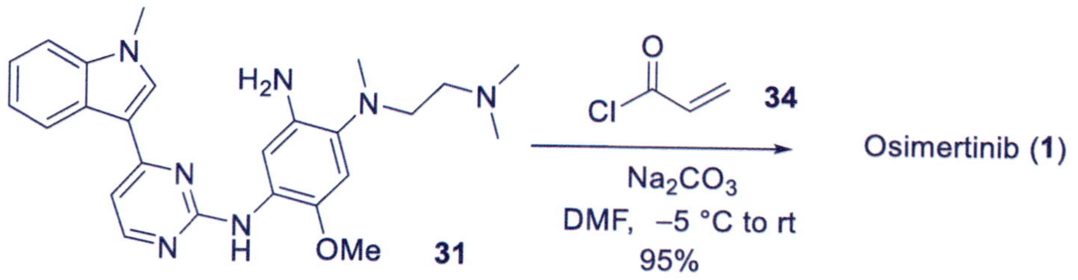

Scheme 3. Synthetic route to osimertinib (**1**).

Another Chinese patent[24] describes a convergent synthesis of route to osimertinib (Scheme 4). Protection of **27** with a Boc group to give **36** in 84% yield, and following S_NAr reaction of **36** with **29** to afford **37** in yield of 98%. After the reduction of nitro group of **37** and following the amidation with acryloyl chloride (**34**) to deliver fragment **33**. Finally, osimertinib (**1**) was obtained from **26** and **33** *via* an S_NAr reaction under mild condition (55 °C) in yield of 93%, which is probably attributes to the higher nucleophilicity of the amine **33** compared with **27**. In summary, this route takes six steps with an overall yield of 65% in more convergent manner. This route has potential application in manufacturing in the future.

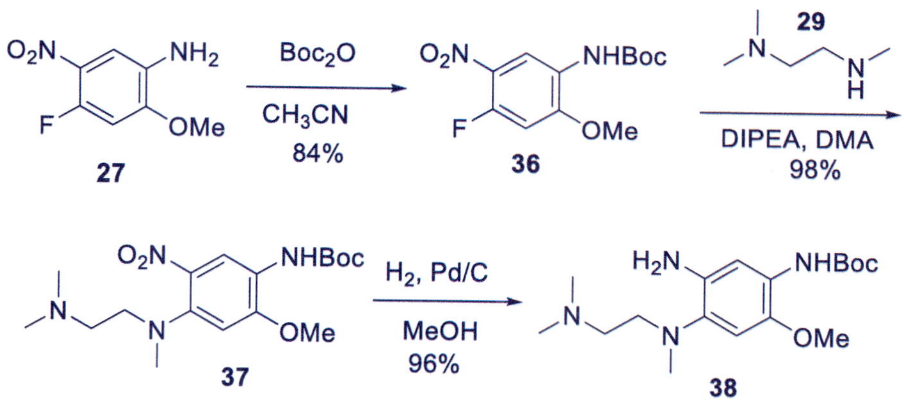

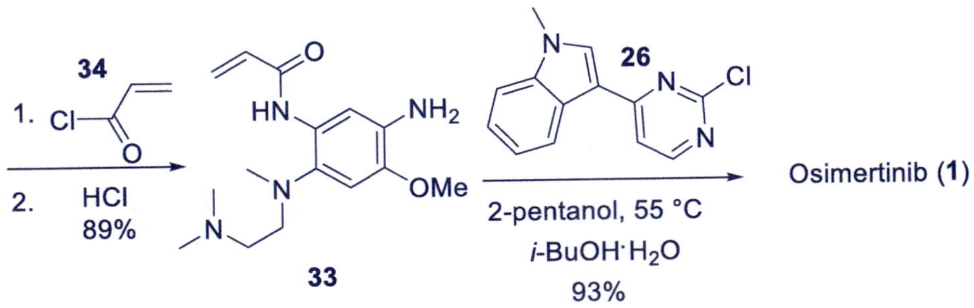

Scheme 4. Alternate route synthesis route of osimertinib (**1**)

9.7. Summary

Osimertinib mesylate (Tagrisso, **1**) is a third-generation EGFR inhibitor that was approved from the FDA in 2015 for the treatment of metastatic NSCLC harboring EGFR T790M mutation. Osimertinib (**1**) displays high cellular potency, good efficacy, acceptable adverse effects, and good selectivity over the wild type EGFR. The story of osimertinib (**1**) is especially fascinating and has been one of the fastest developments cases in the history of drug development. It just takes only 2 years and 8 months from the first patient to the approved by the US FDA. Due to the high efficacy in NSCLC harboring EGFR T790M mutation with acceptable side effects, osimertinib achieved a great success in the treatment in NSCLC harboring EGFR T790M mutation. However, the occurrence of acquired resistance is a major obstacle to achieving long-term remission in lung cancer therapy. Clinic data suggests a subset of patients with NSCLC, become resistant to osimertinib *via* a C797S mutation.[25] The EGFR C797S mutation cannot form a covalent bond with third-generation EGFR inhibitor in a Michael addition manner under physiological conditions, which has been believed to be a key mechanism of resistance. Thus, it is of great urgency that develop a next-generation of EGFR inhibitors to overcome C797S mutation and other mutations in the ATP site of the kinase domain of EGFR. Additionally, identifying of new allosteric inhibitors or development of EGFR protein degraders are alternative approaches to overcome resistance to EGFR TKIs. In a word, there is still a long way to achieve optimal therapies for the treatment of NSCLC harboring EGFR mutations.

9.8. References

1. Gridelli, C.; Rossi, A.; Carbone, D. P.; Guarize, J.; Karachaliou, N.; Mok, T.; Petrella, F.; Spaggiari, L.; Rosell, R. Non-small-cell lung cancer. *Nat. Rev. Dis. Primers* **2015**, *1*, 15009.

2. Singh, M.; Jadhav, H. R. Targeting non-small cell lung cancer with small-molecule EGFR tyrosine kinase inhibitors. *Drug Discov. Today* **2018**, *23*, 745–753.
3. Ohashi, K.; Maruvka, Y. E.; Michor, F.; Pao, W. Epidermal growth factor receptor tyrosine kinase inhibitor-resistant disease. *J. Clin. Oncol.* **2013**, *31*, 1070–1080.
4. Engel. J.; Richters, A.; Getlik, M.; Tomassi, S.; Keul, M.; Termathe, M.; Lategahn, J.; Becker, C.; Mayer-Wrangowski, S.; Grütter, C.; Uhlenbrock, N.; Krüll, J.; Schaumann, N.; Eppmann, S.; Kibies, P.; Hoffgaard, F.; Heil, J.; Menninger, S.; Ortiz-Cuaran, S.; Heuckmann, J.M.; Tinnefeld, V.; Zahedi, R.P.; Sos, M.L.; Schultz-Fademrecht, C.; Thomas, R.K.; Kast, S.M.; Rauh, D. Targeting drug resistance in EGFR with covalent inhibitors: a structure-based design approach. *J Med. Chem.* **2015**, *58*, 6844–6863.
5. Barker, A. J.; Gibson, K. H.; Grundy, W.; Godfrey, A. A.; Barlow, J. J.; Healy, M. P.; Woodburn, J. R.; Ashton, S. E.; Curry, B. J.; Scarlett, L.; Henthorn, L.; Richards, L. Studies leading to the identification of ZD1839 (iressa™): an orally active, selective epidermal growth factor receptor tyrosine kinase inhibitor targeted to the treatment of cancer. *Bioorg. Med. Chem. Lett.* **2001**, *11*, 1911–1914.
6. Pao, W.; Miller, V.; Zakowski, M.; Doherty, J.; Politi, K.; Sarkaria, I.; Singh, B.; Heelan, R.; Rusch, V.; Fulton, L.; Mardis, E.; Kupfer, D.; Wilson, R.; Kris, M.; Varmus, H. EGF receptor gene mutations are common in lung cancers from "never smokers" and are associated with sensitivity of tumors to gefitinib and erlotinib. *Proc. Natl. Acad. Sci. USA.* **2004**, *101*, 13306–13311.
7. Singh, H.; Walker, A. J.; Amiri-Kordestani, L.; Cheng, J.; Tang, S.; Balcazar, P.; Barnett-Ringgold, K.; Palmby, T. R.; Cao, X.; Zheng, N.; Liu, Q.; Yu, J.; Pierce, W. F.; Daniels, S. R.; Sridhara, R.; Ibrahim, A.; Kluetz, P. G.; Blumenthal, G. M.; Beaver, J. A.; Pazdur, R. U.S. food and drug administration approval: neratinib for the extended adjuvant treatment of early-stage HER2-positive breast cancer. *Clin. Cancer Res.* **2018**, *24*, 3486–3491.
8. Shirley, M. Dacomitinib: first global approval. *Drugs* **2018**, *78*, 1947–1953.
9. Ou, S.-H. I. Second-generation irreversible epidermal growth factor receptor (EGFR) tyrosine kinase inhibitors (TKIs): a better mousetrap? a review of the clinical evidence. *Crit. Rev. Oncol. Hemat.* **2012**, *83*, 407–421.
10. Yan, X.-E.; Ayaz, P.; Zhu, S.-J.; Zhao, P.; Liang, L.; Zhang, C. H.; Wu, Y.-C.; Li, J.-L.; Choi, H. G.; Huang, X.; Shan, Y.; Shaw, D. E.; Yun, C.-H. Structural basis of AZD9291 selectivity for EGFR T790M. *J. Med. Chem.* **2020**, *63*, 8502–8511.
11. Hsiao, S.-H.; Lu, Y.-J.; Li, Y.-Q.; Huang, Y.-H.; Hsieh, C.-H.; Wu, C.-P. Osimertinib (AZD9291) attenuates the function of multidrug resistance-linked ATP-binding cassette transporter ABCB1 *in vitro*. *Mol. Pharmaceut.* **2016**, *13*, 2117–2125.
12. Chen, L.; Fu, W.; Zheng, L.; Liu, Z.; Liang, G. Recent progress of small-molecule epidermal growth factor receptor (EGFR) inhibitors against C797S resistance in non-small-cell lung cancer. *J. Med. Chem.* **2018**, *61*, 4290–4300.

13. Bronte, G.; Rizzo, S.; La Paglia, L.; Adamo, V.; Siragusa, S.; Ficorella, C.; Santini, D.; Bazan, V.; Colucci, G.; Gebbia, N.; Russo, A. Driver mutations and differential sensitivity to targeted therapies: a new approach to the treatment of lung adenocarcinoma. *Cancer Treat. Rev.* **2010**, 36, 21–29.
14. Sullivan, I., Planchard, D. Next-generation EGFR tyrosine kinase inhibitors for treating EGFR-mutant lung cancer beyond first line. *Front. Med.* **2017**, *18*, 3–76.
15. Finlay, M. R. V.; Anderton, M.; Ashton, S.; et al. Discovery of a potent and selective EGFR inhibitor (AZD9291) of both sensitizing and T790M resistance mutations that spares the wild type form of the receptor. *J. Med. Chem.* **2014**, *57*, 8249–8267.
16. Ward, R. A.; Anderton, M. J.; Ashton, S.; et al. Structure- and reactivity-based development of covalent inhibitors of the activating and gatekeeper mutant forms of the epidermal growth factor receptor (EGFR). *J. Med. Chem.* **2013**, *56*, 7025–7048.
17. Greig, S. L. Osimertinib: first global approval. *Drugs* **2016**, *76*, 263–273
18. AstraZeneca Pharmaceuticals LP. TagrissoTM (osimertinib) Tablet, for Oral Use: US Prescribing Information. **2015**. http://www.fda.gov. Accessed 10 Dec 2015.
19. Cross, D. A. E.; Ashton, S. E.; Ghiorghiu, S.; et al. AZD9291, an irreversible EGFR TKI, overcomes T790M-mediated resistance to EGFR inhibitors in lung cancer. *Cancer Discov.* **2014**, *4*, 1046–1061.
20. Dickinson, P. A.; Cantarini, M. V.; Collier, J.; Frewer, P.; Martin, S.; Pickup, K.; Ballard, P. Metabolic disposition of osimertinib in rats, dogs, and humans: insights into a drug designed to bind covalently to a cysteine residue of epidermal growth factor receptor. *Drug Metab. Dispos.* **2016**, *44*, 1201–1212.
21. Hughes, D. L. Patent review of manufacturing routes to oncology drugs: carfilzomib, osimertinib, and venetoclax. *Org. Process. Res. Dev.* **2016**, *20*, 2028–2042.
22. Butterworth, S.; Finlay, M. R. V.; Ward, R. A.; Kadambar, V. K.; Chandrashekar, R. C.; Murugan, A.; Redfearn, H. M. 2-(2-,4-,5-Substituted-anilino) pyrimidine derivatives as EGFR modulators useful for treating cancer. WO2013014448A1, **2013**.
23. Qiu, R.; Li, D.; Fu, L.; Zhang, D.; Kanbe, N. Process for preparation of osimertinib AZD9291, CN109134435A (2019).
24. Ji, M.; Li, Y.; Liu, H.; Li, R.; Cai, J.; Hu, H. Synthesis of antitumor drugs. CN104817541A (2015).
25. Thress, K. S.; Paweletz, C. P.; Felip, E.; et al. Acquired EGFR C797S mutation mediates resistance to AZD9291 in non-small cell lung cancer harboring EGFR T790M. *Nat. Med.* **2015**, *21*, 560–562.

10

Sotorasib (LUMAKRAS), An Irreversible Covalent Inhibitor of KRAS^{G12C}

Brian A. Lanman and Andrew T. Parsons

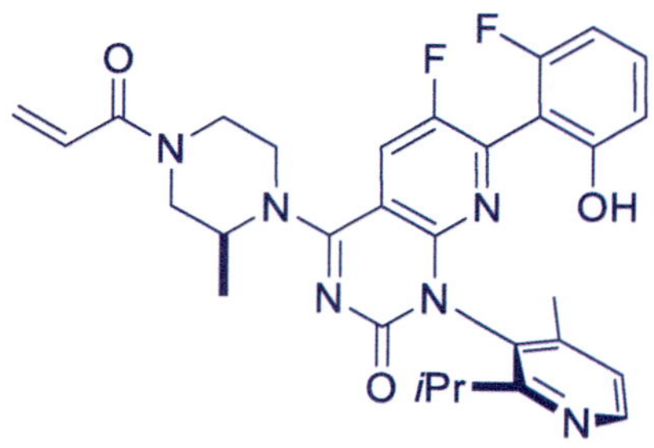

USAN: Sotorasib
Trade name:
LUMAKRAS (US)/LUMYKRAS (UK)
Amgen
Launched: 2021

10.1 Background

In 1967, Kirsten and Mayer isolated a murine retrovirus (Ki-MuSV) that caused the rapid formation of sarcomas in infected rodents and oncogenically transformed cells in tissue culture.[1] Through work in the 1970s and early 1980s, it was determined that this virus had co-opted a normal mammalian gene[2] for a 21 kDa phosphoprotein,[3] and that transfer of a mutant form of this protein to infected cells was responsible for the virus's ability to render infected cells cancerous.[4] The non-mutated cellular homolog of this transforming gene subsequently came to be known as *KRAS* in recognition of its initial identification as the transforming factor of the Kirsten rat sarcoma virus.[5–7] Single point mutations in the *KRAS* gene were determined to be sufficient for its oncogenic transformation.[8]

Subsequent investigations revealed mutations of *KRAS* and its cellular homologs, *HRAS* and *NRAS*, to be common driver mutations in ~30% of all human cancers.[9] *KRAS* mutations are the most prevalent *RAS* mutations, being found in 86% of all *RAS*-driven tumors.[10] Activating mutations of *KRAS* most frequently occur at codon

Current Drug Synthesis, First Edition. Edited by Jie Jack Li.

12, with the *KRAS p.G12C* mutation—in which glycine 12 of the KRAS protein is replaced with a cysteine residue—representing the most common *KRAS* mutation in lung adenocarcinoma (LAC) (42% of *KRAS* mutations, 13% of all LAC patients) and a frequent mutation in colorectal cancer (CRC) (9% of *KRAS* mutations, 4% of all CRC patients).[10] Given the prevalence of the *KRAS p.G12C* mutation in non-small cell lung cancer (NSCLC; a class of lung cancers that includes LAC) and CRC and the poor prognosis of patients harboring these activating *KRAS* mutations,[11] the development of inhibitors of these oncogenic *KRAS* mutations was seen to be of considerable clinical value.

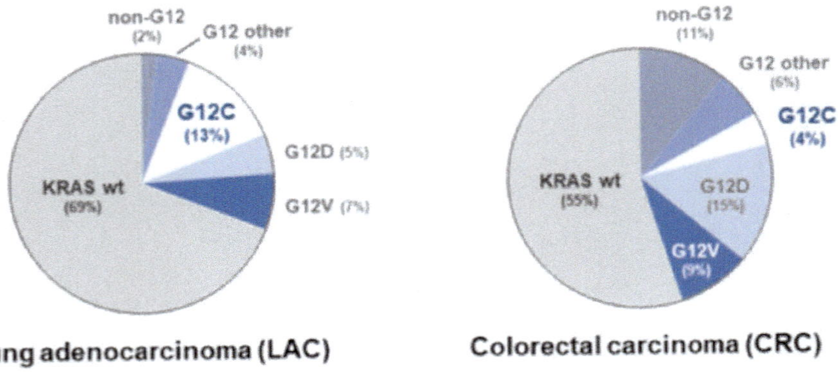

10.2 Pharmacology

Work through the 1980s and 1990s did much to define the signaling pathways whereby *KRAS* regulates cellular growth and proliferation.[9] This work established the KRAS protein to be part of a larger growth-regulatory pathway known as the mitogen-activated protein kinase (MAPK) pathway, which couples extracellular growth signaling to intracellular pro-survival and pro-proliferative signaling. KRAS serves as a "molecular switch" in regulating this pathway, switching between a guanosine diphosphate (GDP)-bound, inactive form and a guanosine triphosphate (GTP)-bound, active form under the action of guanine nucleotide exchange factors (GEFs), which are themselves regulated by receptor tyrosine kinase (RTK)-mediated recruitment to the cell membrane in response to extracellular growth factor signaling.

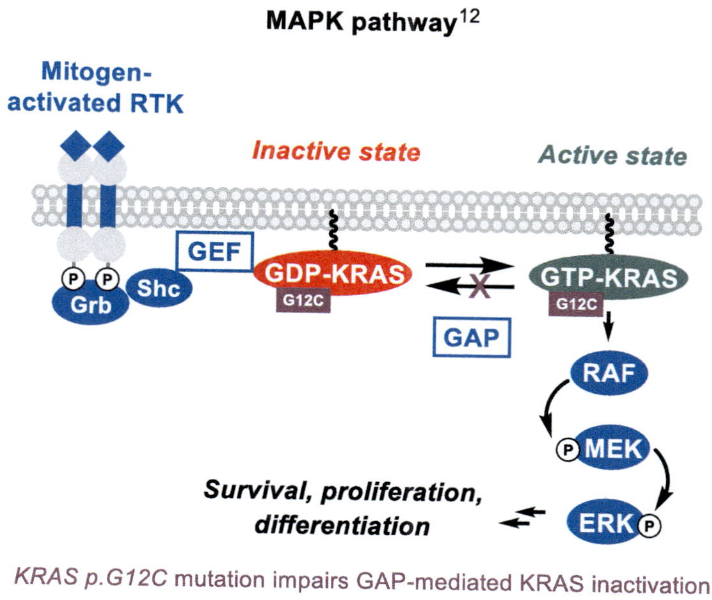

GTP binding to KRAS induces a conformational change in the protein, enabling it to interact with downstream effector proteins such as RAF, thereby initiating a downstream phosphorylation cascade that leads to cellular transcriptional changes that promote cellular survival, proliferation, and differentiation. This pro-proliferative signaling is normally regulated through the action of GTPase-activating proteins (GAPs), which catalyze the hydrolysis of KRAS-bound GTP to GDP, returning KRAS to its inactive state and terminating the pro-proliferative signaling. Mutation of glycine 12 of KRAS to cysteine (resulting in the oncogenic $KRAS^{G12C}$ protein) impairs the ability of GAPs to catalyze this GTP-hydrolysis event, which leads to the accumulation of the GTP-bound form of KRAS and to dysregulated pro-proliferative signaling.

Although work through the 1990s and early 2000s identified MAPK pathway inhibitors both upstream[13] and downstream[14] of KRAS, direct inhibitors of KRAS remained elusive, with RAS proteins gaining a reputation of being "undruggable."[10] The source of this challenge in developing direct inhibitors of the GTP-bound (active) form of KRAS could be attributed to the structure of the RAS proteins, in general. As shown in the following text, the GTP-bound form of oncogenic RAS (e.g., KRAS or HRAS) possesses only one druggable (i.e., deep and well-enclosed) surface pocket for inhibitor binding, the nucleotide binding site (highlighted in blue). This pocket, however, possesses sub-nanomolar affinity for both GDP and GTP, both of which are present in near-millimolar concentrations in cells. These factors combine to make reversible, nucleotide-competitive inhibition of KRAS a nearly intractable challenge.

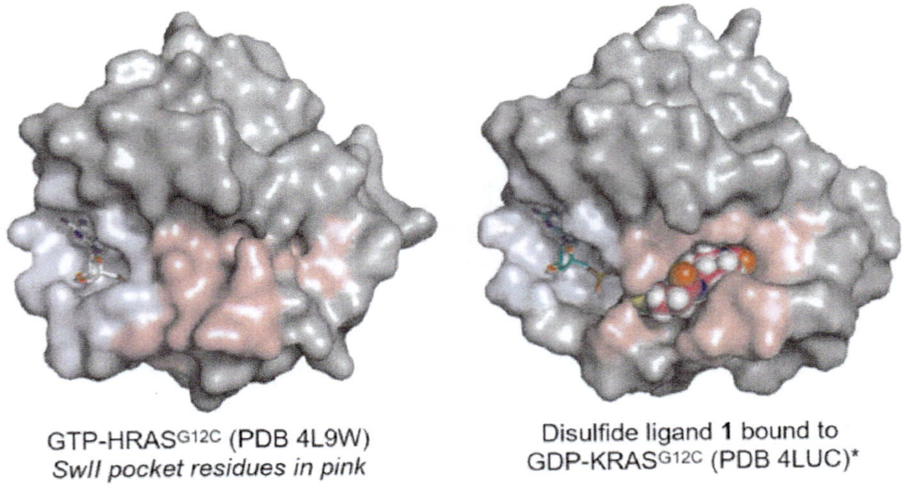

*Glu62 sidechain removed for clarity

Efforts starting in 1990s to find allosteric inhibitors of KRAS failed to identify promising starting points.[15] A significant breakthrough came in 2013 with the report of a covalent inhibitor (disulfide ligand **1**, see Table 1) of the inactive, GDP-bound form of KRAS^{G12C}.[16] This inhibitor bound to KRAS^{G12C} by simultaneously non-covalently engaging the P2[17] or "switch-II" (SwII)[16] pocket of GDP–KRAS^{G12C} (pink, above) while covalently engaging the mutant cysteine residue at position 12. The binding of compound **1** to GDP-KRAS^{G12C} highlighted the significant conformational flexibility of the P2/SwII pocket of KRAS: this ligand-binding pocket was not accessible to compound **1** in the GTP-bound state (at left, above), but only became available upon GDP-binding, which enabled conformational changes in the residues of the P2/SwII pocket to accommodate ligand binding. Subsequent studies with optimized inhibitors revealed that such covalent inhibitors—although binding only to the inactive-state of KRAS^{G12C}—were capable of disrupting subsequent KRAS activation and its interaction with downstream effectors.[18] At the time of this disclosure of compound **1**, related efforts were already underway at Amgen to identify high-affinity covalent inhibitors of KRAS^{G12C}, and these literature findings only heightened our enthusiasm for the potential for this strategy.

10.3 Structure–Activity Relationship (SAR)

Initial covalent ligands for KRAS^{G12C} showed relatively low affinity for KRAS, requiring high concentrations (~100 μM) and long incubation times to achieve significant covalent modification of the protein. To identify higher affinity ligands, we screened custom-designed libraries of low intrinsic-reactivity acrylamide-containing fragments using a combination of mass spectrometry- and fluorescence-based biochemical assays.[19] These efforts identified a series of azetidine acrylamide-based molecules (e.g., phenoxy-acetamide **2**) that were subsequently optimized through iterative screening and structure-based design to provide the potent indoloacetamide inhibitor **3**.

Co-crystallization of compound **3** with $KRAS^{G12C}$ revealed this compound—like disulfide ligand **1**—to bind in the P2 pocket of KRAS (pink; see the following text) and to covalently modify cysteine 12; however, unlike ligand **1**, compound **3** additionally occupied a proximal "cryptic" pocket[20] (blue), whose presence was induced in the presence of bound ligand through the rotation of the histidine 95 sidechain away from the P2 pocket.

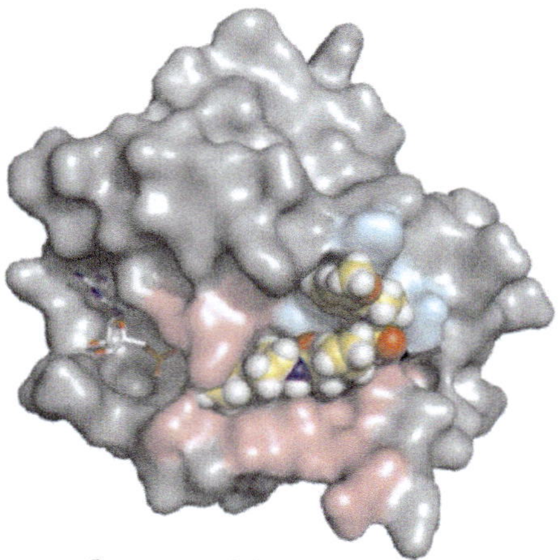

Compound **3** bound to GDP-$KRAS^{G12C}$ (PDB 6P8Z)

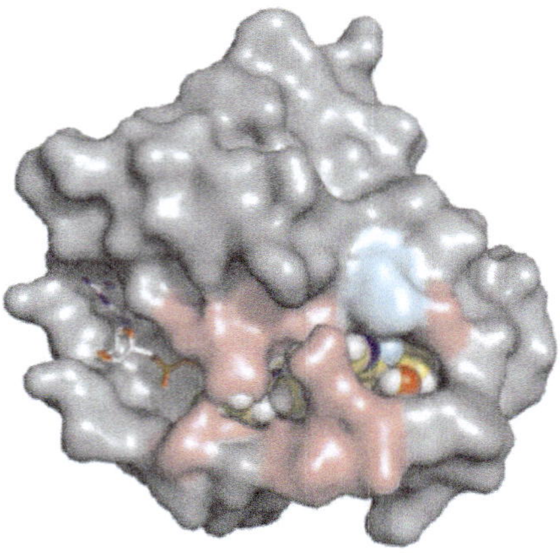

ARS-1620 (Compound **4**) bound to GDP-$KRAS^{G12C}$ (PDB 5V9U)

Occupancy of this cryptic pocket drove exceptional potency in this series of compounds. Unfortunately, these compounds also demonstrated high intrinsic clearance in rodent microsomes and hepatocytes, rapid *in vivo* clearance, and low oral bioavailability.

While this work was in progress, ongoing efforts to optimize the disulfide lead (**1**) at Araxes Pharma had led to the identification of compound **4**.[21] It would be several years before this molecule was subsequently published as ARS-1620,[22] however in the meantime, we prepared compound **4** based on its activity in its initial patent disclosure and found that, while compound **4** was only a moderately potent inhibitor of $KRAS^{G12C}$, it demonstrated significantly greater metabolic stability than compound **3**. A co-crystal structure of compound **4** with $KRAS^{G12C}$ showed that this compound also bound in the P2 pocket of KRAS, but that it did not induce the formation of the cryptic pocket seen in the binding of compound **3**, but bound, instead, with the histidine 95 sidechain (blue) rotated inward toward the P2 pocket and forming a hydrogen bond to N1 of compound **4**.

Table 1. Covalent inhibitors of KRAS^{G12C}

Entry	Structure		2 h p-ERK IC$_{50}$ (μM; Mia PaCa-2)[a]
1	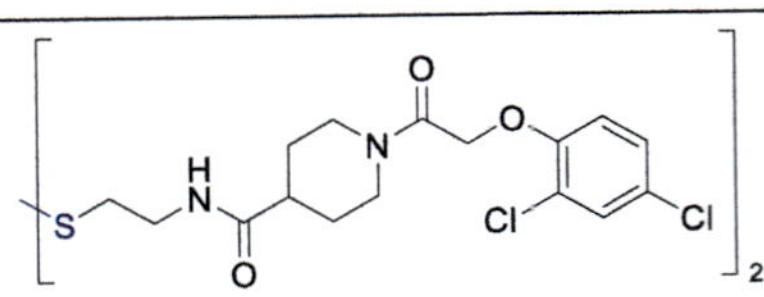		>80% KRAS^{G12C} alkylation (100 μM, 1 h)
2	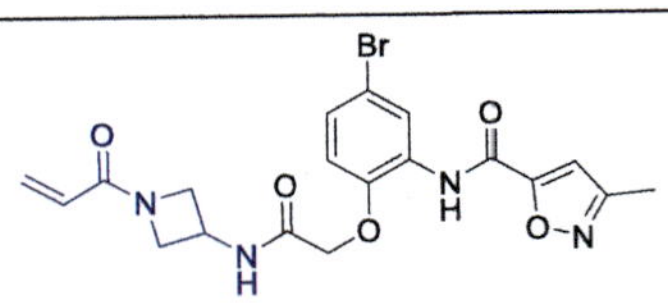		63% KRAS^{G12C} alkylation (10 μM, 20 h)
3	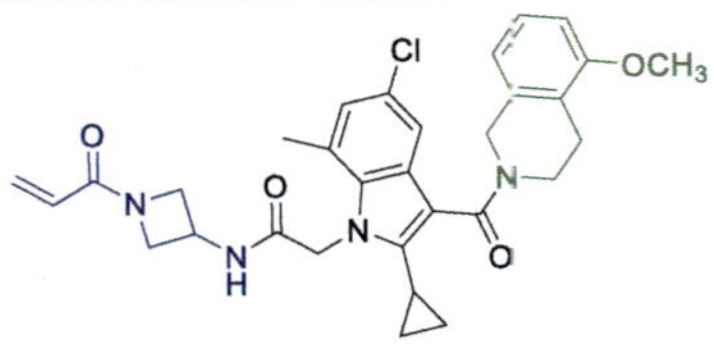		0.220
4	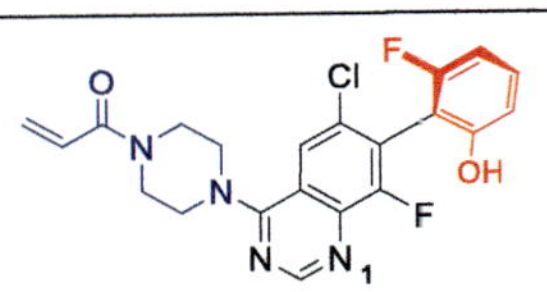		0.831
5	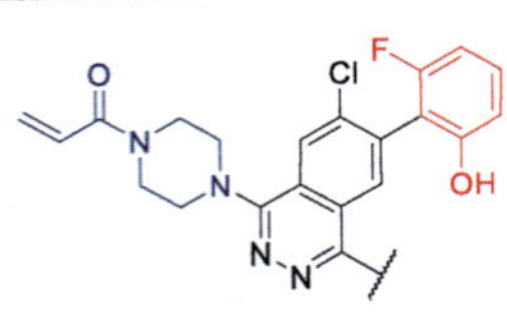	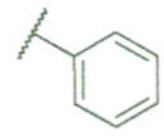	58.0
6		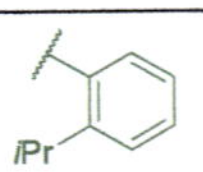	1.80
7	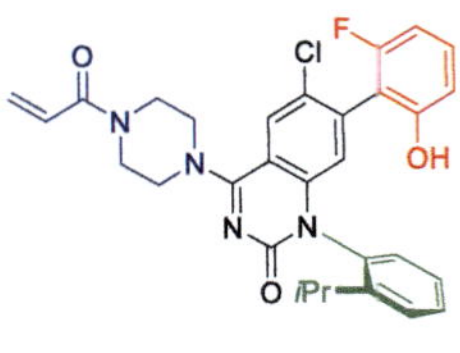		0.335

8	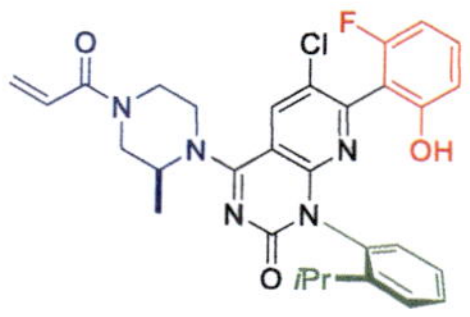	0.044
9	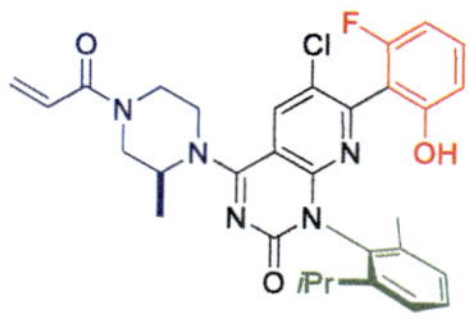	0.028
10	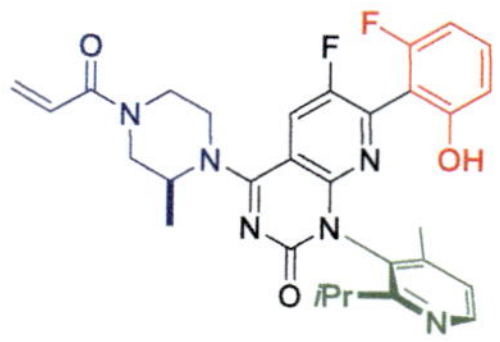 **Sotorasib** (LUMAKRAS™)	0.068

[a] IC_{50} values determined by phospho-ERK1/2 quantification following treatment of EGF-stimulated MIA PaCa-2 cells. blue = cysteine-12-reactive warhead; green = cryptic pocket-engaging group; red = P2/SwII pocket-engaging "tail."

This suggested the possibility of hybridizing the two scaffolds to potentially deliver a compound with the enhanced potency of cryptic pocket-engaging compound **3** and the enhanced metabolic stability of compound **4**. Initial efforts revealed the challenge of successfully implementing this strategy.[23] Crystallographic studies revealed that phthalazine **5** (which substituted the compound **4** scaffold with a phenyl ring) was successful in inducing the formation of the cryptic pocket, but failed to effectively fill the induced pocket, leading to a significant loss in potency relative to compound **4**. Substitution of the newly introduced phenyl ring with an *ortho*-isopropyl substituent, however, more effectively filled in the induced cryptic pocket, leading to a >30-fold boost in potency. Ground-state conformational analysis of the resulting analog, compound **6**, showed that the isopropylbenzene ring and phthalazine core adopted an approximately co-planar conformation rather than the orthogonal conformation observed in the bound state, however. We, therefore, explored substitution of the bicyclic core of compound **6** adjacent to the isopropylbenzene substituent in an effort to enforce the perpendicular orientation of these two ring systems in the ground state. Introduction of a carbonyl group adjacent to the isopropylbenzene ring successfully accomplished this goal, and the resulting quinazolinone analog (**7**) gratifyingly demonstrated an additional fivefold boost in cellular potency.

Further modifications to compound **7**—introduction of an (*S*)-methyl substituent at the 2-position of the piperazine ring and introduction of a nitrogen atom at the 8-

position of the quinazolinone core (see compound **8**)—led to additional improvements in potency and cellular permeability (the latter improvement attributable to intramolecular hydrogen bonding between the phenolic hydroxyl group and the pyridine ring nitrogen).

Interestingly, the introduction of the carbonyl group in compound **7** led to restricted rotation about the adjacent carbon-nitrogen bond, leading to separable atropisomers of the isopropylbenzene ring (i.e., isomers that oriented the isopropyl group either toward or away from the cryptic pocket). Although these atropisomers were separable by chiral supercritical-fluid chromatography, the separated atropisomers were found to slowly interconvert with a half-life of ~8 days. This posed a significant development challenge, as such isomerization could detrimentally impact the long-term stability of formulated drug product derived from this compound.

Several strategies were explored to overcome this atropisomer stability issue, however ortho-methyl substitution of the isopropylbenzene ring (see compound **9**) proved the preferred option. The introduction of this additional methyl group led to a dramatic enhancement in isomeric stability (isomerization half-life of >2000 years) and additionally provided a modest boost in cellular potency.[23]

Disappointingly, crystalline forms of compound **9** proved to have low oral bioavailability in rodent pharmacokinetic studies due to poor aqueous solubility. Two concerted changes—replacing the lipophilic chlorine atom of compound **9** with a more hydrophilic fluorine atom and replacing the greasy isopropyl methylbenzene ring with a more polar (but analogously substituted) pyridine ring—ultimately addressed this issue, resulting in compound **10**.

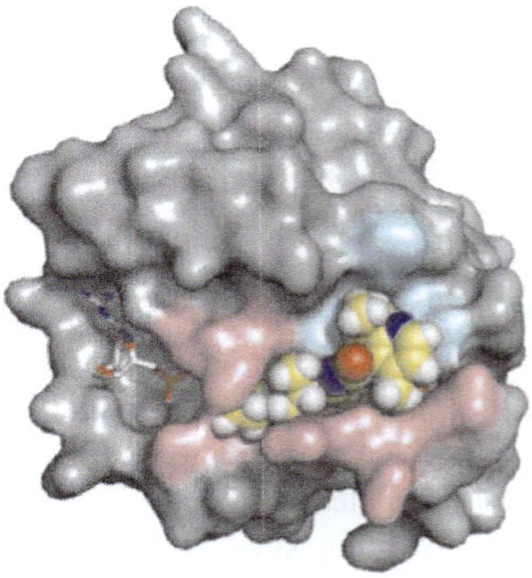

Sotorasib (**10**) bound to GDP-KRAS^{G12C} (PDB 6OIM)

A co-crystal structure of compound **10** with KRAS^{G12C} (see earlier) illustrates how the replacement of the quinazoline core of compound **4** with an isopropylpyridinesubstituted azaquinazolinone provided a novel P2 pocket-binding ligand capable of accessing the cryptic pocket first identified with compounds **2** and **3**. Through its combination of cellular potency, atropisomer configurational stability, and favorable biopharmaceutical properties, compound **10** ultimately successfully navigated pre-clinical investigations to be selected for clinical development as sotorasib (**1**).

10.4 Pharmacokinetics and Drug Metabolism

In preclinical pharmacokinetic studies, sotorasib (**10**) demonstrated good microsomal and hepatic stability across preclinical species, which in turn translated into low-to-moderate clearance and acceptable oral bioavailability (30–34%) in mice, rats, and dogs. *In vitro*, the acrylamide warhead of sotorasib demonstrated moderate reactivity toward glutathione under physiological conditions (5 mM GSH, pH 7.4, 37 °C)[24] and demonstrated a half-life of 200 minutes.[23]

Clinically, at the recommended dose of 960 mg once-daily, sotorasib demonstrates a steady-state apparent clearance of 26.2 L/h and a mean steady-state volume of distribution of 211 L. Sotorasib has an elimination half-life of 5 hours and exhibits 89% protein binding. Sotorasib demonstrates non-linear, time-dependent pharmacokinetics, with similar exposure (AUC_{0-24} and C_{max}) across doses between 180 and 960 mg.[25]

Sotorasib is primarily cleared by non-enzymatic conjugation to glutathione (and subsequent metabolism) as well as by CYP3A-mediated oxidation; as a result, co-administration with strong CYP3A4 inducers can lead to decreased sotorasib exposure (C_{max} & AUC). Sotorasib also demonstrates increased aqueous solubility at low pH (cf, 1.3 mg/mL at pH 1.2 vs. 0.03 mg/mL at pH 6.8), therefore concomitant co-administration with gastric acid-reducing agents such as proton-pump inhibitors (PPI; e.g., omeprazole) or H_2 receptor antagonists (e.g., famotidine) can lead to decreased sotorasib C_{max} and AUC.[25]

10.5 Efficacy and Safety

In preclinical studies, sotorasib (**10**) rapidly inactivated $KRAS^{G12C}$ *via* covalent modification (k_{inact}/K_I = 9900 $M^{-1}s^{-1}$). Such modification proved highly selective. A cysteine proteomic analysis showed the Cys12-containing peptide fragment of $KRAS^{G12C}$ to be the only cysteine-containing peptide to be covalently modified by sotorasib out of 6451 unique peptides tested. Covalent inactivation of $KRAS^{G12C}$ not only inhibited p-ERK signaling in a range of homo- and heterozygous *KRAS p.G12C*-mutant cell lines, but also strongly impaired cellular viability in these cells. No corresponding MAPK signaling or growth inhibitory effects were observed across a range of other *KRAS* mutant or wildtype cell lines.[26]

In *KRAS p.G12C* mouse xenograft models, sotorasib inhibited the growth of MIA PaCa-2 T2 and NCI-H358 tumors at doses ≥10 mg/kg and produced tumor regressions at doses ≥30 mg/kg. In xenograft studies using an NSCLC tumor cell line (NCI-H358), sotorasib also demonstrated strong synergies with clinical standard-of-care therapies such as chemotherapy (carboplatin) and targeted therapies including a MEK inhibitor (PD-0325901), a pan-HER inhibitor (afatinib), a SHP2 inhibitor (RMC-4550), and a CDK4/6 inhibitor (palbociclib). Furthermore, in an immune-competent syngeneic mouse model (CT-26^{KRASG12C}), sotorasib significantly enhanced survival and resulted in durable cures when combined with an anti-PD-1 immune checkpoint inhibitor.[26]

Clinically, the efficacy of sotorasib was demonstrated in a single-arm, open-label study of patients having locally advanced or metastatic *KRAS G12C*-mutated NSCLC (CodeBreaK 100; NCT03600883). Patients in this study had previously progressed following immune checkpoint inhibitor and/or platinum-based chemotherapy and had a least one measurable lesion as defined by RECIST v1.1 criteria.[27] *KRAS G12C*-mutation status was prospectively confirmed through PCR analysis of tumor tissue samples. In the dose-expansion arm of this study, patients were treated with 960 mg sotorasib once daily until disease progression, with treatment efficacy being assessed radiographically by blinded central review according to RECIST v1.1 criteria.[28]

Of 124 evaluable patients, ~2% exhibited complete responses (disappearance of all target lesions) and ~35% demonstrated partial responses (≥30% decrease in the sum of diameters of target lesions relative to baseline) for an objective response rate (ORR) of 36%. The median duration of response was 10.0 months.[25,29]

On the basis of these results, sotorasib (LUMAKRAS) was approved by the US Food and Drug Administration (FDA) for the treatment of *KRAS G12C*-mutated locally advanced or metastatic NSCLC in May 2021. The most common adverse reactions were diarrhea, musculoskeletal pain, nausea, fatigue, hepatotoxicity, and cough, with periodic liver function tests being indicated to monitor for signs of hepatotoxicity (<1.7% of patients) and to guide dose modifications, as necessary.[25]

10.6 Syntheses

10.6.1 Discovery Synthesis

The discovery route to sotorasib was first published in 2020.[23] To support rapid structure–activity relationship (SAR) exploration of the azaquinazolinone core of compound **8**, a modular synthetic route was developed which allowed for the replacement of the N1-aryl substituent (green, below), piperazine spacer (blue), and phenolic "tail" moiety (red), all while supporting changes to the quinazolinone core (black).

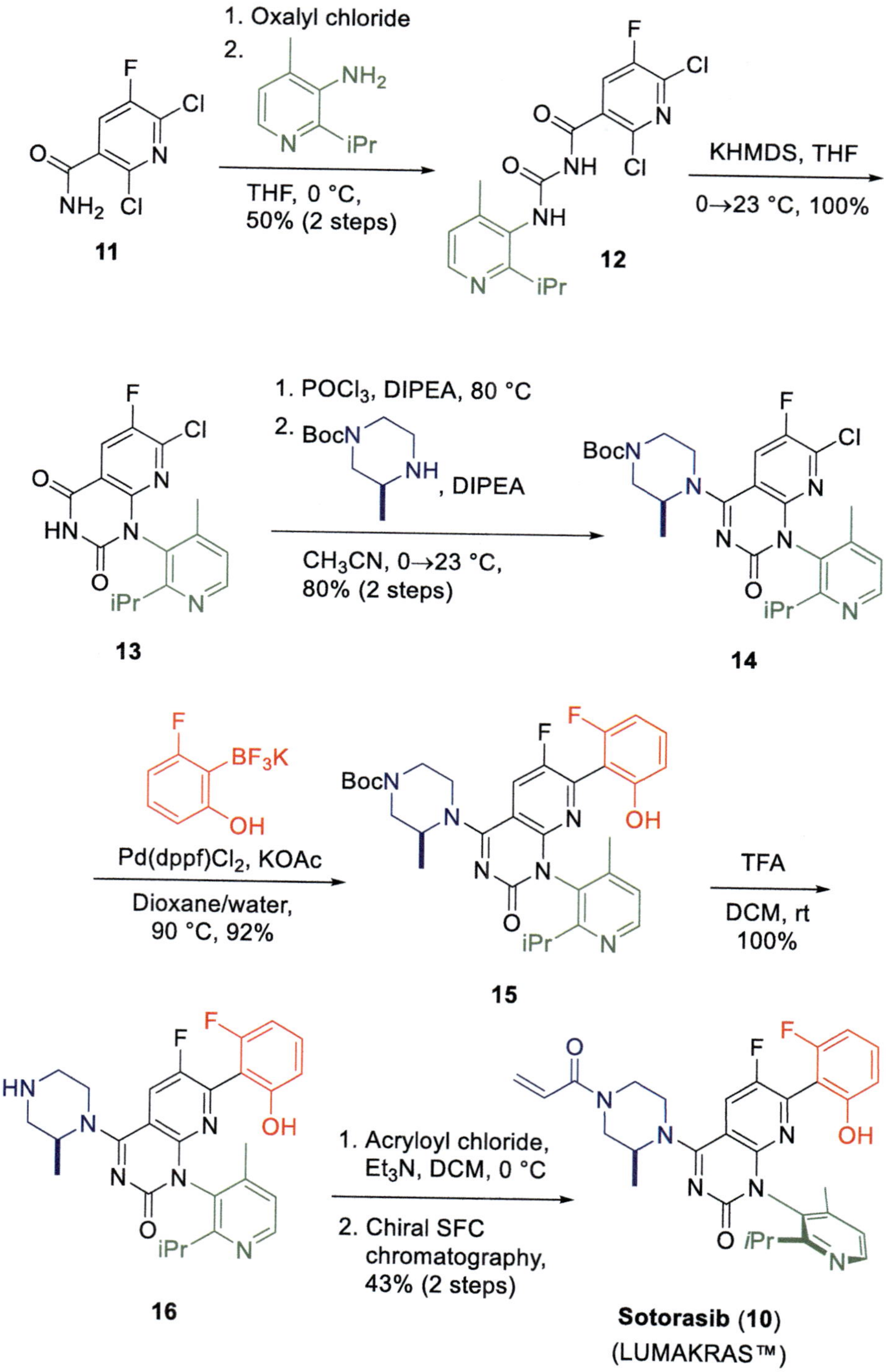
1. Oxalyl chloride
2.
THF, 0 °C,
50% (2 steps)
11
12
KHMDS, THF
0→23 °C, 100%
1. POCl3, DIPEA, 80 °C
2. , DIPEA
CH3CN, 0→23 °C,
80% (2 steps)
13
14
Pd(dppf)Cl2, KOAc
Dioxane/water,
90 °C, 92%
15
TFA
DCM, rt
100%
16
1. Acryloyl chloride,
Et3N, DCM, 0 °C
2. Chiral SFC
chromatography,
43% (2 steps)
Sotorasib (10)
(LUMAKRAS™)

In this route, a nicotinamide building block such as 2,6-dichloro-5-fluoronicotinamide was treated with oxalyl chloride to provide the corresponding acyl isocyanate, which was then combined with an aniline building block (blue) to provide the corresponding *N*-acyl-*N*-aryl urea. Treatment of the resulting urea with potassium hexamethyldisilazide cyclized this urea to the corresponding quinazolinedione, installing the aniline-derived aryl unit at N1.

Selective chlorination of the C4 carbonyl of the resulting scaffold ($POCl_3$, DIPEA, 80 °C) followed by nucleophilic aromatic substitution with a suitably protected diamine (blue) then installed a diamine spacer at C4 of the quinazolinone core. Substitution of the chloropyridine portion of the quinazolinone core by a Suzuki coupling [Pd(dppf)Cl_2, 90 °C] with an aryl trifluoroborate coupling partner (red) allowed for the late-stage installation of a range of differing aromatic "tail" substituents.

Finally, trifluoroacetic acid-mediated cleavage of the Boc protecting group and acylation of the resulting amine with acryloyl chloride allowed for the installation of the cysteine 12-reactive acrylamide warhead. For compounds exhibiting restricted rotation about the N1-aryl bond, atropisomer separation was subsequently accomplished by supercritical-fluid chromatography using a chiral stationary phase. By this route, sotorasib (LUMAKRAS) could be isolated in ~16% overall yield from 2,6-dichloro-5-fluoronicotinamide in an efficient and modular 8-step (5-pot) synthesis.

10.6.2 Process Development Route

Due to the speed with which the program advanced through pre-clinical and clinical studies (~2.5 years from IND to NDA submission), the process route to sotorasib utilized the same general synthetic strategy (bond disconnections) developed by the discovery team. This approach provided a starting point for process development activities as well as preliminary knowledge of process impurities. Two critical challenges relating to the discovery approach were identified, requiring substantial development efforts to render the process commercially viable and to ensure long-term drug substance supply. The first challenge was to eliminate the late-stage chromatographic separation of diastereomers, which resulted in ≥50% mass loss at the end of the synthesis. The second was to substitute the aryl potassium trifluoroborate nucleophile in the Suzuki–Miyaura coupling with a fluoride-free reagent to obviate the need for corrosion-resistant reaction vessels and to improve manufacturing flexibility.

The late-stage chromatographic separation used in the discovery synthesis was necessary to separate diastereomers with different stereochemical configurations at the atropisomeric axis. In the synthesis of sotorasib, the first atropisomeric intermediate is *rac*-**13**, the product of Step 2 and the first isolated intermediate of the process route. Early clinical lots were prepared through the use of chiral chromatographic separation of *rac*-**13**, but the development of a classical resolution was pursued as a more favorable long-term solution to prepare atropisomerically pure (*M*)-**13**. It was found that efficient resolution could be achieved using (+)-2,3-dibenzoyl-D-tartaric acid (DBTA), providing (*M*)-**13** in 40% yield and >99.9% *ee* after freebasing.[30]

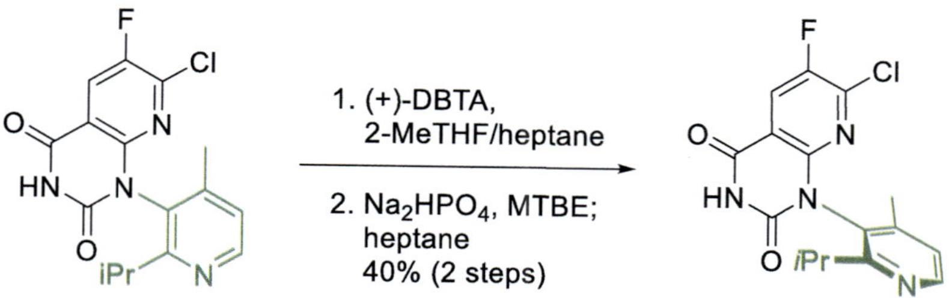

rac-**13** (*M*)-**13**

The Suzuki–Miyaura cross-coupling conditions used to convert intermediate **14** to **15** initially utilized an aryl potassium trifluoroborate coupling partner. While this transformation was high yielding (92%), the batch scale was limited by the size of Hastelloy (corrosion-resistant) reaction vessels available at the manufacturing facility. To increase manufacturing flexibility and scale, alternative coupling partners were explored with the goal of enabling execution in stainless steel or glass-lined vessels. Following experimentation, it was found that boronic acid trimer **17** was a competent coupling partner under optimized reaction conditions.[31] The improved conditions utilized (dpePhos)$PdCl_2$ as the catalyst and substituted 1,4-dioxane with 2-methyltetrahydrofuran as a greener reaction solvent. Using these conditions, the cross coupling was executed with 0.6 mol% Pd (representing a >75% reduction in catalyst versus the initial reaction conditions) while maintaining yields >90%.

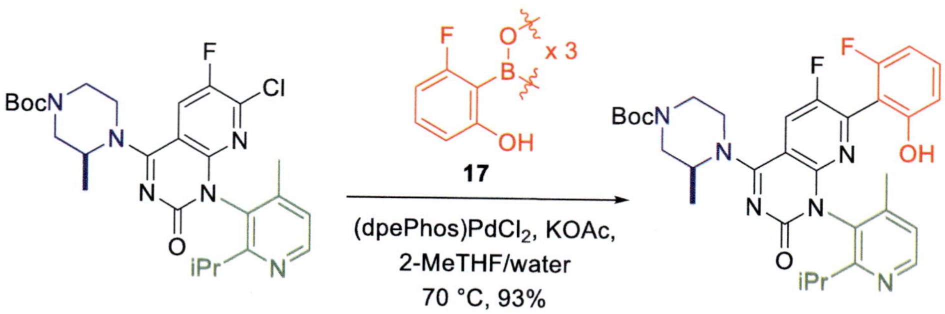

14 **15**

Through implementation of a classical resolution and improved Suzuki–Miyaura coupling conditions (along with other minor modifications to the synthesis and process conditions), substantial improvements to process efficiency (yield and cycle time) and drug substance quality were realized. These improvements enabled commercialization of a manufacturing process derived from the discovery synthesis. This allowed for a compressed development timeline while also ensuring both short-term clinical and long-term commercial supplies of sotorasib drug substance.

10.7. Summary

In the nearly 40 years since its identification as one of the first known human oncogenes, *KRAS* has become one of the most compelling targets in oncology as well as one of the most challenging. The accelerated approval of LUMAKRAS (sotorasib; **10**) for the treatment of *KRAS G12C*-mutant NSCLC by the US FDA in May 2021 (and by the United Arab Emirates' Ministry of Health and Prevention (MOHAP) in June 2021 and the United Kingdom's Medicines and Healthcare Products Regulatory Agency (MHRA) and Health Canada in September 2021) marks the culmination of nearly four decades of research defining the structure and function of the KRAS protein in human cancer.

The discovery of LUMAKRAS brought together advances in structural biology (the identification of critical new binding pockets on the surface of KRAS),[16,19] inhibitor design strategies (the covalent inhibition of non-catalytic surface cysteine residues),[32] and hit-finding strategies (the screening of cysteine-reactive-reactive covalent libraries)[16,19] to identify new starting points for the inhibition of $KRAS^{G12C}$.

The clinical evaluation of LUMAKRAS was likewise facilitated by new technical advances—particularly in the areas of tissue and plasma sequencing (especially next-generation (NGS) sequencing)—which made it possible to rapidly identify patients harboring the *KRAS p.G12C* mutation who could benefit from this new targeted therapy. As the product of a confluence of technological developments, the discovery of LUMAKRAS illustrates that—at least for some cellular targets—“undruggability” is not an intrinsic characteristic, but rather an indication that the correct combination of technologies and strategic approaches have yet to be brought to bear.

10.8. References

1. Kirsten, W. H.; Mayer, L. A. Morphologic responses to a murine erythroblastosis virus. *J. Natl. Cancer Inst.* **1967**, *39*, 311–335.
2. Tsuchida, N.; Gilden, R. V.; Hatanaka, M. Sarcoma-virus-related RNA sequences in normal rat cells. *Proc. Natl. Acad. Sci. U.S.A.* **1974**, *71*, 4503–4507.
3. Shih, T. Y.; Weeks, M. O.; Young, H. A.; Scolnick, E. M. Identification of a sarcoma virus-coded phosphoprotein in nonproducer cells transformed by Kirsten or Harvey murine sarcoma virus. *Virology* **1979**, *96*, 64–79.
4. Ellis, R. W.; Defeo, D.; Shih, T. Y.; Gonda, M. A.; Young, H. A.; Tsuchida, N.; Lowy, D. R.; Scolnick, E. M. The p21 src genes of Harvey and Kirsten sarcoma

viruses originate from divergent members of a family of normal vertebrate genes. *Nature* **1981**, *292*, 506–511.
5. Der, C. J.; Krontiris, T. G.; Cooper, G. M. Transforming genes of human bladder and lung carcinoma cell lines are homologous to the ras genes of Harvey and Kirsten sarcoma viruses. *Proc. Natl. Acad. Sci. U.S.A.* **1982**, *79*, 3637–3640.
6. Chang, E. H.; Gonda, M. A.; Ellis, R. W.; Scolnick, E. M.; Lowy, D. R. Human genome contains four genes homologous to transforming genes of Harvey and Kirsten murine sarcoma viruses. *Proc. Natl. Acad. Sci. U.S.A.* **1982**, *79*, 4848–4852.
7. Coffin, J. M.; Varmus, H. E.; Bishop, J. M.; Essex, M.; Hardy, W. D., Jr.; Martin, G. S.; Rosenberg, N. E.; Scolnick, E. M.; Weinberg, R. A.; Vogt, P. K. Proposal for naming host cell-derived inserts in retrovirus genomes. *J. Virol.* **1981**, *40*, 953–957.
8. Santos, E.; Martin-Zanca, D.; Reddy, E. P.; Pierotti, M. A.; Della Porta, G.; Barbacid, M. Malignant activation of a K-ras oncogene in lung carcinoma but not in normal tissue of the same patient. *Science* **1984**, *223*, 661–664.
9. Downward, J. Targeting RAS signaling pathways in cancer therapy. *Nat. Rev. Cancer* **2003**, *3*, 11–22.
10. Cox, A. D.; Fesik, S. W.; Kimmelman, A. C.; Luo, J.; Der, C. J. Drugging the undruggable RAS: mission possible? *Nat. Rev. Drug Discov.* **2014**, *13*, 828–851.
11. Yu, H. A.; Sima, C. S.; Shen, R.; Kass, S.; Gainor, J.; Shaw, A.; Hames, M.; Iams, W.; Aston, J.; Lovly, C. M.; Horn, L.; Lydon, C.; Oxnard, G. R.; Kris, M. G.; Ladanyi, M.; Riely, G. J. Prognostic impact of KRAS mutation subtypes in 677 patients with metastatic lung adenocarcinomas. *J. Thorac. Oncol.* **2015**, *10*, 431–437.
12. Adapted with permission from Lanman, B. A.; Cee, V. J.; Marx, M. A. Clinical and pre-clinical approaches to the inhibition of KRAS(G12C). In *2020 Medicinal Chemistry Reviews*, Bronson, J. J., Ed.; The Medicinal Chemistry Division of the American Chemical Society: Washington, D.C., **2020**; Vol. 55, pp. 249–271. (© 2020 ACS-MEDI)
13. Ayati, A.; Moghimi, S.; Salarinejad, S.; Safavi, M.; Pouramiri, B.; Foroumadi, A. A review on progression of epidermal growth factor receptor (EGFR) inhibitors as an efficient approach in cancer targeted therapy. *Bioorg. Chem.* **2020**, *99*, 103811.
14. Uehling, D. E.; Harris, P. A. Recent progress on MAP kinase pathway inhibitors. *Bioorg. Med. Chem. Lett.* **2015**, *25*, 4047–4056.
15. Wang, W.; Fang, G.; Rudolph, J. Ras inhibition via direct Ras binding--is there a path forward? *Bioorg. Med. Chem. Lett.* **2012**, *22*, 5766–5776.
16. Ostrem, J. M.; Peters, U.; Sos, M. L.; Wells, J. A.; Shokat, K. M. K-Ras(G12C) inhibitors allosterically control GTP affinity and effector interactions. *Nature* **2013**, *503*, 548–551.
17. Grant, B. J.; Lukman, S.; Hocker, H. J.; Sayyah, J.; Brown, J. H.; McCammon, J. A.; Gorfe, A. A. Novel allosteric sites on Ras for lead generation. *PLoS One* **2011**, *6*, e25711.
18. Patricelli, M. P.; Janes, M. R.; Li, L. S.; Hansen, R.; Peters, U.; Kessler, L. V.; Chen, Y.; Kucharski, J. M.; Feng, J.; Ely, T.; Chen, J. H.; Firdaus, S. J.; Babbar,

A.; Ren, P.; Liu, Y. Selective inhibition of oncogenic KRAS output with small molecules targeting the inactive state. *Cancer Discov.* **2016**, *6*, 316–329.

19. Shin, Y.; Jeong, J. W.; Wurz, R. P.; Achanta, P.; Arvedson, T.; Bartberger, M. D.; Campuzano, I. D. G.; Fucini, R.; Hansen, S. K.; Ingersoll, J.; Iwig, J. S.; Lipford, J. R.; Ma, V.; Kopecky, D. J.; McCarter, J.; San Miguel, T.; Mohr, C.; Sabet, S.; Saiki, A. Y.; Sawayama, A.; Sethofer, S.; Tegley, C. M.; Volak, L. P.; Yang, K.; Lanman, B. A.; Erlanson, D. A.; Cee, V. J. Discovery of *N*-(1-acryloylazetidin-3-yl)-2-(1*H*-indol-1-yl)acetamides as covalent inhibitors of KRAS(G12C). *ACS Med. Chem. Lett.* **2019**, *10*, 1302–1308.

20. Cryptic pockets are binding sites that are not evident in un-ligated protein but form upon ligand binding: Vajda, S.; Beglov, D.; Wakefield, A. E.; Egbert, M.; Whitty, A. Cryptic binding sites on proteins: definition, detection, and druggability. *Curr. Opin. Chem. Biol.* **2018**, *44*, 1–8.

21. Li, L.; Feng, J.; Wu, T.; Ren, P.; Liu, Y.; Liu, Y.; Long, Y. Inhibitors of KRAS G12C. *Patent Application* WO/2015/054572, April 16 (2015).

22. Janes, M. R.; Zhang, J.; Li, L. S.; Hansen, R.; Peters, U.; Guo, X.; Chen, Y.; Babbar, A.; Firdaus, S. J.; Darjania, L.; Feng, J.; Chen, J. H.; Li, S.; Li, S.; Long, Y. O.; Thach, C.; Liu, Y.; Zarieh, A.; Ely, T.; Kucharski, J. M.; Kessler, L. V.; Wu, T.; Yu, K.; Wang, Y.; Yao, Y.; Deng, X.; Zarrinkar, P. P.; Brehmer, D.; Dhanak, D.; Lorenzi, M. V.; Hu-Lowe, D.; Patricelli, M. P.; Ren, P.; Liu, Y. Targeting KRAS mutant cancers with a covalent G12C-specific inhibitor. *Cell* **2018**, *172*, 578–589 e17.

23. Lanman, B. A.; Allen, J. R.; Allen, J. G.; Amegadzie, A. K.; Ashton, K. S.; Booker, S. K.; Chen, J. J.; Chen, N.; Frohn, M. J.; Goodman, G.; Kopecky, D. J.; Liu, L.; Lopez, P.; Low, J. D.; Ma, V.; Minatti, A. E.; Nguyen, T. T.; Nishimura, N.; Pickrell, A. J.; Reed, A. B.; Shin, Y.; Siegmund, A. C.; Tamayo, N. A.; Tegley, C. M.; Walton, M. C.; Wang, H. L.; Wurz, R. P.; Xue, M.; Yang, K. C.; Achanta, P.; Bartberger, M. D.; Canon, J.; Hollis, L. S.; McCarter, J. D.; Mohr, C.; Rex, K.; Saiki, A. Y.; San Miguel, T.; Volak, L. P.; Wang, K. H.; Whittington, D. A.; Zech, S. G.; Lipford, J. R.; Cee, V. J. Discovery of a covalent inhibitor of KRAS(G12C) (AMG 510) for the treatment of solid tumors. *J. Med. Chem.* **2020**, *63*, 52–65.

24. Cee, V. J.; Volak, L. P.; Chen, Y.; Bartberger, M. D.; Tegley, C.; Arvedson, T.; McCarter, J.; Tasker, A. S.; Fotsch, C. Systematic study of the glutathione (GSH) reactivity of N-arylacrylamides: 1. Effects of aryl substitution. *J. Med. Chem.* **2015**, *58*, 9171–9178.

25. LUMAKRAS prescribing information (revised 05/2021). https://www.pi.amgen.com/~/media/amgen/repositorysites/pi-amgen-com/lumakras/lumakras_pi_hcp_english.ashx (last accessed September 5, 2021).

26. Canon, J.; Rex, K.; Saiki, A. Y.; Mohr, C.; Cooke, K.; Bagal, D.; Gaida, K.; Holt, T.; Knutson, C. G.; Koppada, N.; Lanman, B. A.; Werner, J.; Rapaport, A. S.; San Miguel, T.; Ortiz, R.; Osgood, T.; Sun, J. R.; Zhu, X.; McCarter, J. D.; Volak, L. P.; Houk, B. E.; Fakih, M. G.; O'Neil, B. H.; Price, T. J.; Falchook, G. S.; Desai, J.; Kuo, J.; Govindan, R.; Hong, D. S.; Ouyang, W.; Henary, H.; Arvedson, T.; Cee, V. J.; Lipford, J. R. The clinical KRAS(G12C) inhibitor AMG 510 drives anti-tumour immunity. *Nature* **2019**, *575*, 217–223.

27. Eisenhauer, E. A.; Therasse, P.; Bogaerts, J.; Schwartz, L. H.; Sargent, D.; Ford, R.; Dancey, J.; Arbuck, S.; Gwyther, S.; Mooney, M.; Rubinstein, L.; Shankar, L.; Dodd, L.; Kaplan, R.; Lacombe, D.; Verweij, J. New response evaluation criteria in solid tumours: revised RECIST guideline (version 1.1). *Eur. J. Cancer* **2009**, *45*, 228–247.

28. A Phase 1/2, Study Evaluating the Safety, Tolerability, PK, and Efficacy of AMG 510 in Subjects With Solid Tumors With a Specific KRAS Mutation (CodeBreaK 100). https://clinicaltrials.gov/ct2/show/NCT03600883 (last accessed September 5, 2021).

29. Skoulidis, F.; Li, B. T.; Dy, G. K.; Price, T. J.; Falchook, G. S.; Wolf, J.; Italiano, A.; Schuler, M.; Borghaei, H.; Barlesi, F.; Kato, T.; Curioni-Fontecedro, A.; Sacher, A.; Spira, A.; Ramalingam, S. S.; Takahashi, T.; Besse, B.; Anderson, A.; Ang, A.; Tran, Q.; Mather, O.; Henary, H.; Ngarmchamnanrith, G.; Friberg, G.; Velcheti, V.; Govindan, R., Sotorasib for lung cancers with *KRAS p.G12C* mutation. *N. Engl. J. Med.* **2021,** *384*, 2371–2381.

30. Parsons, A. T.; Cochran, B. N.; Powazinik, IV, W.; Caporini, M. A. Improved Synthesis of Key Intermediate of KRAS G12C Inhibitor Compound. *Patent Application* WO/2020/102730, May 22 (2020).

31. Parsons, A. T.; Beaver, M. Improved Synthesis of KRAS G12C Inhibitor Compound. *Patent Application* WO/2021/097212, May 20 (2021).

32. For a review of the re-emergence of covalent inhibitor strategies in the late 1990s and early 2000s, see Singh, J.; Petter, R. C.; Baillie, T. A.; Whitty, A. The resurgence of covalent drugs. *Nat. Rev. Drug Discov.* **2011**, *10*, 307–317.

11

Lorlatinib (Lorbrena), An ALK Inhibitor for Treating NSCLC

Ruby M. Aaron, Hayden K. Low,
Benjamin T. Sokol, and Amy B. Dounay

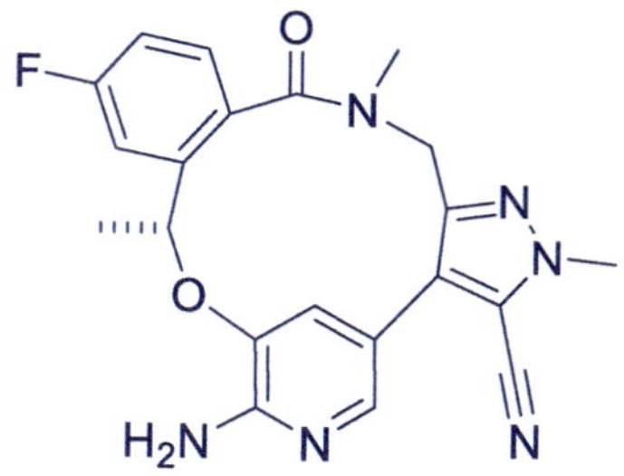

Lorlatinib (Lorbrena, **1**)
Pfizer, 2018
ALK/ROS1 inhibitor

11.1 Background

Lung cancer is the leading cause of global cancer mortality.[1] An estimated 235,760 people will be diagnosed with this indication, and an estimated 131,880 people will die from this disease in the United States in 2021.[2] Despite the advent of new treatments, the outlook for patients remains poor, with an overall 5 year survival rate of 21%.[2] Prolonged tobacco use is the most significant risk factor, followed by exposure to radon gas, secondhand smoke, and asbestos.[1,2] In the United States, the incidence of lung cancer is decreasing for men and stabilizing for women, correlating with historical, gender-based differences in tobacco use.[3] Globally, lung cancer incidence is decreasing or stabilizing in Western countries and increasing in Eastern and developing countries.[3]

Lung cancer presents in two main forms: non-small cell lung cancer (NSCLC), diagnosed in 84% of cases, and small cell lung cancer.[2] Diagnosis and treatment of

Current Drug Synthesis, First Edition. Edited by Jie Jack Li.

NSCLC is stage dependent.[3] Advanced disease is metastatic to multiple organ systems, with more than 40% of patients presenting with brain metastases (BM).[3,4] Advanced-stage patients require systemic treatment with chemotherapy, monoclonal antibodies, immune checkpoint inhibitors, or targeted drugs.[3] Treatment can be further specified to target genetic abnormalities in NSCLC. Anaplastic lymphoma kinase (ALK) and c-ROS oncogene 1 (ROS1) rearrangements, presenting in 3–8% and 1–2% of patients, respectively, are two specific genetic variants targeted by new NSCLC therapeutics.[3,4]

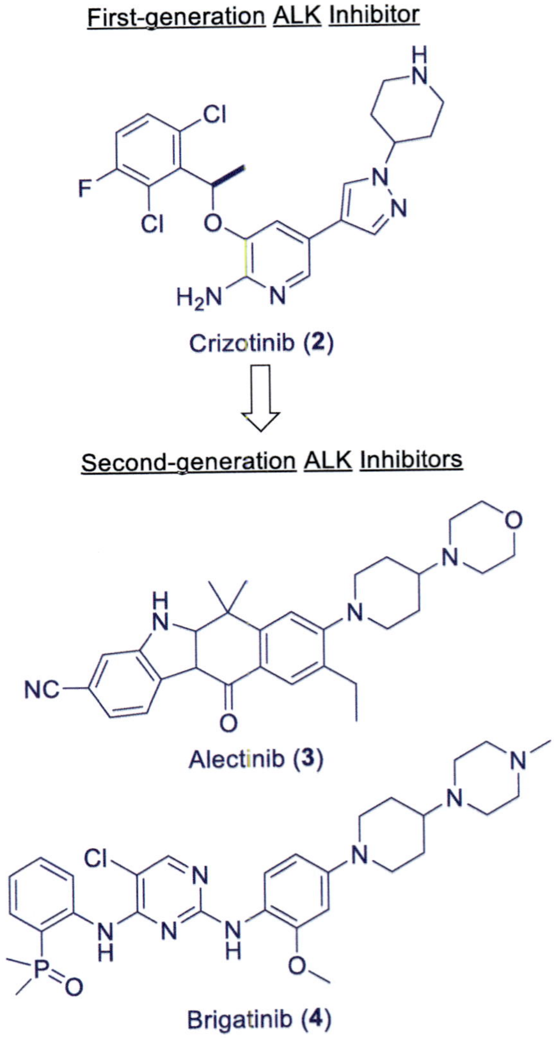

Figure 1. Representative first- and second-generation drugs approved for the treatment of patients presenting with ALK and/or ROS1 positive NSCLC.

The targeting of oncogenic, constitutively active ALK, and/or ROS1 tyrosine kinases represents a novel frontier in the treatment of NSCLC.[5] Crizotinib (**2**, Figure 1) was the first therapeutic approved for the treatment of ALK-positive NSCLC.[4] A multi-targeted tyrosine kinase inhibitor (TKI) affecting ALK, ROS1, and MET, crizotinib (**2**) showed an overall response rate (ORR) of 74%, and median progression free survival (PFS) of 10.9 months in clinical trials of TKI naïve ALK-positive patients.[4] Despite these promising results, high rates of resistance-generating mutations are reported in a majority of patients following 10–12 months of treatment with crizotinib (**2**). Furthermore, crizotinib (**2**) achieves poor blood–brain barrier (BBB) penetration, with a cerebral spinal fluid (CSF) to free plasma ratio of only 0.03.[6] Central nervous system (CNS) progression of NSCLC occurs in 47% of ROS1-positive patients treated with crizotinib (**2**), highlighting its ineffectiveness in treating CNS metastatic disease.[4,6]

Second-generation drugs, including alectinib (**3**) and brigatinib (**4**, Figure 1), have been developed and approved for ROS1 and/or ALK-positive NSCLC treatment. Both **3** and **4** demonstrate better intercranial overall response rates (IC-ORRs) when compared to crizotinib (**2**).[4] Nevertheless, mutation-generated resistance and lack of CNS permeability eventually render these second-generation treatments ineffective. Furthermore, patients who have previously failed crizotinib (**2**) therapy or second-generation therapies show poor ORR and IC-ORR when treated with subsequent drugs.[4] These limitations led Pfizer to launch a research program that ultimately led to the development of lorlatinib (Lorbrena, **1**), a third-generation ALK inhibitor for treatment of NSCLC.[6]

11.2 Pharmacology

ALK fusion genes have been implicated as oncogenic drivers in a variety of cancer types, including ALK-positive NSCLC.[7] In healthy cells, echinoderm microtubule-associated protein-like 4 (EML4) and ALK genes are oriented in opposite directions along the short arm of chromosome 2 (Figure 2A). Paracentric chromosomal inversion in this region results in the fusion of a truncated form of ALK to EML4 (Figure 2B). The truncation of ALK occurs near the 5′ end of exon 20, leading to the exclusion of the cellular domain and transmembrane helix from the EML4-ALK fusion, and inclusion of only the cytoplasmic portion of ALK and the tyrosine kinase (TK)- domain.[8] This fusion creates a constitutively active cytoplasmic TK that causes the cell to proliferate uncontrollably, resulting in cancer growth. The rapid proliferation of ALK positive NSCLC cells is reliant on the EML4-ALK oncogenic fusion protein. This biological vulnerability in ALK positive NSCLC can be exploited with ALK inhibition to block the protein's access to energy, thus deactivating it.

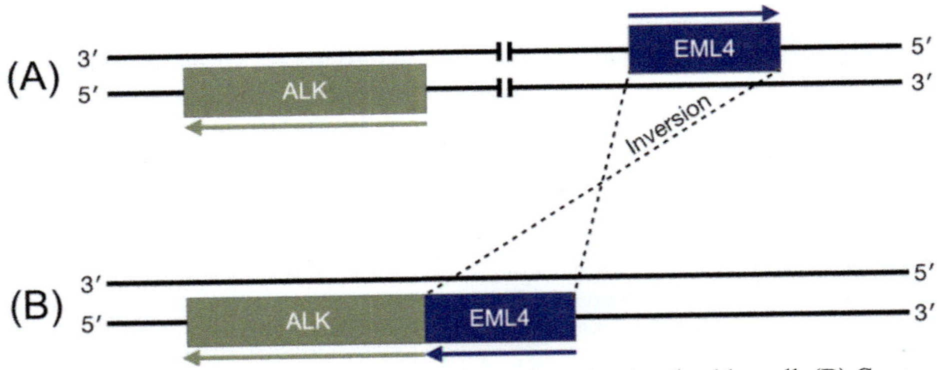

Figure 2. EML4-ALK Gene Fusion. (A) Gene orientation in a healthy cell. (B) Gene orientation in cells with ALK-EML4 gene fusion.

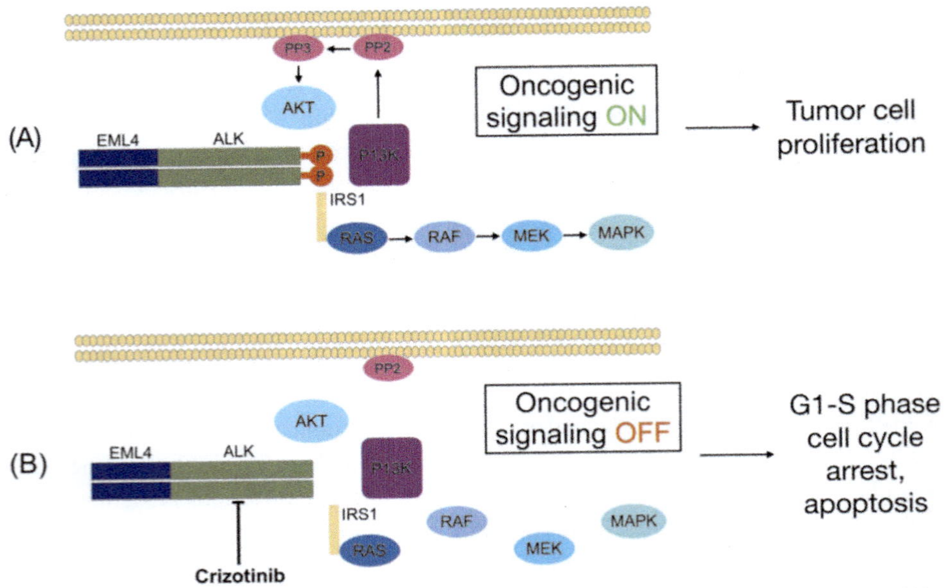

Figure 3. Crizotinib (**2**) therapy interrupts ALK dependent oncogenic signaling. (A) Oncogenic signaling in EML4-ALK fusion leads to tumor cell proliferation. (B) Oncogenic signaling is disrupted by treatment with crizotinib (**2**) therapy, inducing apoptosis. (Based on Reference 9)

Crizotinib (**2**), a first-generation TKI, exhibits competitive binding to the ATP binding site in ALK, ROS1, and MET. Upon binding, crizotinib (**2**) inhibits ALK phosphorylation and signal transduction, which leads to G1-S phase cell cycle arrest and induction of apoptosis in ALK positive NSCLC cells (Figure 3).[9] However, a vast majority of patients treated with crizotinib (**2**) develop resistance within 1 year of treatment, leading to recurrence of disease. Although second-generation ALK inhibitors have shown robust activity against wild type (WT) and activated resistance mutant ALK fusion proteins in clinical trials, they eventually become ineffective due to the emergence of resistance and lack of CNS availability. These limitations prompted research toward a next generation ALK/ROS1 inhibitor with both broad spectrum ALK potency and CNS availability.[7]

11.3 Structure–Activity Relationship (SAR)

11.3.1 Acyclic ALK Inhibitors

Pfizer initiated a next-generation ALK inhibitor program to develop an orally-dosed small molecule drug that combines ALK potency, CNS availability, and favorable absorption, distribution, metabolism, and excretion (ADME) properties.[7,10] A detailed account of the structure–activity relationship (SAR) studies at Pfizer that led to the discovery of lorlatinib (**1**) has been reported previously.[7] Herein, a few key aspects of these SAR studies are highlighted. The primary medicinal chemistry design objectives included increasing the drug's lipophilic efficiency (LipE) and reducing the molecular weight, while incorporating the requisite structural features to overcome resistant mutations and BM.[7] The ALK-L1196M gatekeeper (GK) mutation was identified as the most common crizotinib (**2**)-resistant mutant and was used to assess all new compounds in preliminary cell-based assays. Additionally, the transporter protein P-glycoprotein (P- gp) is largely responsible for the efflux of many cancer drugs, including the first-generation ALK inhibitor crizotinib (**2**). Madin–Darby canine kidney (MDCK–MDR1) cells are engineered to overexpress human P-gp. The MDCK/MDR assay is widely used in drug discovery to identify compounds that are likely to avoid P-gp efflux and achieve adequate drug exposure in the brain. The medicinal chemistry team targeted compounds with low MDR BA/AB ratios (<2.5), indicative of low P-gp efflux, to ensure high probability of achieving efficacious drug concentrations in the CNS.[7]

Initial medicinal chemistry efforts produced a new series of second-generation ALK inhibitors exemplified by compound **5** (PF-06439015, Figure 4).[7] In comparison with crizotinib (**2**), compound **5** increases potency against both WT ALK and the GK mutant (ALK-L1196) by 105-fold and 128-fold, respectively (Figure 4). These potency improvements can be attributed to numerous new interactions between **5** and ALK, which were revealed in co-crystal structures of inhibitors with ALK. Despite its improved potency, compound **5** does not achieve MDR values predictive of CNS availability.[7] Numerous acyclic analogues of **5** were explored to improve CNS penetration through careful attention to lipophilicity and molecular weight. Following this approach, compounds **6** and **7** achieved improvements in MDR ratios, but at the expense of decreased potency against both ALK WT and GK mutant cells (Figure 4). Although compounds **5**, **6**, and **7** each have promising attributes, these compounds illustrate the difficulty of simultaneously optimizing for potency and CNS exposure within a single compound in this acyclic series. Thus, the medicinal chemistry team was prompted to investigate alternate scaffold strategies.[7]

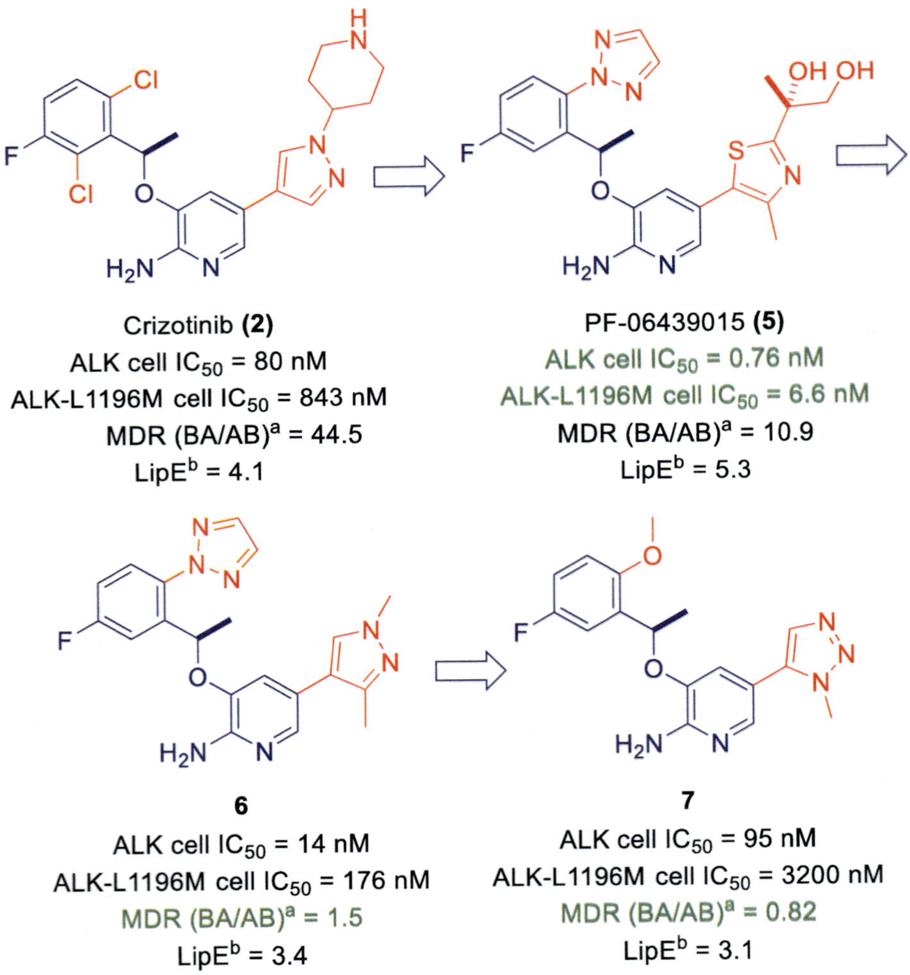

Figure 4. Potency and ADME properties of representative acyclic compounds. [a]MDCK-MDR1 BA/AB efflux ratio at 2 μM drug concentration and pH 7.4. MDCK/MDR cells overexpressing P-gP are cultured as a monolayer in a transwell plate. Test compounds are added to either the apical (A) or basolateral (B) donor chamber and concentrations are measured from the corresponding receiver chamber. The concentration ratio (B to A)/ (A to B) is reported. MDR ratio (BA/AB) values >2.5 are predictive of CNS exclusion in humans. [b]LipE = –[log(pALK – L1196M cell IC_{50})] – log D. Values in green are within targeted range. (Reference 7/American Chemical Society)

11.3.2 Macrocyclic ALK Inhibitors

X-ray crystal structures of ALK protein-inhibitor complexes illuminated the unique binding mode of compounds in this series and led the medicinal chemistry team to design innovative inhibitor scaffolds. One important early structure showed triazole analog **7** in the ATP binding site of ALK. This structure revealed that **7** adopts a U-shaped geometry in the bound conformation (Figure 5), a trend that was observed with numerous additional acyclic compounds crystallized in the binding site.[7] The close proximity of "head" and "tail" (4.1 Å) suggested that these groups could conceivably be covalently linked to form a macrocyclic scaffold (Figure 5). An optimally designed macrocycle could additionally mask the molecule's polarity, creating a more compact, rigid inhibitor in order to maintain potency and selectivity for ALK and to achieve desirable CNS ADME properties.[7]

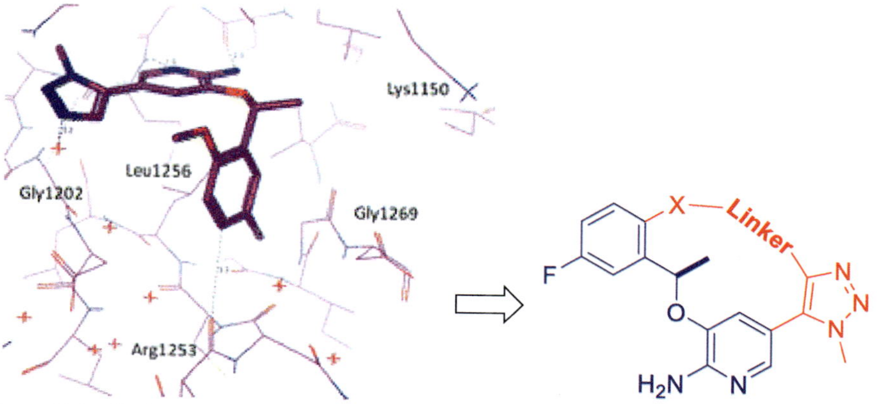

Figure 5. Left: Compound **7** cocrystal structure in the ALK kinase domain (purple, PDB 4CNH) highlighting the U-shaped binding conformation and proximity of the methoxyfluorophenyl headgroup and triazole tail group. (Adapted with permission from Reference 7. Copyright 2014 American Chemical Society) Right: Proposed macrocyclic scaffold.

11.3.3 Discovery of Lorlatinib (PF-06463922)

Initial SAR studies of macrocyclic scaffolds explored ether-linked analogs inspired by methyl ether **7**. A series of macrocycles were prepared by extending the methyl ether linker in **7** to connect to the five-membered heterocycle in the tail.[7] To determine optimal sizing, ether-linked macrocycles ranging in size from 12- to 14-membered rings were prepared. This systematic SAR study revealed that the 12-membered ring system in compound **8** (Figure 6) was most optimal in terms of LipE.[7] Unfortunately, the ether-linked compounds were too lipophilic (log D>3) to provide the desired overlap of potency, ADME properties, and CNS availability.

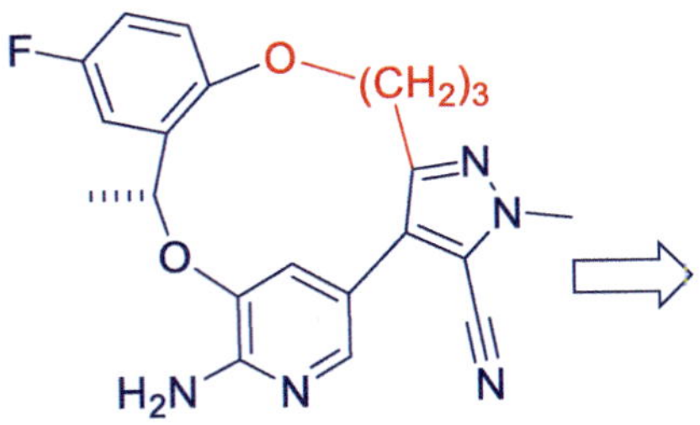

8

ALK cell IC_{50} = 1.0 nM
ALK-L1196M cell IC_{50} = 20 nM
LipE[b] = 4.4

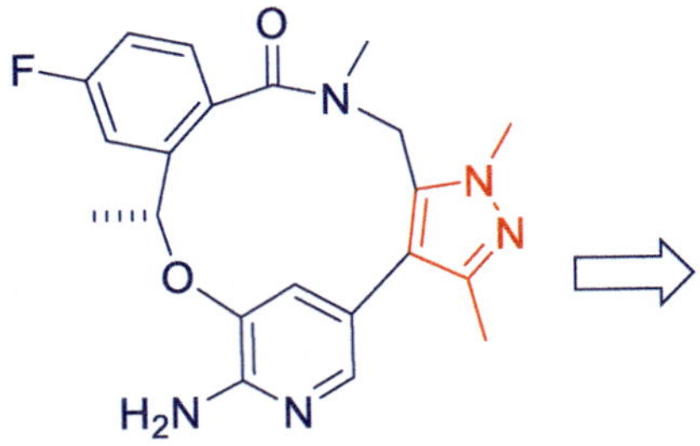

9

ALK cell IC_{50} = 0.70 nM
ALK-L1196M cell IC50 = 14 nM
MDR (BA/AB)[a] = 4.2
LipE[b] = 5.7

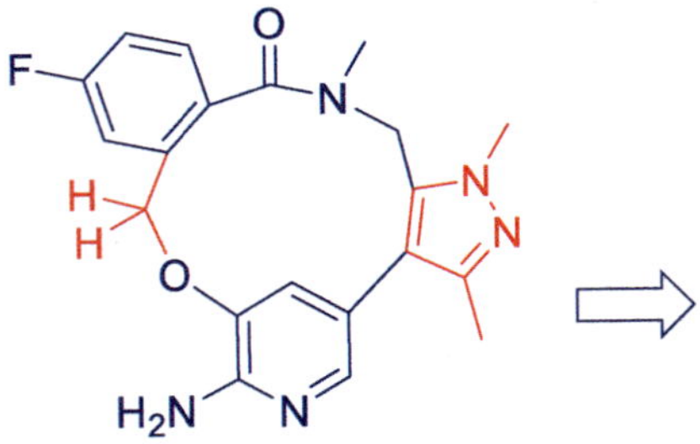

10

ALK cell IC_{50} = 6.4 nM
ALK-L1196M cell IC_{50} = 97 nM
MDR (BA/AB)[a] = 5.5
LipE[b] = 5.0

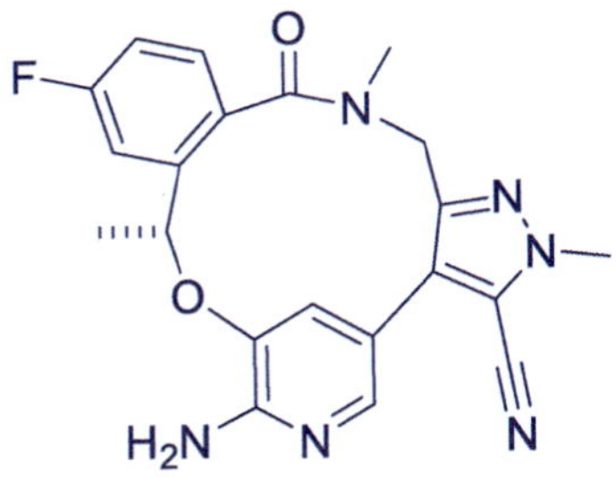

Lorlatinib **(1)**

ALK cell IC_{50} = 1.3 nM
ALK-L1196M cell IC_{50} = 21 nM
MDR (BA/AB)[a] = 1.5

Figure 6. SAR progression from of macrocycles to discovery of lorlatinib (**1**, PF-06463922). [a]MDCK-MDRI BA/AB efflux at 2 μM substrate concentration and pH 7.4. [b]LipE = – [log(pALK – L1196M cell IC_{50})] – log D.

To decrease lipophilicity of the 12-membered macrocyclic ethers, the chemistry team pursued an amide linker. Computational modeling suggested that a single carbon linker between the amide and proximal pyrazole moiety would result in a 12-membered lactam (compound **9**) with minimal ring strain.[7] Additional hydrogen bonding interactions were expected between the amide linker and the ALK protein domain. Most importantly, incorporation of the amide functionality decreases lipophilicity, moving the lactam analogs into an optimal log D range. The co-crystal structure of compound **9** bound to ALK was superimposed on the original docking model (Figure 7). The crystal structure confirms that the *N*-methyl group of **9** is pulled closer toward the G-loop by 1.2 Å, therefore enabling closer contact with the carbonyl group of Leu1122 and the side chains of Leu1122, Gly1123, and Val1130. Additionally, the amide carbonyl interacts with both Lys1150 and His1124 through the formation of a water bridge.[7]

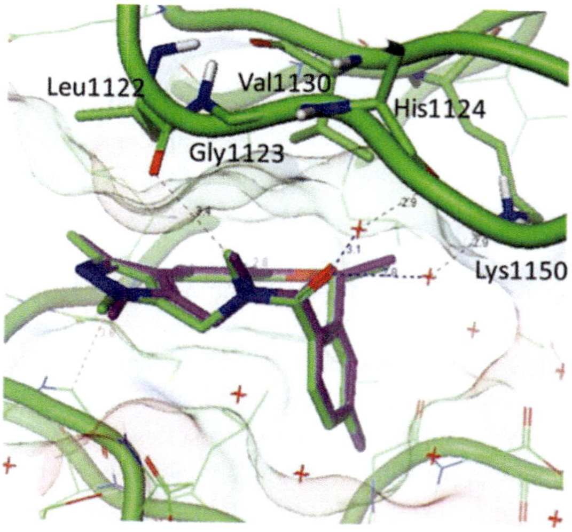

Figure 7. Co-crystal structure of **9** in ALK (green) overlaid with modeled ligand (purple). (Reprinted with permission from Reference 7. Copyright 2014 American Chemical Society.)

Compound **9** addressed several shortcomings of the previous acyclic analogs. Specifically, both cellular potency and LipE against ALK WT and ALK-L1196M were improved. Furthermore, increased permeability and metabolic stability demonstrated the potential of macrocyclic amides as lead compounds for the development of potent ALK inhibitors with CNS availability. Throughout the SAR development on this program, kinase selectivity was optimized against tropomycin receptor kinase B (TrkB). Extensive lead optimization was carried out on the macrocyclic amide scaffold to achieve the best combination of potency, CNS availability, ADME properties, and kinase selectivity in a single compound.[7]

SAR analysis of additional macrolactams highlighted several crucial features in the final optimization toward lorlatinib (**1**, Figure 6).[7]

1. Compound **9** ((*R*)-enantiomer) exhibits a 210-fold increase in potency against the gatekeeper mutant compared with its corresponding (*S*)-enantiomer.
2. Removal of the methyl group from compound **9** to afford compound **10** leads to a ~9-fold decrease in potency in the ALK cell assay.
3. Exploration of heterocyclic replacements demonstrated that the 2-aminopyridine and cyanopyrazole moieties provided the optimal overall potency and properties.

4. The cyanopyrazole moiety was a particularly important feature for kinase selectivity. The nitrile group was hypothesized to make unfavorable contacts with Tyr635 in TrkB.

11.4 Pharmacokinetics and Drug Metabolism

For patients treated with first- or second-generation ALK inhibitors, metastases in the CNS are common (occurring in ~45–70% of patients on crizotinib (**2**)), due to an inability of these drugs to achieve therapeutic concentrations in the CNS. Lorlatinib (**1**) was designed to achieve increased drug exposure in the CNS as well as an improved overall pharmacokinetic profile compared to first- and second-generation ALK inhibitors.[11,12]

Preclinical pharmacokinetic analysis of intravenously (iv) or orally administered lorlatinib (**1**) in rats and dogs demonstrates a low clearance rate, a moderate volume of distribution, and excellent absorption (Table 1).[10] In both species, the oral bioavailability approaches 100% (~100% for rats and 97% for dogs), and maximum plasma lorlatinib (**1**) concentrations are achieved approximately 1 hour post dose (t_{max}). To further predict drug pharmacokinetics in a human patient population, a population pharmacokinetic model for lorlatinib (**1**) was developed using data pooled from clinical trials.[13] This modeling supported the use of lorlatinib (**1**) at 100 mg qd dosing as a dose that achieves safe and efficacious drug levels at steady state.

Table 1. Pharmacokinetic profiling of lorlatinib (**1**) in preclinical rat and dog studies

Parameter	Rats[a]	Dogs[b]
Clearance (ml/min/kg)	15.5	9.1
Volume of distribution (L/kg)	2.6	2.8
$t_{1/2}$ (h)	2.7	4.6
Oral absorption (%F)	~100	97

[a]Rats (n = 2) were administered 1 mg/kg (iv) and 5 mg/kg (oral gavage) in solution formulation. [b]Dogs (n = 2) were administered 1 mg/kg (iv) and 2 mg/kg (oral gavage) in solution formulation. Source: Adapted from Reference 10.

The proliferation of resistant ALK mutants and metastasis to the CNS are the two primary mechanisms by which NSCLC can lose susceptibility to first- and second-generation ALK kinase inhibitors.[12] Lorlatinib (**1**) was specifically designed for improved penetration of the CNS to prevent and treat dangerous cranial metastases. *In vivo* assessments of brain penetration in rats were used to investigate the concentration and distribution of lorlatinib (**1**) in the brain. The ratio unbound (free) drug concentration in brain vs. plasma was ~0.30, an impressive 10-fold improvement in BBB penetration compared to crizotinib (2).[6,12] Autoradiography studies and mass spectrometry analysis showed a relatively uniform distribution of lorlatinib (**1**) in the cerebellum and cerebrum beyond the fenestrations created by brain tumors, indicative of a highly brain penetrant compound.

11.5 Efficacy and Safety

First- and second-generation ALK-targeted therapies can provide immense benefits in the treatment of ALK-positive NSCLC; however, limited treatment options remain after relapse. Preclinical studies showed that lorlatinib (**1**) is highly effective in patient-derived and genetically-engineered cancer models driven by drug-resistant ALK.[12] For example, lorlatinib (**1**) was highly active against the ALK mutant G1202R, a strain with broad resistance to first- and second-generation ALK inhibitors. In biochemical assays on recombinant WT ALK and crizotinib (**2**)-resistant mutants, lorlatinib (**1**) was 30-fold more effective at suppressing ALK-dependent phosphorylation, inducing apoptosis, and inhibiting cellular proliferation than crizotinib (**2**).[12] Lorlatinib (**1**) also demonstrated superior efficacy in disease models using patient-derived tumor xenografts in mice.[12] Both WT ALK and crizotinib (**2**)-resistant tumor xenografts showed greater inhibition from lorlatinib (**1**) than crizotinib (**2**), with durable antitumor effects observed after a four-day treatment of 3 and 10 mg/kg/day of lorlatinib (**1**).[12] These superior antitumor effects were maintained in NSCLC brain metastasis models as well. In addition to its superior efficacy, lorlatinib (**1**) also demonstrated a favorable preclinical safety profile.

Based on its promising preclinical profile, lorlatinib (**1**) was advanced to phase I/II clinical trials to evaluate safety, pharmacokinetics, pharmacodynamics, and patient-reported outcomes in ALK-positive/ROS1-positive, advanced NSCLC patients.[14] Approximately 4% of NSCLC patients are ALK-positive, with younger, non-smoking populations disproportionately represented in this group.[5] In a clinical trial directly comparing lorlatinib (**1**) and crizotinib (**2**) in treatment-experienced individuals, the two primary objectives were to find dose-limiting toxicities (phase I) and to compare the percentage of patients with overall and intracranial objective response (phase II).[15] The overall objective response rate is the percentage of patients with a 30–100% reduction in tumor size. Likewise, the intracranial objective response rate is the percentage of patients with a 30–100% reduction in brain lesion size. Based on the safety and efficacy results from the phase I portion of the study, 100 mg daily was established as the recommended dose for phase II trials.

Phase II studies evaluated the overall and intracranial objective response rate in different patient classes (Table 2).[15] These classes include treatment-naïve patients, treatment-experienced patients relapsing on crizotinib (**2**), and treatment-experienced patients relapsing on one or more lines of ALK inhibitor therapy. Of the 30 treatment-naïve patients receiving lorlatinib (**1**), an overall objective response rate of 90% was observed with an intracranial objective response rate of 75%. The overall and intracranial objective response rates were 74.1% and 58.8%, respectively, in patients relapsing after crizotinib (**2**) therapy, demonstrating the remarkable clinical efficacy of lorlatinib (**1**), even against crizotinib (**2**)-resistant NSCLC strains. Similar results were observed for patients relapsing after treatment on one or more different ALK inhibitor therapies. The substantial efficacy observed after previous ALK inhibitor therapy suggests that lorlatinib (**1**) is highly effective against many clinically relevant drug-resistant ALK mutants. The robust intracranial objective response rates demonstrate that lorlatinib (**1**) also has adequate CNS exposure, which should contribute to increased survival for patients after relapse. Overall, the relatively high dose required for dose-limiting toxicity and the robust efficacy for relapsed NSCLC patients demonstrate that lorlatinib (**1**) has a broad therapeutic window and is effective against resistant tumor strains, making it a valuable tool in the continued treatment of advanced NSCLC.

Table 2. Overall and intracranial objective response to lorlatinib (**1**) in phase II clinical trials.

Patient class	Overall objective response (%)[a]	Intracranial objective response (%)[b]
Treatment-naïve with advanced ALK-positive NSCLC	90.0	75.0
Advanced ALK-positive NSCLC after crizotinib (**2**)	74.1	58.8
Advanced ALK-positive NSCLC after two prior ALK inhibitor therapies	41.5	55.6
Advanced ALK-positive NSCLC after three prior ALK inhibitor therapies	34.8	39.5

[a]A 30–100% reduction in the tumor size for target lesions.
[b]Referring to lesions only within the brain.
Source: Based on Reference 15.

Lorlatinib (**1**) was well-tolerated overall in clinical studies. Most adverse events were related to homeostatic imbalances in lipid levels, such as hypercholesterolemia and hypertriglyceridemia, which are reversible by treatment with statins.[15] Other common (>20%) treatment-related adverse events include edema, cognitive defects, fatigue, weight gain, mood defects, and diarrhea. Serious treatment-related adverse events occurred in only 7% of patients, and only 3% of patients permanently discontinued treatment due to treatment-related adverse events.

Lorlatinib (**1**) is metabolized by both CYP2A and CYP3A and demonstrates moderate potential for drug–drug interactions (DDIs).[14–17] The simultaneous treatment with lorlatinib (**1**) and the CYP3A inducer rifampin caused significant reduction of lorlatinib (**1**) exposure during phase I clinical trials.[16] For this reason, lorlatinib (**1**) is contraindicated for use with strong CYP3A inducers. Conversely, the concomitant administration of lorlatinib (**1**) and the CYP3A inhibitor itraconazole caused higher plasma exposure of lorlatinib (**1**), requiring a reduction in lorlatinib (**1**) dose to maintain therapeutic concentrations.[17] Lorlatinib (**1**) can also produce significant but reversible elevations in the levels of aspartate transaminase and alanine transaminase.[16]

Lorlatinib (**1**) was initially granted accelerated the Food and Drug Administration (FDA) approval in November 2018 as second- or third-line treatment of patients with ALK positive metastatic NSCLC.[18] At that time, the first- and second-generation TKIs were still the front-line standard of care. However, the risk of relapse and CNS metastases with those therapies is high. More recently, a phase III randomized trial (CROWN trial—NCT03052608) demonstrated the efficacy of lorlatinib (**1**) as a first-line treatment for ALK-positive NSCLC compared to crizotinib (**2**).[19] The study showed that 78% of patients were living without disease progression after 1 year of lorlatinib (**1**) treatment compared with only 39% of patients with PFS from the crizotinib (**2**) cohort. For 78 of the 296 patients with CNS metastases at baseline, objective intracranial response was also significantly higher with lorlatinib (**1**) treatment (66%) than with crizotinib (**2**) treatment (20%). Lorlatinib (**1**) was also more effective at delaying CNS progression, as 96% of patients avoided CNS progression over the study's 12-month duration as compared to only 60% in the crizotinib (**2**) cohort. Based on the results of the CROWN trial, in March 2021, the FDA granted expanded regular approval to lorlatinib (**1**) for patients with metastatic NSCLC with ALK-positive tumors.[20]

11.6 Syntheses

11.6.1 Medicinal Chemistry Routes

The synthetic route to lorlatinib (**1**) faced two significant challenges: (1) Efficient construction of the 12-membered macrocyclic lactam and (2) Asymmetric installation of the drug's single chirality center.[7,21] Retrosynthetic analysis (Scheme 1) suggested that the final macrocyclization could be accomplished either *via* lactamization from amino acid **11** or intramolecular direct arylation from bromopyridine **12**.[7]

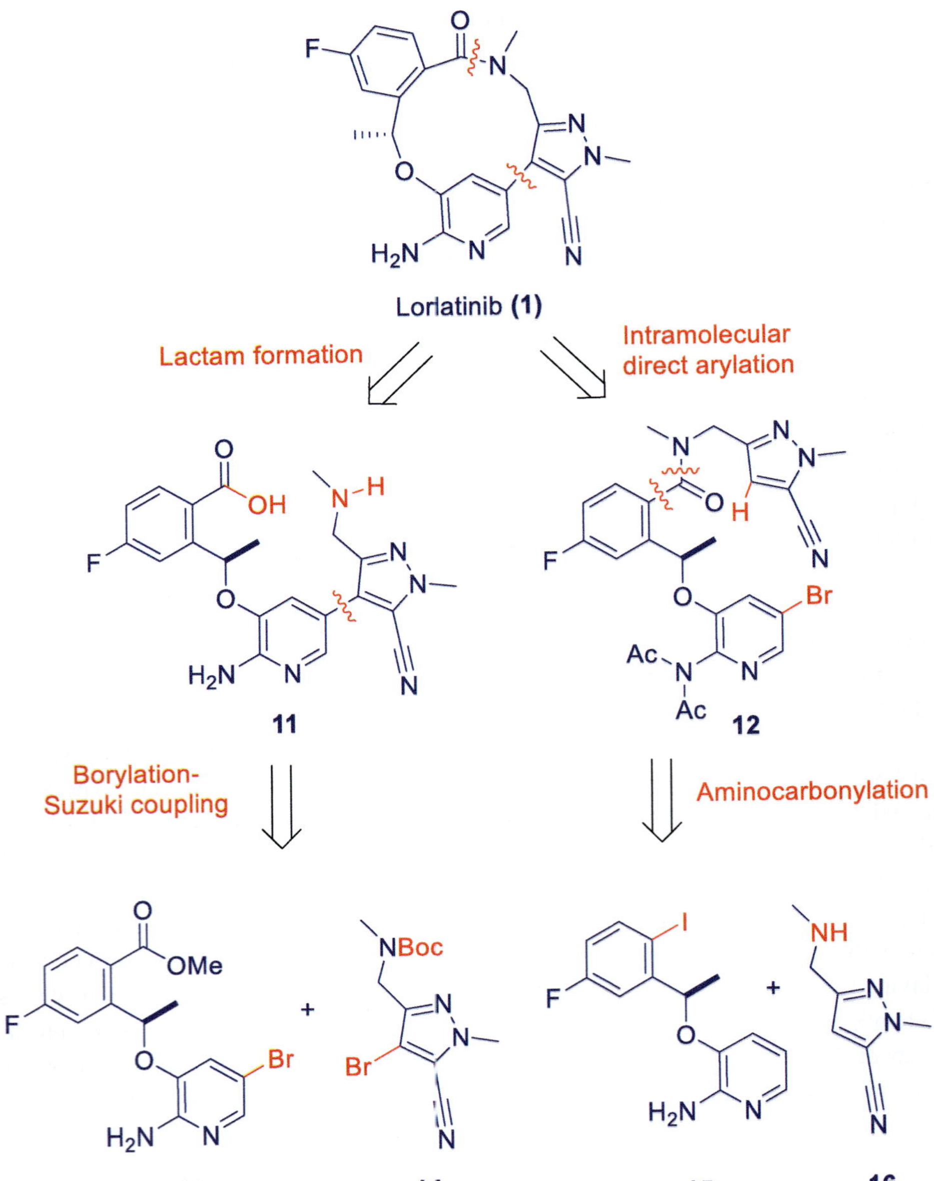

Scheme 1. Retrosynthetic analysis of lorlatinib (**1**).

Accordingly, the macrocyclization precursors could be obtained from appropriately functionalized aminopyridines (**13** or **15**), with a number of possible methods envisioned for installation of the benzylic chirality center. Likewise, parallel approaches could be used for the synthesis of pyrazoles **14** and **16**.

The initial medicinal chemistry route toward lorlatinib (**1**) investigated the late-stage lactamization for closure of the macrocycle. Toward this end, a route toward pyrazole **14** was first developed (Scheme 2).[22] Dibromination of pyrazole **17** *via* a radical pathway was achieved using *N*-bromosuccinimide (NBS) and benzoyl peroxide. Displacement of the primary bromide with methylamine was followed by Boc protection under standard conditions to afford Boc carbamate **19**. Finally, the ethyl ester of **19** was converted to the nitrile using a three-step sequence of saponification, amidation, and dehydration to afford **14** in 23% overall yield.[22]

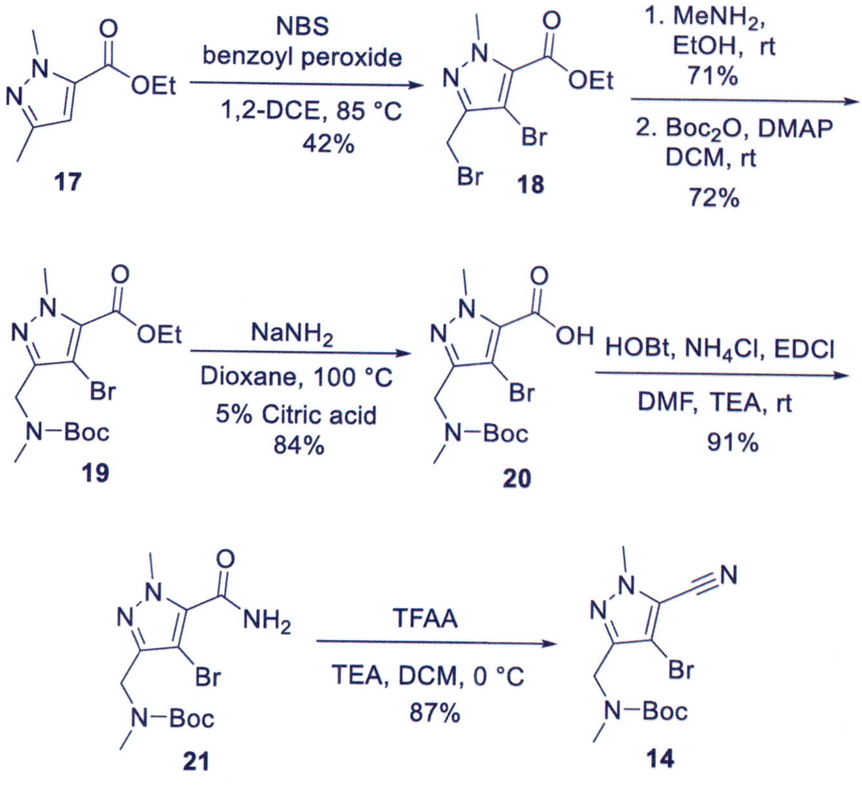

Scheme 2. Synthetic route toward pyrazole **14**. (Reference 22/Google patents)

To address the synthetic challenge of efficiently installing the chirality center, initial medicinal chemistry routes explored the use of chiral boron reducing agents. Both the *B*-diisopinocampheyl chloroborane (DIPCl) reagent and the Corey–Bakshi–Shibata (CBS) catalyst were evaluated in the reduction of ketone **22** to desired (*S*)-alcohol **23** (Scheme 3).[7] However, due to cost considerations, (–)-DIPCl was preferred for initial laboratory scale preparations. Cleavage of the borate intermediate was accomplished using diethanolamine in ethanol and tetrahydrofuran (THF) to yield (*S*)-alcohol **23** in 96% enantiomeric excess (*ee*).

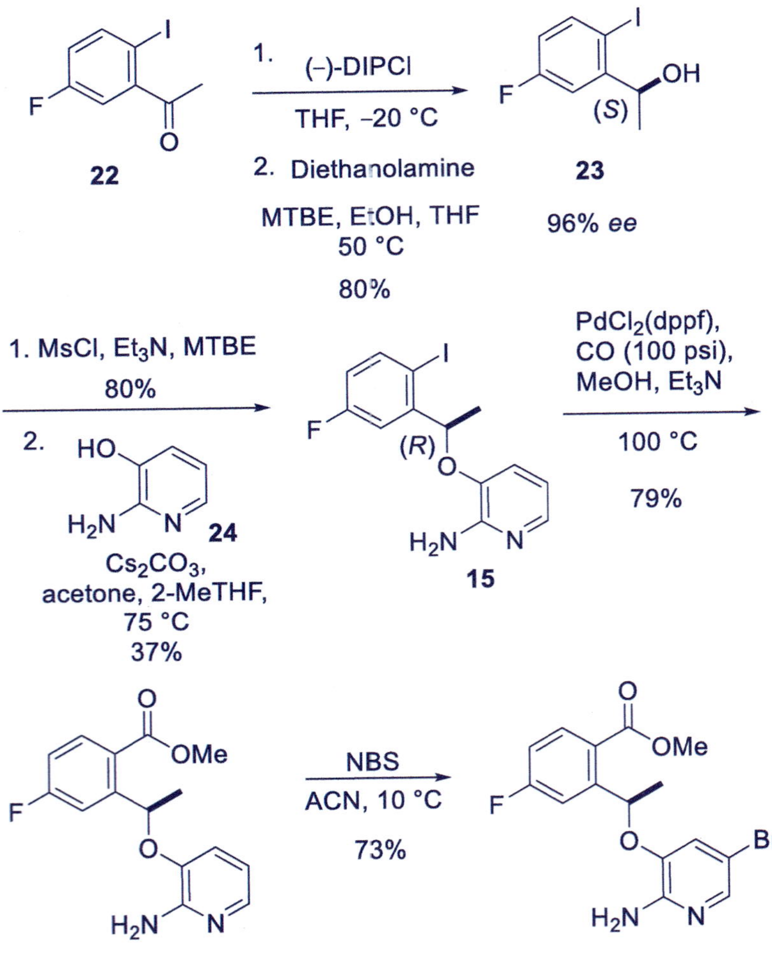

Scheme 3. Synthesis of bromopyridine **13**. (Reference 7/American Chemical Society)

Following the key asymmetric reduction, alcohol **23** was further elaborated to produce intermediates **13** and **15** for the initial medicinal chemistry routes. Toward this end, activation as the mesylate followed by nucleophilic substitution with amine **24** yielded **15** with the desired inversion of configuration at the chirality center.[22] Aryl iodide **15** could be used directly for the late-stage direct arylation cyclization pathway.

Alternatively, to enable the late-stage macrolactamization pathway, esterification of **15** was accomplished using carbon monoxide (CO, 100 psi) and catalytic $PdCl_2$(dppf) to yield methyl ester **25**. Bromination using NBS in acetonitrile provided bromopyridine **13**, which was isolated with high optical purity (99% *ee*).[22]

With intermediates **13** and **14** in hand, the convergent assembly of lorlatinib (**1**) was first investigated following the strategy of late-stage macrolactamization. (Scheme 4).[7] The route commenced with a one-pot borylation–Suzuki coupling in which a solution of **13** and **14** were reacted with bis(pinacolato)diboron (B_2Pin_2), CsF and catalytic $Pd(OAc)_2$ with CataCXiumA to produce **26** in 43% yield. This one-pot protocol provided adequate access to key intermediate **26** for initial medicinal chemistry studies. However, later optimization of this coupling reaction demonstrated improved yields by completing the borylation and Suzuki coupling processes in separate steps. Saponification of methyl ester **26** using sodium hydroxide revealed the carboxylic acid functionality, and subsequent acid-mediated Boc deprotection unmasked the secondary amine in penultimate intermediate **11**. Lactamization was achieved using hexafluorophosphate azabenzotriazole tetramethyluronium (HATU) in DMF to produce lorlatinib (**1**) in 29% yield.[7,22] The presence of the chiral methyl group, which significantly increased potency during SAR studies, also markedly improved lactamization yield relative to the des-methyl compound. Conformational analysis of the methylated and des-methyl analogs illustrated that the allylic methyl group induces $A^{1,3}$ strain. In the resulting low-energy conformations, the amine and carboxylic acid termini are in close proximity, promoting higher lactamization yields.[7]

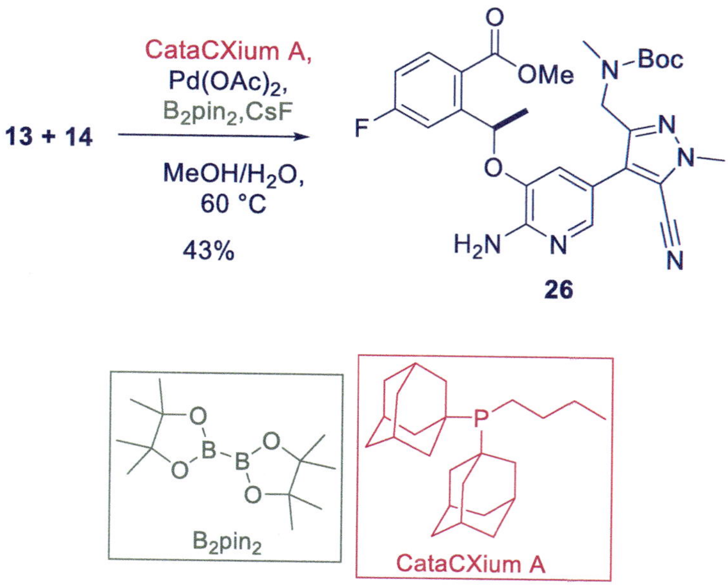

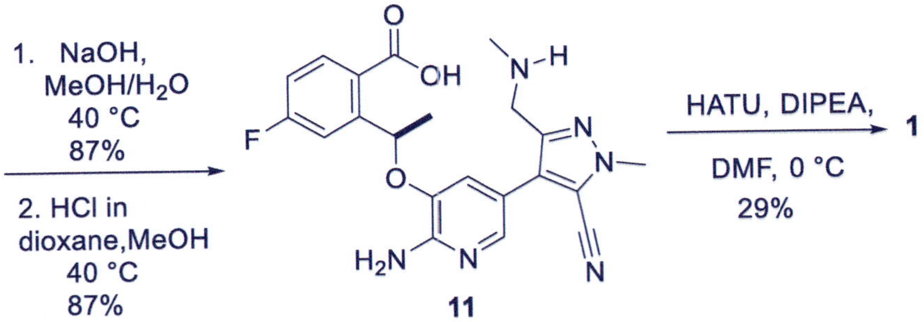

11

Scheme 4. Assembly of lorlatinib (**1**) using macrolactamization approach. (Reference 7/American Chemical Society)

The first discovery synthesis route (Scheme 4) was used to provide the initial 5 g of lorlatinib (**1**) and served as a basis for future scale up approaches.[7] However, a number of limitations were identified that would need to be addressed during process scale optimization. First, in the one-pot borylation–Suzuki coupling, homocoupling of **13** produced by-product **27** in 30% yield (Figure 8). Additionally, competitive hydrolysis of the nitrile during saponification of ester **26** led to formation of a significant amount of amide byproduct **28**. Finally, safety concerns regarding HATU and the requirement for a total of three silica gel chromatography steps presented additional barriers for kilogram scale lorlatinib (**1**) production.[7]

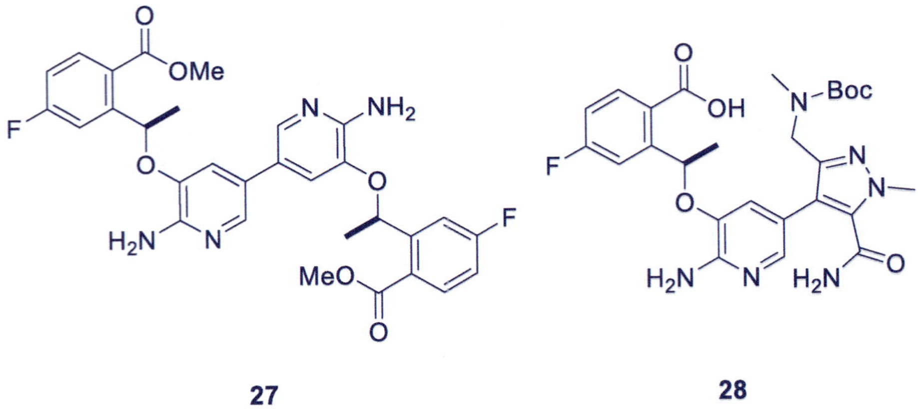

27 **28**

Figure 8. Major by-products formed in synthesis of lorlatinib (**1**) using the macrolactamization route.

In an alternate discovery chemistry approach to lorlatinib (**1**), direct arylation was envisioned as the final, ring-closing step.[7] To initiate this route, aryl iodide **15** was joined to amine **16** *via* a palladium-catalyzed aminocarbonylation reaction using Pd(P*t*-

Bu$_3$)$_2$, diisopropylethylamine (DIPEA), and carbon monoxide (4 bar) to afford **29** in 70% yield (Scheme 5). Selective bromination of the pyridine ring was accomplished at low temperature using 1.1 equivalents of NBS to yield **30**. Attempts at direct intramolecular arylation of **30** led to de-halogenated **29** as the major product, prompting judicious protection of the aminopyridine. Thus, bis-acetamide derivative **12** was synthesized from reaction of **30** in neat acetic anhydride. Direct arylation to effect macrocyclization of aryl bromide **12** was accomplished using CataCXium A, Pd(OAc)$_2$, and potassium acetate (KOAc). Use of *t*-amyl alcohol (*t*-AmOH) as solvent proved to be crucial to the efficiency of this reaction. Acid-mediated deprotection of **31** yielded lorlatinib (**1**) in 42% yield over the final two steps.[7]

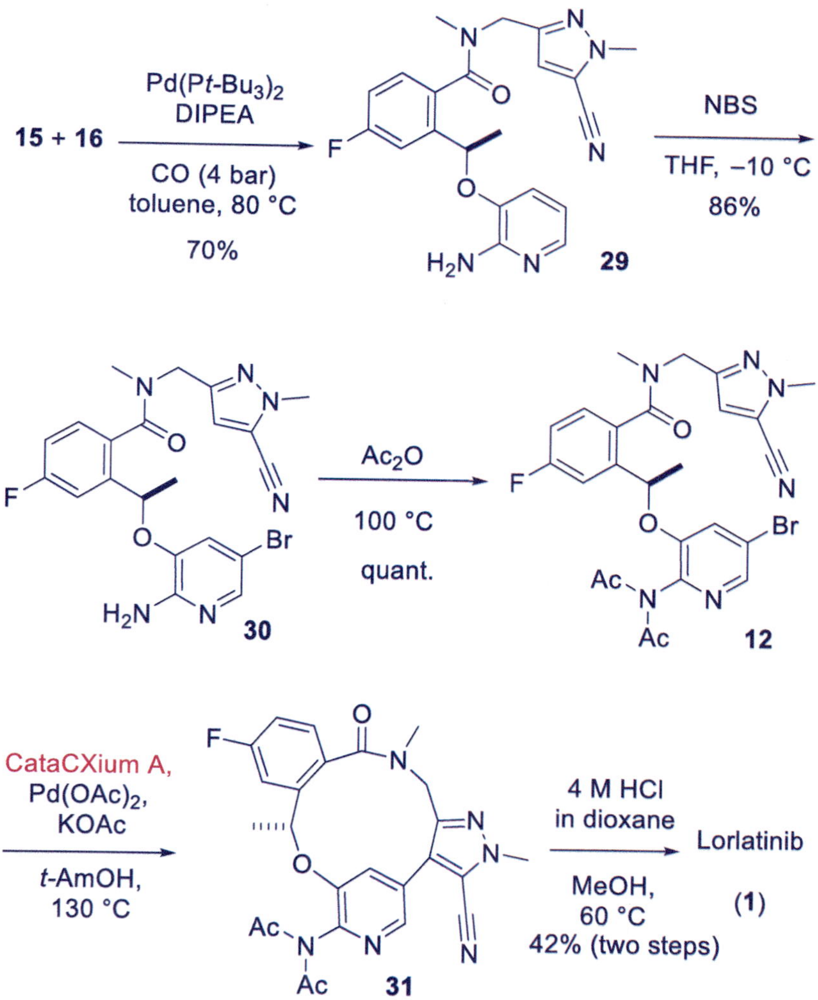

Scheme 5. Assembly of lorlatinib (**1**) using intramolecular direct arylation approach.

The direct arylation provides an innovative alternative synthesis for lorlatinib (**1**). This route provides a marked improvement in efficiency (25% overall yield over five steps from key intermediates **15** and **16**) compared with the original macrolactamization route (11% overall yield over four steps from compounds **13** and **14**). Additionally, the creative application of C–H activation chemistry was the first report of a macrocyclic ring closure accomplished using Pd-catalyzed direct arylation. By avoiding the stoichiometric boron reagents required in Suzuki couplings, the direct arylation method has potential to contribute to a greener and more atom economical overall synthetic process.

11.6.2 Lorlatinib Process Routes

The first scale-up route toward lorlatinib (**1**, Scheme 6) was designed to overcome key limitations of the original medicinal chemistry synthetic approach.[23] First, to address the strong binding affinity of palladium with the 2-aminopyridine moiety of **13**, Boc protecting groups were introduced in intermediate **32**. This modification was designed to reduce the amount of palladium catalyst required for the Suzuki coupling and to facilitate the removal of residual palladium from the product. Additionally, the boronate ester was incorporated into **32** to minimize homo-coupling side products and to improve the yield on the Suzuki coupling. Cleavage of all Boc protecting groups with HCl (g) was followed by chemoselective saponification of the methyl ester using potassium trimethylsilanoate (TMSOK), which was selected to avoid hydrolysis of the nitrile group. The overall yield of 66% over three steps represents a significant improvement over the previous route, which had achieved only 32% yield over the analogous three-step sequence. As an additional benefit, no chromatography was required with this improved sequence, in which crystalline potassium salt **34** precipitated directly out of the reaction mixture in 98% purity.[23]

In the final step of this route, the process chemistry team evaluated alternate coupling reagents for the macrolactamization due to concerns about the use of HATU on large scale. Use of 1-cyano-2-ethoxy-2-oxoethylidenaminooxy)dimethylamino-morpholino-carbenium hexafluorophosphate (COMU), selected as a potentially safer alternative to HATU, provided acceptable yield (46% over the final two steps).[23] In summary, the changes incorporated in this initial scale-up synthesis increased the overall yield from intermediate **14** to **1** from 10% to 30% for the initial 438 g scale-up batch.[23] Ultimately, further safety testing of HATU verified that it could be handled safely on scale. Thus, HATU, which promoted the intramolecular amide formation in higher yield and eliminated the only chromatographic step in the sequence, became the preferred coupling reagent for the large-scale macrocyclization.[23]

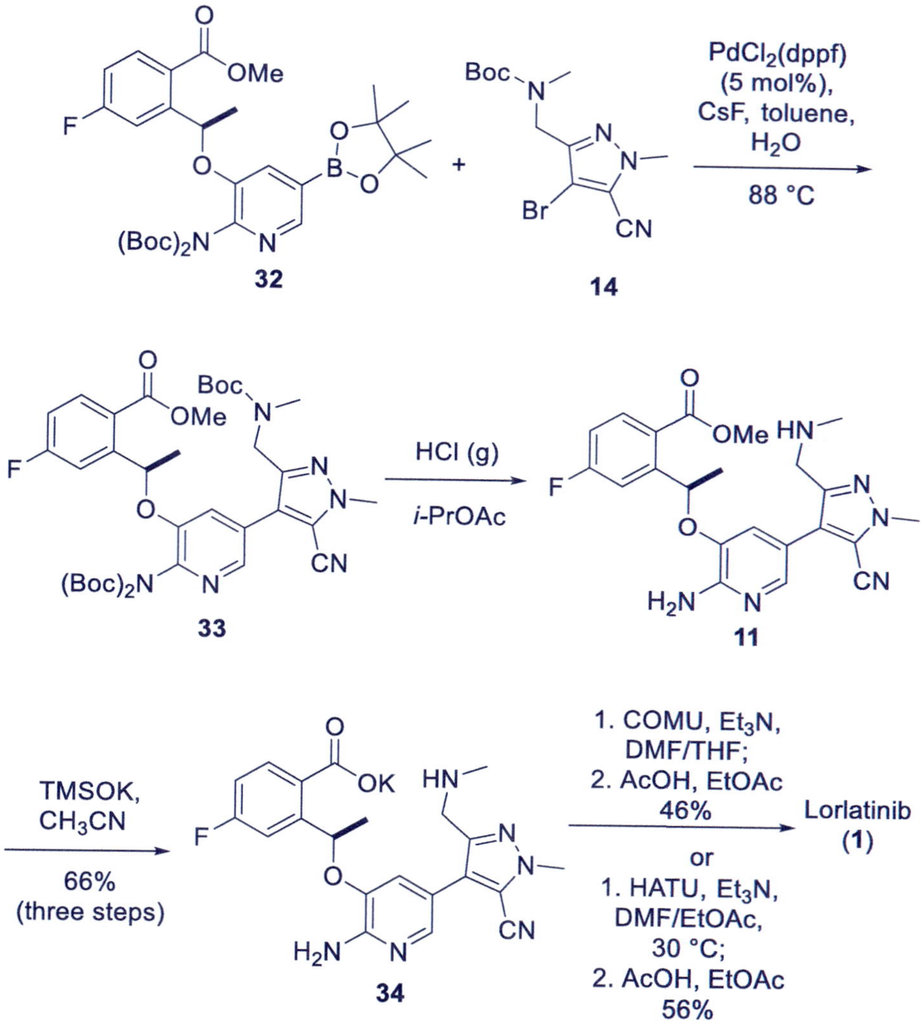

Scheme 6. Initial scale-up synthesis route.

The initial process scale-up route (Scheme 6) was suitable for the preparation of kilogram scale quantities of lorlatinib (**1**) to support clinical studies. However, this route was lengthy and linear, so a more convergent approach was required for commercial manufacturing of lorlatinib (**1**). Additionally, to maximize overall efficiency, the manufacturing route revisited the direct arylation approach for final ring closure, which had shown some advantages during the discovery chemistry investigations.[21] To improve convergence and efficiency, the retrosynthesis was reenvisioned (Scheme 7). *In situ* tosyl transfer from the highly functionalized bromopyridine **36** to chiral alcohol **35** followed by S_N2 substitution would furnish the advanced macrocyclization precursor required for a late-stage direct arylation ring closure. An early-stage amidation between chiral lactone **37** and amine **16** would yield alcohol **35** in convergent fashion.

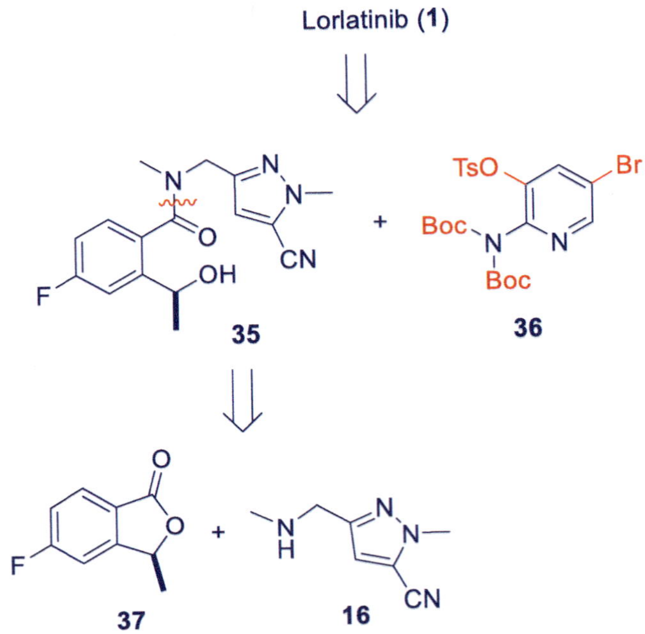

Scheme 7. Manufacturing retrosynthetic strategy.

Optimization of the synthesis of chiral lactone **37** with high chiral purity became a key objective in the commercial process development (Scheme 8).[21] Initial processes toward lactone **37** utilized either CBS or DIPCl for enantioselective reductions of ketone **22**. Although CBS catalyst was used successfully on laboratory scale, it was not cost-effective at industrial scale. Thus, DIPCl was evaluated for use on scale. Reduction of **22** with (–)-DIPCl produced **23** with >97% *ee*; however, the isolation of alcohol **23** after boron-mediated reduction became a significant issue on large scale.[21] Diethanolamine was used to facilitate cleavage of the borate intermediate, but removal of the diethanolamine borate by-product by filtration was challenging on scale. This borate side product was also identified as a potential sensitizer, creating health and safety concerns for its use on scale. Additionally, removal of α-pinene, a by-product of DIPCl, required low-temperature recrystallization, which was inconvenient and reduced the yield to 67% after isolation. These limitations led the process team to investigate alternate methods for the asymmetric reduction of ketone **22**.

Enzymatic reduction emerged as a promising alternative to traditional chemical asymmetric reductions. Screening of enzyme libraries identified 2,4-diketogluconic acid (DkgA) enzyme, which uses NADPH as a cofactor.[21] This remarkably effective bioreduction system produced **23** in >98% chemical purity and >99% *ee*. To recycle the cofactor, isopropanol was used as a co-substrate and *Lactobacillus brevis* alcohol dehydrogenase (ADH LB) as a recycling enzyme. This optimized enzymatic method was successfully implemented on scale up to 350 kg. The bioreduction methodology provided dramatic improvements in product isolation over the original boron reduction methods. The crude product was readily precipitated from the enzymatic reaction mixture and could be easily recrystallized without need for cryogenic conditions to afford **23** in 91% yield and nearly 100% purity. Alcohol **23** was efficiently converted to lactone **37** using Pd-catalyzed carbonylation with $Pd(dppf)Cl_2$ under 50 psi CO. The two-step enzymatic reduction/carbonylation process was used successfully on scale to manufacture up to 1070 kg of **37** with high yield (90%) and purity (>99.8% HPLC, >99.9% *ee*).[21]

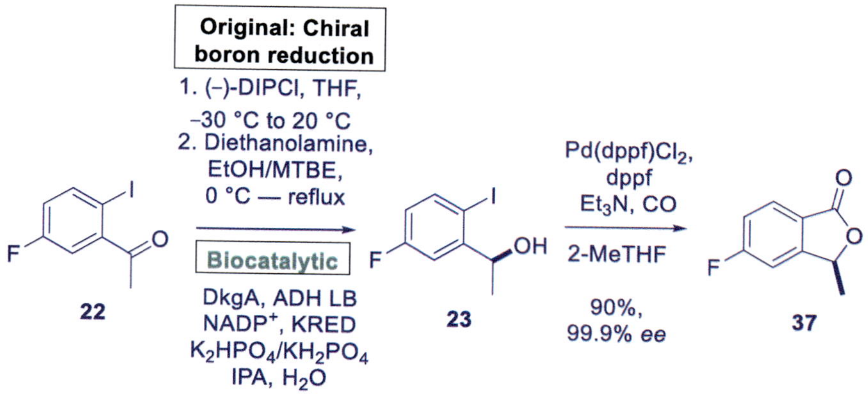

Scheme 8. Alternate scale-up approaches to lactone **37**. (Reference 21/American Chemical Society)

Despite the successful implementation of the enzymatic reduction/carbonylation sequence, the route was deemed too expensive for large-scale production, so an improved route was ultimately required for commercial manufacturing of lorlatinib (**1**).[21] The cost for producing ketone **22**, which was obtained in four to six steps from commodity chemicals, presented one significant concern. Additionally, the cost of enzymes and cofactor for the enzymatic reduction contributed to high overall costs of this route. Finally, the use of toxic CO gas in the carbonylation step became a significant drawback due to worker safety concerns on the manufacturing scale. These limitations led the process chemistry team to develop yet another approach to lactone **37** that would resolve all of these shortcomings. The ideal synthesis of **37** would use an inexpensive, commercially available starting material with the carbonyl group of lactone **37** already intact. Additionally, cost-effective asymmetric reduction conditions would preferably replace the expensive enzymatic reduction. With these criteria in mind, numerous alternative routes and conditions were explored to achieve the most optimal and cost-effective approach for commercial production.

The manufacturing route for lactone **37** began with the amidation of 4-fluorobenzoyl chloride (**38**) with diisopropylamine (Scheme 9).[21] The bulky amide **39** served as an effective substrate for directing *ortho*-metallation with *n*-butyllithium to produce aryllithium intermediate **40**, which was treated with acetaldehyde to afford alcohol **41**. Oxidation with (2,2,6,6-tetramethylpiperidin-1-yl)oxyl (TEMPO) produced ketone **42**, the key intermediate for asymmetric reduction using Ru-cat **43**. Under the optimized transfer hydrogenation conditions, the reaction generated secondary alcohol **44** with superb enantioselectivity (>99.9% ee) and excellent yield (94%). Finally, lactonization of **44** under acidic conditions followed by neutralization yielded the target lactone **37**. Notably, the bulky diisopropylamide group played a pivotal role, both promoting high enantioselectivity in the asymmetric catalytic reduction, while also contributing to the highly crystalline properties of each intermediate in the route, thereby enabling facile isolation and purification without reliance on chromatography. The asymmetric transfer hydrogenation process developed for this manufacturing route was essential for the effective scale-up of lorlatinib (**1**). During the initial production of commercial product, this process was successfully implemented in the manufacture of 400 kg of lactone **37**.[21]

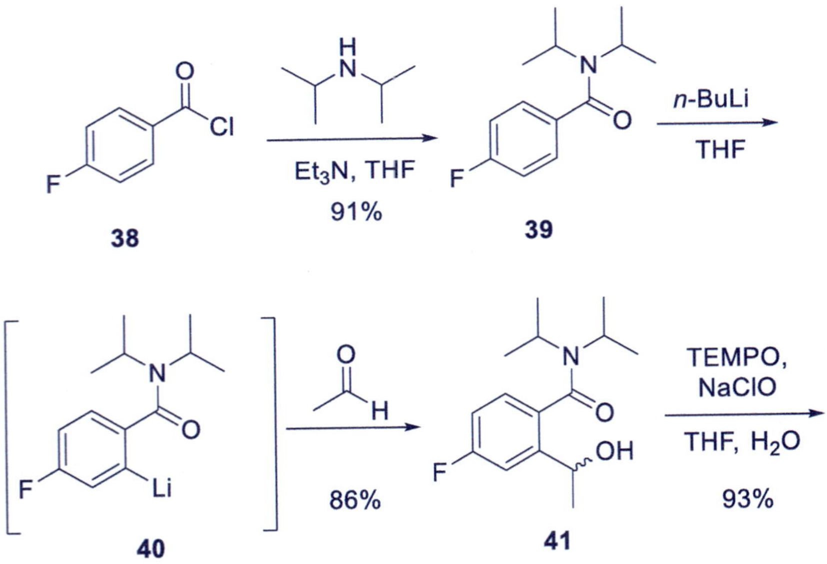

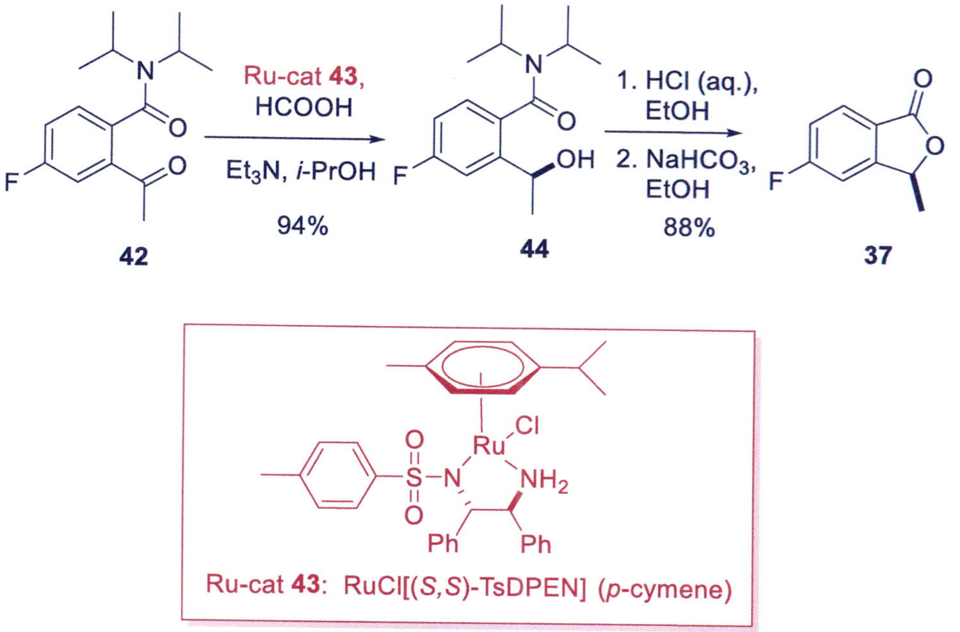

Scheme 9. Manufacturing route to lactone **37**. (Reference 21/American Chemical Society)

In addition to optimizing synthetic access to lactone **37**, the process chemistry team revised the overall assembly of lorlatinib (**1**), incorporating **37** as a key intermediate (Scheme 10).[24] In the remarkably concise commercial route toward lorlatinib (**1**), lactone **37** underwent a facile ring-opening amidation with amine **16**, using $AlCl_3$ as a Lewis acid catalyst, to afford intermediate **35** in nearly 100% conversion. Addition of tosylate **36** and potassium *t*-butoxide to a solution of **35** initiated transfer of the tosyl group to the secondary alcohol, followed by etherification of the resultant phenoxide *via* S_N2 substitution with inversion of configuration at the chirality center. Upon crystallization with *t*-amyl alcohol, cyclization precursor **45** was obtained in high purity with an overall yield of 65–70% over the two-step process. Although previous studies had demonstrated that the planned intramolecular cyclization of **45** could be achieved using $Pd(OAc)_2$ and CataCXium A, alternatives to this expensive, proprietary ligand were sought for the manufacturing process. Screening of additional ligands revealed that di(*n*-butyl)-*t*-butylphosphine provided similar catalytic activity to CataCXium A.[24] For ease of handling, the air-sensitive phosphine ligand was used as its stabilized HBF_4 salt, which was neutralized with NaO*t*-Bu to prepare the active catalyst under the cyclization conditions. In the event, the intramolecular Pd-catalyzed cyclization of **45** produced macrocycle **46** with typical yields of 65–70%. Finally, treatment of **46** with HCl in aqueous methanol removed both Boc groups to afford lorlatinib (**1**) with typical yields of

95% after crystallization. In initial production runs, this process was scaled to produce more than 120 kg of API.[24]

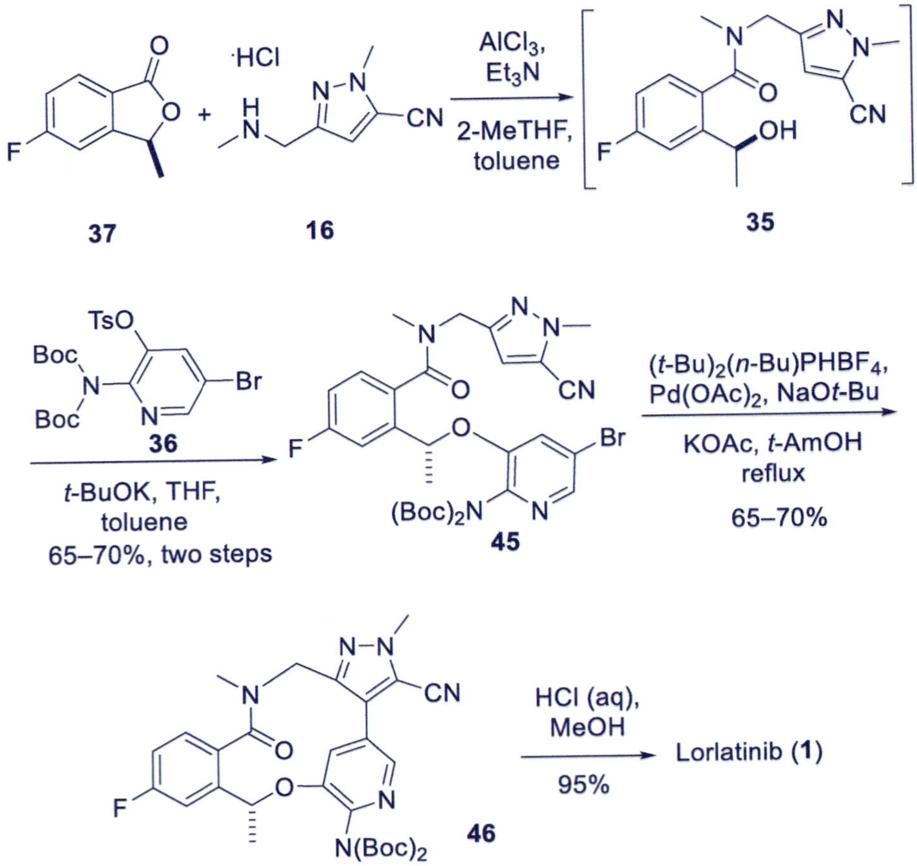

Scheme 10. Commercial route to lorlatinib (**1**). (Reference 24/American Chemical Society)

11.7 Summary

Lorlatinib (**1**, Lorbrena) is a third-generation, macrocyclic ALK/ROS1 kinase inhibitor developed specifically to inhibit drug-resistant ALK mutants and to retain efficacy against CNS metastases.[7] A structure-based drug design approach proved critical in the design of new scaffolds to minimize P-gp efflux and enhance BBB permeability while also avoiding resistance from ALK mutants. Among ALK inhibitors, lorlatinib (**1**) stands out as the most potent inhibitor of WT ALK and ALK mutants resistant to first- and second-generation ALK inhibitors (e.g., crizotinib (**2**)). The implementation of innovative chemistry, including the novel application of the intramolecular direct arylation, was key to the development of an efficient, cost-effective process for commercial manufacturing of lorlatinib (**1**).

Lorlatinib (**1**) is markedly more effective at targeting BM of NSCLC than previous generations of ALK inhibitors. It also has the benefit of once-a-day, 100 mg dosing with high tolerability at this dose. On November 2, 2018, Lorlatinib (**1**) was granted accelerated FDA approval for second- or third-line treatment of patients with ALK-positive metastatic NSCLC.[18] Based on the remarkable patient outcomes in the CROWN trial, on March 3, 2021 the FDA granted an expanded regular approval to this breakthrough medicine to include lorlatinib (**1**) as a first-line treatment of adult patients with metastatic ALK-positive NSCLC.[20]

11.8 References

1. Miller, R. A.; Cagle, P. T. Lung Cancer Epidemiology and Demographics, In *Precision Molecular Pathology of Lung Cancer. Molecular Pathology Library*. Cagle, P. et al. Eds. Springer, Cham., 2018; pp. 15–17.
2. Siegel, R.; Kimberley, M.; Fuchs, H. E.; Jemal, A. Cancer statistics, 2021. *CA Cancer J. Clin.* **2021**, *71*, 7–33.
3. Duma, N.; Santana-Davila, R.; Molina, J. R. Non-small cell lung cancer: epidemiology, screening, diagnosis, and treatment. *Mayo Clin. Proc.* **2019**, *94*, 1623–1640.
4. Remon, J.; Pignataro, D.; Novello, S.; Passiglia, F. Current treatment and future challenges in ROS1- and ALK-rearranged advanced non-small cell lung cancer. *Cancer Treat. Rev.* **2021**, *95*, 102178.
5. Waqar, S. N.; Morgensztern, D. Lorlatinib: a new-generation drug for ALK-positive NSCLC. *Lancet Oncol.* **2018**, *19*, 1555–1557.
6. Richardson, P. J. Discovery and Early Development of the Next-Generation ALK Inhibitor, Lorlatinib (18): Agent for Non-Small-Cell Lung Cancer. In *Drug Discovery and Development*, 3rd, O'Donnell, J. T., O'Donnell, J. J., Somberg, J., Idemyor, V., Eds.; CRC Press: Boca Raton, FL, 2019; pp. 185–218.
7. Johnson, T. W.; Richardson, P. F.; Bailey, S.; Brooun, A.; Burke, B. J.; Collins, M. R.; Cui, J. J.; Deal, J. G.; Deng, Y.-L.; Dinh, D.; Engstrom L. D.; He, M.; Hoffman, J.; Hoffman, R. L.; Huang, Q; Kania, R. S.; Kath, J. C.; Lam, H.; Lam, J. L.; Le, P. T.; Lingardo, L.; Liu, W.; McTigue, M.; Palmer, C. L.; Sach, N. W.; Smeal, T.; Smith, G. L.; Stewart, A. E.; Timofeevski, S.; Zhu, H.; Zhu, J.; Zou, H. Y.; Edwards, M. P. Discovery of (10*R*)-7-amino-12-fluoro-2,10,16-trimethyl-15-oxo-10,15,16,17-tetrahydro-2H-8,4-(metheno)pyrazolo[4,3-h][2,5,11]-benzoxadiazacyclotetradecine-3-carbonitrile (PF-06463922), a macrocyclic inhibitor of anaplastic lymphoma kinase (ALK) and c-Ros oncogene (ROS1) with preclinical brain exposure and broad spectrum potency against ALK-resistant mutations. *J. Med. Chem.* **2014**, *57*, 4720–4744.
8. Bayliss, R.; Choi, J.; Fennell, D. A.; Fry, A. M.; Richards, M. W. Molecular mechanisms that underpin EML4-ALK driven cancers and their response to targeted drugs. *Cell Mol. Life Sci.* **2016**, *73*, 1209–1224.

9. Sahu, A.; Prabhash, K.; Noronha, V.; Joshi, A.; Desai, S. Crizotinib (2): a comprehensive review. *South Asian J. Cancer*. **2013**, *2*, 91–97.
10. Yamazaki, S.; Lam, J. L.; Johnson, T. W. Discovery and Pharmacokinetic–Pharmacodynamic Evaluation of an Orally Available Novel Macrocyclic Inhibitor of Anaplastic Lymphoma Kinase and c-Ros Oncogene 1. In *Practical Medicinal Chemistry with Macrocycles: Design, Synthesis, and Case Studies*, Marsault, E.; Peterson, M. L.; Eds.; John Wiley & Sons, Inc: Hoboken, NJ, 2017; pp. 519–543.
11. Basit, S.; Ashraf, Z.; Lee, K.; Latif, M. First macrocyclic 3rd-generation ALK inhibitor for treatment of ALK/ROS1 cancer: clinical and designing strategy update of lorlatinib. *Eur. J. Med. Chem*. **2017**, *134*, 348–356.
12. Zou, H. Y.; Friboulet, L.; Kodack, D. P.; Engstrom, L. D.; Li, Q.; West, M.; Tang, R. W.; Wang, H.; Tsaparikos, K.; Wang, J.; Timofeevski, S.; Katayama, R.; Dinh, D. M.; Lam, H.; Lam, J. L.; Yamazaki, S.; Hu, W.; Patel, B.; Bezwada, D.; Frias, R. L.; Lifshits, E.; Mahmood, S.; Gainor, J. F.; Affolter, T.; Lappin, P. B.; Gukasyan, H.; Lee, N.; Deng, S.; Jain, R. K.; Johnson, T. W.; Shaw, A. T.; Fantin, V. R.; Smeal, T. PF-06463922, an ALK/ROS1 inhibitor, overcomes resistance to first and second generation ALK inhibitors in preclinical models. *Cancer Cell* **2015,** *28*, 70–81.
13. Chen, J.; Houk, B.; Pithavala, Y. K.; Ruiz-Garcia, A. Population pharmacokinetic model with time-varying clearance for lorlatinib using pooled data from patients with non-small cell lung cancer and healthy participants. *CPT Pharmacometrics Syst. Pharmacol*. **2021**, *10*, 148–160.
14. A Study of PF-06463922 An ALK/ROS1 Inhibitor In Patients With Advanced Non Small Cell Lung Cancer With Specific Molecular Alterations. https://clinicaltrials.gov/ct2/show/NCT01970865.
15. Solomon, B. J.; Besse, B.; Bauer, T. M.; Felip, E.; Soo, R. A.; Camidge, D. R.; Chiari, R.; Bearz, A.; Lin, C. C.; Gadgeel, S. M.; Riely, G. J.; Tan, E. H.; Seto, T.; James, L. P.; Clancy, J. S.; Abbattista, A.; Martini, J. F.; Chen, J.; Peltz, G.; Thurm, H.; Ou, S. I.; Shaw, A. T. Lorlatinib in patients with ALK-positive non-small-cell lung cancer: results from a global phase 2 study. *Lancet Oncol.* **2018**, *19*, 1654–1667.
16. Chen, J.; Xu, H.; Pawlak, S.; James, L.P.; Peltz, G.; Lee, K.; Ginman, K.; Bergeron, M.; Pithavala, Y.K. The effect of rifampin on the pharmacokinetics and safety of lorlatinib: results of a phase one, open-label, crossover study in healthy participants. *Adv Ther.* **2020**, *37,* 745–758.
17. Patel, M.; Chen, J.; McGrory, S.; O'Gorman, M.; Nepal, S.; Ginman, K.; Pithavala, Y.K. The effect of itraconazole on the pharmacokinetics of lorlatinib: results of a phase I, open-label, crossover study in healthy participants. *Invest. New Drugs*. **2020**, *38*, 131–139.
18. FDA approves lorlatinib for second- or third-line treatment of ALK-positive metastatic NSCLC. US Food and Drug Administration. *https://www.fda.gov/drugs/fda-approves-lorlatinib-second-or-third-line-treatment-alk-positive-metastatic-nsclc* (Accessed November 2, 2021).
19. Shaw, A. T.; Bauer, T. M.; de Marinis, F.; Felip, E.; Goto, Y.; Liu, G.; Mazieres, J.; Kim, D.-W.; Mok, T.; Polli, A.; Thurm, H.; Calella, A. M.; Peltz, G.;

Solomon, B. J. First-line lorlatinib or crizotinib in advanced ALK-positive lung cancer. *N. Engl. J. Med.* **2020**, *383*, 2018–2029.

20. FDA approves lorlatinib for metastatic ALK-positive NSCLC. US Food and Drug Administration. US Food and Drug Administration. *https://www.fda.gov/drugs/resources-information-approved-drugs/fda-approves-lorlatinib-metastatic-alk-positive-nsclc* (Accessed November 2, 2021)
21. Duan, S.; Li, B.; Dugger, R. W.; Conway, B.; Kumar, R.; Martinez, C.; Makowski, T.; Pearson, R.; Olivier, M.; Colon-Cruz, R. Developing an asymmetric transfer hydrogenation process for (*S*)-5-fluoro-3-methylisobenzofuran-1(*3H*)-one, a key intermediate to lorlatinib. *Org. Process Res. Dev.* **2017**, *21*, 1340–1348.
22. Jensen, A. J.; Luthra, S.; Richardson, P. F. Solid Forms of a Macrocyclic Kinase Inhibitor. WO 2014/207606 Al. December 31, 2014.
23. Li, B.; Barnhart, R. W.; Hoffman, J. E.; Nematalla, A.; Raggon, J.; Richardson, P.; Sach, N.; Weaver, J. Exploratory process development of lorlatinib. *Org. Process Res. Dev.* **2018**, *22*, 1289–1293.
24. Dugger, R.; Li, B.; Richardson, P. Discovery and Development of Lorlatinib: A Macrocyclic Inhibitor of EML4-ALK for the Treatment of NSCLC. In *Complete Accounts of Integrated Drug Discovery and Development: Recent Examples from the Pharmaceutical Industry, Vol 2.* Pesti, J. A.; Abdel-Magid, A. F.; Vaidyanathan, R.; Eds. ACS: Washington, D.C., 2019; pp. 27–60.

12

Niraparib (Zejula), A Small Molecule, PARP1/2 Inhibitor for Treating Breast, Ovarian, and Pancreatic Cancers

Raymond Ng

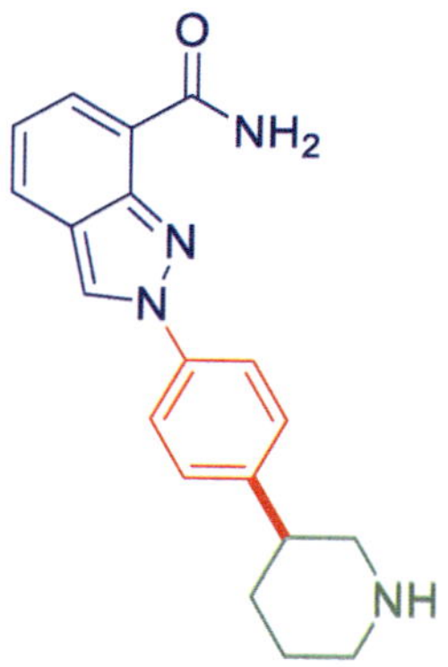

USAN: Niraparib
Trade name: Zejula (**1**)
GSK/Tesaro/Merck
Launched: 2017

12.1 Background

Poly(ADP-ribose) polymerases (PARPs), also known as ADP-ribosyltransferase diphtheria toxin-like proteins (ARTDs), are a family of 17 proteins involved in post-translational modification of proteins (Table 1).[1,2a,2b] PARPs regulate various processes such as DNA damage response, chromatin remodelling, transcription activation and repression,[2c] ubiquitylation, RNA metabolism, and cellular stress responses through a post-translational modification process called PARylation.

Current Drug Synthesis, First Edition. Edited by Jie Jack Li.

Table 1. PARPs (ARTD) as writers of post-translational modification.

Name	Alternative Names	Enzymatic Activity (automodification)	Cellular Localization[1]
ARTD1	PARP1	PARylation, branching	N
ARTD2	PARP2	PARylation	N >> C
ARTD3	PARP3	MARylation	N > C
ARTD4	PARP4, vPARP	MARylation	C > N
ARTD5	PARP5A, tankyrase 1	OARylation	C >> N
ARTD6	PARP5A, PARP6, tankyrase 2	OARylation	C >> N
ARTD7	PARP15, BAL3	MARylation	C
ARTD8	PARP14, BAL2, CoaSt6	MARylation	C > N
ARTD9	PARP9, BAL1	no activity reported	C >> N
ARTD10	PARP10	MARylation	C >> N
ARTD11	PARP11	MARylation	N and C
ARTD12	PARP12, ZC3HDC1	MARylation	C >> N
ARTD13	PARP13, ZC3HAV1, ZAP1	no activity reported	C
ARTD14	PARP7, tiPARP, RM1	MARylation	C and N
ARTD15	PARP16	MARylation	C
ARTD16	PARP8	MARylation	C
ARTD17	PARP6	MARylation	C
ARTD18	TPT1	no activity reported	unknown

Cellular localization during interphase (N = nuclear; C = cytoplasm)
(Adapted from Reference 1 with permission from Elsevier.)

Target proteins are post-translationally modified with ADP-ribose units using β nicotinamide adenine dinucleotide (β-NAD^+) as substrate. The best characterized PARPs, -1, -2, -5A (TNKS1), and -5B (TNKS2), are sometimes called polyPARPs, because of their ability to PARylate or transfer multiple ADP-ribose units, resulting in linear and branched poly(ADP-ribose) (PAR) chains. The remaining members of the PARP family are similarly called monoPARPs, since they MARylate or transfer a single ADP-ribose unit onto target proteins. PARPs (ARTDs) act as writers by adding ADP-ribose units onto proteins, while enzymes such as ADP-ribosyl hydrolases 1 and 3 (ARH1; ARH3), MDO1/2, TARG1, and poly(ADP-ribose) glycohydrolase (PARG) act as erasers by removing ADP-ribose units (Scheme 1).

Scheme 1. ADP-ribosylation.[1]

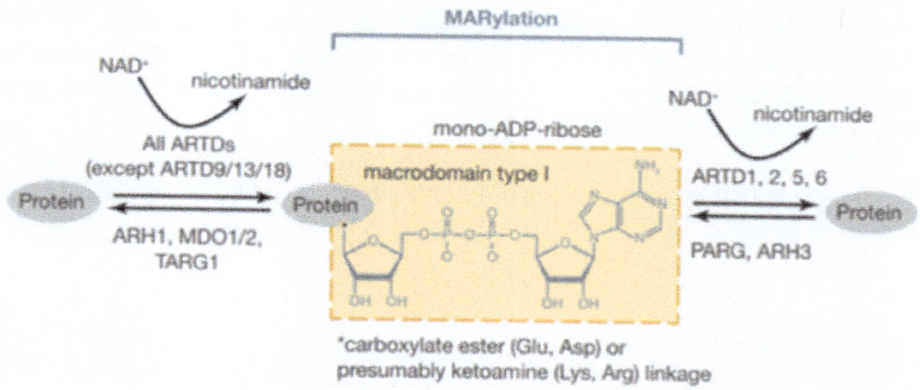

(Adapted from Reference 1 with permission from Elsevier.)

There is limited knowledge about the function(s) of many of the members of the monoPARP family. However, there is growing interest in PARP7,[3a] PARP10,[3b] and PARP16[3c] as targets for oncology indications. In 2020, Ribon Therapeutics disclosed their PARP7 inhibitor (first-in class) entering phase I clinical trials.

Although PARP1 was first discovered[4] in 1966, it took another 50 years before a PARP inhibitor was shown to be the first cancer drug to exploit synthetic lethality.[5] PARP1 and PARP2 are nuclear proteins that contain both a DNA binding domain and a catalytic domain. The DNA binding domain, which contains zinc fingers, enables PARP1 to localize rapidly and bind to DNA at the site of damage. The ensuing conformation change in the protein alters the catalytic domain, thereby increasing its activity up to 500-fold.

Figure 1. Timeline from PARP1 discovery to First Regulatory Approval.[5] (With permission from Clinical Care Options Oncology; clinicaloptions.com/oncology)

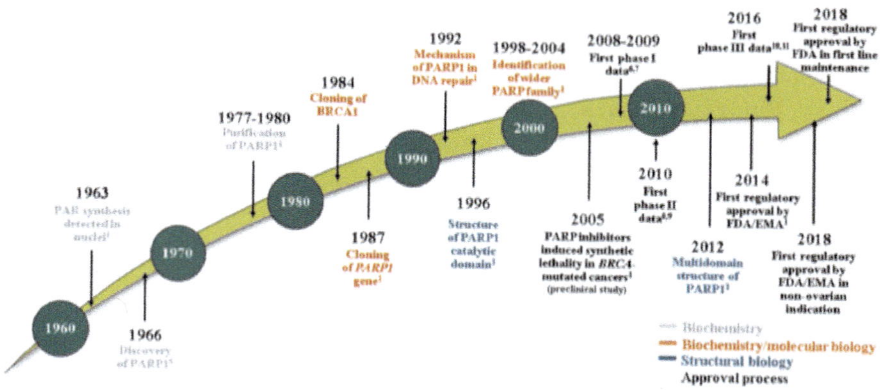

At the time of this writing, there are four approved PARP inhibitor drugs, all having activity at both PARP1 and PARP2.[12] Due to the high structural similarity between PARP1 and PARP2, it has been challenging to discover a PARP1-selective inhibitor. Only recently, Astra-Zeneca reported the discovery of a PARP1-selective inhibitor, AZD5305[13] which may provide a better safety profile with less hematological toxicity than first-generation PARP1 inhibitors.

Figure 2. Biological mechanism of PARP1 inhibition and DNA trapping.

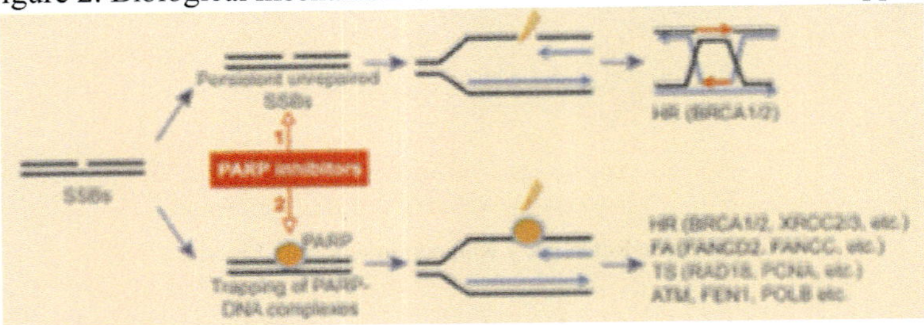

(Adapted from Reference 14 with permission from American Association for Cancer Research)

PARP1/2 inhibitors mediate their antitumor effects through dual cytotoxic mechanisms. The catalytic inhibition (upper pathway; Figure 2) blocks repair of DNA single strand breaks (SSBs) leading to persistent unrepaired SSBs and replication fork damage that requires homologous recombination (HR) and base excision repair (BER).[14] Stalled replication forks can further lead to DNA double strand breaks (DSB), which are more difficult to repair. In addition, inhibitors such as olaparib, veliparib, rucaparib, niraparib (**1**), and talazoparib also act in part as poisons that trap the PARP1 and PARP2 enzymes at damaged DNA (lower pathway; Figure 2). These resulting trapped PARP-DNA complexes also lead to replication fork damage but utilize additional repair pathways including Fanconi pathway, template switching, ATM, polymerase β, and FEN1 (replicative flap endonuclease). Moreover, the potency of inhibitors to trap PARP is not correlated with the catalytic inhibitory properties of each drug. Interestingly, talazoparib has the highest PARP trapping potency, followed by niraparib (**1**), rucaparib, olaparib, and veliparib (the weakest).

Knockout or inhibition of PARP1 impairs BER and sensitizes tumor cells to radiation or cytotoxic agents that induce DNA damage that is normally repaired by BER. Targeting more than a single DNA repair pathway in tumor cells could induce "synthetic lethality."[15] Two genes which loss of function of either alone retains cell viability, however loss of function of both contemporaneously that results in cell death is considered synthetic lethal. This concept has been demonstrated in the selective killing of BRCA-1 or BRCA-2 (known tumor suppressors involved in HR DSB repair) deficient tumor cells.[16] Ultimately these DSBs lead to genomic instability, cell cycle arrest, and apoptosis.

Nearly 90% of all ovarian malignancies are caused by epithelial ovarian cancer. A hallmark of high grade serous ovarian cancer (HGSOC) is a defective DNA damage response, of which PARP inhibitors can exploit through synthetic lethality. Not surprising, approved PARP inhibitors including clinical development candidate veliparib,[17] are indicated for ovarian cancer therapy.[18a] Other cancers with HR repair mutations, such as advanced prostate[18b] and pancreatic[18c] cancer, are also good candidates for treatment with PARPi.

12.2 Pharmacology

PARPi compete with β-NAD^+ at the catalytic site of the PARP enzyme. In addition to catalytic inhibition, these PARPi can produce conformational changes in PARP1 upon binding and stabilize the complex between PARP1 DNA-binding domain and DNA.

Figure 3. Mechanism of ADP-ribose transfer.[19]

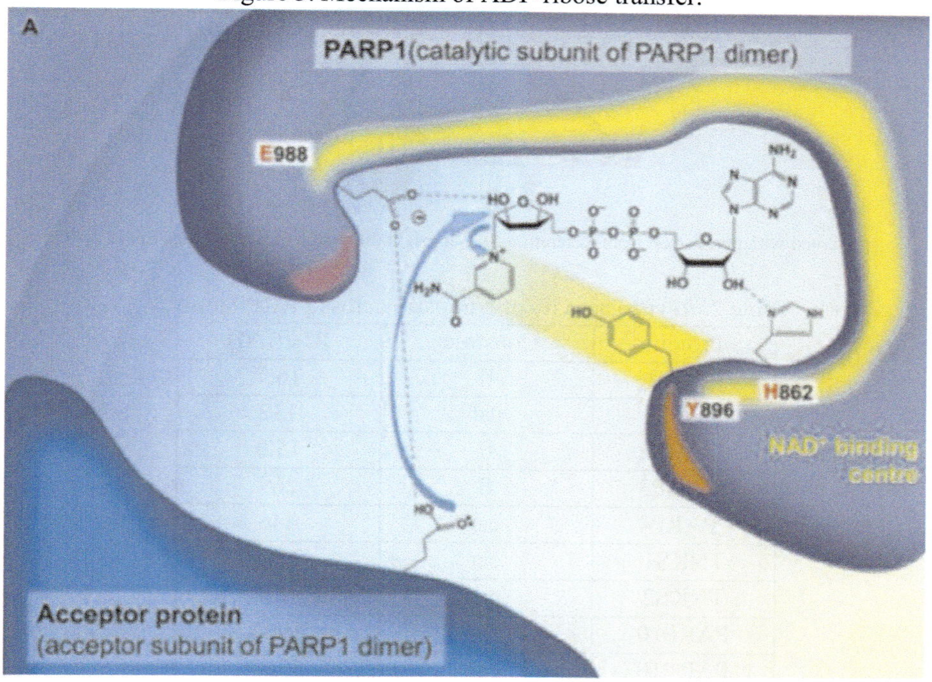

Figure 3 shows the carboxy group from the acceptor protein (in this case, a PARP1 dimer) reacting with β-NAD^+ bound to the catalytic subunit of another PARP1 dimer, thereby transferring an ADP-ribose unit to the acceptor protein.[19] Common amino acids in the acceptor protein that can be PARylated are glutamic acid (E), aspartic acid (D), arginine (R), or lysine (K) residues. Nicotinamide is released during this process.

After reports of PARP inhibition and synthetic lethality triggered great excitement, numerous drug discovery efforts across industry were initiated, including the program at IRBM (Merck Research Laboratories, Rome), which identified niraparib (**1**). As shown in Figure 4 and Table 2, niraparib (**1**) inhibits PARP1 (IC_{50} = 16.7 nM) and PARP2 (IC_{50} = 15.3 nM) with selectivity over other members of the PARP family.[20]

Figure 4. *In vitro* potency of niraparib (**1**) across PARP family members.

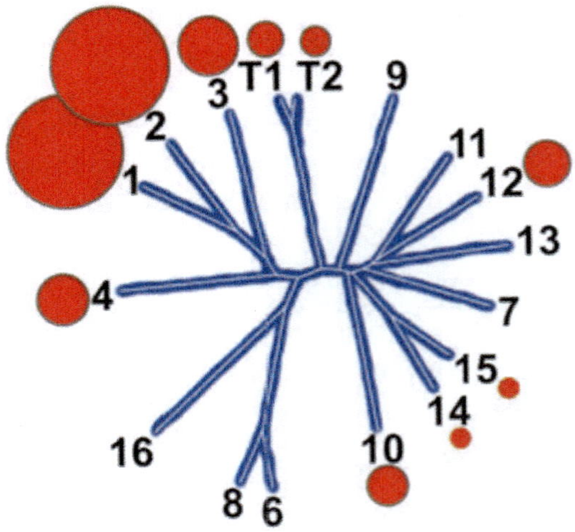

Table 2. *In vitro* activity of Niraparib across PARP family.

Protein	Construct	IC_{50} (nM)
PARP1	fl	16.7
PARP1	cat	132
PARP2	fl	15.3
PARP3	fl	269
PARP4	cat	446
TNKS1	cat	2355
TNKS2	cat	5130
PARP10	fl	1900
PARP10	cat	3300
PARP14	cat	17300
PARP15	cat	29200

The first PARPi approved was olaparib (2014), followed by rucaparb (2016), niraparib (2017), and talazoparib (2018). All of these first generation PARPi show selectivity for polyPARPs, such as PARP1 and PARP2, and show much weaker activity towards monoPARPs.

Figure 5. Structures of PARP Inhibitors.

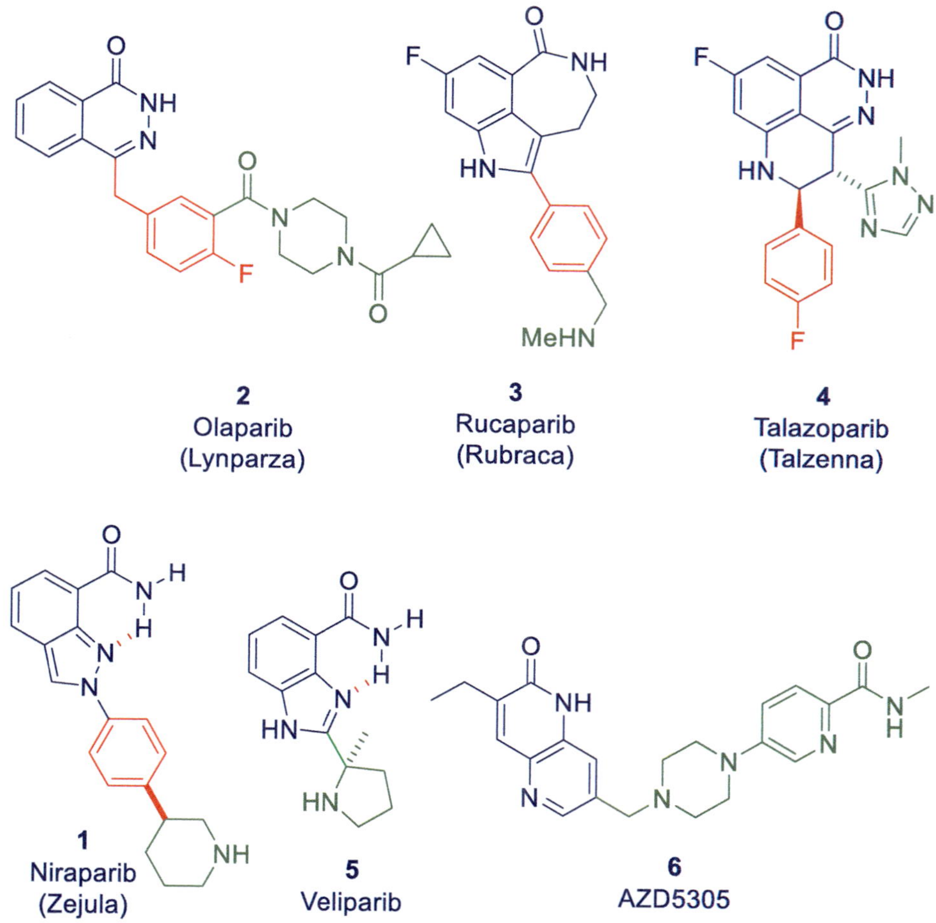

Shown in Figure 5 are the chemical structures of olaparib (**2**), rucaparib (**3**), and talazoparib (**4**). Since the strategy for PARPis are to compete with β-NAD^+ at the nicotinamide binding site, it is not surprising that all the PARPi, including niraparib (**1**), contain structural mimics of nicotinamide (heterocycle in blue). The weak potency of nicotinamide, as a PARP inhibitor, is partly due to the free rotation of the amide bond. Olaparib (**2**), rucaparib (**3**), and talazoparib (**4**) lock the carboxamide group into the desired anti-conformation by cyclizing into a ring. An alternate strategy employed in veliparib (**5**) and niraparib (**1**) is to incorporate a nitrogen within a heteroaromatic ring to form a six-membered H-bond with the carboxamide.

Although PARP inhibitors have achieved regulatory approval for HR repair deficient tumors including BRCA mutation, there are reports of overlapping hematological toxicities when dosed in combination with first-line chemotherapies. It has been hypothesized that the lack of selectivity of these first generation PARPi for PARP1 over PARP2 may contribute to the observed toxicity. Recently Astra Zeneca has reported a second-generation PARP1-DNA trapper, AZD5305 (**6**) with high selectivity for PARP1 over other PARPs, including PARP2.[13] PARPi, **1–5** show PARP1/2-fold selectivity of 51, 1, 25, 3, and 18, respectively. AZD5305 (**6**) has PARP1/2-fold selectivity of 460, which translated to 46% viability of remaining proliferating and differentiating hematopoietic stem/progenitor cells (HSPCs) at 100 nM without significantly further viability loss at concentrations up to 10 μM.

12.3 Structure–Activity Relationship (SAR)

With known X-ray crystal structures and modelling studies showing the amide of nicotinamide making three key hydrogen bonds to the hydroxyl group of Ser904 and the amide backbone of Gly863, the medicinal chemistry group at IRBM/Merck Research labs in Rome started screening four [6,5]-bicyclic heteroaromatic carboxamide derivatives **7–10** (Table 3).[21,22] All four of these designs incorporate a nitrogen within the heteroaromatic ring to lock the carboxamide group into a six-membered intramolecular H-bond. Indazole (**7**) with IC_{50} = 24 nM showed higher affinity than **8–10** and also showed cellular activity in a PARylation assay.

Table 3. *In vitro* activity of heterocyclic carboxamides.

Compound	Structure	PARP1 IC_{50} (nM)	PARylation EC_{50} (nM)	PARylation EC_{90} (nM)
7		24	3,700	6,200
8		71	ND	ND

9	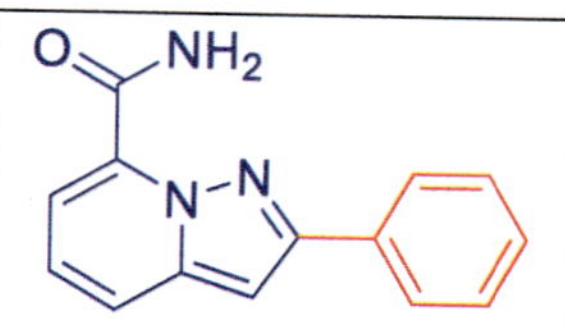	55	ND	ND
10		270	ND	ND

Source: Based on References 21 and 22.

Indazole **7** was also evaluated *in vivo* and the pharmacokinetics (PK) of **7** showed moderate plasma clearance in rat with volume of distribution of 1.8 L/kg and terminal half-life of 5.1 hours. Additionally, the acceptable oral bioavailability, $F = 41\%$, of **7** makes it a good starting point for further optimization.

Figure 6. Profile of Indazole **7**.[22]

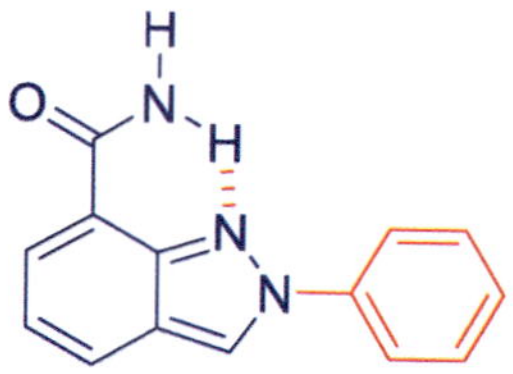

Indazole 7
PARP1 IC_{50} = 24 nM
HeLa cervical cells, EC_{50} = 3.7 μM
Rat LM: 123 (μL/min)/mgP
Human LM: 138 (μL/min)/mgP
Rat PK, Cl = 30 (mL/min)/kg
V_{ss} = 1.8 L/kg
$T_{1/2}$ = 5.1 h
F = 41%

Substitution at the *para*-position of the phenyl ring improved both enzymatic activity and cellular activity (Table 3). Polar groups were explored to potentially improve solubility with the possibility of picking up additional binding interactions in the adenosine binding pocket. The addition of a dimethylaminomethyl group at the *para*-position of the phenyl ring to give **11**, improved the enzymatic activity by 6-fold over unsubstituted indazole **7**. The solubility of **11** (3 mg/mL in phosphate buffer) was also an improvement over the sparingly soluble **7** (26 μg/mL in phosphate buffer). More importantly, **11** showed a significant improvement in the cell-based PARP inhibition assay (PARylation EC_{50} = 110 nM), which may be attributed to better cell penetration.[21a]

Pyrrolidine **12** and piperidine **13** show similar PARP inhibitory activity, whereas the less basic morpholine **14** was less active. Secondary amines are also tolerated, where methylamine **15** shows similar enzymatic activity as **11**, but improved activity in cells. Larger alkyl chains on the secondary amine, extending the distance between the aryl ring and the basic amine, or moving the basic group from the *para*-position to the *meta*-position all led to dramatic loss in cellular activity. The high correlation of the EC_{90} for inhibition of PARylation with the CC_{50} for the inhibition of the proliferation of BRCA-1 silenced HeLa cells suggests that strong sustained PARP inhibition is required to inhibit cell proliferation.

Compound **11** was radio-labeled, dosed intravenously to rats, and urine and bile were examined. The majority of the dosed radioactivity was excreted as the metabolite carboxylic acid and its acylglucuronide.[21a] The oxidation was determined to be driven by CYP1A1, which is expressed in the lung. Constraining the amine in a ring such as in **19** reduced clearance in both rat liver microsomes and CYP1A1, while maintaining good potency.

Table 4. *In vitro* activity of 3- and 4-substituted phenyl on N2-phenyl indazole carboxamides.

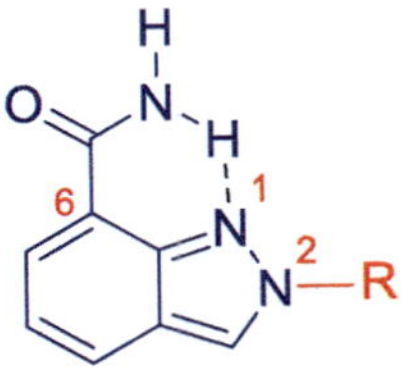

Compound	R	PARP1 IC_{50} (nM)	PARylation EC_{50} (nM)	PARylation EC_{90} (nM)	BRCA1-CC_{50} (nM)
11	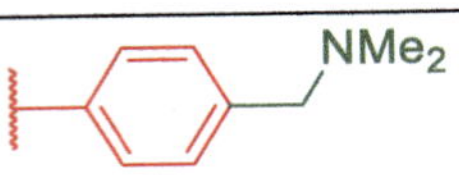	3.7	110	630	520
12	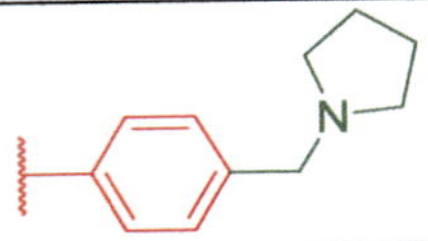	4.8	94	450	1,200
13	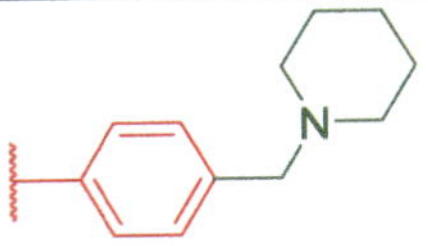	1.9	180	970	880
14	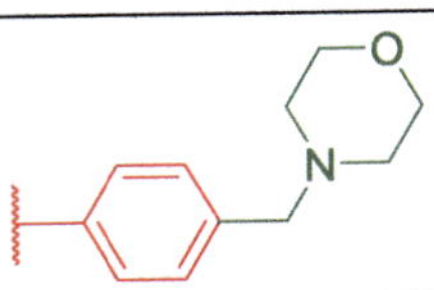	17	1,200	6,200	5,300
15	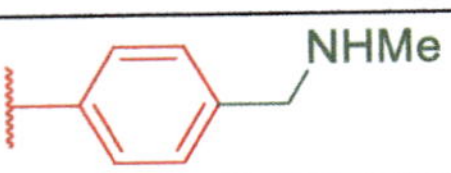	3.8	68	740	460

16	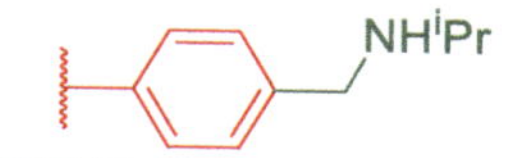	6.7	700	2,000	2,000
17	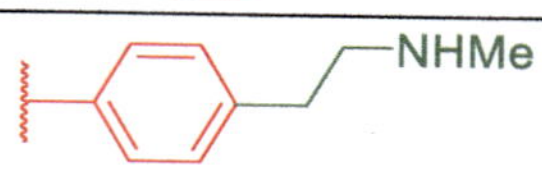	200	ND	ND	19,000
18	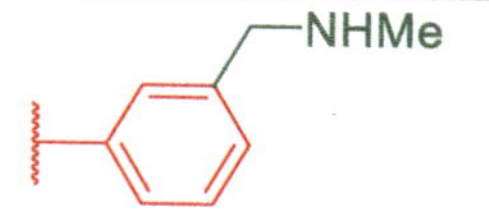	5.3	ND	ND	10,300
19	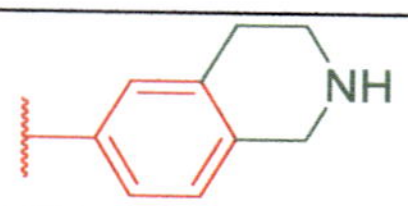	1.4	13	140	270

An alternate strategy of appending saturated nitrogen heterocycles was fruitful as it yielded the 3-substituted piperidine **21**, which showed good enzyme and cell potency, reduced liver microsome and CYP1A1 *in vitro* clearance, and moderate *in vivo* plasma clearance in rat (Table 4). The resolution provided **1** (*S*-isomer) and **23** (*R*-isomer), which both showed excellent inhibition of PARP1. The *S*-enantiomer **1** was more potent in the cell-based assays (PARylation EC_{50} = 4.0 nM; BRCA1-HeLa CC_{50} = 34 nM). Additionally, the *in vitro* turnover in human liver microsomes was similar (HLM Cl_{int}, **1** = 3, **23** = 4 μL/min/mg).

Table 5. *In vitro* activity of 4-substituted phenyl on N2-phenyl indazole carboxamides.

Compound	R	PARP1 IC_{50} (nM)	PARylation EC_{50} (nM)	PARylation EC_{90} (nM)	BRCA1- CC_{50} (nM)
20	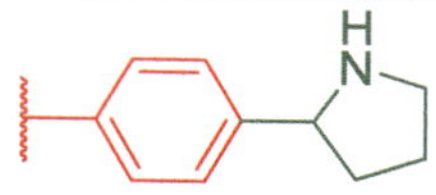	3.1	31	430	190
21	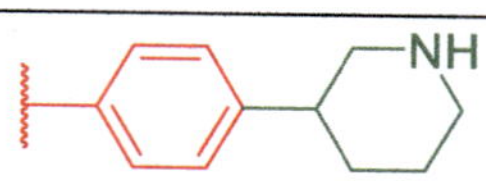	3.2	24	220	72
22	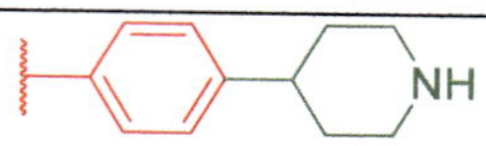	9	ND	ND	410
1	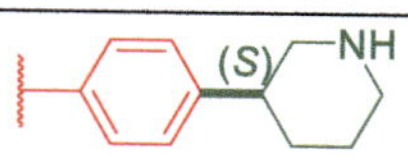	3.2	4.0	45	34
23	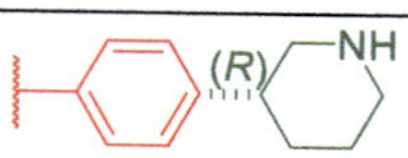	2.4	30	280	470

During preclinical development, the issue of species-specific PK needed to be addressed. In rat, **1** showed good bioavailability (65%) and a high volume of distribution (V_{dss} = 6.9 L/kg), resulting in a reasonably long terminal half-life ($t_{1/2}$ = 3.4 h). However, the dog PK of **1** showed high clearance (CLp = 31 mL/min/kg). Experiments in hepatocytes showed high turnover in dogs to be due to amide hydrolysis of **1** to the corresponding carboxylic acid. Confirmation with *in vivo* dog experiments showed 52% of the radioactivity of [^{14}C]-**1** dosed intravenously to bile duct cannulated dogs was recovered as the metabolite carboxylic acid. This metabolite is inactive in the PARP1 assay since it is missing the key H-bond donor needed for binding in the nicotinamide site. The co-crystal structure of niraparib (**1**) with PARP1 (shown in Figure 7) show the critical H-bond of the amide with Gly863. Nevertheless, based on the low metabolic turnover in human *in vitro* systems, the human PK was anticipated to be good, which was borne out in the clinic. This illustrates the value of conducting mechanistic PK studies.

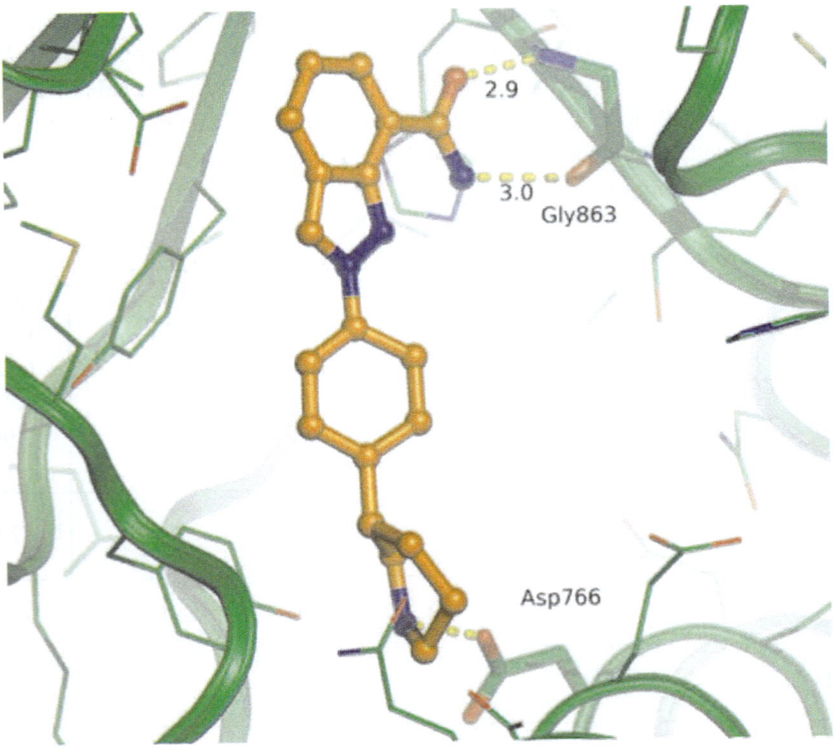

Figure 7. Co-Xtal structure of niraparib (**1**) and PARP1.

(Generated from PDB: 4R6E (2.2 Å res.) by Marc Adler at ChemPartner)

12.4 Pharmacokinetics and Drug Metabolism

Niraparib's (**1**) PK have been studied at 10 dose levels between 30 and 400 mg/day in 60 patients with advanced solid tumors.[21a,23] The maximum tolerated dose was determined by reversible grade 4 thrombocytopenia. Notably, the terminal half-life of 36 hours allowed for once daily dosing, an improvement over previously approved PARPi. Niraparib shows 73% bioavailability and 83% plasma protein binding. The drug is rapidly absorbed and the T_{max} is achieved with 3 hours.[23] Target coverage, as determined by xenograft studies, is achieved at all doses of 40 mg/day or above.

Niraparib (**1**) is primarily metabolized by carboxylesterases, followed by glucuronidation. At the Food and Drug Administration (FDA) approved dose, approximately 47.5% of the drug is excreted in urine and about 38.8% in feces over 21 days.

12.5 Efficacy and Safety

In the randomized, placebo-controlled, phase III trial (ENGOT-OV16/NOVA) conducted by the European Network for Gynecological Oncological Trial groups and investigators in the United States, Canada, and Hungary, the objective was to evaluate the efficacy and safety of niraparib vs. placebo as maintenance treatment in a broad population of patients with platinum-sensitive, recurrent ovarian cancer.[11] During the earlier phase I dose escalation study, the maximum dose of 300 mg per day of niraparib resulted in an objective clinical response in patients with ovarian cancer and was associated with a low frequency of high-grade toxic effects. For the ENGOT-OV16/NOVA trial, the primary end point of the duration of progression-free survival (PFS) was defined as the time from treatment randomization to the earliest date of disease progression or death from any cause. After the occurrence of disease progression or death in 103 patients in the g*BRCA* cohort and in 101 in the HRD-positive subgroup of the non-g*BRCA* cohort, the efficacy analysis was performed. During the study, 213 of such events had occurred in the overall non-g*BRCA* cohort. The median rate of compliance in the niraparib group was approximately 90% in the two cohorts and compliance in the placebo group was also high (> 99%).

The niraparib group demonstrated significantly longer duration of PFS than that in the placebo group in all three primary efficacy populations ($p < 0.001$). In the g*BRCA* cohort (with germline BRCA mutation), the median duration of PFS was 21.0 months in the niraparib group and 5.5 months in the placebo group (hazard ratio, 0.27; 95% confidence interval [CI], 0.17 - 0.41). Niraparib treatment also resulted in significantly longer PFS vs. placebo in the HRD-positive subgroup of the non-g*BRCA* cohort (median, 12.9 months vs. 3.8 months; hazard ratio, 0.38; 95% CI, 0.24 - 0.59). Interestingly, longer PFS in the niraparib-treatment group vs. placebo in the overall non-g*BRCA* cohort (median, 9.3 months vs. 3.9 months; hazard ratio, 0.45; 95% CI, 0.34 - 0.61) was also observed.

Following this presentation of ENGOT-OV16/NOVA, niraparib became the first PARPi to receive full approval by the US FDA for the maintenance treatment of recurrent ovarian cancer, regardless of a patient's germline or somatic mutational status. This approval is notable since the earlier approved olaparib and rucaparib required a confirmatory trial in conjunction with companion diagnostics to show efficacy in patients with a somatic BRCA mutated advanced ovarian cancer. The ENGOT-OV16/NOVA trial was a randomized, placebo-controlled, double-blind phase III trial that enrolled 553 patients that had previously received at least two prior treatments of platinum-based chemotherapy and were in complete or partial response to most current chemotherapy treatment, which produced survival data rather than response rates.

In PRIMA, a phase III, randomized, double-blind, placebo-controlled study evaluating niraparib as maintenance treatment following platinum-based chemotherapy, the most common adverse events observed after niraparib dosing are anemia, nausea, thrombocytopenia, constipation, fatigue, platelet count decrease, and neutropenia.

Emerging mechanisms of PARPi resistance in advanced disease continue to complicate their clinical use. Examples of possible drug resistance mechanisms are: (1) regained HR capacity through restored BRCA1/2 function; (2) upregulation of the P-glycoprotein drug efflux transporter; and (3) mutation or down-regulation of PARP1.[24]

6 Syntheses

Here shown are the initial Merck discovery route (Scheme 2) to niraparib **(1)**, subsequent improvements for pilot plant scale-up, and a highlight of the transaminase route (Scheme 3) developed by the Merck Process group.

Scheme 2. Merck Discovery Route

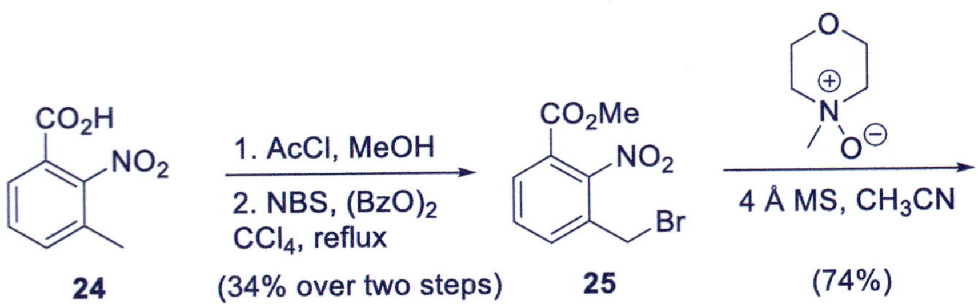

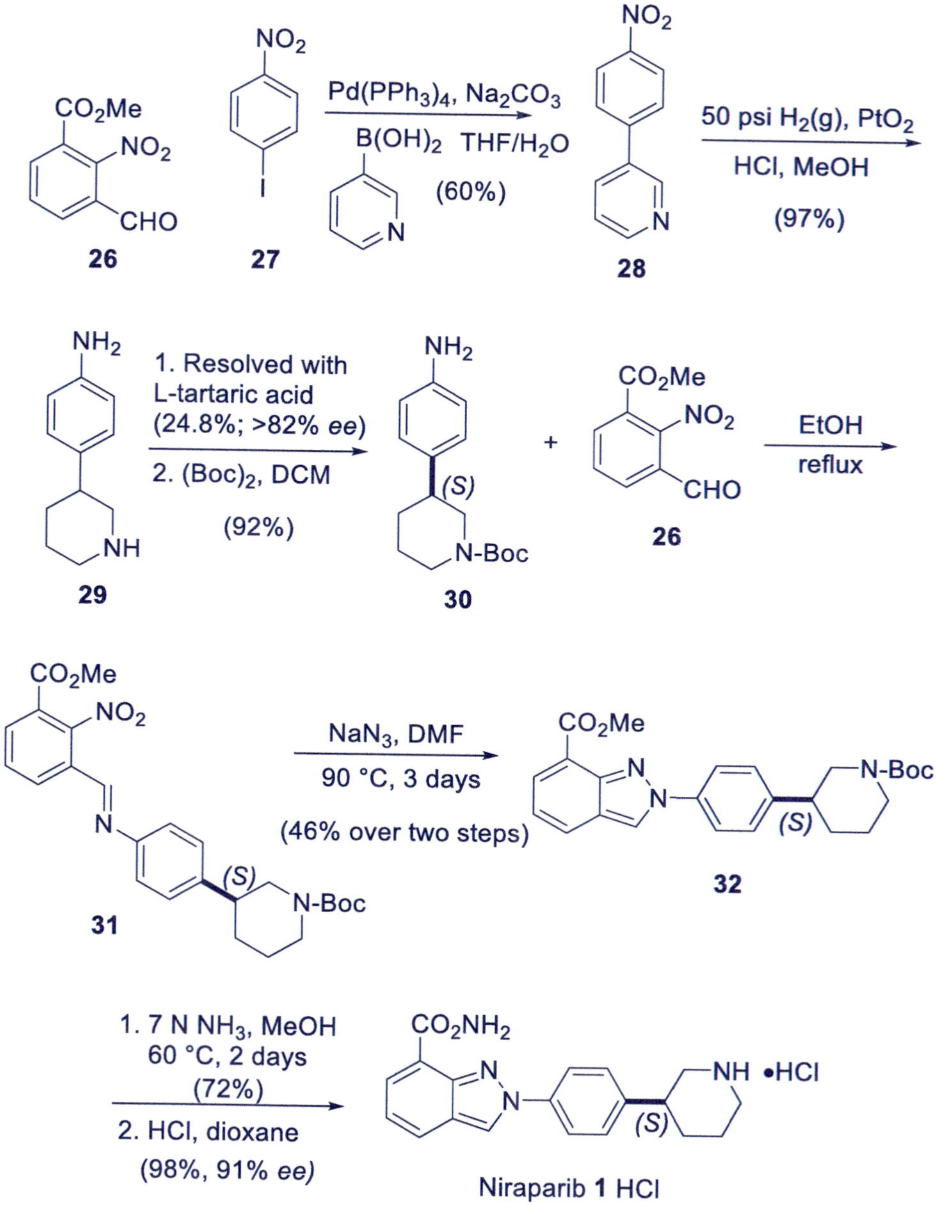

The first kilogram-scale synthesis[25] of API incorporated some improvements over the Merck discovery route such as (1) utilizing chiral HPLC separation to achieve higher enantiopurity and (2) telescoping reactions to avoid chromatographic separations. Generally, the first non-Good Manufacturing Practices (GMP) kilogram scale-up of API to support Investigational New Drug (IND)-enabling studies only requires modest improvements in the process in order to implement in larger reactors safely and does not necessarily need the level of efficiency and impurity quantification necessary in a GMP manufacturing route. This first-generation process route (not shown) afforded niraparib **1** tosylate monohydrate in 11 total steps with 11% overall yield.

A second-generation route[25] was developed by Merck Process group to address several shortcomings: (1) chiral resolution by salt formation or chiral chromatographic separation, (2) the use of azide reagents, especially under heating conditions, (3) use of metals and high-pressure hydrogenation equipment, and (4) low occupational exposure limits for the penultimate intermediate and API.

Scheme 3. Highlight of transaminase process route

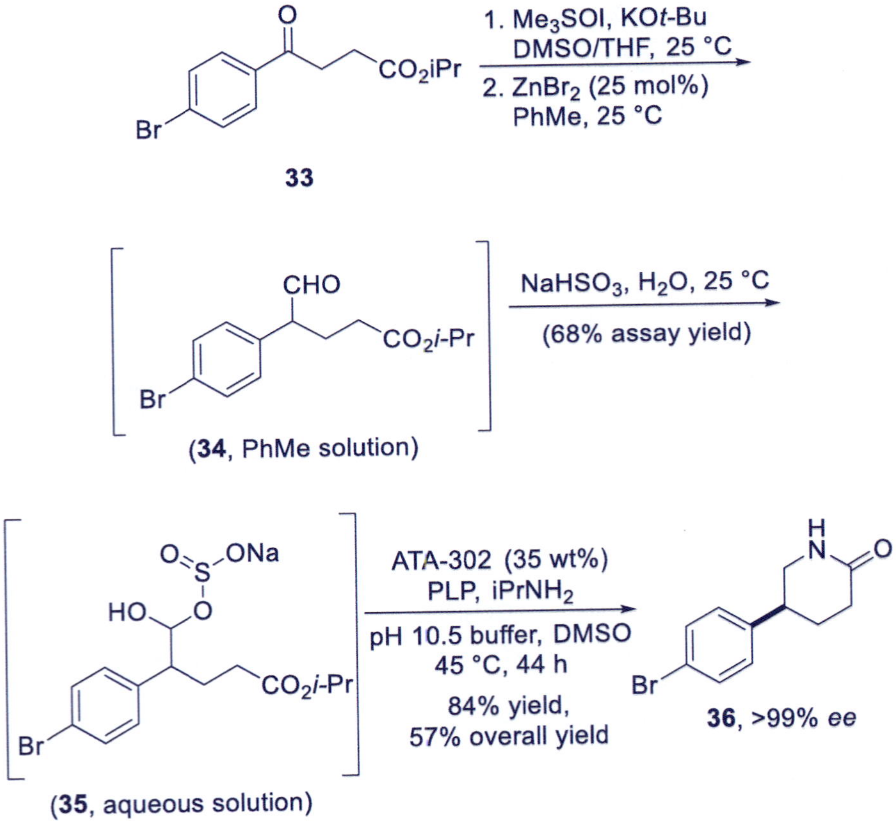

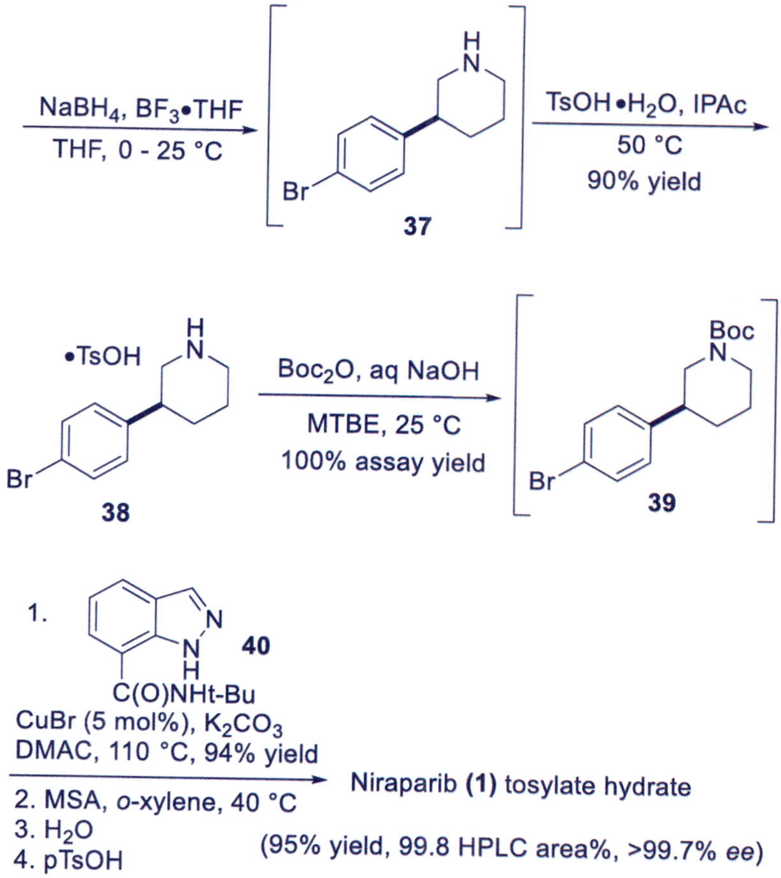

To obviate potent-compound handling of the penultimate intermediate, the primary amide was replaced with *t*-butyl-protected amide **40**. The additional steric bulk of the *t*-butyl-protected amide at C7 further improved the N-arylation regioselectivity favoring N2 over N1 arylation on indazole **40**. To avoid the catalytic hydrogenation step and the chiral resolution in the earlier routes, a transaminase-mediated dynamic kinetic resolution (DKR) approach was successfully employed. Selection of the desired (*S*)-aldehyde enantiomer by the transaminase enzyme, with concomitant racemization of the unreacted (*R*)-aldehyde, would result in a DKR process that theoretically converts all the aldehyde to desired chiral amine without the need for discarding the unwanted enantiomer during a classical resolution. The chiral amine was intramolecularly trapped by the suitably positioned isopropyl ester to afford **36** in >99% *ee*. Finally, optimized N2-selective arylation between **39** and **40** using CuBr (5 mol%) and 8-hydroxyquinoline (10 mol%) in DMAC, followed by salt exchanges provided niraparib **(1)** tosylate[26] monohydrate (99.8% HPLC assay area); > 99.7% *ee*.

12.7 Summary

Niraparib (**1**) and other PARPi represent a new class of oncology drugs that exploit synthetic lethality. As the third US FDA-approved PARP inhibitor, niraparib (**1**) was a significant improvement over olaparib and rucaparib since it was approved for patients without a germline or somatic BRCA mutation with single daily dosing. This approval expanded the population of patients who could receive and benefit from a PARP inhibitor. Niraparib (**1**) does pose a greater risk of thrombocytopenia, which may be attributed to inhibition of PARP2. The recently reported PARP1-selective AZD5305 (**6**)[13] may demonstrate it is possible to dial out the hematological toxicity risk associated with first generation PARPi.

Perhaps pleiotrophic effects across multiple members of the PARP family may lead to more efficacious cancer therapeutics with less propensity to develop drug resistance. Notably inhibition of PARP16 has recently been found to contribute to the polypharmacology of talazoparib **(4)**.[27]

The story of exploiting synthetic lethality to target cancers is just beginning. There are many other products of commonly altered genes in cancer that remain untargeted. The magnitude and robustness of synthetic lethal effects that would translate to efficacy in the clinic is a critical consideration. The field is moving towards investigating simultaneous exploitation of multiple synthetic lethal effects and identifying higher-order synthetic lethal effects, which may lead to combination therapies not previously conceived.[28]

12.8 References

1. Hottiger, M. O. SnapShot: ADP-ribosylation signaling. *Mol. Cell* **2015**, *58*, 1134.
2. (a) Kraus, W. L. PARPs and ADP-Ribosylation: 50 years … and counting. *Mol. Cell* **2015**, *58*, 902–910. (b) Langelier, M.-F.; Eisemann, T.; Riccio, A. A.; Pascal, J. M. PARP family enzymes: regulation and catalysis of the poly (ADP-ribose) posttranslational modification. *Curr. Opin. Struct. Biol.* **2018**, *53*, 187–198. (c) Pahi, Z. G.; Borsos, B. N.; Pantazi, V.; Ujfaludi, Z.; Pankotai, T. PARylation during transcription: insights into the fine-tuning mechanism and regulation. *Cancers* **2020**, *12*, 183–201.
3. (a) Gozgit, J. M.; Vasbinder, M. M.; Abo, R. P.; Kunii, K.; Kuplast-Barr, K. G.; Gui, B.; Lu, A. Z.; Molina, J. R.; Minissale, E.; Swinger, K. K.; Wigle, T. J.; Blackwell, D. J.; Majer, C. R.; Ren, Y.; Niepel, M.; Varsamis, Z. A.; Nayak, S. P.; Bamberg, E.; Mo, J.-R.; Church, W. D.; Mady, A. S. A.; Song, J.; Utley, L.; Rao, P. E.; Mitchison, T. J.; Kuntz, K. W.; Richon, V. M.; Keilhack, H. PARP7 negatively regulates the type I interferon response in cancer cells and its inhibition triggers antitumor immunity. *Cancer Cell* **2021**, *39*, 1–13. (b) Venkannagari, H.; Verheugd, P.; Koivunen, J.; Haikarainen, T.; Obaji, E.; Ashok, Y.; Narwal, M.; Pihlajaniemi, T.; Lüscher, B.; Lehtiö, L. Small-molecule

chemical probe rescues cells from mono-ADP-ribosyltransferase ARTD10/PARP10-induced apoptosis and sensitizes cancer cells to DNA damage. *Cell Chem. Biol.* **2016**, *23*, 1251–1260. (c) Challa, S.; Khulpatteea, B. R.; Nandu, T.; Camacho, C. V.; Ryu, K. W.; Chen, H.; Peng, Y.; Lea, J. S.; Kraus, W. L. Ribosome ADP-ribosylation inhibits translation and maintains proteostasis in cancers. *Cell* **2021**, *184*, 1–16.
4. Chambon, P.; Weill, J. D.; Doly, J.; Strosser, M. T.; Mandel, P. On the formation of a novel adenylic compound by enzymatic extracts of liver nuclei. *Biochem. Biophys. Res. Commun.* **1966**, *25*, 638–643.
5. Ledermann, J. A.; Coquard, I. R.; Marme, F. Key questions on PARP inhibitors in ovarian cancer: experts evaluate the latest evidence. *Clinical Care Options* 2021, released online: clinicaloptions.com
6. Plummer, R.; Jones, C.; Middleton, M.; Wilson, R.; Evans, J.; Olsen, A.; Curtin, N.; Boddy, A.; McHugh, P.; Newell, D.; Harris, A.; Johnson, P.; Steinfeldt, H.; Dewji, R.; Wang, D.; Robson, L.; Calvert, H. Phase I study of the poly (ADP-ribose) polymerase inhibitor, AG014699 in combination with temozolomide in patients with advanced solid tumors. *Clin. Cancer Res.* **2008**, *14*, 7917–7923.
7. Fong, P. C.; Boss, D. S.; Yap, T. A.; Tutt, A.; Wu, P.; Mergui-Roelvink, M.; Mortimer, P.; Swaisland, H.; Lau, A.; O'Connor, M. J. Inhibition of poly(ADP-ribose) polymerase in tumors from *BRCA* mutation carriers. *N. Engl. J. Med.* **2009**, *361*, 123–134.
8. Audeh, M. W.; Carmichael, J.; Penson, R. T.; Friedlander, M.; Powell, B.; Bell-McGuinn, K. M.; Scott, C.; Weitzel, J. N.; Oaknin, A.; Loman, N.; Lu, K.; Schmutzler, R. K.; Matulonis, U.; Wickens, M.; Tutt, A. Oral poly(ADP-ribose) polymerase inhibitor Olaparib in patients with BRCA1 or BRCA2 mutations and recurrent ovarian cancer: a proof-of-concept trial. *Lancet* **2010**, *376*, 245–251.
9. Tutt, A.; Robson, M.; Garber, J. E.; Domchek, S. M.; Audeh, M. W.; Weitzel, J. N.; Friedlander, M.; Arun, B.; Loman, N.; Schmutzler, R. K.; Wardley, A.; Mitchell, G.; Earl, H.; Wickens, M.; Carmichael, J. Oral poly(ADP-ribose) polymerase inhibitor Olaparib in patients with BRCA1 or BRCA2 mutations and advanced breast cancer: a proof-of-concept trial. *Lancet* **2010**, *376*, 235–244.
10. Bang, Y.; Boku, N.; Chin, K.; Lee, K.; Park, S. H.; Qin, S.; Rha, S. Y.; Shen, L.; Xu, N.; Im, S.; Locker, G.; Rowe, P.; Shi, X.; Hodgson, D.; Liu, Y.; Xu, R.; 2742 – Olaparib in combination with paclitaxel in patients with advanced gastric cancer who have progressed following first-line therapy: phase III GOLD study. OncologyPRO 2016, ESMO Abstract.
11. Mirza, M. R.; Monk, B. J.; Herrstedt, J.; Oza, A. M.; Mahner, S.; Redondo, A.; Fabbro, M.; Ledermann, J. A.; Lorusso, D.; Vergote, I.; Ben-Baruch, N. E.; Marth, C.; Madry, R.; Christensen, R. D.; Berek, J. S.; Dørum, A.; Tinker, A. V.; duBois, A.; González-Martin, A.; Follana, P.; Benigno, B.; Rosenberg, P.; Gilbert, L.; Rimel, B. J.; Buscema, J.; Balser, J. P.; Agarwal, S.; Matulonis, U. A. Niraparib maintenance therapy in platinum-sensitive, recurrent ovarian cancer. *N. Engl. J. Med.* **2016**, *375*, 2154–2164.
12. Underhill, C.; Toulmonde, M.; Bonnefoi, H. A review of PARP inhibitors: from bench to bedside. *Ann. Oncol.* **2011**, *22*, 268–279.

13. Johannes, J. W.; Balazs, A.; Barratt, D.; Bista, M.; Chuba, M. D.; Cosulich, S.; Critchlow, S. E.; Degorce, S. L.; Di Fruscia, P.; Edmondson, S. D.; Embrey, K.; Fawell, S.; Ghosh, A.; Gill, S. J.; Gunnarsson, A.; Hande, S. M.; Heightman, T. D.; Hemsley, P.; Illuzi, G.; Lane, J.; Larner, C.; Leo, E.; Liu, L.; Madin, A.; Martin, S.; McWilliams, L.; O'Connor, M. J.; Orme, J. P.; Pachl, F.; Packer, M. J.; Pei, X.; Pike, A.; Schimpl, M.; She, H.; Staniszewska, A. D.; Talbot, V.; Underwood, E.; Varnes, J. G.; Xue, L.; Yao, T.; Zhang, K.; Zhang, A. X.; Zheng, X. Discovery of 5-{4-[(7-Ethyl-6-oxo-5,6-dihydro-1,5-naphthyridin-3-yl)methyl]piperazin-1-yl}-*N*-methylpyridine-2-carboxamide (AZD5305): a PARP1-DNA trapper with high selectivity for PARP1 over PARP2 and other PARPs. *J. Med. Chem.* **2021**, *64*, 14498–14512.

14. Murai, J.; Huang, S. N.; Das, B, B.; Renaud, A.; Zhang, Y.; Doroshow, J. H.; Ji, J.; Takeda, S.; Pommier, Y. Differential trapping of PARP1 and PARP2 by clinical PARP inhibitors. *Cancer Res.* **2012**, *72*, 5588–5599.

15. Lord, C. J.; Ashworth, A. PARP inhibitors: synthetic lethality in the clinic. *Science* **2017**, *355*, 1152–1158.

16. (a) Bryant, H. E.; Schultz, N.; Thomas, H. D.; Parker, K. M.; Flower, D.; Lopez, E.; Kyle, S.; Meuth, M.; Curtin, N. J.; Helleday, T. Specific killing of BRCA2-deficient tumours with inhibitors of poly(ADP-ribose) polymerase. *Nature* **2005**, 434, 913–917. (b) Farmer, H.; McCabe, N.; Lord, C. J.; Tutt, A. N. J.; Johnson, D. A.; Richardson, T. B.; Santarosa, M.; Dillon, K. J.; Hickson, I.; Knights, C.; Martin, N. M. B.; Jackson, S. P.; Smith, G. C. M.; Ashworth, A. Targeting the DNA repair defect in BRCA mutant cells as a therapeutic strategy. *Nature* **2005**, *434*, 917–921.

17. Boussios, S.; Karihtala, P.; Moschetta, M.; Abson, C.; Karathanasi, A.; Zakynthinakis-Kyriakou, N.; Ryan, J. E.; Sheriff, M.; Rassy, E.; Pavlidis, N. Veliparib in ovarian cancer: a new synthetically lethal approach. *Invest. New Drugs* **2020**, *38*, 181–193.

18. (a) González-Martin, A.; Pothuri, B.; Vergote, I.; Depont Christensen, R.; Graybill, W.; Mirza, M. R.; McCormick, C.; Lorusso, D.; Hoskins, P.; Freyer, G.; Baumann, K.; Jardon, K.; Redondo, A.; Moore, R. G.; Vulsteke, C.; O'Cearbhaill, R. E.; Lund, B.; Backes, F.; Barretina-Ginesta, P.; Haggerty, A. F.; Rubio-Peréz, M. J.; Shahin, M. S.; Mangili, G.; Bradley, W. H.; Bruchim, I.; Sun, K.; Malinowska, I. A.; Li, Y.; Gupta, D.; Monk, B. J. Niraparib in patients with newly diagnosed advanced ovarian cancer. *N. Engl. J. Med.* **2019**, *381*, 2391–2402. (b) Nizialek, E.; Antonarakis, E. S. PARP inhibitors in metastatic prostate cancer: evidence to date. *Cancer Manag. Res.* **2020**, 12, 8105–8114. (c) Zhu, H.; Wei, M.; Xu, J.; Hua, J.; Liang, C.; Meng, Q.; Zhang, Y.; Liu, J.; Zhang, B.; Yu, X.; Shi, S. PARP inhibitors in pancreatic cancer: molecular mechanisms and clinical applications. *Mol. Cancer* **2020**, *19*, 49–63.

19. Alemasova, E. E.; Lavrik, O. I. Poly(ADP-ribosyl)ation by PARP1: reaction mechanism and regulatory proteins. *Nucleic Acids Res.* **2019**, *47*, 3811–3827.

20. Thorsell, A.-G.; Ekblad, T.; Karlberg, T.; Löw, M.; Pinto, A. F.; Trésauges, L.; Moche, M.; Cohen, M. S.; Schüler, H. Structural basis for potency and promiscuity in poly (ADP-ribose) polymerase (PARP) and tankyrase inhibitors. *J. Med. Chem.* **2017**, *60*, 1262–1271.

21. (a) Jones, P.; Wilcoxen, K.; Rowley, M.; Toniatti, C. Niraparib: a poly(ADP-ribose) polymerase (PARP) inhibitor for the treatment of tumors with defective homologous recombination. *J. Med. Chem.* **2015**, *58*, 3302–3314 (b) Jones, P.; Ontoria, J. M.; Scarpelli, R.; Schultz-Fademrecht, C. Amide Substituted Indazoles as Poly(ADP-Ribose)polymerase (PARP) Inhibitors. WO 2008/084261

22. Jones, P.; Altamura, S.; Boueres, J.; Ferrigno, F.; Fonsi, M.; Giomini, C.; Lamartina, S.; Monteagudo, E.; Ontoria, J. M.; Orsale, M. V.; Palumbi, M. C.; Pesci, S.; Roscilli, G.; Scarpelli, R.; Schultz-Fademrecht, C.; Toniatti, C.; Rowley, M. Discovery of 2-{4-[(3*S*)-piperidin-3-yl]phenyl}-2*H*-indazole-7-carboxamide (MK-4827): a novel oral poly(ADP-ribose)polymerase (PARP inhibitor efficacious in BRCA-1 and -2 mutant tumors. *J. Med. Chem.* **2009**, *52*, 7170–7185.

23. (a) Longoria, T. C.; Tewari, K. S. Pharmacokinetic drug evaluation of niraparib for the treatment of ovarian cancer. *Expert Opin. on Drug Metab. & Toxicol.* **2018**, *14*, 543–550. (b) Sandhu, S. K.; Schelman, W. R.; Wilding, G.; Moreno, V.; Baird, R. D.; Miranda, S.; Hylands, L.; Riisnaes, R.; Forster, M.; Omlin, A.; Kreischer, N.; Thway, K.; Gevensleben, H.; Sun, L.; Loughney, J.; Chatterjee, M.; Toniatti, C.; Carpenter, C. L.; Iannone, R.; Kaye, S. B.; de Bono, J. S.; Wenham, R. M.; The poly(ADP-ribose) polymerase inhibitor niraparib (MK4827) in BRCA mutation carriers and patients with sporadic cancer: a phase 1 dose-escalation trial. *Lancet Oncol.* **2013**, *14*, 882–892.

24. Michelena, J.; Lezaja, A.; Teloni, F.; Schmid, T.; Imhof, R.; Altmeyer, M. Analysis of PARP inhibitor toxicity by multidimensional fluorescence microscopy reveals mechanisms of sensitivity and resistance. *Nat. Commun.* **2018**, 9, 2678–2693.

25. (a) Chung, C. K.; Bulger, P. G.; Kosjek, B.; Belyk, K. M.; Rivera, N.; Scott, M. E.; Humphrey, G. R.; Limanto, J.; Bachert, D. C.; Emerson, K. M. Process development of C-N cross-coupling and enantioselective biocatalytic reactions for the asymmetric synthesis of niraparib. *Org. Process Res. Dev.* **2014**, *18*, 215–227. (b) Chung, C. K.; Scott, M. E.; Bulger, P. G.; Belyk, K. M.; Limanto, J.; Humphrey, G. R. Regioselective N-2 arylation of indazoles *US Pat. 2017/0137403 A1.* (c) Hughes, D. L. Patent Review of Manufacturing Routes to Recently Approved PARP Inhibitors: Olaparib, Rucaparib, and Niraparib. *Org. Process Res. Dev.* **2017**, *21*, 1227–1244.

26. Foley, J.; Wilson, R. D. Pharmaceutically Acceptable Salts of 2-{4-[(3S)-Piperidin-3-yl]phenyl}-2H-indazole-7-carboxamide. WO 2009/087381 A1

27. Palve, V.; Knezevic, C. E.; Bejan, D. S.; Luo, Y.; Li, X.; Novakova, S.; Welsh, E. A.; Fang, B.; Kinose, F.; Haura, E. B.; Monteiro, A. N.; Kooman, J. M.; Cohen, M. S. The non-canonical target PARP16 contributes to polypharmacology of the PARP inhibitor talazoparib and its synergy with WEE1 inhibitors. *Cell Chem. Bio.* **2022**, *29*, 1–13.

28. Ashworth, A.; Lord, C. J. Synthetic lethal therapies for cancer: what's next after PARP inhibitors. *Nat. Rev. Clin. Oncol.* **2018**, *15*, 564–576.

13

Selinexor (Xpovio), An XPO1 Inhibitor and A New Class of Therapeutics for Treating Multiple Myeloma

John Mancuso

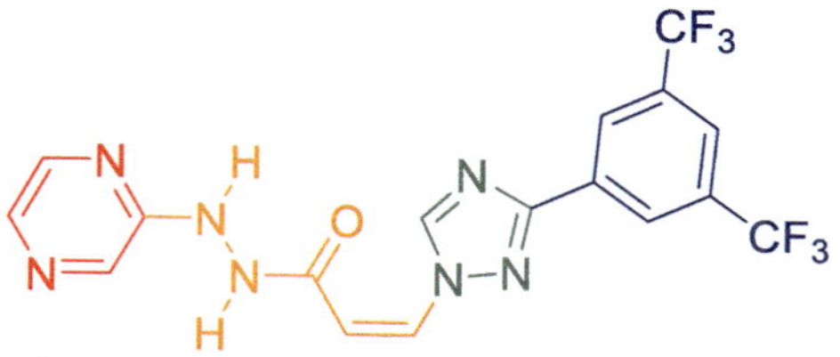

1

USAN: Selinexor
Trade name: Xpovio
Karyopharm Tx
Launched: 2018

13.1 Exportin1 (XPO1)

Within the cell, the nucleus encapsulates genetic material and the nuclear envelope aids cellular organization by forming a barrier restricting passage of proteins between the nucleus and cytoplasm. There is a tightly regulated system of transfer of materials in and out of the nucleus which permits transport of specific proteins and mRNAs in and out of this organelle. Small molecules (under 40 kDA) can passively move between nucleus and cytoplasm. Large molecules (>40 kDA) require specific transport proteins (chaperones) in order to traverse the nuclear pore complex. Signaling sequences are encoded within the proteins themselves, which dictate transportation to the correct location.[1,2]

Karyopherins are a main group of transport receptor chaperones. They consist of 20 members that include karyopherin alpha (KPNA) 1–6, karyopherin beta (KPNB) 1, and exportin-1 (XPO1), also termed the chromosome region maintenance-1 (CRM1). Transport signs within protein cargo redirects in (nuclear localization signals; importins)

Current Drug Synthesis, First Edition. Edited by Jie Jack Li.

or out (nuclear export signals; exportins) of the nucleus. This process is supported *via* energy from the RanGTPase complex.[3]

XPO1 is a ring-shaped karyopherin that uses its outer groove to bind protein cargos.[4] Proteins transported by XPO1 contain a 10–15 residue NES that bind to a hydrophobic groove on the convex surface of XPO1.[5]

XPO1 is the main exporter of leucine-rich proteins from the nucleus to the nuclear pore complex to the cytoplasm. The XPO1 cargo proteins include nearly all tumor suppressor proteins such as p53, Rb, p73; cell cycle regulators such as p21 and galactin-3; immune response regulators (e.g., IkB); chemotherapeutic targets (e.g., DNA topoisomerase).[2]

Overexpression and/or mutations of XPO1 are associated with both solid (lung, pancreatic, cervical, breast, and ovarian cancer, glioma, and sarcoma) and hematologic and lymphoma malignancies including multiple myeloma (MM). This results in enhanced transport of cargos out of the nucleus thereby leading to deactivation of the apoptotic pathway; deregulation of the cell cycle; resistance to chemotherapeutics; and disruption to cellular growth signaling (see Figure 1).

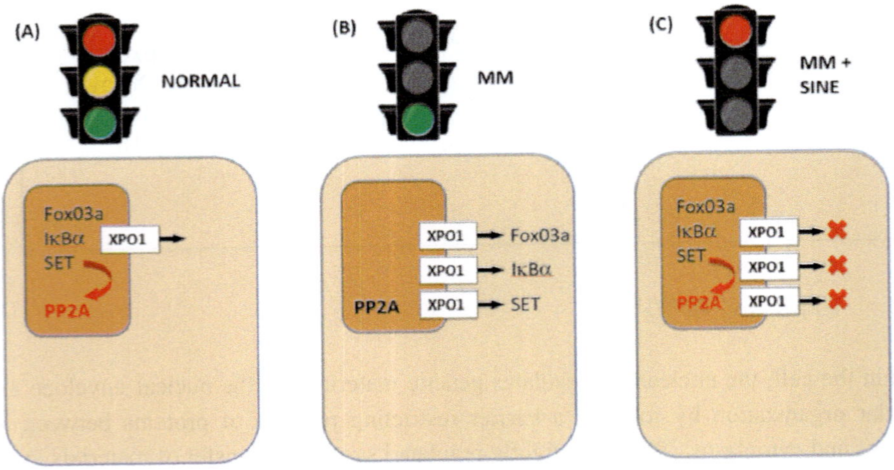

Figure 1. Normal cells maintain function by regulation of intracellular traffic (ions, small molecules, proteins). (A) Proteins located in the nucleus function normally. SET is able to activate its target protein phosphatase 2A (PP2A), a tumor suppressor. (B) In the disease condition, such as multiple myeloma, overexpression of XPO1 gives the "green light" to re-localize tumor suppressor proteins to the cytosol, including IKBa and Fox03a. Protein activity is altered and PP2A is no longer activated. (C) Treatment with a SINE (selective inhibitor of nuclear export) compound forces a "red light" condition in which proteins are no longer diverted from the nucleus and can resume their normal functions. (Adapted from Reference 1).

Inhibition of XPO-1 with a selective inhibitor of nuclear export (SINE) compound results in nuclear retention and accumulation of tumor suppressor proteins thus amplifying their role in the apoptosis pathway and reducing oncoprotein synthesis in cells with damaged genetic material. This ultimately leads to induction of apoptosis in cancer cells while largely sparing normal cells. Therefore, XPO1 is of interest as a therapeutic target for the treatment of cancer in general and specifically MM.

Several XPO1 targeting compounds have been tested as potential cancer therapeutics including:

- Leptomycin B (LBM, CI-940), natural and semisynthetic analogues
- SINE compounds:
  - KPT-185
  - KPT-251
  - KPT-276
  - KPT-335 (verdinexor)
  - KPT-8602 (eltanexor)
  - KPT-330 (selinexor, **1**)

For MM, preclinical antitumor activity has been demonstrated with ratjadone, CBS9106, KPT-276, and selinexor (**1**).[3]

13.2. Overview of Multiple Myeloma

MM is defined as a cancer of plasma cells, a type of white blood cell that produces antibodies. Specific causes are open to conjecture. Several risk factors have been identified and include obesity, radiation exposure, family history, and chemical exposure. The disease is chronic and despite several therapeutic options, resistance and relapse are frequent outcomes.[6]

MM is the second most frequent malignancy of the blood in the United States after non-Hodgkin lymphoma.[7] Median age at diagnosis is about 62 years for men and 61 for women (range 20–92). This disease does not predominantly afflict the younger population (only 2% patients under 40 years). The incidence varies globally from 1 per 100,000 people in China to 4 per 100,000 in most developed countries.

Symptoms include bone lesions, excess bone marrow plasma cells, renal disease, and decline in immune function. Studies have indicated that the bone marrow microenvironment of tumor cells has a key role in myeloma pathogenesis. Median survival after conventional treatments is 3–4 years; high-dose treatment followed by autologous stem-cell transplantation can extent median survival to 5–7 years.[8]

The first successful treatment was a combination of melphalan and prednisone that was introduced in the late 1960s and was further improved by high-dose drug regimens with autologous stem-cell transplantation in the 1980s. The new era of treatment was initiated in the late 1990s with the introduction of thalidomide and analogue lenalidomide in combination with bortezomib. These drugs have several dose-limiting side effects and most patients relapse over time.

MM therapy has improved over the past 20 years. As several patients have been treated with multidrug combinations, a number of refractory cases of MM to current therapeutics agents have occurred.[3]

- Double class (DCR): refractory to IMiD (thalidomide-class molecules) and a protease inhibitor (PI) such as bortezomib
- Triple-class (TCR): myeloma refractory to DCR agents with the addition of CD38-targeted therapies such as daratumumab
- Quad-refractory MM: refractory to at least two IMiDs and two PIs
- Penta-refractory MM: refractory to two IMiDs, two PIs and CD-38 targeted therapy

Treatments for penta-refractory MM patients are very limited. Options depend on the patent history but may involve use of multi-agent cytotoxic therapies, cellular therapies (stem-cell transplant), or inclusion into a clinic study with new agents. There is a need for new potent therapies, both single and multiple-agent, that can extend survival and minimize serious side effects.[7]

13.3. Development of Selinexor

Inhibition of XPO1 had been demonstrated by the natural product leptomycin B, an unsaturated branched fatty acid with antifungal and antibiotic properties isolated from *Streptomyces* spp., however, its toxicity in clinical studies made it unsuitable for further development.[1,4]

The group of Daelemans at the Katholieke Universiteit Leuven performed studies in 2002 on PKF050638, an *N*-azolyl acrylate and an effective and selective inhibitor of HIV-1 Rev, which found additional inhibitory activity against XPO1 in the micromolar range.[9] A subsequent and extensive SAR study in 2008 using PKF050638 as a starting point further defined structural elements required for XPO1 inhibition in this scaffold.[10]

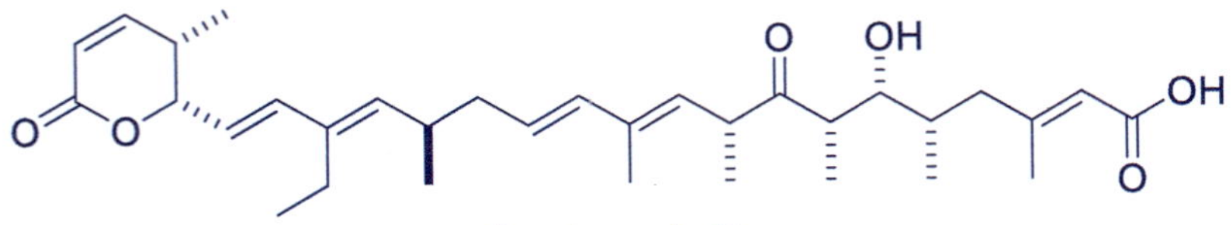

Leptomycin B

PKF050-638 KPT-185 KPT-251

A patent awarded to Karyopharm in 2011 included the structure of KPT-185.[11] KPT-251 was identified through an *in silico* molecular modeling strategy by Karyopharm, based on a docking-and-binding mode analysis using a published crystal structure of XPO1[12] with a small virtual library of compounds against the NES groove.[13] Replacement of the isopropyl ester of KPT-185 with the more stable oxadiazole in KPT-251 was found to improve bioavailability and metabolic stability and maintain electrostatic interactions within the active site. Of particular note, KPT-251 was significantly active against human acute myeloid leukemia (AML) cells in a xenograft model, with only minimal toxicity to normal tissues. The same study also noted that an analogue of KPT-251, KPT-330 (selinexor (**1**) had superior pharmacokinetic properties and had entered phase I clinical trials in humans with advanced cancers.

13.4. Pharmacology and Mechanism

Selinexor (**1**) selectively inhibits XPO1 by interaction with Cys528 in the cargo binding pocket (NES binding groove), forming a slowly reversible covalent bond.[4,6] This results in the accumulation of tumor suppressing proteins in the nucleus. This ultimately leads to a drop of levels of oncoproteins and results in cell cycle arrest and apoptosis. As there is generally overexpression of XPO1 in cancer cells,[14] this process generally spares normal cells.[15]

This compound was approved by the Food and Drug Administration for fast-track designation for treatment of penta-refractory MM in April 2018.[16] In combination with dexamethasone, Selinexor (**1**) received accelerated approval in the United States in 2019 for the treatment of MM. The approval was directed for adult patients with refractory or relapsed MM that showed resistance to at least four prior therapies and was refractory to at least two proteosome inhibitors and at least two immunomodulatory agents and an anti-CD38 mAb.[3,6]

13.5. Pharmacokinetics, Pharmacodynamics and Drug Metabolism

Structurally, selinexor (**1**) follows the guidelines established by Lipinski et al.[17] The compound has been demonstrated to be orally bioavailable in rats and monkeys (60–70%); absorption is moderately rapid with a median T_{max} of 2–4 hours.[18] Peak plasma concentrations were unaffected by administration in either fed or fasted states. The estimated apparent volume of distribution is 125 L with >95% protein bound in the plasma; there is evidence from animal studies that the compound does penetrate the central nervous system (CNS) blood–brain barrier (BBB, *vide infra*). This brain penetrance has indications for potential treatment of CNS and meningeal manifestation of MM but may also provide a reason for observed CNS-related side effects such as nausea and dizziness.[3] There was no evidence to indicate that repeat dosing of selinexor (**1**) leads to accumulation.[7,19]

In a study on a cohort of 189 patients with disease progression and prior therapy (73% had three or more previous lines of therapies), peak serum concentration had a medium time of 2–4 hours with a terminal half-life of 6–7 hours. Analysis of patient samples indicated glucuronidation was the primary mode of drug metabolism.[8] The compound also undergoes oxidation *via* cytochrome P450 (CYP3A4) generating the most common circulating metabolite (<5% of peak of parent levels) as the *trans*-isomer of selinexor (**2**, designated as KPT-375). This isomer, resulting **1** from olefin isomerization is minimally active, demonstrating 10% of XPO1 inhibition as compared to the parent compound.

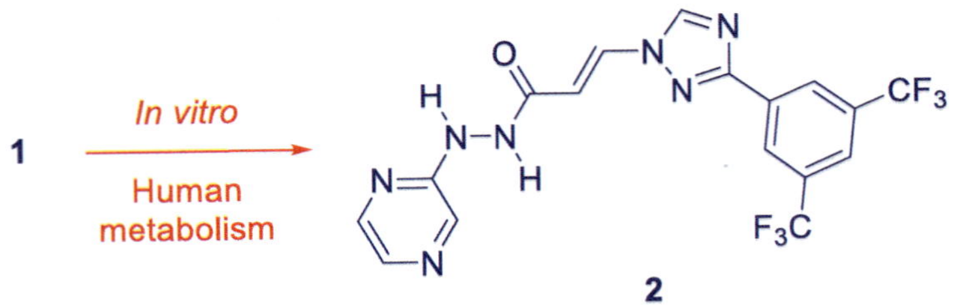

2
KPT-375

No definitive study on the elimination process of selinexor (**1**) has been performed; however, hepatobiliary secretion into feces seems to be the primary route of excretion based on radiolabeled studies in rats with urinary excretion being a minor route.[20]

The original studies with leptomysin B demonstrated that the downstream effects of nuclear export inhibition can persist longer than the actual nuclear export block.[21] Observation of increased XPO1 mRNA expression in leucocytes through a positive loop feedback has been found in patents with hematological malignancies and solid tumors, which provides evidence of XPO1 inhibition. XPO1 transcription increased dose proportionally from 3 to 28 mg/m^2, which reached an induction plateau of 5.3 ± 2.9-fold at doses ≥ 28 mg/m^2. Thus, the half-life of pharmacodynamic activity based on circulated leucocytes is at least 48 hours and is consistent with twice-a-week dosing.[8]

Similarly, treatment with selinexor (**1**) in lymphoid and myeloma cells results in disruption of three-dimensional nuclear organization of telomeres. Normal cells were predominantly unaffected.[22]

13.6. Efficacy and Safety

Selinexor (**1**) is a highly selective and slowly reversible covalent inhibitor of XPO1. A long pharmacokinetic half-life of approximately 48 hours was observed which permits twice-weekly dosing.

As combination or monotherapy, selinexor (**1**) had an acceptable tolerability profile in several clinical trials. The most frequently observed (> 20%) adverse events associated with the drug were dose and schedule-dependent and could be optimized by modifying the dosing frequency.[4] Symptoms included thrombocytopenia, fatigue, nausea, anemia, decreased appetite, decreased weight, diarrhea, vomiting, hypernatremia, neutropenia, leukopenia, constipation, dyspnea, and upper respiratory tract infection in patients with myeloma.[5] It has been demonstrated that some symptoms (weight loss, fatigue, anorexia) could be improved through the use of appetite stimulants (megesterol with olanzapine) and antinausea agents (odanstron).[4]

13.7. Syntheses

The patent disclosed in 2013[23] by Karyopharm proposed generation of the compound though conversion of an aryl nitrile **3** to thioamide **4** by magnesium chloride-mediated addition of sodium thiolate (Scheme 1). Thioamide **4** is condensed with hydrazine and formic acid to form the 1,2,4-triazole **5**. Subsequent addition/elimination reaction with the *Z*-iodoacrylate gives intermediate **6**, which is then saponified with lithium hydroxide. At this stage, the *Z*-isomer of intermediate **7** is the predominate regioisomer (90%) and a minor amount (8%) of the *E*-isomer can be separated out *via* chromatography.

The final step involves an amide bond forming reaction with 2-hydrazinopyrazine using the T3P coupling reagent to give selinexor (**1**) in 48% yield.

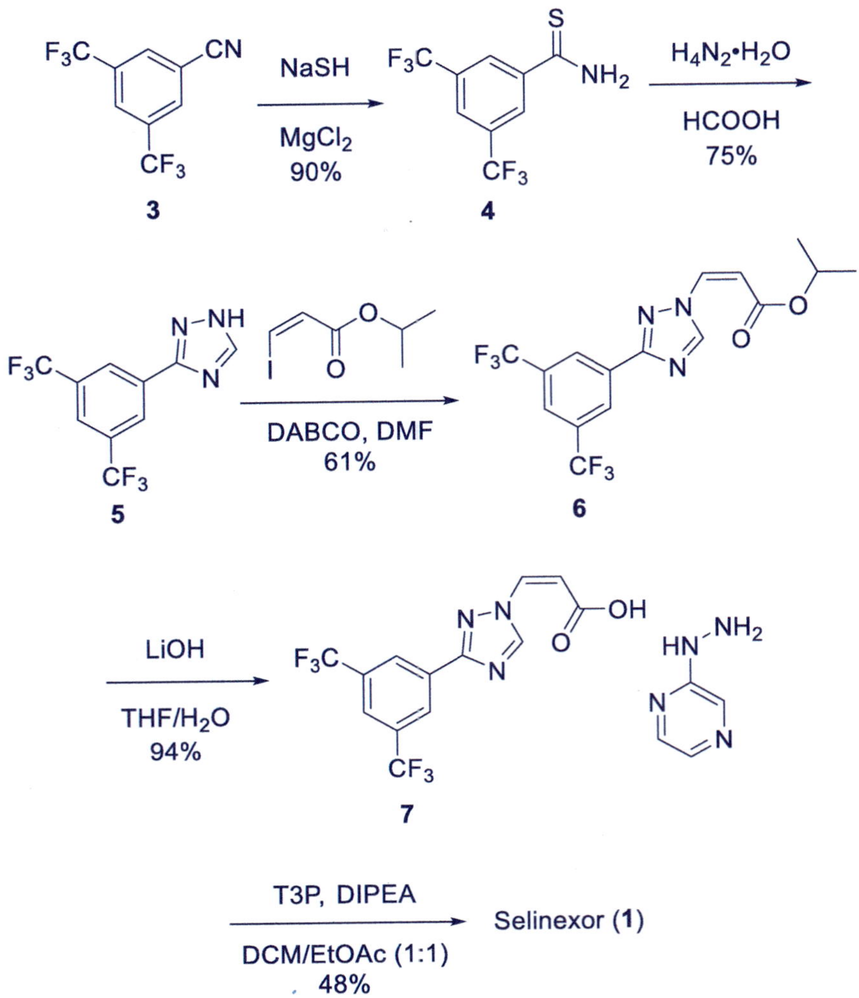

Scheme 1: Karyopharm's synthesis of Selinexor.

A patent filed by Watson Laboratories in 2018[24] covers novel crystalline forms of selinexor (**1**) and describes a modified synthesis (Scheme 2).

Compound **3** undergoes hydration under basic conditions to give **4**, which is then converted to acylformamidine **8**. Cyclization then occurs under acidic conditions and, compared with the original route, has the advantage of avoiding the liberation of toxic fumes such as H_2S.

Concerning conversion of intermediate **5** to **9**, the *Z*-methyl iodoacrylate is added slowly to keep the temp within –5 to +5 °C and water is added to precipitate the product at the end of the reaction. The yield increases significantly compared with the original route (88% vs. 61%), and the use of the potassium carbonate increases the rate of reaction which reduced the overall *E*-isomer content to 0.89% without any purification.

The use of *tert*-butanol during the subsequent saponification step is also beneficial and seems to prevent any undesired isomerization of **7** to the *E*-isomer,[25] thus another purification step can be avoided.

The last step substitutes the more cost effective EDCI•HCl for T3P. The advantage is that the former does not require cryogenic conditions as opposed to the –40 to –20 °C range needed for the latter reagent. Room temperature conversion leads to an improved isolated yield of 90–96% of compound **1**.

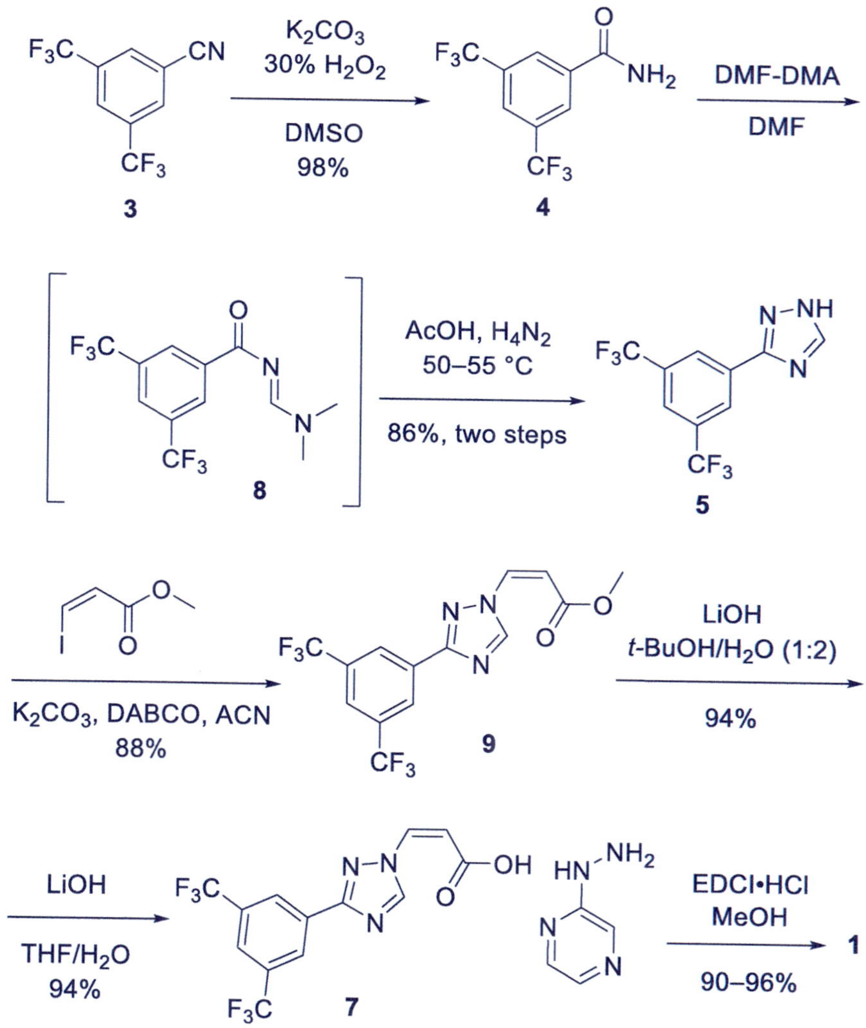

Scheme 2: Watson Laboratories' synthesis of Selinexor.

13.8. Summary and Future

Selinexor (**1**) is currently in several clinical trials as a single or combination agent for a wide range of hematological and solid cancers (over 70 trials, with 8 for MM). The BOSTON phase III study with examined a combination of weekly selinexor (**1**) with dexamethasone and once weekly bortezomib was found to be superior to standard twice weekly bortezomib and dexamethasone across all efficacy endpoints in patients with MM who had received 1–3 prior therapies (NCT03110562). Significant reduced rates of peripheral neuropathy were also observed which is an improvement for older and/or frail patients due to the toxicity associated with long term bortezomib treatment.[26] In terms of non-oncology applications, there is evidence that selinexor (**1**) interferes with SARS-CoV-2-virus replication by blocking ACE-2 (angiotensin converting enzyme-2) transport to the cytosol. Reduction of ACE-2 receptors on the cell surface protects cells from viral infection.[27]

Further work is currently focused on ameliorating the tolerability and increasing the therapeutic index of SINE compounds. The second-generation SINE compound, KPT-8602 (eltanexor), shows similar target binding and inhibitory properties as selinexor (**1**) but with reduced brain penetration and a mildly increased degree of reversible binding. The associated drop in toxicity permits a more prolonged and frequent dosing schedule in two murine models of hematological malignancies with respect to selinexor (**1**).[4] This is an exciting development for further exploration of this novel mode of cancer therapy and may open up treatment to other indications for which current therapy options are limited.

13.9. References

1. Abraham, S. A.; Holyoake, T. L. Redirecting traffic using the XPO1 police. *Blood* **2013**, *122*, 2926–2928.
2. Gerecitano, J. SINE (selective inhibitor of nuclear export) – translational science in a new class of anti-cancer agents. *J. Hematol. Oncol.* **2014**, *7*, 67.
3. Podar, K.; Shah, J.; Chari, A.; Richardson, P. G.; Jagannath, S. Selinexor for the treatment of multiple myeloma. *Expert Opin. Pharmacother.* **2020**, *21*, 1–10.
4. Fung, H. Y. J.; Chook, Y. M. Atomic basis of CRM1-cargo recognition, release and inhibition. *Semin. Cancer Biol.* **2014,** *27*, 52–61.
5. Hing, Z. A.; Fung, H. Y. J.; Ranganathan, P.; Mitchell, S.; El-Gamal, D.; Woyach, J. A.; Williams, K.; Goettl, V. M.; Smith, J. L.; Yu, X.; Meng, X.; Sun, Q.; Cagatay, T.; MacMillan, J. B.; Lehman, A. M.; Lucas, D. M.; Baloglu, E.; Shacham, S.; Kauffman, M.; Byrd, J. C.; Chook, Y. M.; Garzon, R.; Lapalombella, R. Next generation XPO1 inhibitor shows improved efficacy and in vivo tolerability in hematologic malignancies. *Blood* **2015**, *126*, 317–317.

6. Gandhi, U. H.; et al. Clinical implications of targeting XPO1-mediated nuclear export in multiple myeloma. *Clin. Lymphoma Myeloma Leuk.* **2018**, *18*, 335–345.
7. Raab, M. S.; Podar, K.; Breitkreutz, I.; Richardson, P. G.; Anderson, K. C. Multiple myeloma. *Lancet* **2009**, *374*, 324–339.
8. Munshi, N. C. Plasma cell disorders: an historical perspective. *Hematol. Am. Soc. Hematol. Educ. Program* **2008**, *2008*, 297.
9. Daelemans, D.; et al. A synthetic HIV-1 Rev inhibitor interfering with the CRM1-mediated nuclear export. *Proc. Natl. Acad. Sci.* **2002**, *99*, 14440–14445.
10. Neck, T. V.; et al. Inhibition of the CRM1-mediated nucleocytoplasmic transport by N-azolylacrylates: structure–activity relationship and mechanism of action. *Bioorg. Med. Chem.* **2008**, *16*, 9487–9497.
11. Shechter, S.; Kauffman, M.; Sandanayaka, V. P.; Shacham, S. (2011) Nuclear Transport Modulators and Uses Thereof [Karyopharm Therapeutics, US] WO2011109799A1.
12. Dong, X.; et al. Structural basis for leucine-rich nuclear export signal recognition by CRM1. *Nature* **2009**, *458*, 1136–1141.
13. Etchin, J.; et al. Antileukemic activity of nuclear export inhibitors that spare normal hematopoietic cells. *Leukemia* **2013**, *27*, 66–74.
14. Senapedis, W. T.; Baloglu, E.; Landesman, Y. Clinical translation of nuclear export inhibitors in cancer. *Semin. Cancer Biol.* **2014**, *27*, 74–86.
15. Syed, Y. Y. Selinexor: first global approval. *Drugs* **2019**, *79*, 1485–1494.
16. https://www.targetedonc.com/view/selinexor-granted-fast-track-designation-by-fda-for-pentarefractory-multiple-myeloma. Accessed Sept 20, 2021.
17. Lipinski, C. A.; Lombardo, F.; Dominy, B. W.; Feeney, P. J. Experimental and computational approaches to estimate solubility and permeability in drug discovery and development settings. *Adv. Drug Deliver. Rev.* **2001,** *46*, 3–26.
18. Fda.gov; ODAD Briefing document on Selinexor (KPT-330), Karyopharm Therapeutics Inc (fda.gov/media/121669/download). Accessed Sept 20, 2021.
19. Razak, A. R. A.; et al. First-in-class, first-in-human phase I study of selinexor, a selective inhibitor of nuclear export, in patients with advanced solid tumors. *J. Clin. Oncol.* **2016**, *34*, 4142–4150.
20. Peterson, T. J.; Orozco, J.; Buege, M. Selinexor: a first-in-class nuclear export inhibitor for management of multiply relapsed multiple myeloma. *Ann. Pharmacother.* **2020,** *54*, 577–582.
21. Mutka, S. C.; et al. Identification of nuclear export inhibitors with potent anticancer activity in vivo. *Cancer Res.* **2009**, *69*, 510–517.
22. Taylor-Kashton, C.; et al. XPO1 inhibition preferentially disrupts the 3D nuclear organization of telomeres in tumor cells. *J. Cell. Physiol.* **2016**, *231*, 2711–2719.
23. Sandanayaka, V. P.; Shacham, S.; McCauley, D.; Shechter, S. Hydrazide Containing Nuclear Transport Modulators and Uses Thereof [Karyopharm Therapeutics, US] WO2013019548A1 (2013).
24. Muthusamy, A. R.; Kanniah, S. L.; Ravi, A.; Das, T. C.; Chemate, R. P.; Singh, A. K.; Wagh, Y.D. Novel Crystalline Forms of Selinexor and Process for their Preparation [Watson Lab Inc, IN] WO2018129227A1 (2018).
25. The inventors observed only 0.66% content of *E*-isomer at this step.

26. Auner, H. W.; Gavriatopoulou, M.; Delimpasi, S.; et al. Effect of age and frailty on the efficacy and tolerability of once-weekly selinexor, bortezomib, and dexamethasone in previously treated multiple myeloma. *Am. J. Hematol.* **2021**, *96*, 708–718.
27. Kashyap, T.; et al. Selinexor, a novel selective inhibitor of nuclear export, reduces SARS-CoV-2 infection and protects the respiratory system in vivo. *Antivir. Res.* **2021**, *192*, 105115.

III

CNS DRUGS

14

Sage 217 (Zuranolone) for Treatment of Major Depressive Disorder

Richard T. Beresis

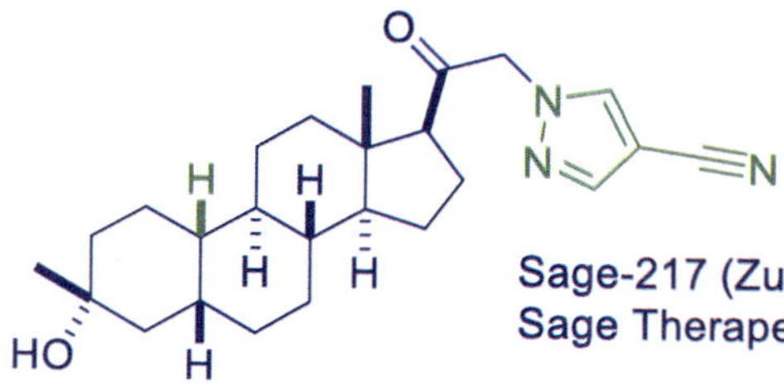

Sage-217 (Zuranolone, **1**)
Sage Therapeutics

14.1 Background

Major depressive disorder (MDD) and its accompanying related mood disorders constitute the most frequently diagnosed mental illness with ~10% of US adults suffering from a mood disorder event in the past year.[1] Primary mood disorder states may be broken down as depression (depressed mood), manic (elevated mood), and bipolar (a mix of depressed and elevated moods), and there exists a variety of disorder subtypes that may vary in length of episode and severity.

MDD patients are typically treated with antidepressant medication in some combination with professional counseling with a period of remission as the primary therapeutic goal. MDD has been historically treated with drugs, which include selective serotonin reuptake inhibitors (SSRIs), monoamine oxidase inhibitors (MAOIs), and tricyclic antidepressants (TCAs).[2]

The older antidepressant medications TCAs and MAOIs are currently less prescribed relative to newer medications, as they tend to feature a more severe side-effect profile, interact with specific foods, and possess known drug–drug interactions (DDIs).

Current Drug Synthesis, First Edition. Edited by Jie Jack Li.

However, these drugs may still show greater efficacy compared with the more modern therapeutics for certain patients.

SSRIs are the most commonly prescribed antidepressants today and have become the MDD frontline treatment. This class of drugs can be represented by fluoxetine (**2**), sertraline (**3**), paroxetine (**4**), and citalopram (**5**).

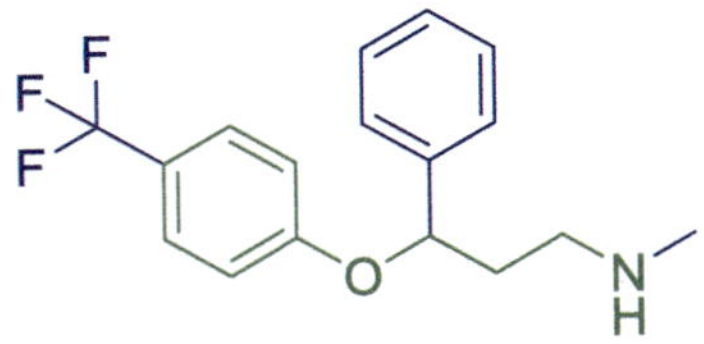

2

Fluoxetine (Prozac)
Lilly, 1986
SSRI

3

Sertraline (Zoloft)
Pfizer, 1991
SSRI

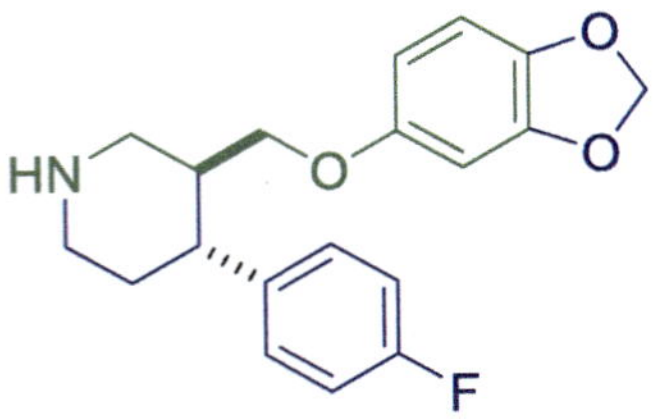

4

Paroxetine (Paxil)
GlaxoSmithKline, 1992
SSRI

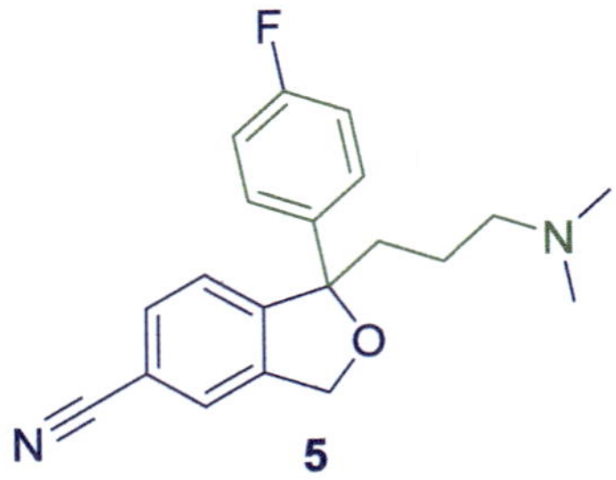

5

Citalopram (Celexa)
Lundbeck, 1998
SSRI

SSRIs function by selectively blocking the reuptake of serotonin to increase the amount of serotonin available in the brain synapses and tend to have lower frequency and milder side effects when compared to MOAIs and TCAs. There does not appear to be significant clinical efficacy differences among the set of Food and Drug Administration (FDA) approved SSRIs for MDD treatment, and in this regard these antidepressant medications work equally well. Thus, patients who do not tolerate a particular SSRI are frequently switched to another SSRI or other type of antidepressant.

As an additional treatment option, the related serotonin–norepinephrine reuptake inhibitors (SNRI) sertraline (**6**, Zoloft)[3] and venlafaxine (**7**, Effexor) have found wide use for certain patients where these SNRIs may be as efficacious as any of the SSRIs.

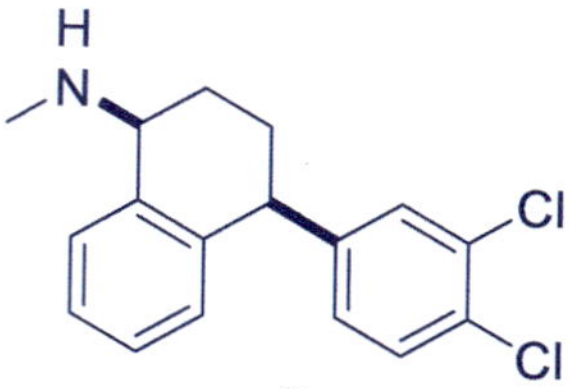

6

Sertraline (Zoloft)
Pfizer, 1991
SNRI

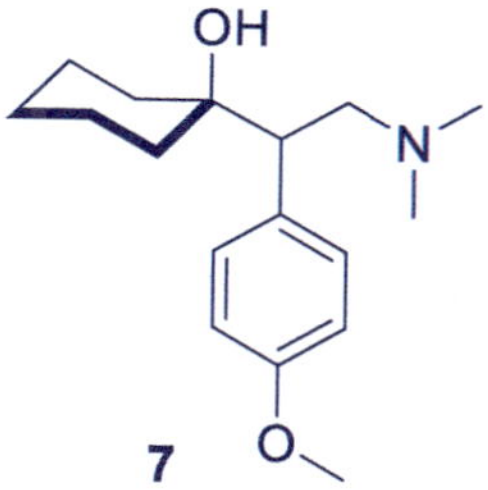

7

Venlafaxine (Effexor)
Wyeth, 1993
SNRI

The antidepressant class of drugs strongly influence the overall balance of the primary neurotransmitters dopamine, norepinephrine, and serotonin in the brain, which in turn regulate emotion, stress response, libido, appetite, and sleep. Given that antidepressant drugs affect the fundamental neurotransmission pathways in concert with the complex interrelated nature of central nervous system (CNS) biology, it comes as no surprise that therapeutic gaps remain in the treatment of mood disorders. These include insufficient patient response, poorly tolerated side-effect profiles leading to compliance issues, and significant efficacy lag times. Subsequently, there exists a variety of disorder subtypes that are currently unmet by the current generation of antidepressant medications.

Thus, researchers remain focused on the discovery of novel treatment strategies and a new generation of effective medications for MDD and other mood disorders. Recently, the *N*-methyl-*D*-aspartate receptor (NMDA) antagonist class of drugs has emerged, wherein initially existing drugs such as ketamine (**8**) and its *S*-enantiomer esketamine (**9**) have been demonstrated to be clinically efficacious in the treatment of mood disorders.[4]

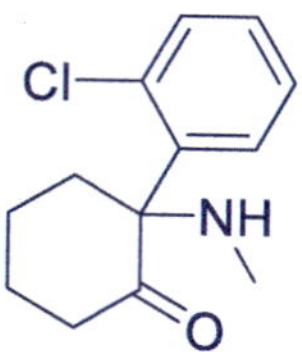

8

Ketamine

9

Esketamine

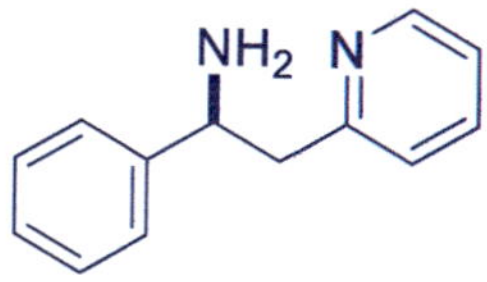

10

Lanicemine AZD6765

Lanicemine AZD6765 (**10**) was advanced into the clinic by AstraZeneca as a low-trapping NMDA receptor antagonist; "low-trapping" refers to the compound's antidepressant efficacy without the undesired psychomimetic effects.[5] AV-101 (**11**) is an orally active pro-drug of 4-chloronurenic acid that was advanced by VistaGen into phase II treatment-resistant MMD before showing a lack of efficacy.[6] Allergan's IV-administered rapastinel GLYX-13 (**12**) acts as a selective partial agonist acting on the allosteric glycine site of the NMDA receptor complex.[7] This agent was granted FDA Breakthrough Therapy status and was advanced into phase III for MDD before placebo equivalency resulted in its discontinuation.[8]

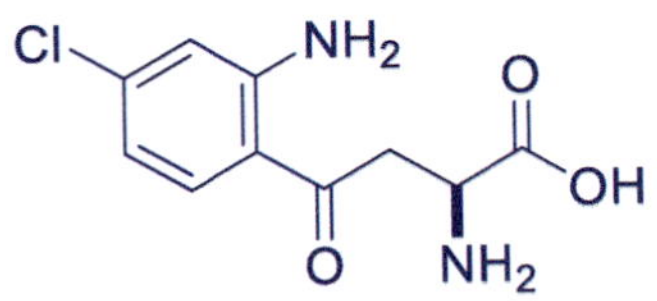

11

4-Chlorokynurenine AV-101

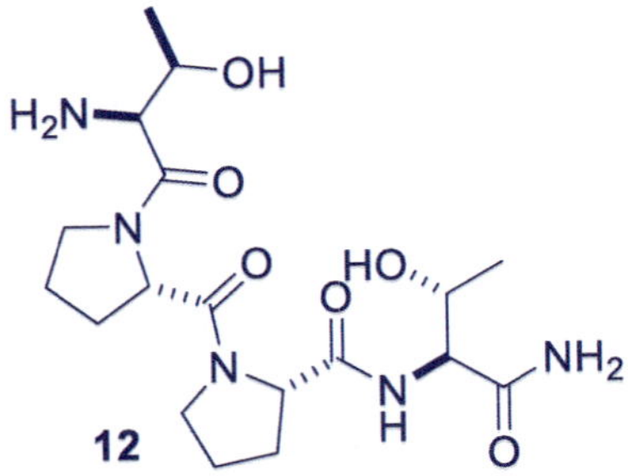

12

Rapastinel GLYX-13

Most recently, discoveries made in the past 25 years point to the Glutamate and $GABA_A$ amino acid neurotransmitter system as key players in MDD disorders. They have been thus actively targeted by academics and industry to address this unmet medical need. Multiple agents are progressing into phase II and III, and this recent progress has been expertly reviewed by Wilkinson and Sanacora.[9] Of particular focus and note are neuro-active steroid (NAS) class, which impacts CNS function as positive allosteric modulators (PAM) of the $GABA_A$ receptor.[10] A successful outcome stemming from this novel and exciting therapeutic approach will be discussed in this chapter.

14.2 Pharmacology

$GABA_A$ is the primary inhibitory neurotransmitter in the CNS, playing a vital role in homeostasis of neuronal activity.[11] Expressed in ~40% of all brain neurons, the $GABA_A$ receptor family is an abundant and ubiquitous neurotransmitter system influencing a wide range of brain circuitry central to the modulation a variety of behavioral states. $GABA_A$ dysfunction may occur in several different brain regions and *via* the imbalance of neuronal excitation vs. inhibition, and a number of disease states may arise, such as schizophrenia, mood and sleep disorders, seizure disorders, and epilepsy.[12]

In physical form, the $GABA_A$ receptor consists of a penta-meric ion channel formed by assembly of two alpha subunits ($\alpha1-\alpha6$), two beta subunits ($\beta1-\beta3$), and one additional subunit ($\gamma1-\gamma3$, δ, ε, π, or θ). The subunit composition determines the biophysical and pharmacological characteristics of the channel and influences its location at synaptic or extra-synaptic sites.[13]

Endogenous NASs such as pregnanolone (**14**) and allopregnanolone (**15**) are synthesized in the brain from cholesterol and have been shown to impact CNS function through positive allosteric modulation of the $GABA_A$ receptor ($GABA_A$-R). This allosteric modulation increases receptor efficacy and potency of the endogenous ligand γ-aminobutyric acid (**13**).[14] Despite the shared common steroid structural motif, NASs are distinct from glucocorticoids, mineralcorticoids, estrogens, and other steroidal systems and selectively target only $GABA_A$ at both the extra-synaptic and synaptic receptor sites. In general, the development of NAS chemical matter has fortunately not been compromised by undesired cross-steroidal off-target activity.

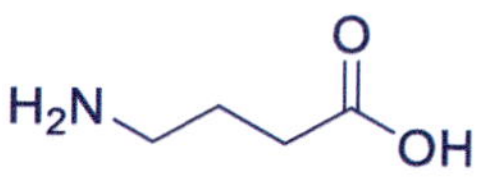

13
γ-Aminobutyric acid
(GABA)

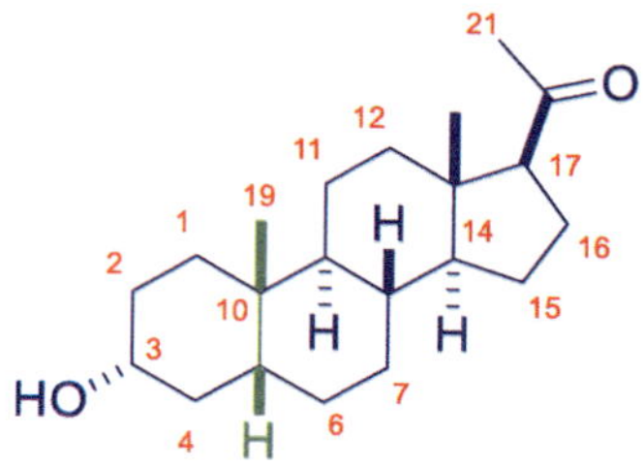

14
Pregnenolone (*cis*-)
steroid core conventional numbering

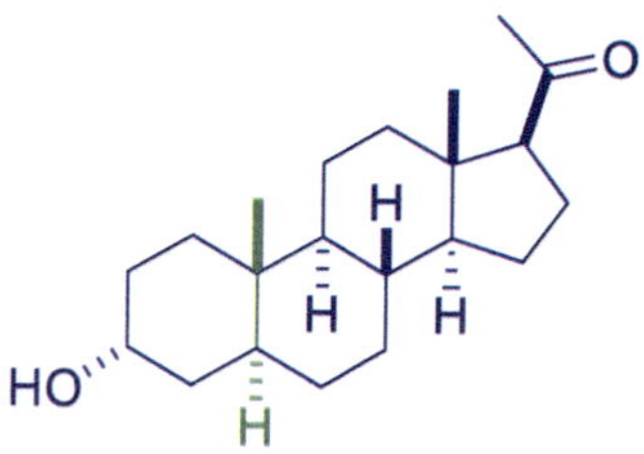

15
Allopregnanolone (*trans*-)
brexanolone (Zulresso)

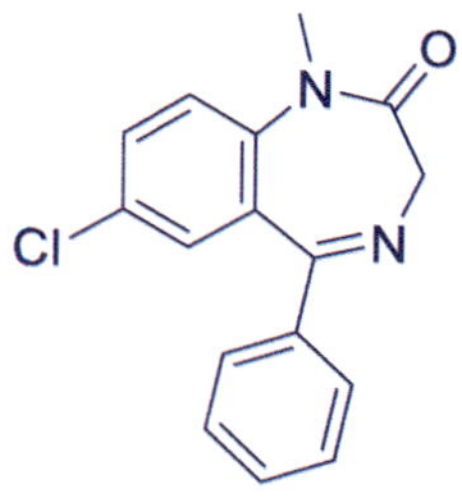

16
Diazepam (Valium)
benzodiazepine

Generally, NASs have the potential to treat a wide range of CNS disorders due to their fundamental inhibitory effects on brain circuitry central to a variety of behavioral states. NASs present the opportunity to differentiate from the classic $GABA_A$-R PAM benzodiazepines (BDs) represented by diazepam (**16**) by targeting separate populations of $GABA_A$-R subtypes. BDs bind to the allosteric α/γ subunit interface distinct from the

GABA binding site, while NASs are capable of also targeting extra-synaptic $GABA_A$-Rs, including the δ subunit.[15]

BD therapy suffers from a tolerance issue, as BDs act only at synaptic $GABA_A$–Rs, which are in turn down-regulated by BDs themselves, and subsequently BD efficacy is diminished/lost due to a reduction of the synaptic receptor population.[16]

Thus specifically, NASs have the potential to treat seizures that are BD resistant, an example being status epilepticus (SE), which can only be targeted at extra-synaptic $GABA_A$-Rs. In regard to MDD, clinical evidence points toward $GABA_A$ involvement, as decreased levels of both $GABA_A$-R and endogenous NASs are observed in the brain originating from patients suffering from depression.[17] Encouragingly, therapeutic modulation to positively affect NAS levels has resulted in demonstrated efficacy in *in vivo* models for depression,[18] and has provided the impetus for the industry to evaluate clinically endogenous NASs and explore modifications to improve their drug-like properties and ultimate therapeutic potential.

14.3 Structure–Activity Relationship (SAR)

$GABA_A$ PAM endogenous ligands pregnanolone (**14**) and allopregnanolone (**15**) have historically been assessed clinically for a variety of indications. However low oral bioavailability and high clearance due to rapid **C3** hydroxyl oxidation requires IV dosing. Pregnanolone (**14**) has been investigated for clinical uses as diverse as general anesthesia, bipolar disorder (BPD), back pain, and schizophrenia, but unfortunately noted side effects and poor pharmacokinetic properties have halted any advancement. Allopregnanolone[19] (brexanolone, Zulresso) (**15**), a naturally occurring steroid, was recently FDA approved and marketed by Sage Therapeutics for the treatment of Postpartum Depression (PPD)[20] where it is IV administered over a long infusion period. PPD pathophysiology is theorized to be driven primarily by the greatly reduced levels of this endogenous NAS following childbirth. An immediate relief of PPD's serious adverse mood state is effectively treated with (**15**) by simply returning the allopregnanolone levels to their optimal endogenous concentrations.

Moving beyond the endogenous steroid structure with the intent of uncovering novel drug molecules with increased efficacy and oral delivery potential, researchers have employed proven medicinal chemistry strategies to improve NAS drug-like properties.

Marinus Pharma is currently advancing ganaxolone[21] (**17**) through clinical trials for drug-resistant partial onset seizure epilepsy and Fragile X Syndrome, but, due to its limited %F, (**17**) dosing ranges from 600 to 900 mg (tid or bid). C3 metabolism-blocking methyl group improves metabolic stability but does not provide robust oral bioavailability. Currently, two phase II studies are progressing, the first *via* IV administration, the second by an orally bio-available form.[22]

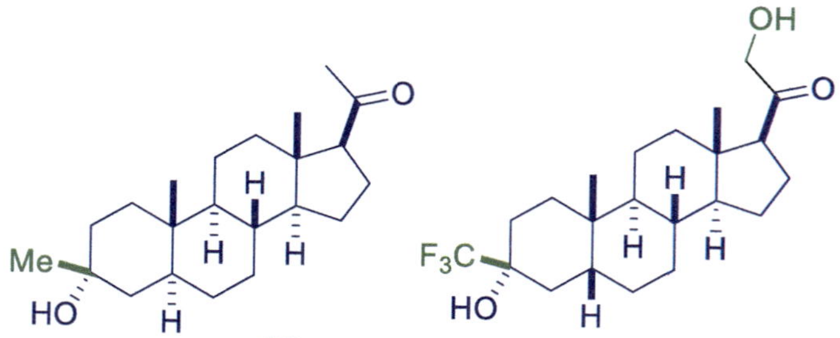

17 **18**

Ganaxolone (Marinus Pharma) Co26749 (Cocensys)

Cocensys demonstrated anticonvulsant and anxiolytic *in vivo* model efficacy with orally dosed Co26749 (**18**)[23] and Co134444 (**19**)[24]. While substitution at C3 (CF_3, CH_2OMe) and C23 (OH, heterocycle) was shown to improve adsorption, distribution, metabolism, and excretion (ADME) properties sufficiently for potential oral dosing, (**18**) and (**19**) were not ultimately advanced into the clinic.

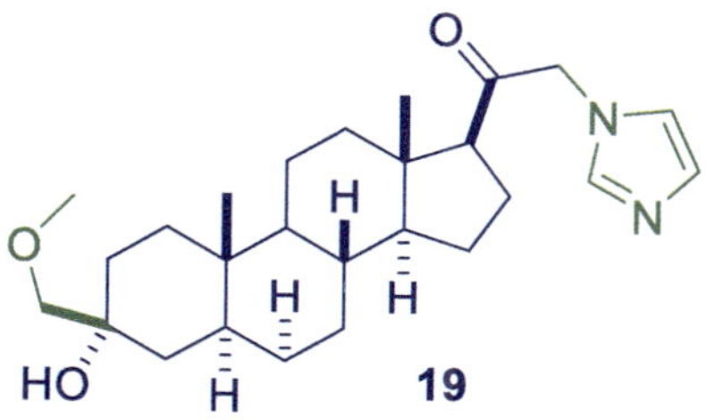

19

Co134444 (Cocensys)

Thus currently, there remains an unmet clinical need for GABA$_A$-R PAM acting agents with greater potency, improved ADME properties, and superior oral delivery, and a new set of researchers have taken on this challenging discovery goal.[25] As Sage and collaborator ChemPartner developed an structure activity relationship (SAR) strategy to address this medical need, the focus became a facile identification of an oral proprietary GABA$_A$-R PAM candidate based on the **(14)**/**(15)** scaffold with a good safety profile and a low tolerance/sedation liability. Figure 3 illustrates opportunities to leverage SAR lessons gleaned from prior GABA$_A$-R efforts.

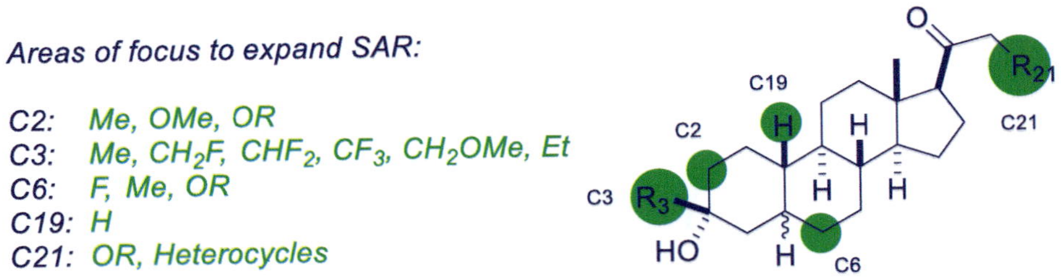

Figure 3 Areas of focus to expand the SAR identified.

Intellectual property (IP) space to work in was established by the development of both the C19 des-methyl NOR-19 in the *cis*- and *trans*-cores combined with novel functionality in other regions of the target molecules.

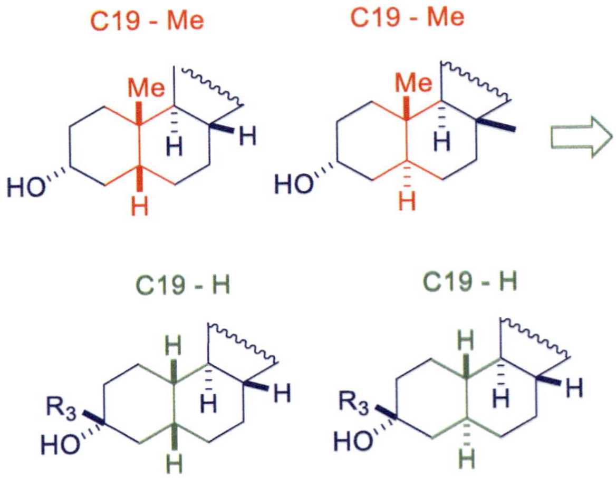

Manual and Q-patch assays were identified as the key research drivers to advance the chemical matter and provide greater understanding of the electro-physiologic activity and selectivity with the differing $GABA_A$-R subtypes. These functional assays constitute electro physiology experiments that are evaluated in both recombinant $\alpha1\beta2\gamma2$ $GABA_A$-R in LTK cells and $\alpha4\beta3\delta$ $GABA_A$-R in CHO cells using a Q-patch or manual patch assay protocol, respectively. Herein, low EC_{50} combined with a high E_{max} is desired.

Table 1 summarizes the initial exploration centered on diverse substitutions at various steroid core positions with the C17 ketone kept constant. Ultimately, analogs bearing changes at C2, C3, C6, and C10 offered no advantage over the reference example allopregnanolone (**15**) in terms of $GABA_A$ activity.

Table 1. Exploration of steroid core diverse substitution.

Compound / Structure	α1 (nM)/E_{max}	α4 (nM)/E_{max}	Liver microsome M/R/H
15 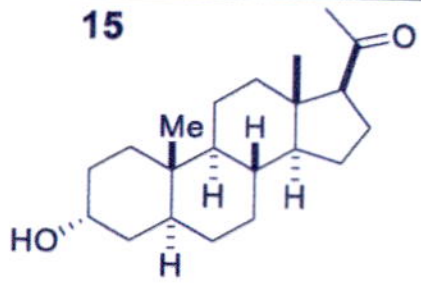	180/475	80/418	415/992/16
20 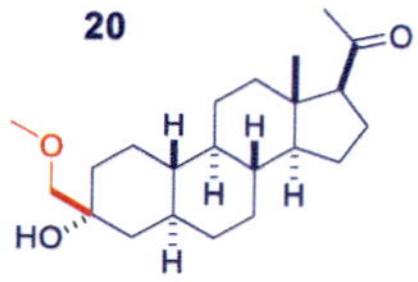	490/734	90/359	260/538 /12
21 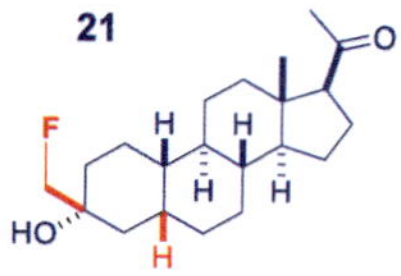	170/373	120/361	280/235/8.3
22 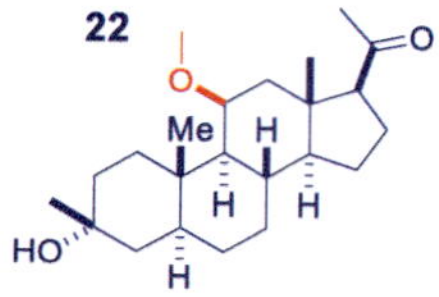	1340/1352	–	608/91/13
23 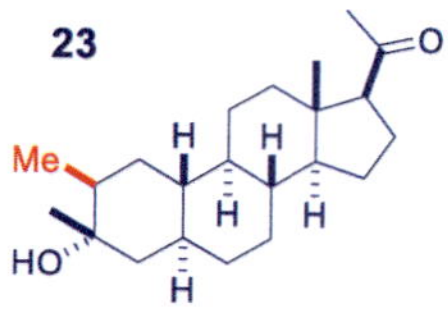	1330/308	308/770	211/53/8.6
24 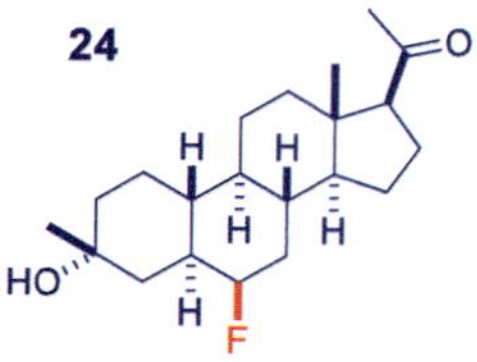	350/326	–	62/25/5.9

SAR focus then shifted to analogs bearing C21 functionalization, as encapsulated in Table 2. Increased polarity in conjunction with decreased basicity at C21 resulted in improvements in the functional activity at both $GABA_A$-Rs measured. C21 heterocyclic analogs such as the N-linked triazole (**26**) demonstrated very good α1 and α4 potency in combination with reasonable solubility and rodent microsomal stability when compared to the first generation NASs.[26]

Table 2. Substituted C21 heterocycles.

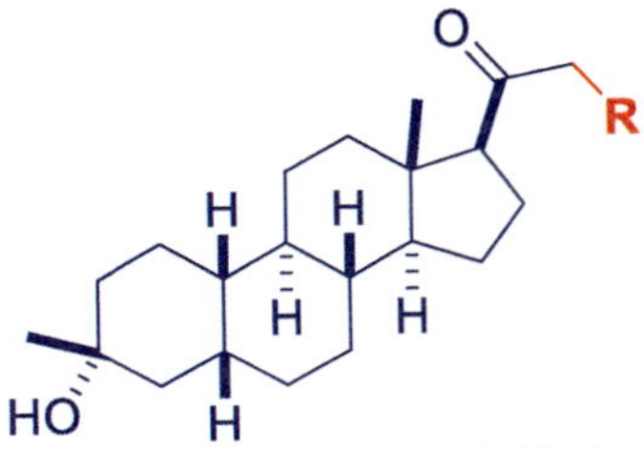

Compound / R	α1 (nM)/E_{max}	α4 (nM)/E_{max}	Microsome M/R/H	PTZ MED mpk
15	180/475	80/418	415/992/16	>10
25 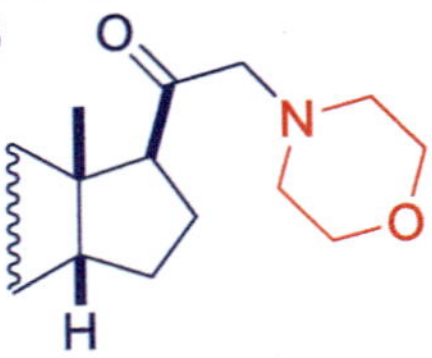	2275/514	3000/460	84/4 /5.7	–
26 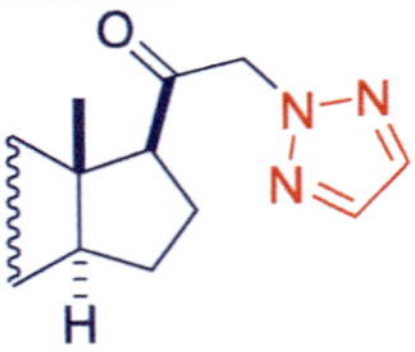	120/652	190/474	149/69/4.6	3
27 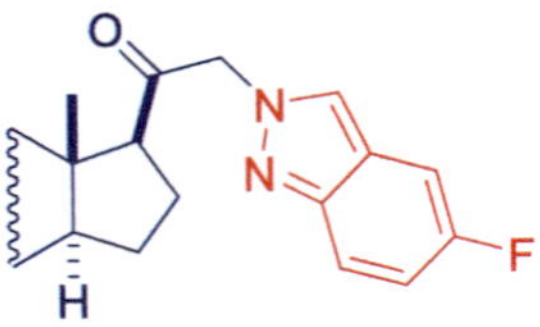	910/976	590/238	130/207/11	–
28 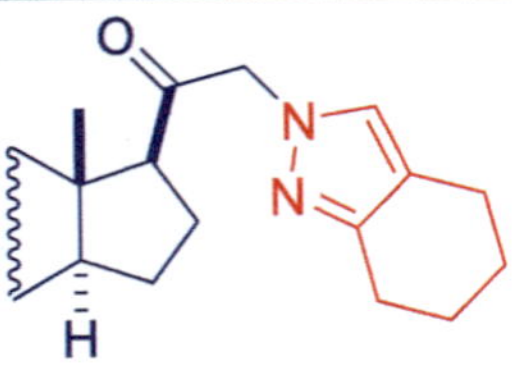	724/467	178/455	4/3.3/0.6	–

29 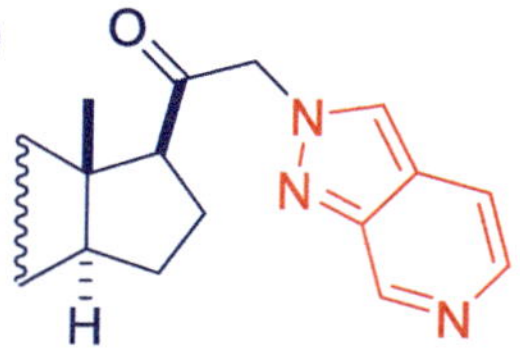	>3000 / >744	160/838	33/137/5.7	1

C21 bicyclic analogs such as (**29**) demonstrated excellent potency and GABA subunit selectivity for α4. (Figure 3a). Selective targeting of δ-containing receptors specifically enhance tonic GABAergic inhibition with the potential to positively affect disorders specific to tonic dysfunction including Fragile X, Autism spectrum disorder, anxiety and sleep disorders.

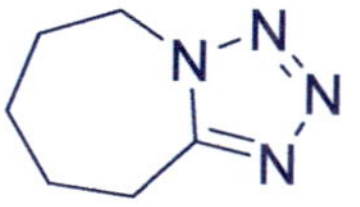

Pentylenetetrazol

PTZ

The PTZ Induced Seizure Model was the *in vivo* animal model chosen as the key project driver to select potential preclinical compound candidates for nomination. Pentylenetetrazol (PTZ) is a selective $GABA_A$ antagonist that upon administration will cause generalized CNS stimulation in CD-1 mice resulting in tonic-clonic seizures, and a compound's ability to suppress these seizures over a defined time period is assayed with minimum effective concentration (MED), the primary readout. (**29**) demonstrated a PTZ MED of 1 mpk, a >10-fold efficacy improvement over (**15**). Subsequent mouse PK studies showed good bioavailability (40%) for (**29**), but unfortunately high clearance (Cl > 3 L/h/kg) was also seen.

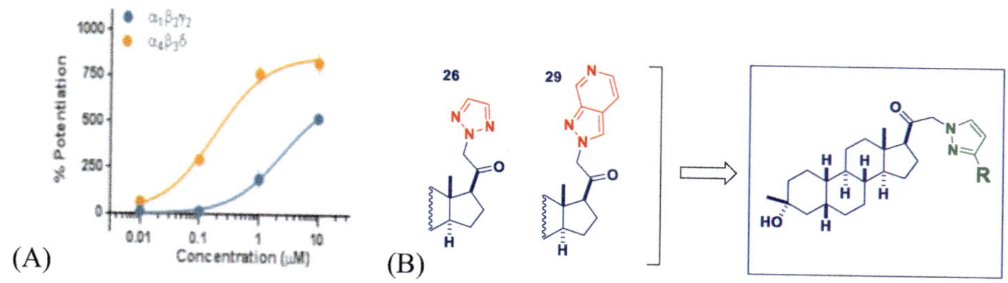

Figure 1 (A) Compound **29** patch clamp $GABA_A$ subtypes selectivity. (B) Medicinal chemistry design for substituted pyrazole series.

Considering prior SAR to continue heterocycle exploration at C21, opportunities became apparent to interrogate structural hybrids of the lead project compounds. Thus, with the goal of leveraging the receptor potency of the triazole (**26**) and the *in vivo* efficacy of (**29**), a focused series of substituted 5-membered heterocycles analogs were designed and synthesized. (Figure 1 B).

From an initial large set of five- and siv- membered heterocycles surveyed at the C21 position (data not shown), the pyrazole emerged as the superior example. Table 3 gives biodata for a subset of examples from the subsequent survey of pyrazole R-substitutions. The initial methyl substituted analog (**30**), while active, suffered from poor liver microsome stability and was then expanded on through a further examination of polar substitutions guided by logD and solubility considerations.

Table 3. Substituted C21 pyrazoles.

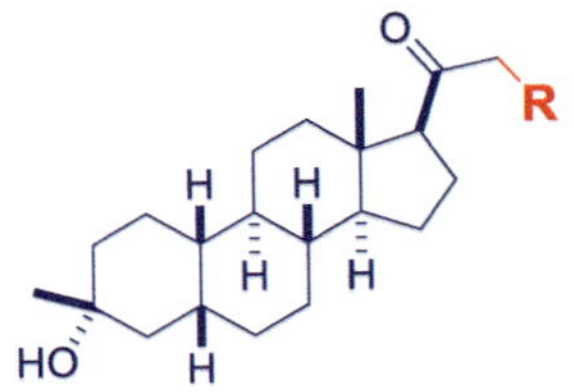

Compound / R	α1 (nM) /E_{max}	α4 (nM) /E_{max}	Microsome M/R/H	PTZ MED mpk
15	180/475	80/418	415/992/16	>10
30 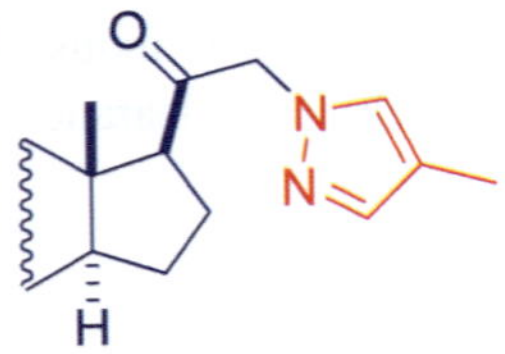	250/412	1230/824	409/204/14.6	–
31 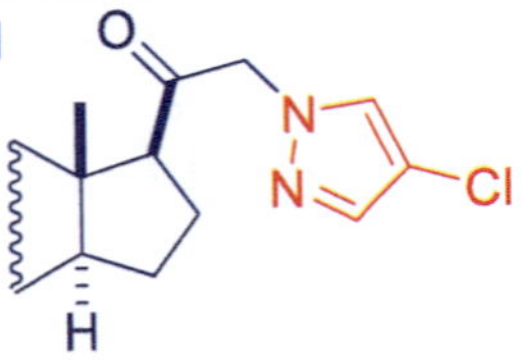	1199/564	773/1045	231/65/7.9	–
32 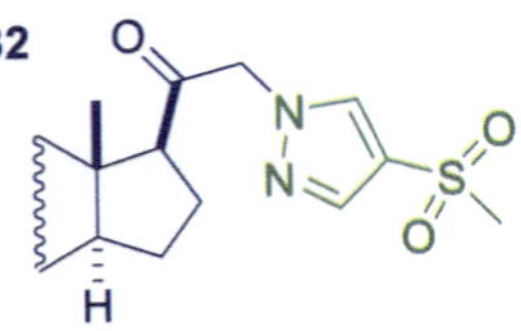	>3000/>352	675/410	19/47/2.4	<1

1 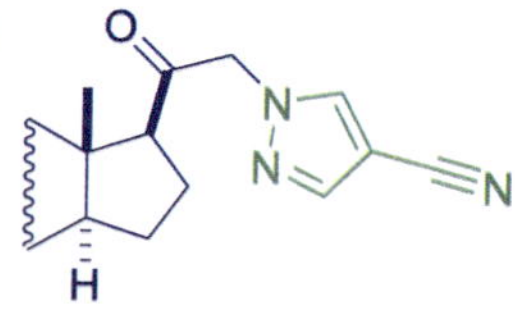	374/1041	163/640	77/44/6	0.1

The sulfone functionalized compound (**32**) is notable for its α4 selectivity, good metabolic stability, and PTZ efficacy. Based on these properties, (**32**) has been evaluated as a potential clinical candidate for α4 specific clinical indications. The nitrile functionalized compound zuranolone (**1**), which ultimately emerged from this effort, demonstrated very good cross-species liver microsome stability and excellent dual α1 and α4 $GABA_A$ nM potency combined with noted high E_{max} values in the electro physiology assays.[27]

14.4 Pharmacokinetics and Drug Metabolism

Zuranolone (**1**) initially demonstrated an encouraging rodent pharmacokinetics (PK) profile with good 24-hour exposure by both IV and PO in the mouse (Figure 2).[28] Similar good 24-hour exposure was then recapitulated in the secondary development species, the dog.

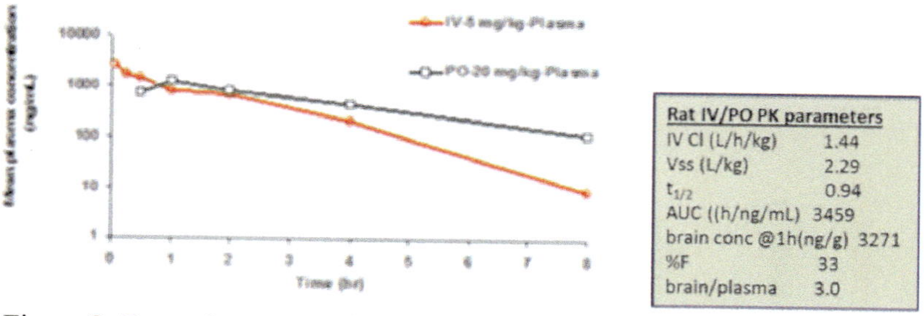

Figure 2 Zuranolone rat IV / PO plasma exposure and PK parameters from reference 28.

Zuranolone (**1**) was found to be highly orally bioavailable in humans with half-life of 16–21 hours and a C_{max} of ~1 hour providing an optimized oral PK profile, which was determined to support daily oral dosing in subsequent clinical trials (Figure 3).[29]

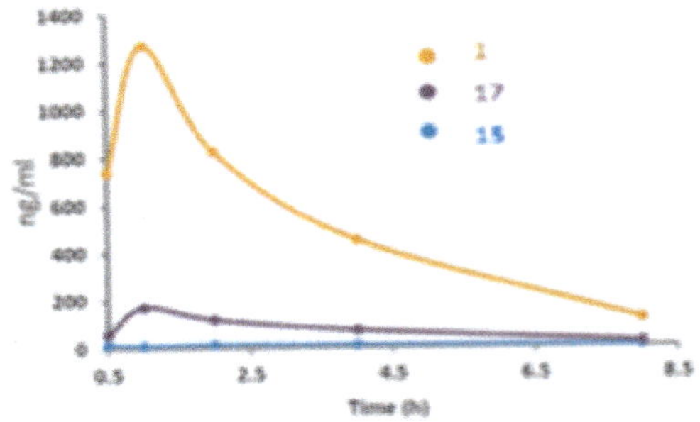

Figure 3. Human oral exposure comparison between zuranolone (**1**), ganaxolone (**17**), and allopregnanolone (**15**).

The comparison of (**1**), (**17**), and (**15**) human oral bio-availability over time (Figure 3) nicely encapsulates the evolving progression made in developing the endogenous NAS motif into molecules with improved drug-like properties and oral dosing potential.[30]

14.5 Efficacy and Safety

In addition to the excellent receptor potency and ADME profile, zuranolone (**1**) showed superiority in the PTZ efficacy model (MED 0.1 mpk) in comparison with all project analogs evaluated, notably suppressing clonic/tonic convulsions and death rate as well as seizure latency time. In this experiment, a remarkable brain concentration at 1 hour and brain/plasma ratio (3.0) was measured indicating excellent CNS exposure (Figure 4).

(A)

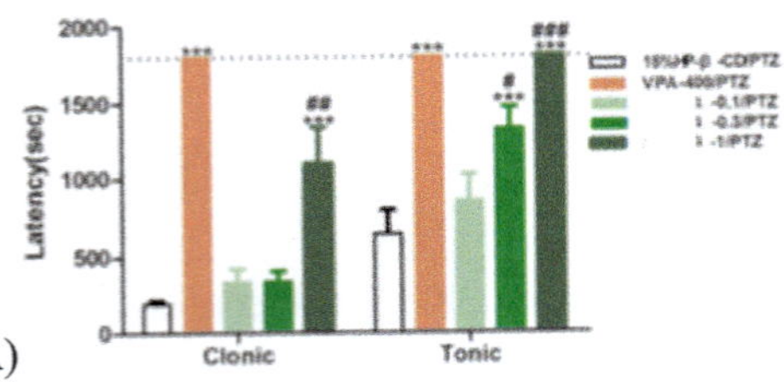

(B)

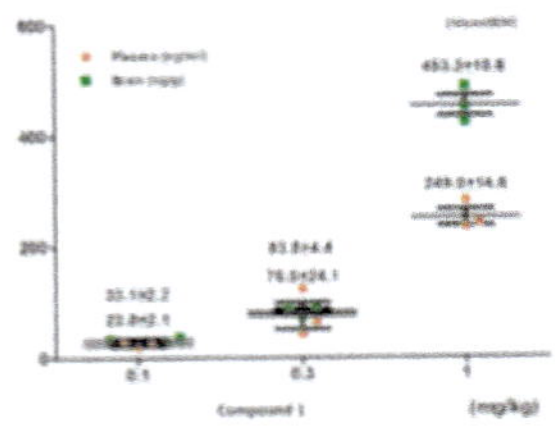

Figure 4 (A) Zuranolone (**1**) PTZ model efficacy. (B) Corresponding Zuranolone (**1**) brain/plasma levels.

Correspondingly, good efficacy was demonstrated in a battery of complimentary CNS animal models with (**1**), including locomotor activity (LMA), pharmaco-resistant ESE, mesio-temporal lobe epilepsy, and the pharmaco-resistant lithium pilocarpine.[31] Zuranolone (**1**) has also demonstrated efficacy across numerous preclinical models of seizure represented by the seizure suppression in a chronic epilepsy model as well as having shown useful activity in a genetic epilepsy model.[32] The encouraging results from these diverse *in vivo* CNS models truly speak to the very broad therapeutic potential of successful $GABA_A$-R PAM acting molecule such as zuranolone (**1**).

In terms of safety profile, zuranolone (**1**) was shown to be without hERG cardiac ion channel or CYP liabilities. No nuclear hormone receptor (NHR) cross-reactivity was seen in functional assays measuring agonistic and antagonistic effects in a 22-member panel, and no significant hits were found in an off-target CEREP panel evaluation for zuranolone (**1**). In subsequent human clinical trials, zuranolone (**1**) was generally very well tolerated with a minimal side effect profile observed at 25–50 mpk qd oral dosing.

Zuranolone (**1**) has demonstrated a solid safety profile to date, with no indications of withdrawal, weight gain, sexual dysfunction, euphoria, or sleep disruption symptoms, which are typically the cause of treatment discontinuation with current standard antidepressant medications.

6 Synthesis

The Sage/ChemPartner discovery chemistry synthesis route for zuranolone (**1**) is described in the following text.[33] Starting from the commercially available advanced steroid intermediate (+)-19-Norandrost-4-ene-3,17-dione (**33**), Pd/C hydrogenation of the C4/5 double bond selectively generated the desired core *cis*-orientation of (**34**). The regio- and stereo-selective Grignard methyl addition to the C3 ketone in the presence of the C17 ketone providing (**35**) was accomplished in high yield and enantiomeric excess through complexation with the sterically large ligand methyl aluminum bis-[2,6-di-*tert*-butyl-4-methylphenoxide] (MAD).

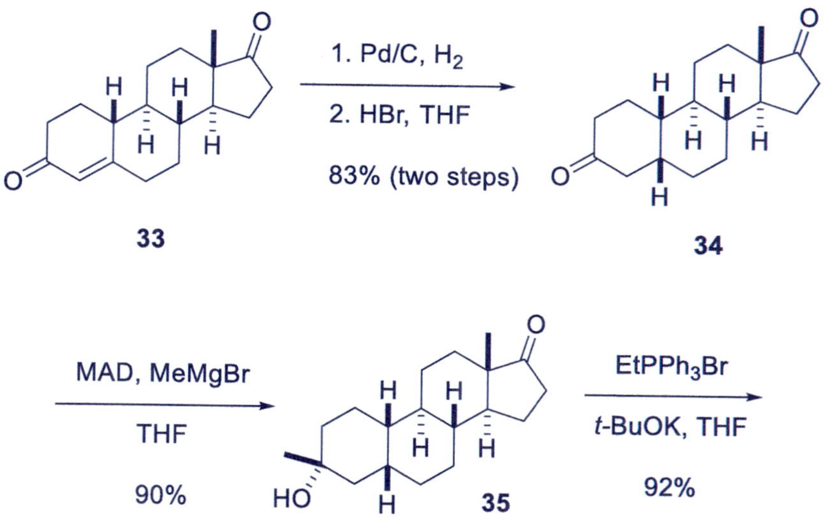

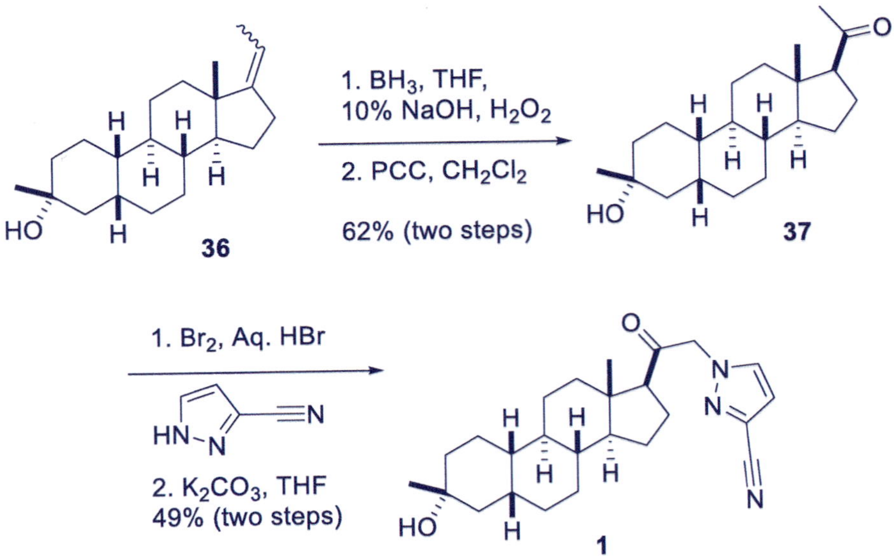

Intermediate (**35**) was then primed for Wittig olefination to provide (**36**) as a mixture of *E*- and *Z*- olefins. This was followed in the reaction sequence by selective hydroxylation with borane, then simple oxidation with PCC to generate the ketone (**37**) in reasonable yield. Selective bromination at the C21 position afforded the reactive common synthetic intermediate used throughout the described SAR exploration, and, in the case of zuranolone (**1**), the addition of 3-cyano-pyrazole to this alpha keto bromide intermediate provided the final product in 49% for the last two transformations.

14.7 Summary

In 2013, zuranolone (**1**) was successfully submitted as a first-in-class novel extra-synaptic $GABA_A$ oral modulator Investigational New Drug (IND), now known also as SAGE-217. Zuranolone (**1**) was granted Breakthrough Therapy Status by the FDA, and the success of this molecule assisted Sage to complete an IPO in 2014.

Zuranolone (**1**) is currently progressing through phase III clinical trials for postpartum depression[34] and MDD[35] and entering phase II for resistant depression, generalized anxiety disorder, and bipolar depression with potential opportunities for an additional diverse array of CNS-based therapies.[36]

While generally well tolerated with a minor side effect profile in the clinic, some setbacks were encountered in the initial MDD clinical trials with zuranolone (**1**) in terms demonstrating statistical efficacy relative to placebo.[37] Subsequently, Sage has partnered with Biogen to provide the resources necessary to complete redesigned clinical trials.

Referencing data from more recent trials,[38] in October 2021 Sage reported consistent clinically meaningful data with zuranolone (**1**) in the LANDSCAPE and NEST clinical development programs. These studies show significant improvements in depressive mood with zuranolone-treated patients, as well as rapid onset of measurable efficacy by Day 3, which is sustained at Day 42.[39] This notable immediate onset of action favorably contrasts typically used antidepressant medications that feature a long treatment period to achieve full efficacy. 2022 is targeted for zuranolone (**1**) new drug application (NDA) submission and FDA approval.

14.8 References

1. NIH National Institute of Mental Health 2017 National Survey on Drug Use and Health, **2017**
2. Dupuy, J. M.; Ostacher, M. J.; Huffman, J.; Perlis, R. H.; Nierenberg A. A. A critical review of pharmacotherapy for major depressive disorder. *Int. J. Neuropsychopharmacol.* **2011**, *14*, 1417–1431.
3. Patel, K.; Sophie, A.; Haque, M. N.; Angelescu, I.; Baumeister, D.; Tracy, D. K. Bupropion: a systematic review and meta-analysis of effectiveness as an antidepressant. *Ther. Adv. Psychopharmacol.* **2016**, *6*, 99–144.
4. Pochwat, B.; Nowak, G.; Szewczyk, B.; An update on NMDA antagonists in depression. *Expert Rev. Neurother.* **2019,** *19*, 1055–1067.
5. Sanacora, G.; Johnson, M. R.; Khan, A.; Atkinson, S. D.; Riesenberg, R. R.; Schronen, J. P.; Burke, M. A.; Zajecka, J. M.; Barra, J.; Su, H-L.; Posener, J. A.; Bui, K. H.; Quirk, M. C.; Piser, T. M.; Mathew, S. J.; Pathak, S.; Adjunctive lanicemine (AZD6765) in patients with major depressive disorder and history of inadequate response to antidepressants: a randomized, placebo-controlled study, *Neuropsychopharmacology,* **2017**, *42*, 844–853.
6. Park, L. T.; Kadriu, B.; Gould, T. D.; Zanos, P.; Greenstein, D.; Evans, J. W.; Yuan, P.; Farmer, C. A.; Oppenheimer, M.; George, J. M.; Adeojo, L. W.; Snodgrass, H. R.; Smith, M. A.; Henter, I. D.; Machado-Vieira, R.; Mannes, A. J.; Zarate, C. A.; A randomized trial of the N-methyl-d-aspartate receptor glycine site antagonist prodrug 4-chlorokynurenine in treatment-resistant depression., *Int. J. Neuropsychopharmacol.* **2020**, *23*, 417–425.
7. Moskal, J. R.; Jeffrey S. Burgdorf, J. S.; Stanton, P. K.; Kroes, R. A.; Disterhoft, J. F.; Burch R. M. Khan, M. A.; The development of rapastinel (formerly GLYX-13); a rapid acting and long-lasting antidepressant., *Curr. Neuropharmacol.* **2017**, *15*, 47–56.
8. Preskorn, S.; Macaluso, M.; Vishaal Mehra, D. O.; Zammit, G.; Moskal J. R.; Burch, R. M.; Randomized proof of concept trial of GLYX-13, an N-methyl-D-aspartate receptor glycine site partial agonist, in major depressive disorder nonresponsive to a previous antidepressant agent. *J. Psychiatr. Pract.* **2015**, *21*, 140–149.
9. Wilkinson, S. T.; Sanacora, G. A new generation of antidepressants: an update on the pharmaceutical pipeline for novel and rapid-acting therapeutics in mood

disorders based on glutamate/GABA neurotransmitter systems. *Drug Discov. Today* **2019**, *24*, 606–615.
10. Kleinman, R. A.; Schatzberg, A. F. Understanding the clinical effects and mechanisms of action of neurosteroids. *Am. J. Psychiatry* **2021**, 178, 221–223.
11. Engin, E.; Benham, R. S.; Rudolph, U. An emerging circuit pharmacology of GABAA receptors. *Trends Pharmacol. Sci.* **2018**, *39*, 710–732.
12. Patel, A. B.; Sarawagi, A.; Soni, N. D. Glutamate and GABA homeostasis and neurometabolism in major depressive disorder. *Front. Psychiatry* **2021**, *12*, doi:10.3389/fpsyt.2021.637863
13. Rudolph, U.; Mohler, H. GABA-based therapeutic approaches: GABAA receptor subtype functions. *Curr. Opin. Pharmacol.* **2006**, *6*, 18–23.
14. Maguire, J.; MacKenzie, G. Neurosteroids and GABAergic signaling in health and disease. *Biomol. Concepts* **2013**, *4*, 29–42.
15. (a) Olsen, R. W. GABAA receptors: subtypes provide diversity of function and pharmacology. *Neuropharmacology* **2009**, *56, 141–148.* (b) Modgil, A. Endogenous and synthetic neuroactive steroids evoke sustained increases in the efficacy of gabaergic inhibition via a protein kinase C-dependent mechanism. *Neuropharmacology* **2017**, *113*, 314–322.
16. Vinkers, C. H.; van Oorschot, R.; Nielsen, E. Ø.; Cook, J. M.; Hansen, H. H.; Groenink, L.; Olivier, B.; Mirza, N. M. GABAA receptor α subunits differentially contribute to diazepam tolerance after chronic treatment. *Neuropharmacol.* **2012**, *PLoS ONE,* 7, e43054. doi./10.1371/journal.pone.0043054.
17. Luscher, B.; Shen, Q.; Sahir, N. The GABAergic deficit hypothesis of major depressive disorder. *Mol. Psychiatry* **2011**, *16*, 383–406.
18. Schule, C.; Nothdurfter, C.; Rupprecht, R. The role of allopregnanolone in depression and anxiety. *Prog. Neurobiol.* **2014**, *113*, 79–87.
19. Melcangi, R. C.; Panzica, G. C. Allopregnanolone: state of the art. *Prog. Neurobiol.* **2014**, *113*, 1–5, introduces series of articles contained in this special issue.
20. Paul, S. M.; Pinna, G.; Guidottide, A. Allopregnanolone: from molecular pathophysiology to therapeutics. a historical perspective. *Neurobiol. Stress* **2020**, *12*, 100215.
21. Carter, R. B.; Wood, P. L; Wieland, S.; Hawkinson, J. E.; Belelli, D.; Lambert, J. J.; White, H. S.; Wolf, H. H.; Mirsadeghi, S.; Tahir, S. H.; Bolger, M. B.; Lan, N. C.; Gee, K. W. Characterization of the anticonvulsant properties of ganaxolone (CCD 1042; 3α-hydroxy-3β-methyl-5α-pregnan-20-one), a selective, high-affinity, steroid modulator of the γ-aminobutyric acid A receptor. *J. Pharmacol. Exp. Ther.* **1997**, *280*, 1284–1295.
22. FDA Clinical Trials NCT01725152 and NCT04391569.
23. Vanover, K. E.; Rosenzweig-Lipson, S.; Hawkinson, J. E.; Lan, N. C.; Belluzzi, J. D.; Stein, L.; Barrett, J. E.; Wood, P. L.; Carter, R. B. Characterization of the anxiolytic properties of a novel neuroactive steroid, Co 2-6749 (GMA-839; WAY-141839; 3alpha, 21-dihydroxy- 3beta-trifluoromethyl-19-nor-5beta-pregnan-20-one), a selective modulator of gamma-aminobutyric acid (A)

receptors. *J. Pharmacol. Exp. Ther.* **2000**, *295*, 337−345 and WO1994027608A1.

24. Vanover, K.; Hogenkamp, D.; Lan, N.; Gee, K.; Carter, R. B. Behavioral characterization of Co134444 (3alpha-hydroxy-21-(1′-imidazolyl)-3beta-methoxymethyl-5alpha- pregnan-20-one), a novel sedative-hypnotic neuroactive steroid. *Psychopharmacol.* **2001**, *155*, 285−291.

25. Burton, G.; Veleiro, A. S. Structure-activity relationships of neuroactive steroids acting on the GABAA receptor. *Curr. Med. Chem.* **2009**, *16*, 455–472.

26. Martinez Botella, G.; Salituro, F. G.; Harrison, B. L.; Beresis, R. T.; Bai, Z.; Shen, K.; Belfort, G. M.; Loya, C. M.; Ackley, M. A.; Grossman, S. J.; Hoffmann, E.; Jia, S.; Wang, J.; Doherty, J. J.; Robichaud, A. J.; Neuroactive steroids. 1. positive allosteric modulators of the (γ-aminobutyric acid) a receptor: structure−activity relationships of heterocyclic substitution at C-21. *J. Med. Chem.* **2015**, *58*, 3500−3511.

27. Martinez Botella, G.; Salituro, F. G.; Harrison, B. L.; Beresis, R. T.; Bai, Z.; Blanco, M-J.; Belfort, G. M.; Cai, J.; Loya, C. M.; Ackley, M. A.; Althaus, A. L.; Grossman, S. J.; Hoffmann, E.; Doherty, J. J.; Robichaud, A. J. Neuroactive steroids. 2. 3α-Hydroxy-3β-methyl-21-(4- cyano-1H-pyrazol-1′-yl)-19-nor-5β-pregnan-20-one (SAGE-217): a clinical next generation neuroactive steroid positive allosteric modulator of the (γ-aminobutyric acid) A receptor. *J. Med. Chem.* **2017**, *60*, 7810−7819.

28. Althaus, A. L.; Ackley, M. A.; Belfort, G. M.; Gee, S. M.; Dai, J.; Nguyen, D. P.; Kazdoba, T. M.; Modgil, A.; Davies, P. A.; Moss, S. J.; Salituro, F. G.; Hoffmann, E.; Hammond, R. S.; Robichaud, A. J.; Quirk, M. C.; Doherty, J. J. Preclinical characterization of zuranolone (SAGE-217), a selective neuroactive steroid GABAA receptor positive allosteric modulator. *Neuropharmacol.* **2020**, *181*, 108333.

29. Hoffmann, E.; Nomikos, G. G.; Kaul, I.; Raines, S.; Wald, J.; Bullock, A.; Sankoh, A. J.; Doherty, J.; Kanes, S. J.; Colquhoun, H. SAGE-217, a novel GABA receptor positive allosteric modulator: clinical pharmacology and tolerability in randomized phase I dose-finding studies. *Clin. Pharmacokinet.* **2020**, *59*, 111−120.

30. Sage Therapeutics. Novel Medicines for Life-Altering CNS Disorders, Q3 2015 Financial Results. Posted November 5, **2015**. https://investor.sagerx.com/static-files/0892c0bb-ef58-44e6-a429-ddac4191f25f.

31. Sage Therapeutics. R&D Day 2016. Posted December 13, **2016**. https://investor.sagerx.com/static-files/487e787c-b1a6-433c-a345-9e0d80b5359b.

32. Hammond, R. S.; Belfort, G. M.; Robichaud, A. J.; Doherty, J. J. Efficacy of a second-generation neuroactive steroid, SAGE-217, in a mouse model of chronic medial temporal lobe epilepsy. Poster presented at the 2015 Society for Neuroscience Annual Meeting, Chicago, IL, October **2015**. Program#/Poster#: 497.15/H27.

33. 19-Nor-C3,3-disubstituted C21-N-pyrazolyl Steroids as GABA Receptor Modulators and Their Preparation, WO2014169833.

34. Frieder, A.; Fersh, M.; Hainline, R.; Deligiannidis, K. M. Pharmacotherapy of postpartum depression: current approaches and novel drug development. *CNS Drugs* **2019**, *33*, 265–282.
35. Gunduz-Bruce, H.; Silber, C.; Kaul, I.; Rothschild, A. J.; Riesenberg, R.; Sankoh, A. J.; Li, H.; Lasser, R.; Zorumski, C. F.; Rubinow, D. R.; Paul, S. M.; Jonas, J.; Doherty, J. J.; Kanes, S. J. Trial of SAGE-217 in patients with major depressive disorder. *N. Engl. J. Med.* **2019**, *381*, 903–911.
36. Neuroactive Steroids and Their Methods of Use for Treating CNS Related Disorder WO2019051264. 19-Nor-C3,3-disubstituted C21-N-pyrazolyl Steroid and Methods of Use Thereof for Treating Depression Such as Major Depression WO2021113786. A 19-nor C3,3-disubstituted C21-N-pyrazolyl Steroid and Methods of Use Thereof for Treating Sleep Disorders WO2019055764. Method of Treating CNS Disorders with Neurosteroids and Gabaergic Compounds WO2019075362. Neuroactive Steroids, Compositions, and Uses Thereof for Treating Tremor, Depression and Anxiety Disorder WO2017156103. Bullock, A.; Kaul, I.; Li, S.; Silber, C.; Doherty, J.; Kanes, S. J. Zuranolone as an oral adjunct to treatment of parkinsonian tremor: a phase 2, open-label study. *J. Neurol. Sci.* **2021**, *421*, 117277. Methods for the Treatment of Perimenopause and Menopause Using GABA-A Receptor Positive Allosteric Modulators WO2020180955.
37. ten Doesschate, F.; van Waarde, J. A.; van Wingen, G. A. Non-superiority of Zuranolone (SAGE-217) at the longer-term. *J. Affect. Disord.* **2021**, *291*, 329–330.
38. Arnaud, A.; Suthoff, E.; Stenson, K.; Werneburg, B.; Hodgkins, P.; Bonthapally, V.; Jonas, J.; Meyer, K.; O'Day, K. Number needed to treat and number needed to harm analysis of the Zuranolone phase 2 clinical trial results in major depressive disorder. *J. Affect. Disord.* **2021**, *285*, 112–119.
39. Sage Therapeutics. LANDSCAPE and NEST Clinical Development Programs. Presented at the 34th European College of Neuropsychopharmacology (ECNP) Congress, Lisbon, Portugal, October 2021.

15

Risdiplam (Evrysdi), A Small Molecule, *SMN2*-Directed RNA Splicing Modifier for Treating Spinal Muscular Atrophy

Jie Jack Li

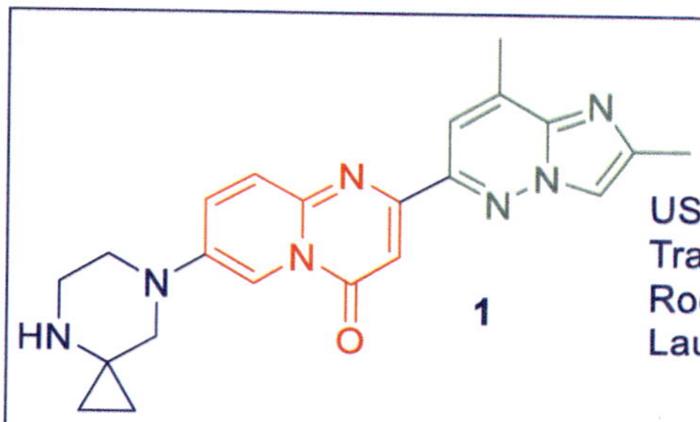

1

USAN: Risdiplam
Trade name: Evrysdi
Roche/PTC/SMA Foundation
Launched: 2020

15.1 Background

Enzymes, receptors, and ion channels are all protein drug targets, and the whole pharmaceutical industry has been founded by protein modulators. Yet we have had little experience in targeting ribonucleic acids (RNAs). Not long ago, the concept of selectively targeting RNAs with small molecules was perceived as a formidable scientific challenge.

RNAs as therapeutic targets are actually very attractive. It has been uncovered that only 1.5% of our genome is ultimately translated into proteins, while 70–90% is transcribed to RNAs! However, using small molecules to target RNA comes with its own perils. For one thing, RNA is inherently dynamic and has limited chemical diversity. For another, promiscuous RNA-binding ligands are often identified during screening campaigns. Different from proteins, which comprise a total of 22 proteinogenic amino acids, RNAs consist of only four primary nucleotide monomeric units, adenine (A), uracil

Current Drug Synthesis, First Edition. Edited by Jie Jack Li.

(U), guanine (G), and cytosine (C), as building blocks. This may be why RNA-targeting small molecule drugs on the market are few and far between.[1]

As it so happens, a disease is a perfect candidate for targeting RNA splicing. That is spinal muscular atrophy (SMA).

SMA is one of the most common inherited neuromuscular disorders, affecting 1 in every ~10,000 live births. Even though it is a rare disease, SMA is devastating, causing selective and progressive loss of motor neurons and progressive muscle wasting. The hallmark of SMA is the progressive degeneration of α-motor neuron in the brain stem and spinal cord that causes muscle atrophy and disease-related complications that can impact survival. Left untreated, SMA may result in infant deaths although the symptoms are milder if the disease onset occurs at later life.

The primary etiology of SMA is low levels of survival motor neuron (SMN) protein as the consequence of homozygous deletion or mutation in the *SMN1* gene on chromosome 5q13. Humans carry two paralogous *SMN1* and *SMN2* genes that are ubiquitously expressed. The *SMN1* full-length mRNA codes SMN protein, an essential protein for normal development and functional homeostasis in all species. The *SMN2* gene is almost identical to the *SMN1* gene with exception of a cytosine (C) to tyrosine (T) substitution that results in the exclusion of exon 7 during splicing of the precursor messenger RNA (pre-mRNA) in 85% of the transcripts. In terms of terminology, *exon* stands for the coding sequence while *intron* is the non-coding stretch of the RNA sequence. Gene splicing is a post-transcriptional modification in which a single gene can code for multiple proteins. It is done, prior to mRNA translation, by differential inclusion or exclusion of pre-mRNA.

Since SMA is characterized by the deficiency of SMN protein, all existing therapies work by boosting SMN protein levels. This can be achieved either by altering the *SMN2* splicing process to increase exon 7 inclusion or by re-expressing *SMN1*. Although lack of *SMN1* gene is the underlying cause of the disease, *SMN2* is the primary disease-modifying gene in SMA.

Biogen's nusinersen (Spinaraza), available since 2016, is an antisense oligonucleotide (ASO). It modulates alternative splicing of the *SMN2* gene, functionally converting it into the *SMN1* gene, thus increasing the level of SMN protein in the central nervous system (CNS). Nevertheless, nusinersen has to be administered intrathecally (by direct injection into the spinal column). This requires anesthesia. Reports have circulated that some parents have elected to pull their children/SMA patients from school for a month beforehand to prevent any illness that might compromise the procedure.[1]

Another treatment for SMA is AveXis/Novartis' adeno-associated virus 9 (AAV9)-based gene therapy, onasemnogene (Zolgensma). Approved in 2019, it is a recombinant adeno-viral vector that encodes the SMN protein. This approach is theoretically not dependent on the *SMN2* gene. Onasemnogene is given as a one-time injection into a vein with at least two months of corticosteroids to protect the liver. Since it costed 2.15 million per treatment in 2019, Zolgensma earned the dubious distinction of being the most expensive medicine in the world![1]

Roche's risdiplam (Evrysdi, **1**) is a small-molecule drug that modulates the splicing of the related *SMN2* gene, which results in the production of functional SMN protein.[2–4]

15.2 Pharmacology

Francis Crick, the co-discoverer of the deoxyribonucleic acid (DNA) double helix structure, advanced the *central dogma* for the flow of genetic information in the late 1950s. Central dogma describes that the flow of genetic information in a two-step process of transcription and translation in a uni-directional vector:

$$\text{DNA} \rightarrow \text{mRNA} \rightarrow \text{Protein}$$

In essence, after DNA is transcribed to mRNA, which is then translated to protein (the SMN protein in this particular case). Recently, mRNA vaccines by Pfizer and Moderna have achieved great success to generate spike proteins of SAR-CoV-2. mRNA-based vaccines saved the day during the COVID-19 pandemic.

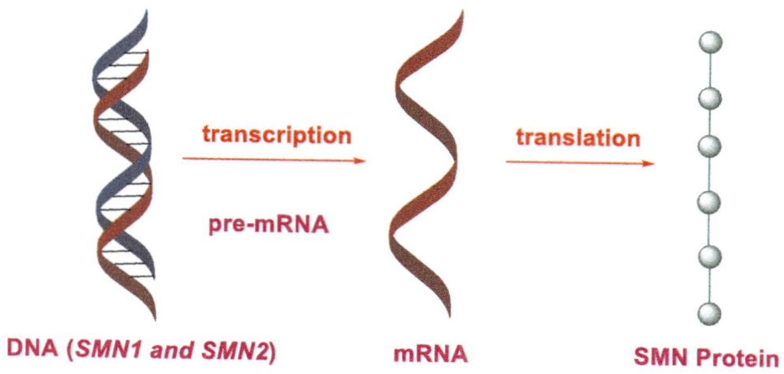

During RNA slicing, RNAs may have multiple splice isoforms. As alluded earlier, loss-of-function of *SMN1* results in loss of muscle function and early death. Humans have a paralog gene *SMN2*, but the resulting transcript from this gene lacks the 7th exon, critical for protein stability, thus does not rescue loss of *SMN1* function. If we could find drugs as *SMN2* splicing modifiers that could increase exon 7 inclusion, those drugs would be promising treatments of SMA.[5–7]

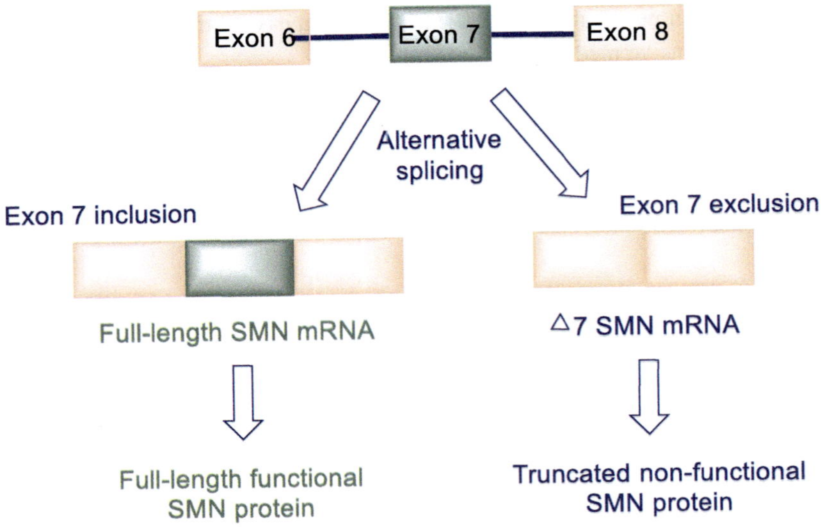

On the molecular level, risdiplam (**1**) modifies splicing by acting as a "molecular glue". It binds the RNA:RNA duplex formed between U1 RNA and *SMN2* pre-mRNA thus stabilizing the interaction between 5′ splice site of exon 7 and the U1 small nuclear ribonucleoprotein (snRNP). The end results are that risdiplam (**1**) interacts with *SMN2* pre-mRNA and enhances exon 7 inclusion to increase the level of functional SMN protein. Ironically, PTC Therapeutics obtained their initial hits including coumarin **3** employing a phenotypic high-throughput screening (HTS) campaign, which eventually led to the discovery of risdiplam (**1**). Its precise mechanism of action was elucidated later.[1]

15.3 Structure–Activity Relationship (SAR)

In 2015, Novartis reported their discovery of branaplam (**2**), an *SMN2* splice modulator that enhanced U1-pre-mRNA association and rescue SMA mice. Its SMN ELISA half-maximum effective concentration ($EC_{50\ PRO}$) was 20 nM. Novartis arrived at branaplam (**2**) using a disubstituted pyridazine hit from their HTS campaign.[8]

Branaplam (**2**)
EC_{50} PRO = 20 nM

PTC Therapeutics carried out a phenotypic HTS campaign to identify small molecule compounds that favor the inclusion of exon 7 through interacting with the RNA components of the splicing machinery during *SMN2* pre-mRNA splicing. Using a mini-gene assay that contained an *SMN2* gene fragment (from 6 to the 5′ region of exon 8), 200,000 compounds were screened. Among several hits, coumarin **3** was identified as a promising hit. Although it had an EC_{50} of 220 nM in a mini-gene luciferase assay, it was inactive in functional assay, inducing no measurable increase in the level of SMN protein in SMA patient-derived fibroblast due to its weak activity.[9]

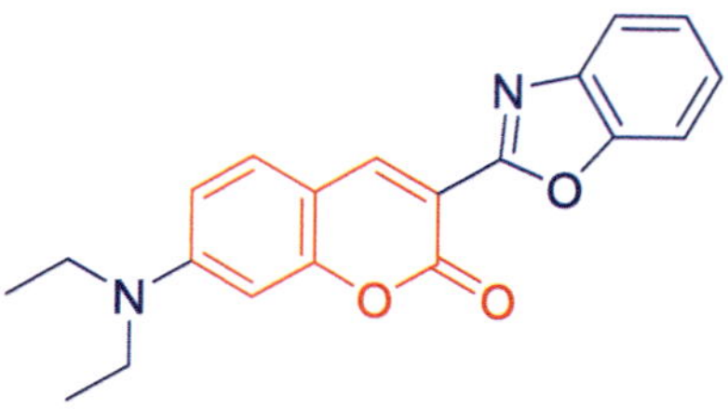

HTS hit, coumarin **3**
SMN2 splicing $EC_{1.5x\ RNA}$ = 0.22 μM
SMN protein $EC_{1.5x\ PRO}$ > 32 μM

The structure–activity relationship (SAR) of the C-3 substituent R apparently depends upon the relative orientation of the Me/N pharmacophore as shown in Table 1. When the relative energy for rotation about coumarin–imidazopyrazine bond such as **4** is calculated, the lowest energy conformation gives a dihedral angle of 180°. In this planar orientation the key nitrogen is located *above* the key methyl substituent. However, when the 5-6 di-heterocyles switched to 6-5 di-heterocyle as pyrazolopyrazine **8**, the 180° dihedral angle is still highly favored but the key nitrogen is located *below* the key methyl when drawn in a similar planar fashion. Empirically, compounds containing heterocycles such as **4**–**7** that favor the "nitrogen above" orientation are highly potent, whereas compounds such as **8** and **9** that highly favor the "nitrogen blow" orientation are nearly inactive.[9]

Table 1. *In vitro* activity of 3-substituted coumarins.

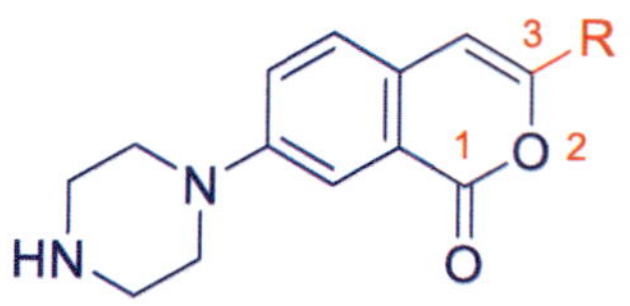

Compound	R	RNA mini-gene $EC_{1.5\times}$ RNA (nM)	SMN protein $EC_{1.5\times\ PRO}$ (nM)

4	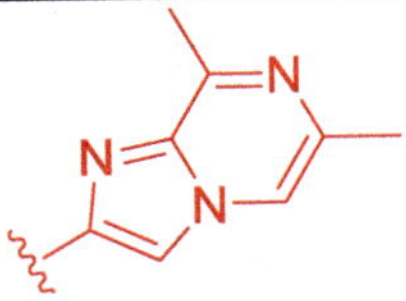	2	15
5	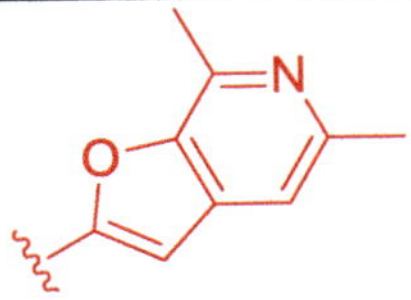	4	16
6	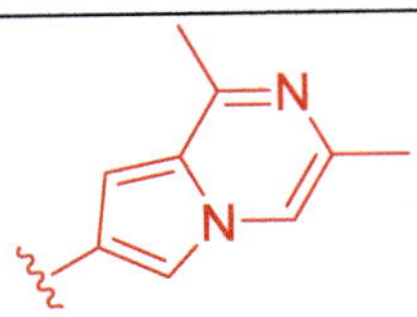	4	29
7	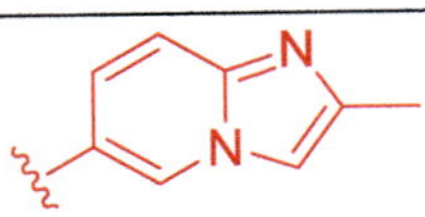	5	66
8	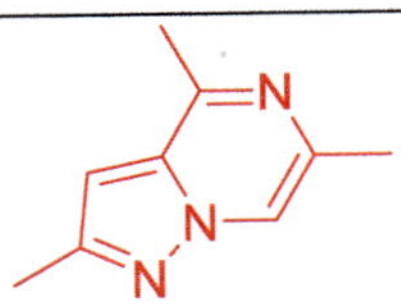	>10,000	>10,000
9	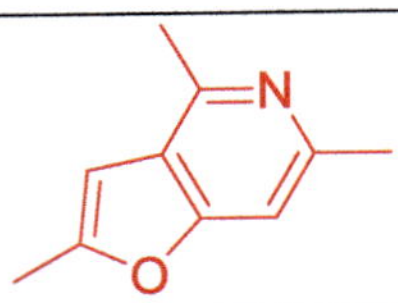	1,500	>10,000

We have learned from experiences that SAR can be really capricious from time to time.

Additional SAR investigations revealed that isocoumarin **10** was potent in functional assay with an EC_{50} of 120 nM. More encouragingly, compound **10** had a good absorption, distribution, metabolism, and excretion (ADME) profile with a brain/plasma (B/P) distribution ratio of 11.[9]

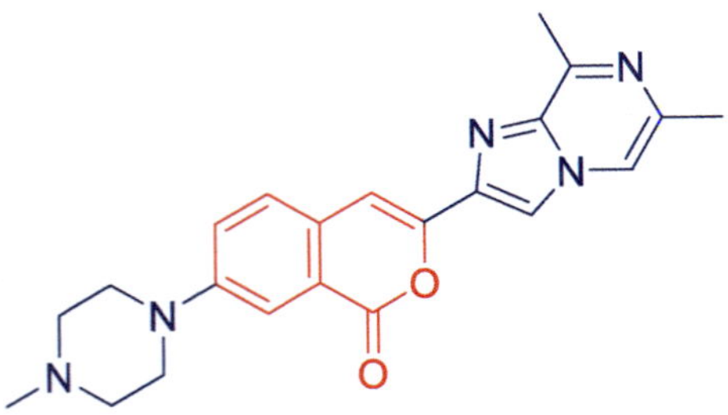

Isocoumarin **10**
SMN $EC_{1.5x\ PRO}$ = 120 nM
Caco-2: P_{app} = 1.2 x 10^{-6} cm/s
B–A/A–B = 1.3
PPB = 93%
Rat PK, $AUC_{0-6\ h}$ = 0.79 mg•h/mL
B/P = 11

But isocoumarin **10** was tested positive for the Ames test, indicating a potential of mutagenicity. It was strongly implied from a collection of data that the right-hand imidazopyrazine might be the culprit *for isocoumarins*. In addition, both coumarins and isocoumarins have been notoriously associated with phototoxicity. To make things worse, isocoumarin **10** exhibited pronounced human plasma instability. In fact, only 55% of the parent molecule **10** remained after a 5 h incubation in human plasma. The nucleophilic ring-opening reaction of the lactone moieties on coumarins and isocoumarins was the likely culprit of their plasma instability.

Just when the project looked hopeless, clever medicinal chemistry came to rescue. Scaffold hopping from coumarin and isocoumarin to many bioisosteres led to many core structures that were not active. However, when pyridopyrimidin-4-one was selected as the core structure, the analogues began to show activities. More importantly, pyridopyrimidinones no longer had the plasma protein instability issue that plagued both coumarins and isocoumarins.

In Table 2, pyridopyrimidinones **11** and **12** were not as potent *in vitro* as their coumarin counterparts **4** and **7**, respectively. Interestingly, while coumarin **8** was completely inactive, its pyridopyrimidinone analog **13** was quite potent. While this phenomenon may be explained by the lowest conformations of **8** and **13**, the more consequential outcome was that pyridopyrimidinones no longer had the liability toward plasma proteins.[9]

Unfortunately, even though pyridopyrimidin-4-ones **14** was tested active in both RNA and protein assays, it was still tested positive for the Ames assay. The structure–toxicology relationship (STR) as shown in Table 3 strongly suggests that the outcome of the Ames assay was dependent on the right-hand-side fragments.[10] Miraculously, pyridopyrimidin-4-one **18** *with merely an additional methyl substitution* was tested negative for the Ames assay while the corresponding analogue **17** was tested positive.[10] Chemistry is magical. It never ceases to amaze!

Table 2. *In vitro* activity of isocoumarins and pyridopyrimidinones

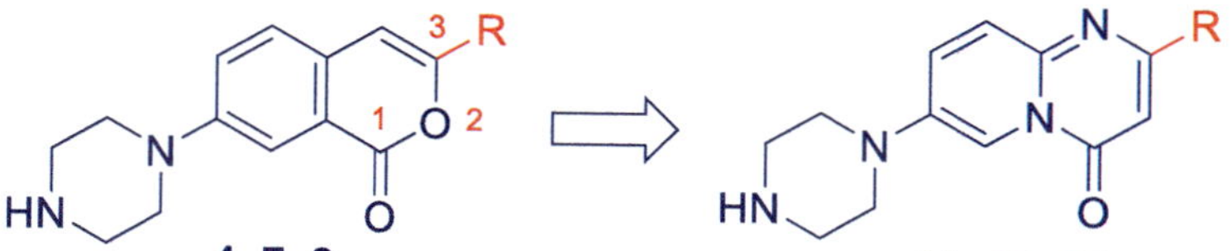

4, 7, 8 **11, 12, 13**

Compound	R	RNA mini-gene $EC_{1.5\times}$ RNA (nM)	SMN Protein $EC_{1.5\times}$ PRO (nM)

4	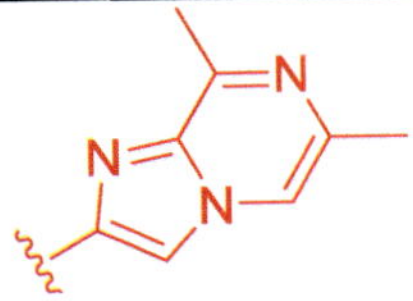	2	15
7	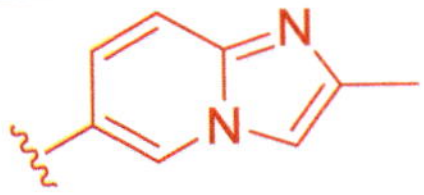	5	66
8	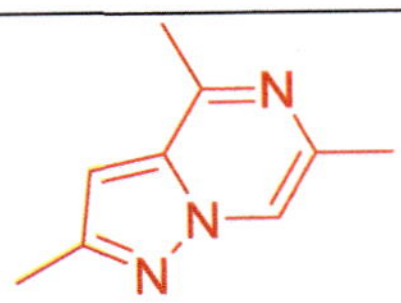	>10,000	>10,000
11	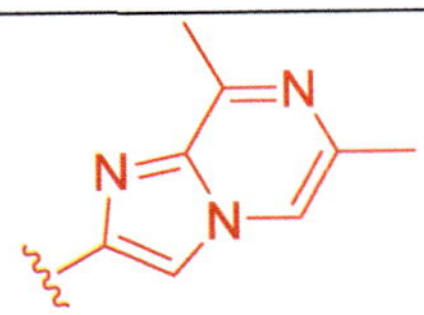	180	430
12	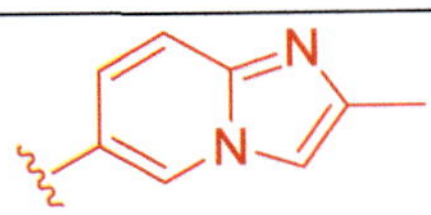	48	31
13	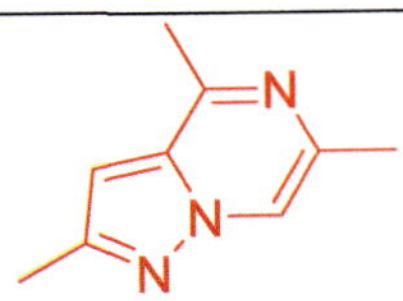	6	31

In collaboration with Roche and the SMA Foundation, PTC continued their lead optimization. In particular, special attention was paid to bioavailability. In order to design efficacious compounds leading to increased SMN protein levels in multiple tissues, the peripheral and CNS *in vivo* are included.

Table 3. The structure–toxicology relationship (STR) of pyridopyrimidinones.

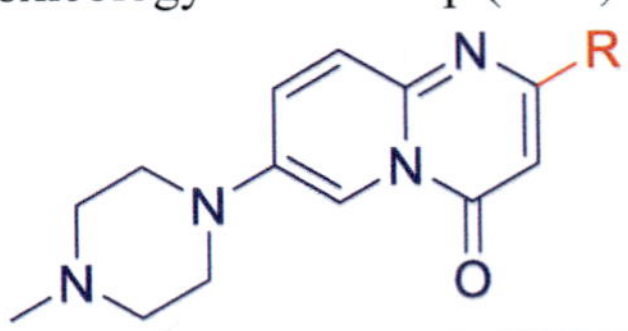

Compound	R	Activity in Ames assay	RNA and SMN $EC_{1.5\times}$ (nM)

14	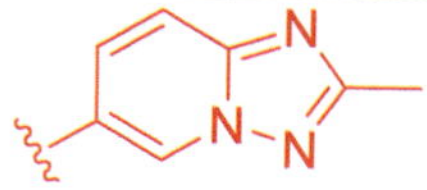	Positive	314/820
15		Positive	3/13
16	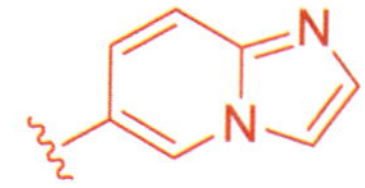	Positive	582/3400
17	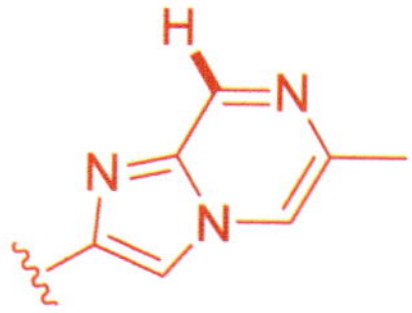	Positive	182/760
18	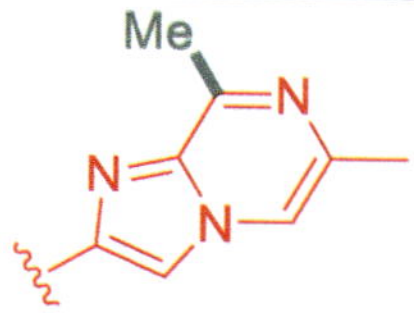	Negative	7/14
19	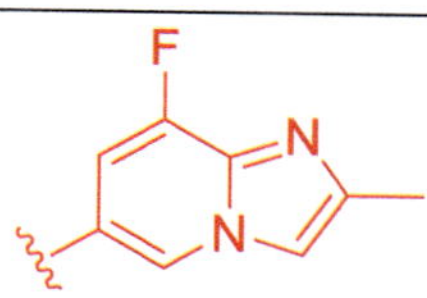	Negative	32/40
20	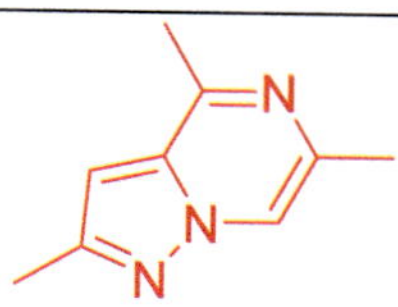	Negative	92/170

Based on reference 10.

To prepare compounds that were not Pgp substrates, they observed that lipophilicity was a good predictor of efflux ration (ER), a measure of Pgp. Compounds were not or were only weak Pgp substrates when the measure of log *D* was greater than 1.6. Pyridopyrimidin-4-one **21** (RG7800) was obtained from the lead optimization with good efficacy, drug metabolism and pharmacokinetics (DMPK) profile, and acceptable safety profile in animal models including cynomolgus monkeys (cyno). RG7800 (**21**) was the first small molecule *SMN2* splicing modifier to enter human clinical trials.[10]

Meanwhile, an effort was taken to replace the pyridopyrimidin-4-one core structure with simpler benzamide. It led to the discovery of a novel class of *SMN2* splicing modifiers. But the series was not further pursued, presumably due to the success of the pyridopyrimidin-4-one derivatives.[11]

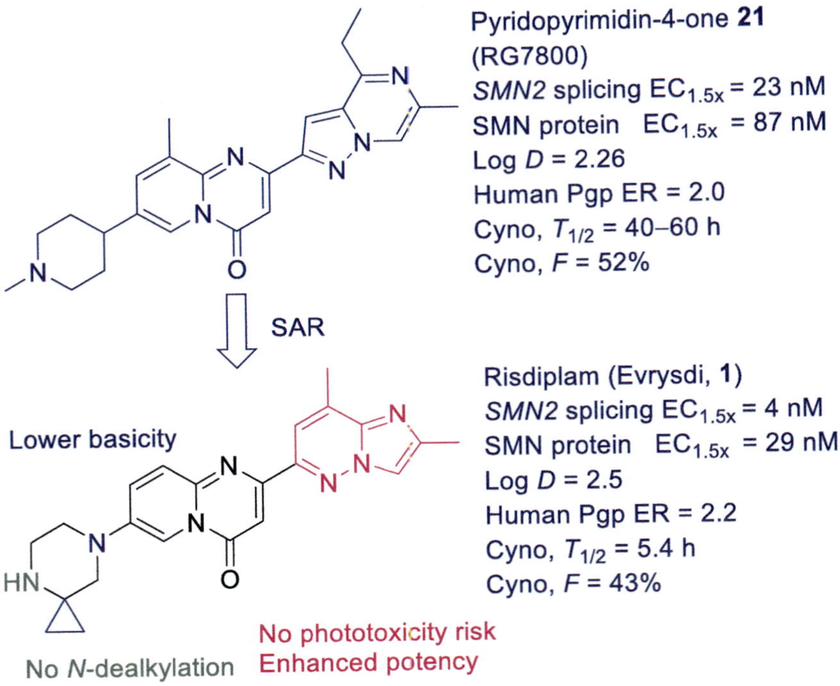

Regrettably, the clinical study of RG7800 (**21**) was put on hold as a precautionary measure due to safety findings in cynomolgus monkeys in the long term (39 weeks) toxicity studies. Non-reversible histological findings in the retina were observed. It was discovered that RG7800 (**21**) also had some selectivity and human ether-à-go-go (hERG, Kv11.1) issues. Its IC_{50} for the hERG potassium ion channel was 1.8 µM, a weak modulator. Moreover, the drug also had potential to cause phototoxicity.[12]

Further optimization of RG7800 (**21**) lowered activities against off-target genes while significantly increased the on-target potency thus reduced the efficacious doses and improved the therapeutic window. Gratifyingly, the selection of a basic amine moiety with the lowest pK_a was possible to maintain its potency and prevented any hERG inhibition or phospholipidosis. The choice of imidazopyrazine as the right-hand substituent removed the phototoxicity potential and boosted the potency, even though isocoumarin–imidazopyrazine **11** was tested positive in the Ames assay. In addition, since *N*-dealkylation to remove the *N*-methyl group readily took place *in vivo*, the *N*-methyl was removed from the drug to relieve metabolism burden of the liver. At the end, the fruit of their labor was risdiplam (**1**) with excellent pharmacokinetic profile (in terms of volume of distribution and half-life) and desired systemic tissue distribution.[13]

15.4 Pharmacokinetics and Drug Metabolism

Being a small molecule fully compliant of Lipinski's rule of five (Ro5), risdiplam (**1**) is "drug-like" with excellent physiochemical properties.

Risdiplam (**1**) has been shown to have good bioavailability into CNS and peripheral tissues when administered orally. In terms of drug distribution of risdiplam (**1**), its total drug levels were similar in the plasma, muscle, and brain in mice, rats, and monkeys. Since it has high passive permeability and is not a human multidrug resistance protein 1 (MDR1) substrate, its cerebrospinal fluid (CSF) levels reflected free drug concentration in plasma in monkeys. The collective *in vitro* and *in vivo* preclinical data strongly suggested that functional protein increase seen in patient's blood following risdiplam (**1**) treatment should reflect similar increase in functional SMN protein in the CNS, muscle, and other peripheral tissues. The prediction was confirmed in clinical trials.

In humans, the apparent clearance of risdiplam (**1**) was 2.1 L/h for a 14.9 kg patient. In healthy adults, the terminal elimination half-life was ~50 h.

Risdiplam (**1**) is largely metabolized by flavin monooxygenase 1 and 3 (FMO1,3), and by CYP1A1, CYP2J2, CYP3A4, and CYP3A7. The parent drug accounted for the majority (83%) of drug-related material in plasma. Metabolism of risdiplam (**1**) has been studied *in vitro* using human liver microsomes (HLMs) and human hepatocytes. The major metabolite was identified as *N*-hydroxylated derivative **22** albeit at a low level of 3.8% and 1.7% in HLMs and human hepatocytes, respectively.[14]

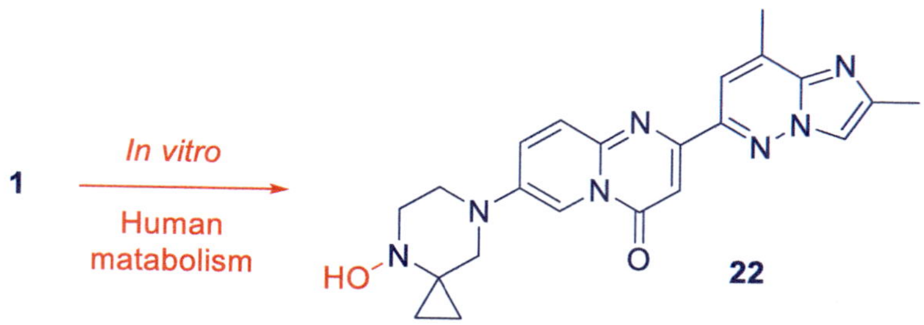

15.5 Efficacy and Safety

Risdiplam (**1**) is a highly potent *SMN2* splicing modifier that increases exon 7 inclusion in *SMN2* mRNA transcript in *in vitro* assays and in transgenic mouse model of SMA. Several *SMN2* splicing modifiers have been shown to improve motor function and longevity in mice with SMA. In clinical trials, risdiplam (**1**) has been shown to increase SMN protein in a dose- and time-consistent manner. After one month of exposure, patients given the maximal dose had a median ratio increase of SMN protein of about 2-fold.

The first clinical drug RG7800 (**21**) had several safety concerns including hERG, phototoxicity, and especially retinal toxicity. Intense scrutiny was focused on risdiplam (**1**)'s retinal toxicity. Gratifyingly, extensive ophthalmologic monitoring conducted in studies in patients with SMA confirmed that risdiplam (**1**) does not induce ophthalmologic toxicity in pediatric or adult patients with SMA at the therapeutic dose. These results suggest that safety ophthalmologic monitoring is not needed in patients receiving risdiplam (**1**), as reflected in the US Prescribing Information for risdiplam (**1**).

At high concentrations, risdiplam (**1**) produced off-target effect on splicing of several genes, including *STRN3*, *FOXM1*, *APLP2*, *MADD*, and *SLC25A17*, which potentially would result in undesired adverse effects. But clinical data indicated that negative impact the low selectivity has on the safety of risdiplam (**1**) is low because no serious adverse event related to the drug was reported, and no ophthalmological findings attributable to risdiplam (**1**) exposure were reported. All in all, risdiplam (**1**) has proven to be a rather safe drug on top of its life-saving efficacy.[15,16]

15.6 Syntheses

PTC/Roche's discovery chemistry route for synthesizing risdiplam (**1**) was published in 2018.[2] An S_NAr reaction between dichloropyridazine **23** and ammonia was not selective, giving rise to a mixture of two regioisomers, which was not separated at this stage. A subsequent N-alkylation of the mixture with 1-bromo-2,2-dimethoxypropane was followed by an intramolecular cyclization to assemble imidazopyridazine **24**. The desired regioisomer was separated and converted to the corresponding boronic ester **26** by the Suzuki coupling with the Miyaura diboron reagent **25**. Meanwhile, a condensation between 5-fluoropyridin-2-amine (**27**) and dimethyl malonate, followed by chlorination with $POCl_3$, afforded chloro-intermediate **28**. A Suzuki coupling between **28** and boronic ester **26** provided adduct **29**. Finally, another S_NAr reaction between fluoride **29** and 4,7-diazaspiro[2,5]octane (**30**) delivered risdiplam (**1**).

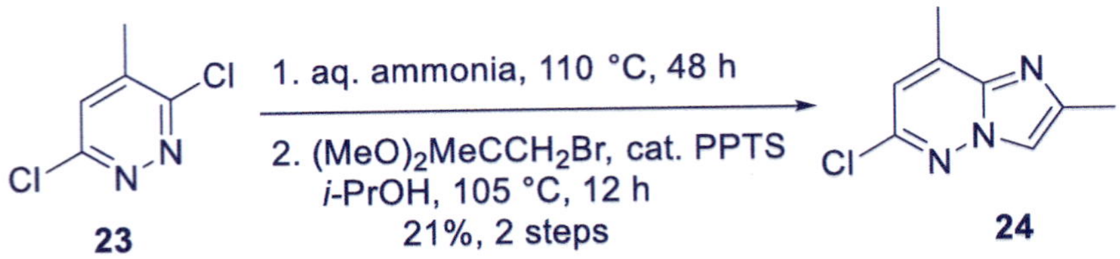

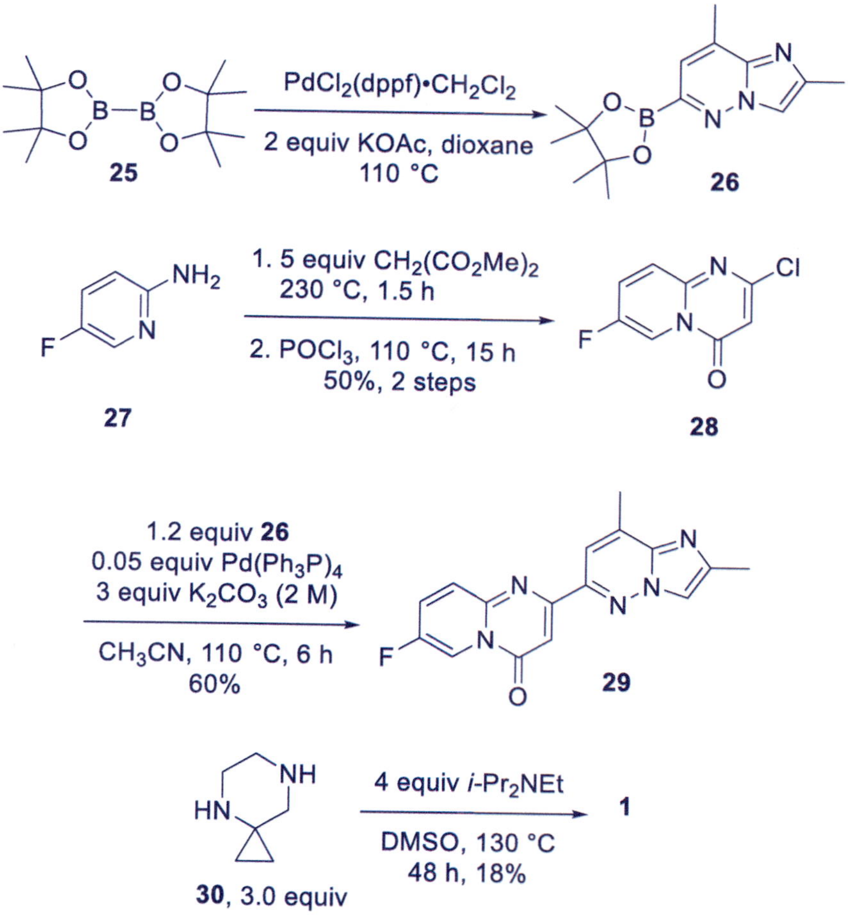

Process chemists at Roche Basel disclosed their large-scale synthesis of risdiplam (**1**) in 2019. Process chemistry has different goals from discovery chemistry. The synthetic routes for discovery chemistry should be modular, allowing quick access of as many and as diverse compounds as possible to probe the SAR. In contrast, the process chemistry routes should shorter, cheaper, easier and more robust for large-scale operations. The Roche process route began with a Buchwald–Hartwig coupling reaction between piperidine **31** and 5-bromo-2-chloropyridine (**32**) to selectively afford the bromo-coupling adduct **33**. Another Buchwald–Hartwig coupling reaction of **33** now took advantage of the α-chlorine atom and furnished the corresponding aniline **34** in the presence of ammonia with the aid of a special ligand *t*-BuBrettPhos. Condensation of aminopyridine **34** with *tert*-butyl malonate installed **35** with the core pyridopyrimidin-4-one structure. Notice that only 1.2 equivalent of the high-boiling *tert*-butyl malonate was needed while five equivalents of low-boiling methyl malonate were required for the discovery route. Intermediate **35** was converted to its corresponding tosylate **36**, which would undergo the Suzuki coupling with organoboronate **26** to afford the adduct. Simple removal of the Boc protection then delivered risdiplam (**1**).[17]

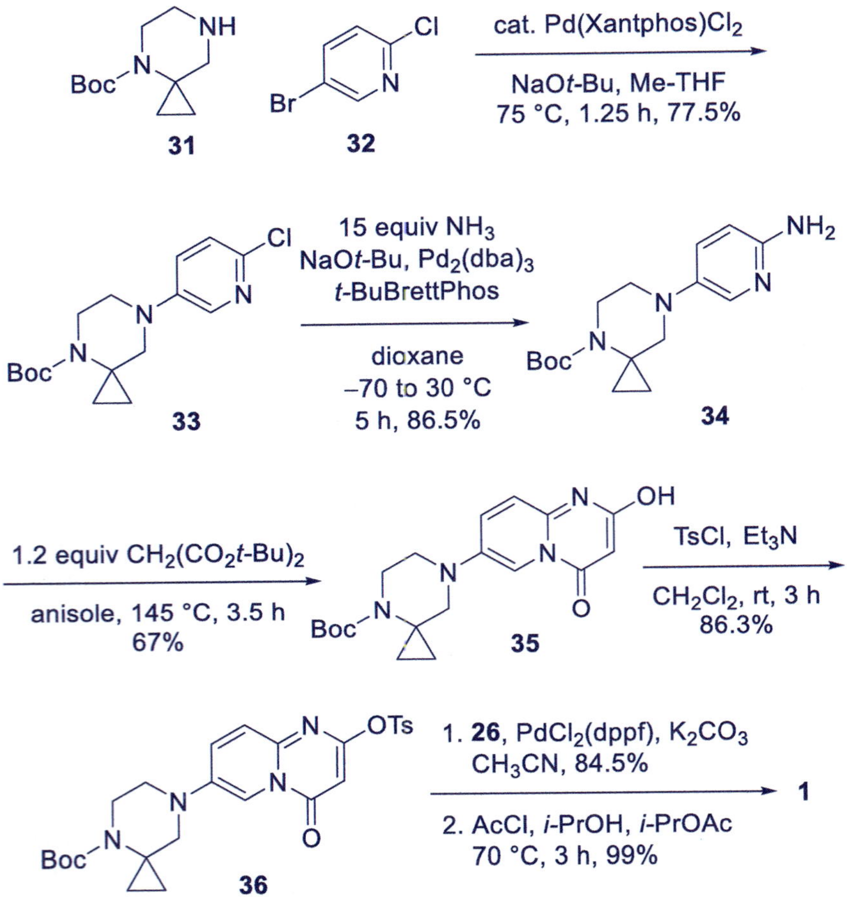

15.7. Summary

Case studies of marketed drugs are interesting because there are always lessons learned. The story of risdiplam (**1**) is especially fascinating. It is the first small-molecule drug on the market to modulate the splicing of the *SMN2* gene, which results in the production of functional SMN protein. The approval of risdiplam (**1**) added a more convenient treatment of SMA considering the previous two treatments were both cumbersome and expensive. From identification of hits from an HTS campaign to extensive lead optimization, the discovery of risdiplam (**1**) encountered many issues involving a typical drug discovery program. They include potency, selectivity, Ames assay/mutagenicity, Pgp substrate, hERG/cardiotoxicity, plasma protein stability, phototoxicity potential, bioavailability, etc. The story of risdiplam (**1**) is a cornucopia of knowledge in drug design and medicinal chemistry.

Risdiplam (Evrysdi, **1**) is important because it is a watershed. With the legitimate proof-of-concept (PoC), more and more small-molecule drugs targeting RNAs are surely to follow, tackling many drug targets that have not been on our radar screen thus far.

It is exciting to gain access to treatment for a disease that has been considered as incurable for more than a century. Nevertheless, despite having three remarkable drugs to treat SMA, clinical outcomes in patients are variable. There is still a long way to go to achieve optimal therapies for the treatment of this devastating genetic disease.

15.8. References

1. Cormac, S. First small-molecule drug targeting RNA gains momentum. *Nat. Biotechnol.* **2021**, *39*, 6–8.
2. Ramdas, S.; Servais, L. New treatments in spinal muscular atrophy: an overview of currently available data. *Exp. Opin. Pharmacother.* **2020**, *21*, 307–315.
3. Dhillon, S. Risdiplam: First Approval *Drugs* **2020**, *80*, 1853–1858.
4. Singh, R. N.; Ottesen. E. W; Singh, N, N. The First Orally Deliverable Small Molecule for the Treatment of Spinal Muscular Atrophy. *Neurosci. Insights* **2020**, *15*, 633105520973985.
5. (a) Yu, A.-M.; Choi, Y. H.; Tu, M.-J. RNA drugs and RNA targets for small molecules: principles, progress, and challenges. *Pharmacol. Rev.* **2020**, *72*, 862–898. (b) Wang, J.; Schultz, P. G.; Johnson, K. A. Mechanistic studies of a small-molecule modulator of SMN2 splicing. *PNAS* **2018**, *115*, E4604–E4612. (c) Campagne, S.; Boigner, S.; Rudisser, S.; Moursy, A.; Gillioz, L.; Knorlein, A.; Hall, J.; Ratni, H.; Clery, A.; Allain, F. H.-T. Structural basis of a small molecule targeting RNA for a specific splicing correction. *Nat. Chem. Biol.* **2019**, *15*, 1191–1198.
6. Ravi, B.; Chan-Cortés, M. H.; Sumner, C. J. Gene-Targeting Therapeutics for Neurological Disease: Lessons Learned from Spinal Muscular Atrophy *Ann. Rev. Med.* **2021**, *72*, 1–14.
7. Falese, J. P.; Donlic, A.; Hargrove, A. E. Targeting RNA with small molecules: from fundamental principles towards the clinic. *Chem. Soc. Rev.* **2021**, *50*, 2224–2243.
8. (a) Palacino, J.; Swalley, S. E.; Song, C.; Cheung, A. K.; Shu, L.; Zhang, X.; Van Hoosear, M.; Shin, Y.; Chin, D. N.; Keller, C. G.; et al. SMN2 splice modulator that enhanced U1-pre-mRNA association and rescue SMA mice. *Nat.*

Chem. Biol. **2015**, *11*, 511–517. (b) Cheung, A. K.; Hurley, B.; Kerrigan, R.; Shu, L.; Chin, D. N.; Shen, Y.; O'Brien, G.; Sung, M. J.; Hou, Y.; Axford, J.; et al. Discovery of Small Molecule Splicing Modulators of Survival Motor Neuron-2 (SMN2) for the Treatment of Spinal Muscular Atrophy (SMA). *J. Med. Chem.* **2018**, *61*, 11021–11036. (c) Axford, J.; Sung, M. J.; Manchester, J.; Chin, D.; Jain, M.; Shin, Y.; Dix, I.; Hamann, L. G.; Cheung, A. K.; Sivasankaran, R.; et al. Use of Intramolecular 1,5-Sulfur-Oxygen and 1,5-Sulfur-Halogen Interactions in the Design of N-Methyl-5-aryl-N-(2,2,6,6-tetramethylpiperidin-4-yl)-1,3,4-thiadiazol-2-amine SMN2 Splicing Modulators. *J. Med. Chem.* **2021**, *66*, 4744–4761.

9. Woll, M. G.; Qi, H.; Turpoff, A.; Zhang, N.; Zhang, X.; Chen, G.; Li, C.; Huang, S.; Yang, T.; Moon, Y.-C.; et al. Discovery and Optimization of Small Molecule Splicing Modifiers of Survival Motor Neuron 2 as a Treatment for Spinal Muscular Atrophy. *J. Med. Chem.* **2016**, *59*, 6070–6085.

10. Ratni, H.; Karp, G. M.; Weetall, M.; Naryshkin, N. A.; Paushkin, S. V.; Chen, K. S.; McCarthy, K. D.; Qi, H.; Turpoff, A.; Woll, M. G.; et al. Specific Correction of Alternative Survival Motor Neuron 2 Splicing by Small Molecules: Discovery of a Potential Novel Medicine to Treat Spinal Muscular Atrophy. *J. Med. Chem.* **2016**, *59*, 6086–6100.

11. Pinard, E.; Green, L.; Reutlinger, M.; Weetall, M.; Naryshkin, N. A.; Baird, J.; Chen, K. S.; Paushkin, S. V.; Metzger, F.; Ratni, H. Discovery of a Novel Class of Survival Motor Neuron 2 Splicing Modifiers for the Treatment of Spinal Muscular Atrophy. *J. Med. Chem.* **2017**, *60*, 4444–4457.

12. (a) Ratni, H.; Ebeling, M.; Baird, J.; Bendels, S.; Bylund, J.; Chen, K. S.; Denk, N.; Feng, Z.; Green, L.; Guerard, M.; et al. Discovery of Risdiplam, a Selective Survival of Motor Neuron-2 (SMN2) Gene Splicing Modifier for the Treatment of Spinal Muscular Atrophy (SMA). *J. Med. Chem.* **2018**, *61*, 6501–6517. (b) Ratni, H.; Scalco, R. S.; Stephan, A. H. Risdiplam, the First Approved Small Molecule Splicing Modifier Drug as a Blueprint for Future Transformative Medicines. *ACS Med. Chem. Lett.* **2021**, *12*, 874–877.

13. Ratni, H.; Mueller, L.; Ebeling, M. Rewriting the (tran)script: Application to spinal muscular atrophy. *Prog. Med. Chem.* **2019**, *58*, 119–156.

14. Naryshkin, N. A.; Weetall, M.; Dakka, A.; Narasimhan, J.; Zhao, X.; Feng, Z.; Ling, K. K. Y.; Karp, G. M.; Qi, H.; Woll, M. G.; et al. SMN2 splicing modifiers improve motor function and longevity in mice with spinal muscular atrophy. *Science* **2014**, *345*, 688–693.

15. (a) Poirier, A.; Weetall, M.; Heinig, K.; Bucheli, F.; Schoenlein, K.; Alsenz, J.; Bassett, S.; Ullah, M.; Senn, C.; Ratni, H.; et al. Risdiplam distributes and increases SMN protein in both the central nervous system and peripheral organs. *Pharmacol. Res. Perspect.* **2018**, *6*, e00447/1–e00447/12. (b) Ando, S.; Suzuki,

S.; Okubo, S.; Ohuchi, K.; Takahashi, K.; Nakamura, S.; Shimazawa, M.; Fuji, K.; Hara, H. Discovery of a CNS penetrant small molecule SMN2 splicing modulator with improved tolerability for spinal muscular atrophy. *Sci. Rep.* **2020**, *10*, 17472.

16. Sergott, R. C.; Amorelli, G. M.; Baranello, G.; Barreau, E.; Beres, S.; Kane, S.; Mercuri, E.; Orazi, L.; SantaMaria, M.; Tremolada, G.; et al. Risdiplam treatment has not led to retinal toxicity in patients with spinal muscular atrophy. *Ann. Clin. Transl. Neurol.* **2021**, *8*, 54–65.

17. Adam, J.-M.; Fantasia, S. M.; Fishlock, D. V.; Hoffmann-Emery, F.; Moine, G.; Pfleger, C.; Moessner, C. Process for the preparation of 7-(4,7-diazaspiro-[2.5]octan-7-yl)-2-(2,8-dimethylimidazo[1,2-*b*]pyrida-zin-6-yl)pyrido[1,2-*a*]-pyrimidin-4-one derivatives. US Pat. 2020-0216472 (2019).

IV

MISCELLANEOUS DRUGS

16

Esaxerenone (Minnebro), An Oral, Non-steroidal, Selective Mineralocorticoid Receptor Blocker for the Treatment of Essential Hypertension

Narendra B. Ambhaikar

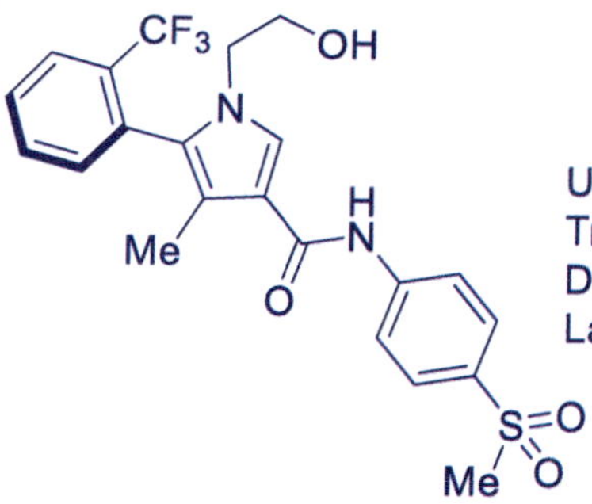

USAN: Esaxerenone
Trade name: Minnebro (**1**)
Daichi Sankyo/Exelixis
Launched: 2020

16.1 Background

Hypertension is a widely prevalent human condition when the systolic blood pressure is > 140 mm Hg and/or diastolic blood pressure is > 90 mm Hg. At present, a significant number of patients suffering from hypertension, about a third, remains undiagnosed. Out of the ones that are diagnosed, half of these patients do not consume an anti-hypertensive medicine. According to the World Health Organization (WHO), high blood pressure has a direct or an indirect impact on at least nine million people worldwide.[1a] Not surprisingly, hypertension is seen as one of the most significant risk factors for morbidity and mortality, with many people receiving sub-optimally treatment. "Essential hypertension" is a common disorder that does not appear to have a specific identifiable cause but impacts about 85% of hypertension patients globally. Specific causes of

Current Drug Synthesis, First Edition. Edited by Jie Jack Li.

hypertension can be determined in only a fraction of cases, while the etiology in a vast majority (~ 90%) of the cases cannot be determined. It is the single most prevalent contributing factor to global burden of diseases and mortality.[1b] Typical cardiovascular diseases and renal disorders are mostly due to hypertension caused by lifestyle, obesity, and genetic factors. These aspects ultimately lead to myocardial infarction, stroke, and heart failure. The human body is conditioned to ingest and excrete < 1 g of sodium (corresponding to 2.5 g salt) per day. When the kidneys are unable to eliminate this ingested sodium volume, essential hypertension develops unless the blood pressure rises. Thus, a well-established influential link exists between a chronically high sodium intake and essential hypertension, particularly when the balance between factors that tend to increase blood pressure and those that try to normalize it gets sufficiently disturbed. Additionally, other factors like obesity, diabetes, ageing, emotional stress, a sedentary life, and low potassium intake aggravate hypertension.[2] Very importantly, hypertension is not only a primary cause of cardiovascular and renal diseases, but also a risk factor that ultimately results in their progression when initiated by other mechanisms like atherosclerosis, ischemic heart disease, diabetes, polycystic kidney diseases, and chronic glomerulon-ephritis to name a few.[3]

The mineralocorticoid receptor (MR), also called aldosterone receptor, belongs to the nuclear receptor family that has equal affinity for mineralocorticoids and glucocorticoids. The last 60 years of mineralocorticoid receptor antagonist (MRA) research have witnessed three key phases: 1) the discovery of aldosterone and its role in promoting sodium conservation in the kidney, salivary glands, sweat glands, and colon;[4] 2) the identification of more specific steroidal anti-mineralocorticoids than spironolactone by companies like Searle, Ciba–Geigy, Roussel Uclaf, and Schering AG; and 3) after the cloning of cDNA of MR a decade later in 1987, companies initiating discovery programs to ultimately seek novel *non-steroidal MRAs* with defined pharmacokinetic and pharmacodynamic properties for use as safe and efficacious drugs for a broad spectrum of diseases.

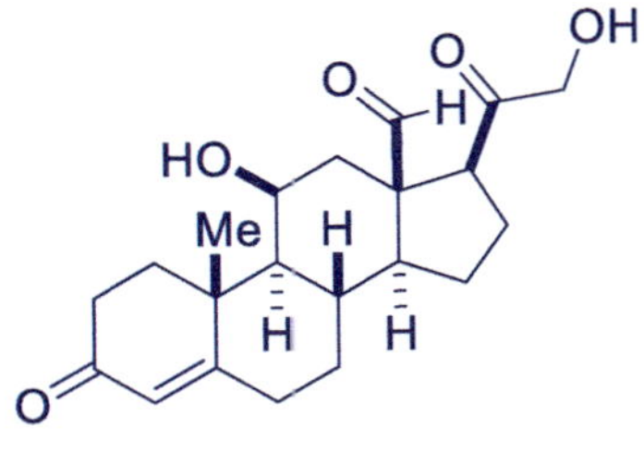

2: Aldosterone

Aldosterone was identified to be an agent controlling renal sodium and potassium excretion in the 1950s. When excessively activated, it plays an important role in development of hypertension, progression of nephropathy, and cardiovascular diseases. With the cloning of cDNA of MR back in 1987, the field has undergone significant advancements. MR is now well understood to be activated by endogenous ligands like aldosterone (compound **2**).[5a] Following this discovery, efforts focused on the discovery of aldosterone antagonists or AAs, that later became known as MRAs. In the search for inhibitors of aldosterone activity in animals and humans in the 1950s, spironolactone (compound **3**) was found to protect rats from aldosterone-induced cardiac necrosis and launched in 1960 as a potassium-sparing diuretic, thanks to the pioneering work of Hans Selye that unfortunately had not been considered clinically significant for decades before. Spironolactone was a first generation MRA to treat essential hypertension.

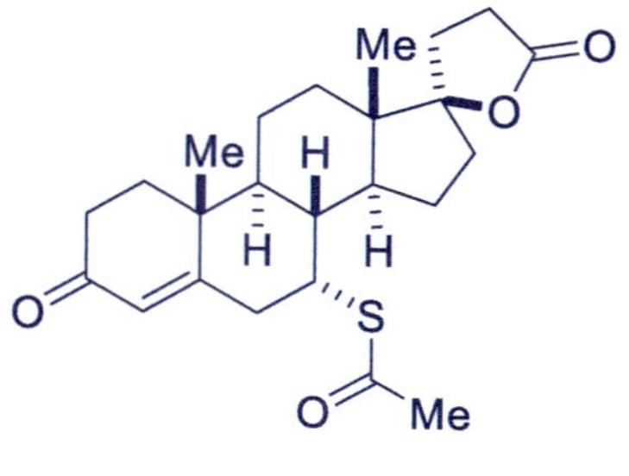

3: Spironolactone
SC-9420

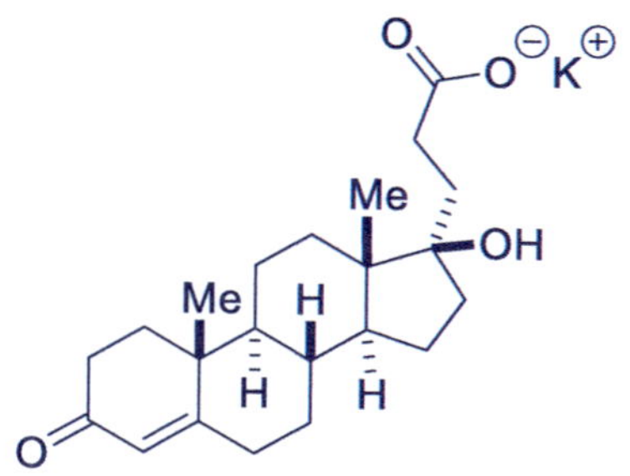

4: Potassium canrenoate
SC-14226

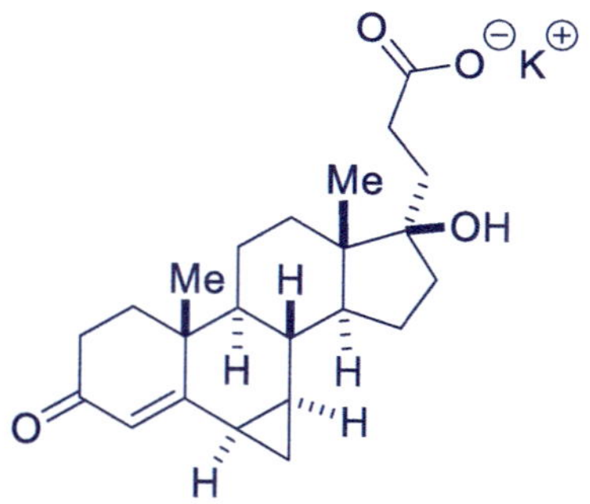

5: Potassium prorenoate
SC-23992

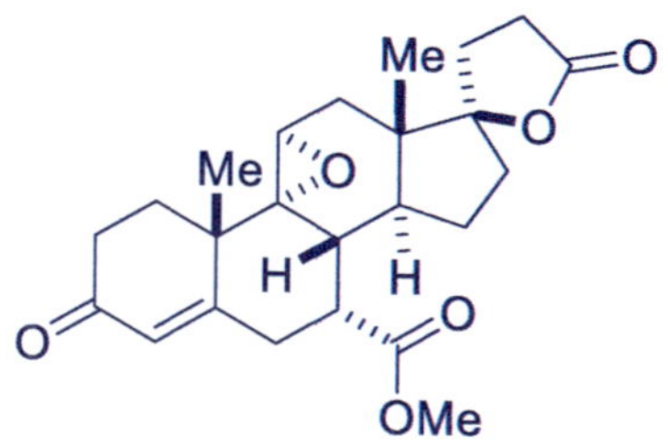

6: Eplerenone
CGP-30083

Over time, the field of genomics underwent a revolutionary change that enhanced understanding of MR and led to subsequent discoveries related to MRAs. Decades of research in the field led to many drugs being introduced in the market, such as potassium canrenoate (SC-14226, **4**), potassium prorenoate (SC-23992, **5**), and notably eplerenone (**6**) that was launched in the United States in 2002.[5b]

These second-generation MRAs were reported to much more specific than spironolactone (**3**) and still based on the initial steroidal 17-spirolactone construct. However, a key risk associated with them is hyperkalemia, a typical side-effect of MR blockade by steroidal MRAs, particularly in combination with other renin-angiotensin-system (RAS) blockers. Therefore, a need for non-steroidal drugs that would avoid undesired consequences of well-established steroidal therapies was rather understandable. Drug discovery programs over the past two decades have increasingly focused on finding novel non-steroidal MRAs, the goal being to mitigate the undesired *side-effects yet retaining the activity of spironolactone.*[5c] Recent promising novel candidates in late phase clinical trials have been esaxerenone (**1**), apararenone, and finerenone. Esaxerenone (**1**) is a groundbreaking therapeutic small molecule that truly represents this effort demonstrating the value of an organized drug discovery program at Exelixis in the United States. Launched in 2020 in Japan by the Daiichi Sankyo Company and prescribed to patients with essential hypertension, this compound is also undergoing clinical trials for other therapies including hypertension with moderate renal impairment, diabetic nephropathies, and hypertension with type 2 diabetes and albuminuria. In the context of the aforementioned background and history of MRAs, the introduction of esaxerenone (Minnebro, **1**) as the first non-steroidal MRA to treat essential hypertension is a defining moment.

16.2 Pharmacology

There are many ways to reduce blood pressure in the human body. In order to understand them, it is important to look at factors that elevate blood pressure with the help of a schematic picture below in Figure 1.[6]

Elevated blood pressure can be treated in many ways through various inhibitors illustrated in the figure below, for which many classical targets that have been traditionally considered. The renin-angiotensin-aldosterone system (RAAS) in the human body needs to be kept under control. Any imbalance in the RAAS may result in occurrence of hypertension directly. Angiotensinogenase, also called renin, is an aspartic protease protein and receptor secreted by the kidneys that participates in RAAS. It mediates extracellular fluid and arterial vasoconstriction; and thereby regulates mean arterial blood pressure.[6] Renin Inhibitors, Angiotensin Converting Enzyme Inhibitors (ACEI), and Aldosterone Synthase Inhibitors are all classes of drugs that can be administered to control increased blood pressure. Aldosterone, a mineralocorticoid secreted by the globular zone of the adrenal cortex, promotes reabsorption of sodium and chloride ions; and increases excretion of potassium and hydrogen ions in renal tubules. It exerts this effect by binding with MR in the kidneys causing development of hypertension. Additionally, MR activation in the extrarenal tissues like heart and blood vessels increases NADPH oxidase levels and reactive oxygen species, promoting development of hypertension and cardiovascular diseases. MR antagonists (MRAs) antagonize the action of aldosterone on MR. It is worth reiterating that elevated aldosterone plasma levels in humans are generally related to hypertension, facilitated through its effect on sodium retention and blood volume.[7] Activation of RAAS triggers elevated levels of renin, angiotensin II, and aldosterone. Angiotensin II, as a vasoconstrictor promotes aldosterone production and stimulates norepinephrine release from sympathetic nerve terminals to elevate heart rate. Aldosterone works to increase blood volume and consequently blood pressure through its action in the kidneys thus retaining sodium. Esaxerenone (**1**) has been well-demonstrated to have potent and long-lasting antihypertensive and cardio-renal protective effects, using rat models in pharmacology studies and in humans in several clinical trials.

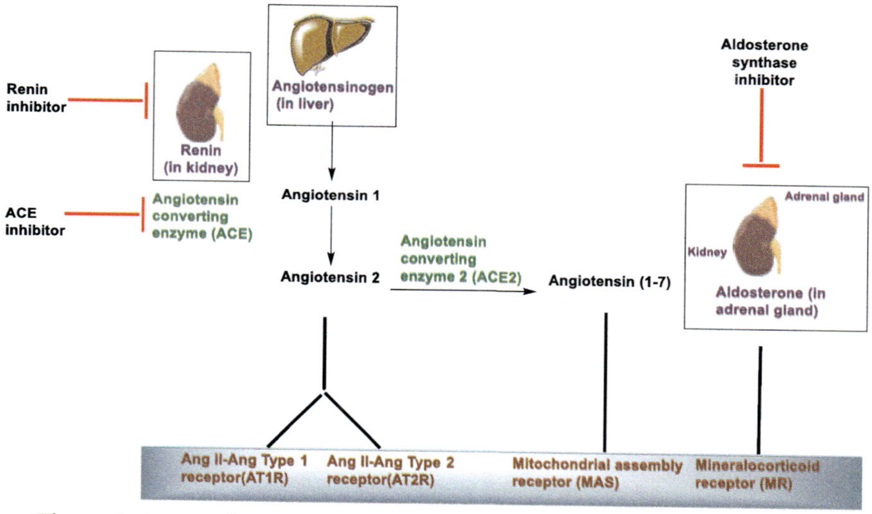

Figure 1. Approaches to reduce blood pressure (modified from Reference 6)

16.3 Structure–Activity Relationship (SAR)

Encoded by the gene NR3C2, MR is a steroid nuclear hormone that belongs to the Greater Nuclear Hormone Receptor (NHR) family, a class that is significantly challenging to the medicinal chemists. Finding high selectivity, maintaining a balance of agonists/antagonist activity involves intensive screening.[8a,b] In the search for novel non-steroidal MRAs, ligands that need to be discovered can possibly induce different MR conformational changes, leading to specific coregulator interactions of value. Both aldosterone and glucocorticoids can activate MR in a cell type-specific fashion. However, aldosterone works to increase blood volume and hence blood pressure, through its action in kidney to retain sodium. An important consideration for the design of new non-steroidal compounds should include a thorough evaluation of the expected physical chemical property space. It should comprise not only the Nuclear Hormone Target Class, but also lessons learnt from the marketed steroidal MRAs and all chemotypes from the patent literature. Several pharmaceutical companies including Pfizer, Bayer, Takeda, and other have entered this space to find non-steroidal MRAs in the past two decades.[9a]

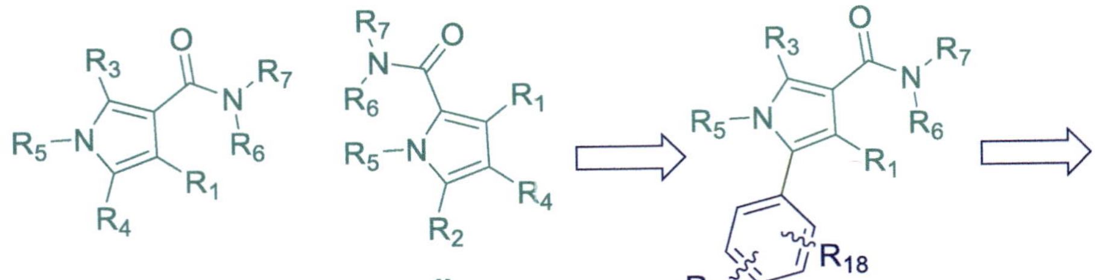

I II

I, II: Substituted pyrroles screened for MR antagonistic activity

III: Scaffold selected for studying activity with wide range of substitutents

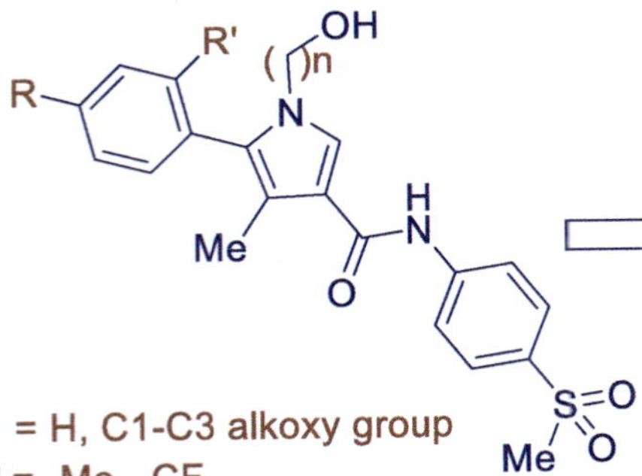

R = H, C1-C3 alkoxy group
R' = -Me, $-CF_3$
n = 1, 2, or 3

7: Atropisomeric compounds selected as MR antagonists

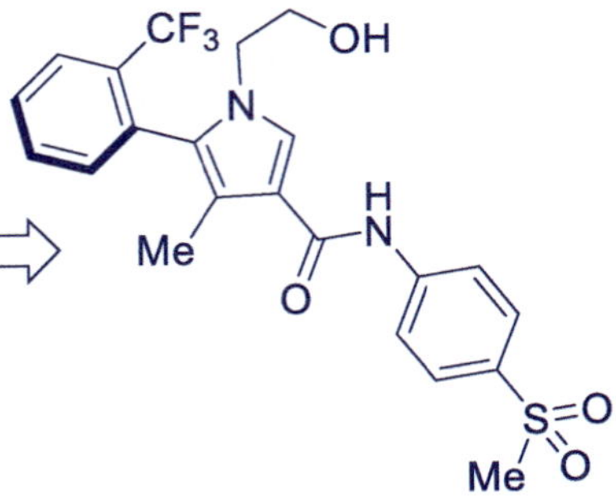

1: Lead candidate XL550 (esaxerenone)

Analysis by Sutherland and Vieth has suggested that marketed NHR ligands have a mean molecular weight of 382 and a mean clogP of 4.1. In 2004, the Exelixis medicinal chemistry team disclosed in their patent on "Pyrrole Derivatives as Pharmaceutical Agents," a series of compounds, compositions, and methods to modulate receptor activities, in a nuclear receptor family that comprises > 150 proteins. Classical receptors include aldosterone receptor (MR), estrogen receptor (ER) (α and β), androgen receptor (AR), progesterone receptor (PR), and the glucocortoid receptor (GR). Exelixis revealed heterocyclic compounds, in which a good majority were phenylpyrrole amides.

In vitro MR activity was reported for a variety of representative compounds. Average IC_{50} values of antagonist activity in a regulatory protein based GAL4-MR assay were as follows: <0.5 μM, 0.5-1 μM and 1-5 μM. Representative molecules, namely, **I** and **II,** are shown in the following text, where substituents R_1, R_2, R_3, R_4, R_5, R_6, and R_7 cover a very wide range of groups. (*Note:* All substituents are not described here owing to the vast body of work and listing all of them would be outside the scope of this chapter.) Exelixis further narrowed down to phenylpyrrole amides as represented in **III**. They found three atropisomeric compounds that they selected for further investigation. Potent MR antagonist property of IC_{50} 2.4 nM, functional assay for the active isomer was reported. A highly selective and potent MR antagonist was thus identified as XL550 for the treatment of hypertension, congestive heart failure, and end organ protection.[9b,c] Medicinal chemistry work described in this patent literature was remarkably extensive. It resulted in a lead molecule XL550 that was later out-licensed to Daiichi-Sankyo, moved into clinical development and ultimately made it to an approved drug in Japan for the treatment of essential hypertension.

16.4 Pharmacokinetics and Drug Metabolism

Esaxerenone (**1**) is a small molecule that complies well with Lipinski's Rule of 5 and is "drug-like." Pharmacokinetics of esaxerenone (**1**) are largely unaffected by food, with no difference seen in C_{max}. In clinical studies conducted by the Daiichi Sankyo Company, it has been observed to "dose-dependently" reduce blood pressure in patients with hypertension, accompanied by its renal protective effects.[10] The elimination half-life at the terminal phase ($t_{1/2}$) of the drug in plasma is longer than that of spironolactone (**3**) and eplerenone (**6**). The superior half-life of esaxerenone (**1**) contributes to its enduring pharmacological effects. Pharmacokinetics reveal no differences observed in the maximum concentration C_{max}, in the body, and area under the curve (AUC) following a single oral dose of 5 mg esaxerenone (**1**) in healthy adult males. Additionally, the time to maximum plasma concentration is about 3 hours, while the elimination half-life is 20 hours. It is important to note that moderate renal impairment does not appear to affect pharmacokinetics of esaxerenone (**1**). Likewise, mild to moderate hepatic impairment also has no impact either. At the same time, the high absorption ratio and various metabolic pathways involved appear to indicate limited drug–drug interaction potential of esaxerenone (**1**) with enzyme inhibitors.

Esaxerenone (**1**) is mainly cleared by metabolism, due to low excretion ratio of unchanged form in urine and feces. Metabolism profiling indicates several oxidized forms, *O*-glucuronide and glucuronides of amide-bond hydrolysate. It is understood to occur in the body by CYP3A4, CYP3A5, and multiple uridine 5'-diphospho-glucuronosyltransferase (UDP) - glucuronosyl-transferase isoforms. Metabolic pathways include oxidation, glucuronidation, and hydrolysis. About 30% of the total body clearance results from oxidation. Unchanged fecal esaxerenone (**1**) is about 18%. The total excretion rate of the excreted drug in the feces and urine was found to be 92.5% with 54.0% and 38.5%, respectively.[1C,11] Proposed major metabolic pathways of esaxerenone (**1**) in humans have been shown below with the help of structures of metabolites M1, M2, M3, M4, M5, and M11 validated against authentic standards. While esaxerenone (1) is recommended to be administered once daily, the dosage can be 1.25, 2.5, or 5.0 mg. Usual dose recommended in adults is 2.5 mg, through oral administration.[12]

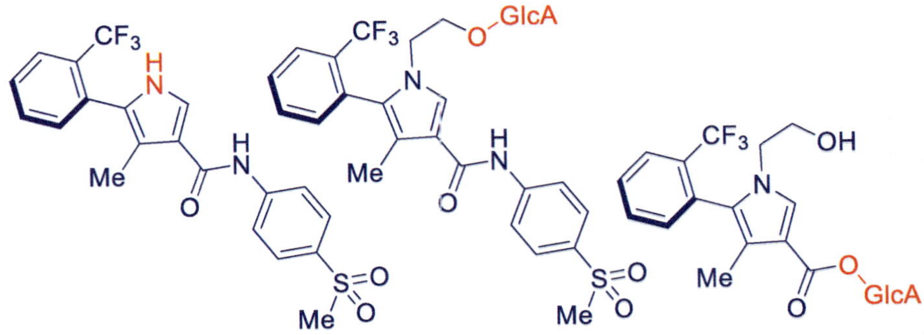

M1: Deshydroxyethyl form by CYP3A4/5

M4: From multiple UGT isoforms; GlcA: glucoronic acid

M11: Acyl glucuronide of amide bond hydrolysis

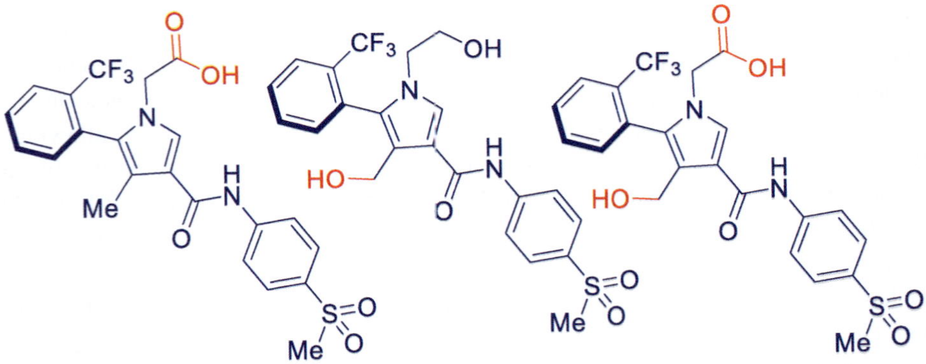

M2: Carboxylic acid form by CYP3A4/5

M3: Oxidized form by CYP3A4/5

M5: Hydroxymethyl carboxylic acid form

16.5 Efficacy and Safety

Esaxerenone (**1**) exhibits a reasonably good dose-dependent antihypertensive effect compared with placebo, as demonstrated in clinical studies. Efficacy is indicated by substantial reductions in the sitting blood pressure values of patients that were given dosages of 2.5 and 5.0 mg/day. A blood pressure decrease was evident in all doses (1.25, 2.5 and 5.0 mg/day). Extrapolation studies further indicate that the antihypertensive effect of 2.5–5.0 mg/day doses of esaxerenone (**1**) is similar to that produced by existing steroidal therapy with 50–100 mg/day of the sMR antagonist eplerenone.[13] Importantly, in Japanese patients with essential hypertension, declines in blood pressure resulting from the treatment of esaxerenone (**1**) 2.5 and 5.0 mg/day for 12 weeks were comparable with 50 mg/day treatment with eplerenone. Additionally, esaxerenone (**1**) 2.5 mg/day dose is *non-inferior* to eplerenone 50 mg/day: A 5.0 mg/day dose of the non-steroidal MRA is superior to the 2.5 mg/day dose. The observation comes from a published report on a phase II, randomized, double-blind, placebo-controlled, open label comparator study in a large group of Japanese patients with essential hypertension.

Significant dose response reductions in systolic and diastolic blood pressure were observed compared with placebo, according to clinical studies conducted by Daichii–Sankyo Multiple Japanese phase III trials. A stable antihypertensive effect with esaxerenone (**1**) across different patient populations was evident, including those with severe hypertension, as well as hypertensive patients with moderate renal impairment or type 2 diabetes mellitus and albuminuria. Similar beneficial effect is also seen with longer-term esaxerenone (**1**) therapy, either alone or in combination with an Angiotensin-II Receptor Blocker (ARB) or an ACEI for a finite period in essential hypertension patients.[14]

Esaxerenone (**1**) demonstrates a *good safety profile* with no major contrary effects overall. A phase I study that was conducted among healthy male Japanese subjects, established the tolerability of esaxerenone (**1**) upon single and multiple dose escalations.[15] When they were given either single doses (5–200 mg), multiple doses (10–100 mg over 10 days) or placebo under fasting conditions, exposures were observed to be generally dose proportional. There were no major critical safety concerns observed in this study. Notably, esaxerenone (**1**) demonstrates a safety profile with no adverse effects related to sex hormones. It stems from the fact that earlier *in vitro* studies of esaxerenone (**1**) as a non-steroidal MR antagonist have proven its selective inhibition of aldosterone to MR, with no agonistic or antagonistic effects on glucocorticoid, androgen or progesterone receptors even at high concentrations.[16] In larger phase II studies, no marked differences in incidence of adverse effects were detected between patients taking placebo and esaxerenone (**1**, 1.25, 2.5, or 5.0 mg/day). Adverse effects that are typically mild to moderate in severity are known to be: nasopharyngitis, upper respiratory tract inflammation, pharyngitis, headaches, elevation blood uric acid, triglycerides, blood creatinine phosphokinase and K^+ levels, reduced $eGFR_{creat}$ (estimated glomerular filtration rate creatinine), and back pain as well as musculoskeletal stiffness. In a phase III trial involving a pooled analysis of Japanese clinical trials, 13% patients given esaxerenone (**1**) had adverse events mainly associated with high serum potassium, blood uric acid, and hyperuricaemia. Despite these adverse effects, none was found to be clinically significant. Orally administered doses of 2.5 mg and 5 mg once daily esaxerenone (**1**) are generally well tolerated for 12 weeks.

16.6 Syntheses

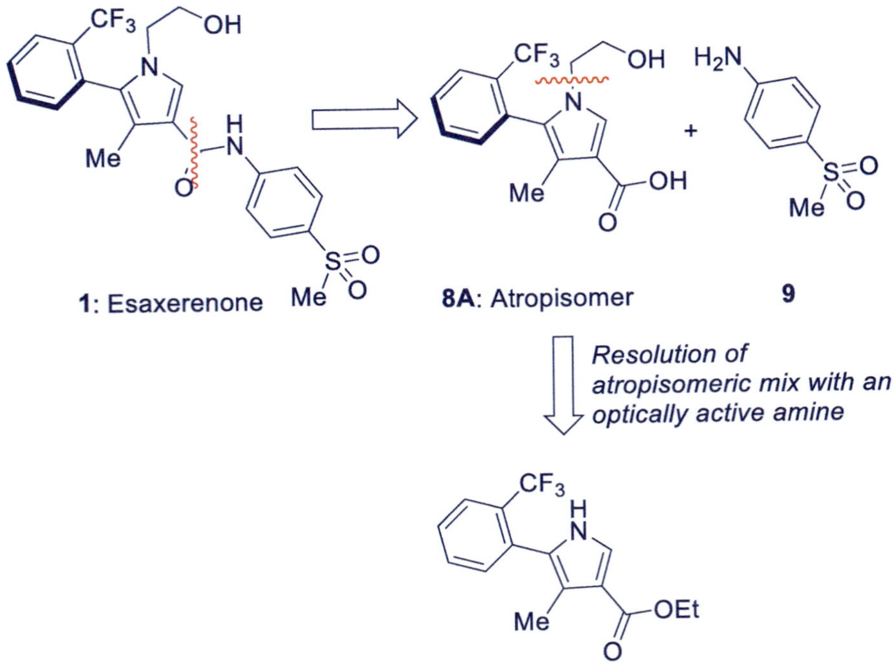

10: Carboxylic ester

Esaxerenone (**1**) has been reported to be synthesized *via* a linear strategy. An important and challenging aspect of this target molecule is its axial chirality requiring the isolation of an atropisomer. As shown in the retrosynthetic scheme earlier, the construction of esaxerenone (**1**) molecule involves three major parts: 1) synthesis of a key carboxylic ester intermediate (compound **10**), 2) a chemical resolution of the atropisomeric mix **8** to achieve axially chiral quinine salt (**8A**), and 3) finally assembly of **8A** with 4-(methylsulfonyl)aniline **9** to yield esaxerenone (**1**). Extensive process research conducted by the Daiichi Sankyo Company and published in 2016 led to the development of a practical process to meet large supplies of the active pharmaceutical ingredient (API) to be formulated.

The synthesis of esaxerenone (**1**) begins with 1-[2-(trifluoromethyl)phenyl]-propan-1-one (compound **8**) shown in the scheme. Ketone **11** is subjected to an α-bromination using bromine in a solution of *t*-butyl methyl ether (MTBE) as the solvent. Incipient product bromoketone **12** formed is not isolated and carried into the next step *in situ* as an MTBE solution. It undergoes reaction with ethyl cyanoacetate to form an intermediate cyanoketoester **13**, another compound that moves *in situ* as a toluene solution and is telescoped into the subsequent cyclization step. Treatment of compound **13** in toluene with thionyl chloride, followed by blowing hydrogen chloride gas at reduced temperature brings about cyclization to generate a chloro-substituted pyrrole (compound **14**), isolated as a crystalline solid.

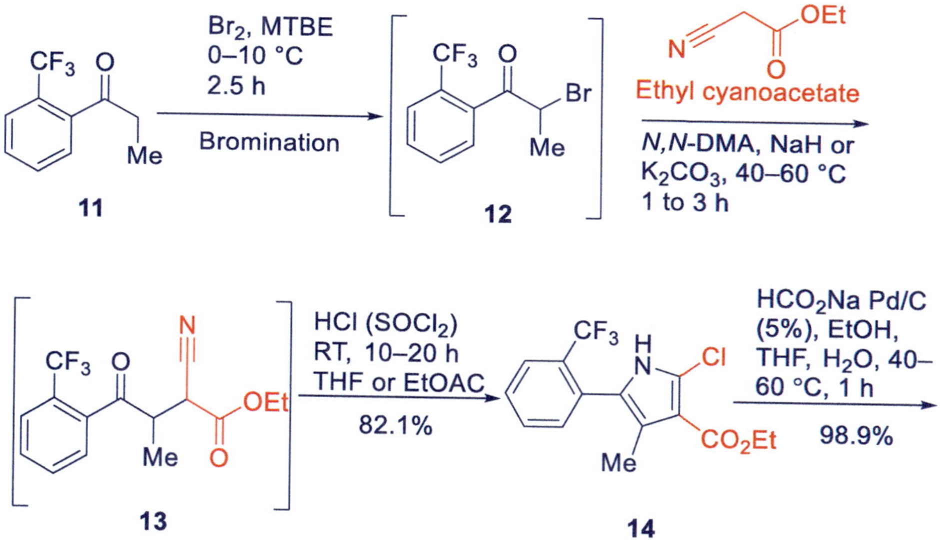

Chloropyrrole **14** is subjected to palladium catalyzed transfer hydrogenation using sodium formate as the source of hydride to yield the des-chloropyrrole **15**, a compound isolated as a crystalline solid. Pyrrole **15** undergoes *N*-alkylation with an ethylene carbonate as the alkylating agent at elevated temperatures to form compound **16**, that is once again not isolated. It is telescoped as an ethanolic solution into the subsequent ester hydrolysis step to yield a key advanced intermediate carboxylic acid compound **8.** This racemic atropisomeric mix is a crystalline compound that serves as the substrate for the key chemical resolution step to separate the desired atropisomer from the undesired one.

The carboxylic acid **8** is essentially considered to be an important and novel intermediate for the large-scale production of esaxerenone (**1**) API (Active Pharmaceutical Ingredient). Extensive process research work was undertaken to seek a scalable technology that would allow to produce large quantities of the desired pure atropisomer **8A**.

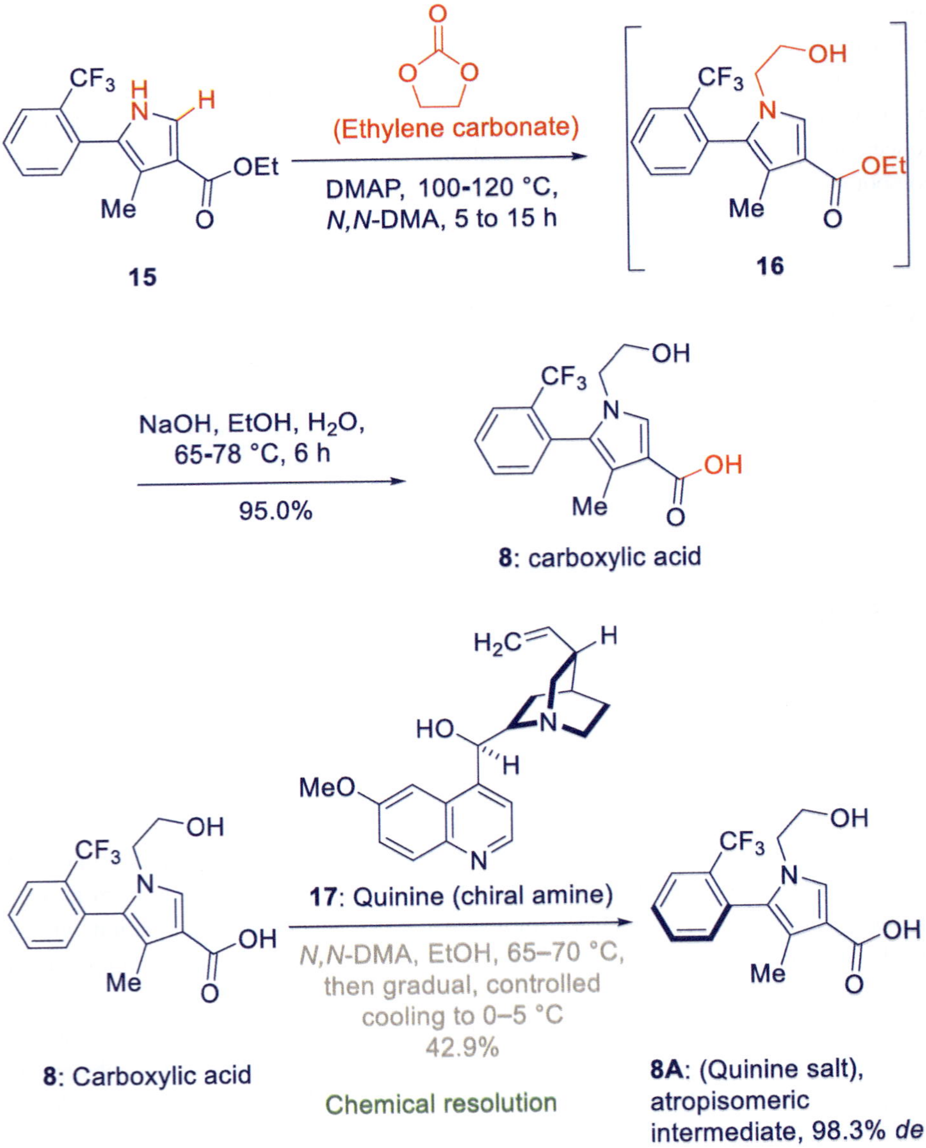

A method for efficient resolution was found to provide the novel atropisomeric pyrrole intermediate in high quality and high yield, by using an industrially advantageous operation. The chemical resolution to generate the pure atropisomer involves the following. To the racemic carboxylic acid compound **8**, *N,N*-dimethylacetamide and ethyl acetate are added at room temperature to dissolve. A solution of quinine is prepared in *N,N*-dimethyacetamide, ethyl acetate and water are added. The mixture heated to 65–70 °C, to which then the solution of racemic **8** is added dropwise at same temperature followed by stirring for an additional hour. Subsequently, gradual cooling to 0–5 °C using a controlled cooling rate, followed by stirring for an additional 30 minutes leads to the formation of crystalline material that can be filtered to yield 42.9%, with a diastereomeric excess of 98.3 (*de*) upon isolation. This is how the atropisomeric pure carboxylic acid in the form of a quinine salt **8A** is generated as a key intermediate for the synthesis of esaxerenone (**1**). The amine salt of compound **8A** obtained this way can also be converted into the free acid form using any typical acid such as hydrochloric acid.

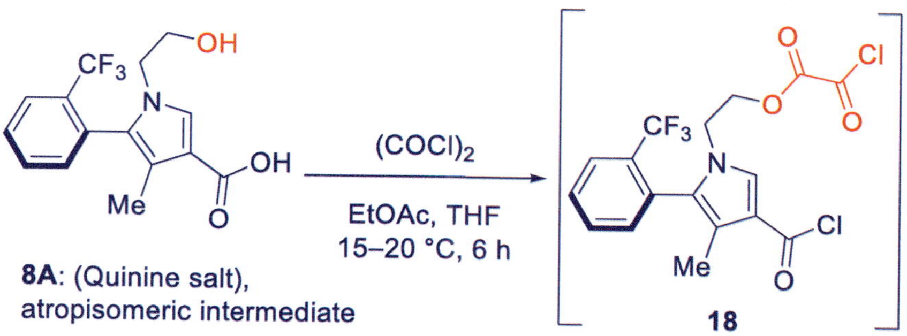

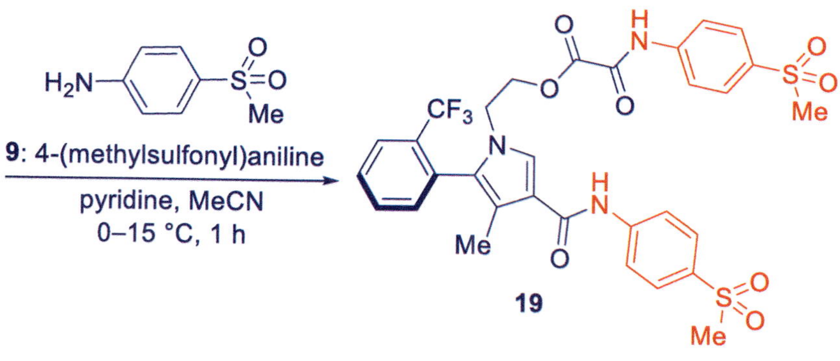

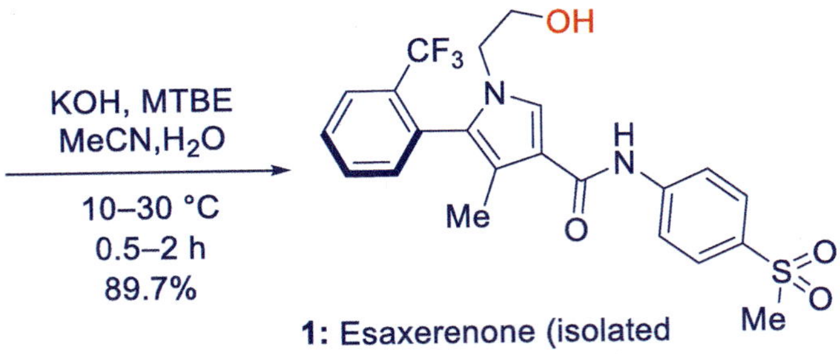

1: Esaxerenone (isolated as crystalline solid)

In the endgame of the synthesis, compound **8A** is subjected to reaction with oxalyl chloride. The carboxylic acid undergoes formation of the corresponding acid chloride, while the hydroxyethyl group undergoes coupling with oxalyl chloride yielding compound **18**. This acid chloride derivative **18** undergoes coupling with 2 equivalents of 4-(methylsulfonyl)aniline (**9**). Hydrolytic work-up yields pure crystalline exaxerenone (**1**).

16.7 Summary

Esaxerenone (**1**) represents the first non-steroidal MRA, reaching patients in Japan that suffer from essential hypertension. It received its first global approval in January 2019. It is the consequence of a long evolution process starting from the time when role of aldosterone was first demonstrated in hypertension in the 1950s. Several developments over the past 60 years clearly suggest that despite the best efforts to design steroidal analogues, they continue to pose limitations and risks owing to side effects and cannot remain a viable widely prescribed therapy. Esaxerenone (**1**) offers a fresh perspective to mitigate these risks. It is an outcome of deep understanding of the mechanism of action and a rigorous effort by the drug discovery community, that was systematically utilized over the past two decades to ultimately bring a non-steroidal small molecule into development.

This is a molecule that can be administered routinely as a tablet to patients with essential hypertension. Adverse effects of this drug are few and typically include increase in serum potassium levels as well as blood uric acid, and hyperuricaemia. Its pharmacokinetics are mostly unaffected by food with no difference seen in C_{max}, leading to a significant decrease in the systolic and diastolic blood pressure. On the synthetic front, the target molecule is structurally interesting that requires a chemically resolved atropisomer as the key intermediate, which can be practically achieved in high yield and purity using a chiral amine with minimal wastage of reagents. This is a challenging technology that has been realistically and sustainably reduced to practice at large scales. Esaxerenone (**1**) is a first-in-class therapy to enter the market and address a grim medical need, that can potentially overcome side-effects of existing steroidal therapy. With many biotech and pharmaceutical companies working on similar strategies with increased understanding of disease, more molecules are expected to eventually reach market and transform patients' lives.

16.8 References

1. (a) Kitt, J.; Fox, R.; Tucker, K. L.; McManus, R. J. New approaches in hypertension management: a review of current and developing technologies and their potential impact on hypertension care. *Curr. Hypertens. Rep.* **2019**, *21*, 44–52.
   (b) Bolívar, J. J. Essential hypertension: an approach to its etiology and neurogenic pathophysiology, Hindawi Publishing Corporation, *Int. J. Hypertens.* **2013**, Article ID 547809, 11 pages. https://www.hindawi.com/journals/ijhy/2013/547809/
2. (a) Messerli, F. H.; Williams, B.; Ritz, E. Essential hypertension. *Lancet* **2007**, *370*, 591–603. (b) Storessen, J. A.; Wang, J.; Bianchi, G.; Birkenhäger, W. H. Essential Hypertension. *Lancet* **2003**, *361*, 629–1641.
3. Hostetter, T. H.; Ibrahim, H. N. Aldosterone in chronic kidney and cardiac disease. *J. Am. Soc. Nephrol.* **2003**, *14*, 2395–2401.
4. Arai, K.; Papadopoulou-Marketou, N.; Chrousos, G. P. *Aldosterone Deficiency and Resistance.* Feingold, K. R.; Anawalt, B.; Boyce, A.; et al., eds. Endotext [Internet]. South Dartmouth, MA: MDText.com, Inc.; 2000–2020.
5. (a) Kolkhof, P.; Bärfacker, L. Mineralocorticoid receptor antagonists: 60 years of research and development. *J. Endocrinol.* **2017**, *234*, T125–T140. (b) Garthwaite, S. M.; McMahon, E. G. The evolution of aldosterone antagonists. *Mol. Cell. Endocrinol.* **2004,** *217*, 27–31. (c) Ottow, E., Weinmann, H. *Nuclear Receptors as Drug Targets*. John Wiley & Sons. (2008), p. 410.
6. Gao, Q.; Xu, L.; Cai, J. New drug targets for hypertension: a literature review. *BBA – Mol. Basis Dis.* **2021**, *1867*, 166037.
7. White P. C. Aldosterone: direct effects on and production by the heart *J. Clin. Endocrinol. Metab.* **2003**, *88*, 2376–2383.
8. (a) Gronemeyer, H.; Gustafsson, J.-Å.; Laudet, V. Principles for modulation of the nuclear receptor superfamily. *Nat. Rev. Drug Discov.* **2004**, *3*, 950–964. (b) Moore, J. T.; Collins, J. L.; Pearce, K. H. The nuclear receptor superfamily and drug discovery. *ChemMedChem* **2006**, *1*, 504–523.
9. (a) Piotrowski, D. W. Mineralocorticoid receptor antagonists for the treatment of hypertension and diabetic nephropathy. *J. Med. Chem.* **2012**, *55*, 7957–7966. (b) Patent WO 2006/012642 (US Publication No. US 2008- 0234270). (c) Patent WO 2008/056907 (US Publication No. US 2010- 0093826).
10. Yamada, M.; Mendell, J.; Takakusa, H.; Shimizu, T.; Ando, O. Pharmacokinetics, metabolism, and excretion of [14C]esaxerenone, a novel mineralocorticoid receptor blocker in humans *Drug Metab. Dispos.* **2019**, *47*, 340–349.
11. Duggan, S. Esaxerenone: First global approval. *Drugs* **2019**, *79*, 477–481.

12. Daiichi Sankyo. Esaxerenone (1) (Minnebro): Japanese prescribing information; 2019. http://www.pmda.go.jp/PmdaS earch /iyaku Detail/43057 4_21490 B6F10 26_1_02#CONTRAINDICATIONS.
13. Ito, S.; Itoh, H.; Rakugi, H.; Okuda, Y.; Yamakawa, S. Efficacy and safety of esaxerenone (CS-3150) for the treatment of essential hypertension: a phase 2 randomized, placebo-controlled, double-blind study *J. Human Hypertens.* **2019**, *33*, 542–551.
14. (a) Ito, S.; Itoh, H.; Rakugi, H.; Okuda,Y.; Yoshimura, M.; Yamakawa, S; Double-blind randomized phase 3 study comparing esaxerenone (1) (CS-3150) and eplerenone in patients with essential hypertension (ESAX-HTN Study) *Hypertens.* **2020**, *75*, 51–58. (b) Ito, S.; Ito, H. Rakugi, H.; Okuda, Y.; Yoshimura, M.; Yamakawa, S. A double-blind phase 3 study of esaxerenone (CS-3150) compared to eplerenone in patients with essential hypertension (ESAX-HTN study) *J. Hypertens.* **2018**, *36*, e239.
15. Kato, M.; Furuie, H.; Shimizu, T.; Miyazaki, A.; Kobayashi, F.; Ishizuka, H. Single- and multiple-dose escalation study to assess pharmacokinetics, pharmacodynamics and safety of oral esaxerenone in healthy Japanese subjects. *Brit. J. Clin. Pharmacol.* **2018**, *84,* 1821–1829.
16. (a) Arai, K.; Tsuruoka, H.; Homma, T. CS-3150, a novel non-steroidal mineralocorticoid receptor antagonist, prevents hypertension and cardiorenal injury in Dahl salt-sensitive hypertensive rats. *Eur. J. Pharm.* **2015**, *769*, 266–273. (b) Arai, K.; Morikawa, Y.; Ubukata N.; Tsuruoka, H.; Homma T. CS-3150, a novel nonsteroidal mineralocorticoid receptor antagonist, shows preventive and therapeutic effects on renal injury in deoxycorticosterone acetate/salt-induced hypertensive rats. *J Pharm. Exp. Ther.* **2016**, *358*, 548–557.

17

Voclosporin (Lupkynis), A Macrocyclic Peptide Inhibitor of Calcineurin for the Treatment of Lupus Nephritis

Yan Wang

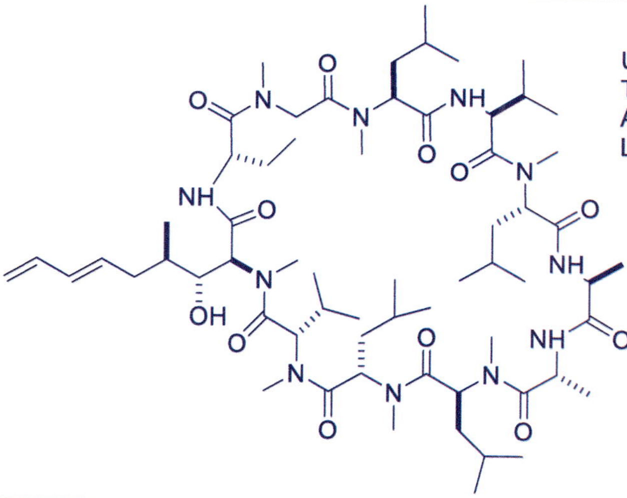

USAN: INN
Trade name: LUPKYNIS
Aurinia Pharmaceuticals
Launched: 2021

17.1 Background

Systemic lupus erythematosus is a chronic autoimmune disease characterized by loss of tolerance against nuclear autoantigens, lymph proliferation, polyclonal autoantibody production, immune complex disease, and multiple organ tissue inflammation. In this disease, the immune system of the body attacks its own tissues, causing widespread inflammation and tissue damage in the affected organs. Lupus nephritis (LN) is one of the most serious and common complications of Systemic lupus erythematosus. In many cases, LN is the presenting manifestation that results in the diagnosis of systemic lupus

Current Drug Synthesis, First Edition. Edited by Jie Jack Li.

erythematosus. Overall, patients with systemic lupus erythematosus of younger age, male sex, and African, Asian, or Hispanic ethnicity are more likely to develop LN. Mortality associated with lupus is significantly higher in those with LN compared with those without LN, and death directly attributable to kidney disease occurs in 5−25% of patients with proliferative LN within 5 years of onset. Furthermore, 10−30% of patients with LN progress to kidney failure requiring kidney replacement therapy.[1–4]

At present, kidney biopsy is used to establish a diagnosis of LN or other processes involving the kidneys in a patient with lupus, and to determine the extent of acute and chronic kidney injury, which has therapeutic implications. The 2003 ISN/RPS classification which is currently the most widely used system, explicitly states that concurrent tubular atrophy, interstitial inflammation and fibrosis, and arteriosclerosis or other vascular lesions should be reported and graded as mild, moderate, or severe. In this system, LN was classified into six classes (I−VI) based on clinical findings.[1,2]

The immune suppression medications that are used to treat LN include glucocorticoids (**1**, **2**), cyclophosphamide (**3**), or mycophenolate mofetil (MMF, **4**). The initial treatment for LN is corticosteroids, such as prednisone (**1**) and prednisolone (**2**). The steroids may be given orally or intravenously. Immunosuppressive therapy is mostly used for the treatment of classes III and IV LN. However, conventional immunosuppressive treatments are not uniformly effective, and even in patients who respond, 35% may relapse. Furthermore, 5–20% of patients with LN develop the end-stage renal disease (ESRD) within 10 years from the initial event. And drug-induced toxicity also remains a concern. Combination therapy using immune-suppression medications and the next generation of calcineurin inhibitor, voclosporin (VCS, **5**), was investigated. The Food and Drug Administration (FDA) has approved Lupkynis in combination with a background immunosuppressive therapy regimen to treat adult patients with active LN in 2021. Lupkynis is the first FDA-approved oral therapy for LN. It is a semi-synthetic structural analog of cyclosporine (CsA, **6**). It has demonstrated a more stable pharmacokinetic and pharmacodynamic relationship than CsA, a higher potency than CsA, and an improved metabolic profile when compared with older calcineurin inhibitors.[1–6]

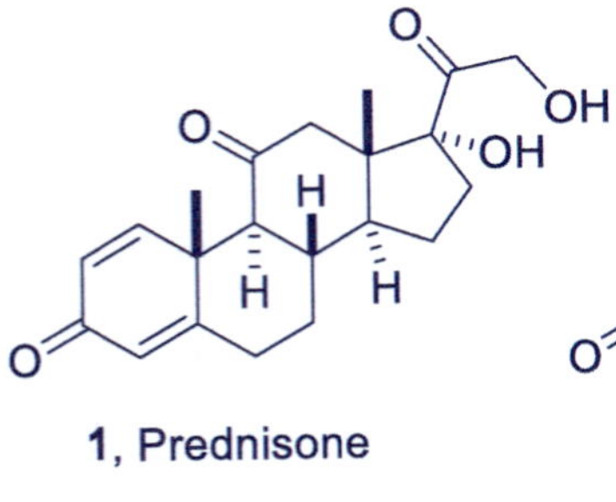

1, Prednisone

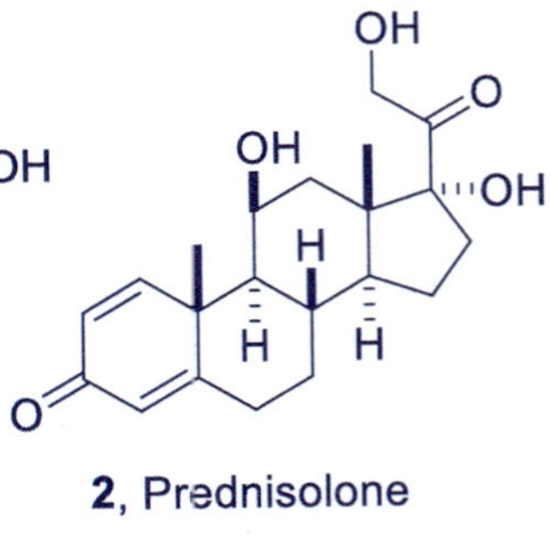

2, Prednisolone

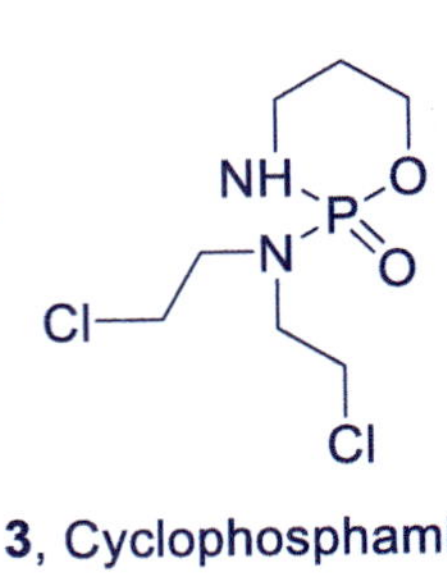

3, Cyclophosphamide

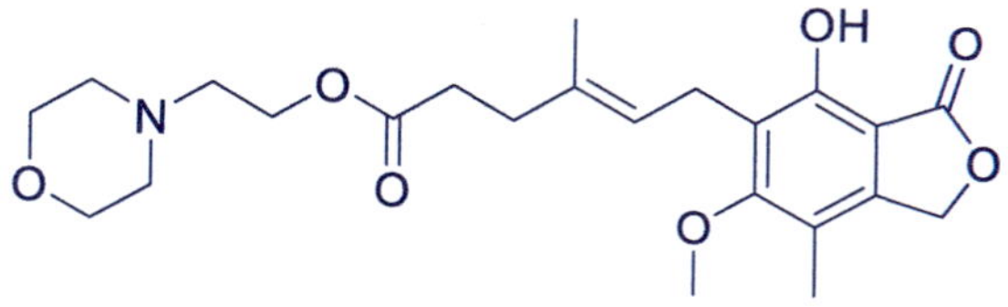

4, Mycophenolate mofetil (MMF)

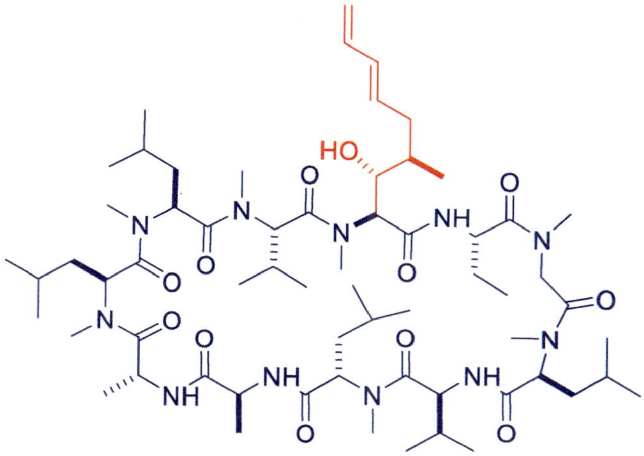

5, Voclosporin (VCS)

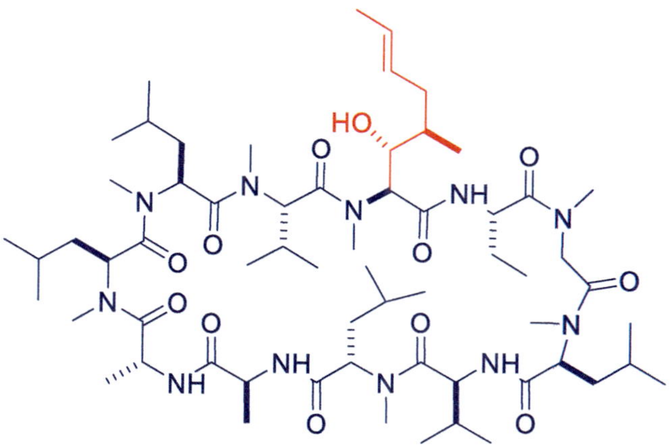

6, Cyclosporin A (CsA)

17.2 Pharmacology

Calcineurin inhibitors are potent immunosuppressants that reversibly inhibit T-cell proliferation and prevent the production and release of pro-inflammatory cytokines by blocking the activity of the calcium-regulated serine-threonine phosphatase calcineurin. Calcineurin inhibitors also inhibit fibroblast proliferation and expression of vascular endothelial growth factor (VEGF). After entering the lymphocyte, calcineurin inhibitors bind to intracellular immunophilins and form complexes that subsequently bind to and inhibit calcineurin. This prevents translocation of the cytoplasmic component of the nuclear factor of activated T cells to the nucleus, which impairs transcription of the genes encoding IL-2 and other cytokines.[7]

CsA reversibly inhibits T-lymphocytes by binding to cyclophilin. It also inhibits lymphokine production and release. As a prototypical calcineurin inhibitor, CsA has been reported to be effective in preventing allograft rejections for a variety of organs in different species. It was introduced to the market in 1983 for the prevention of organ rejection under the trade name SANDIMMUNE.[8]

VCS is a novel calcineurin inhibitor immuno-suppressant, with a synergistic and dual mechanism of action, blocking IL-2 expression and T-cell-mediated immune responses. It is structurally similar to CsA (**6**), only the side chain of amino acid NMeBmt is modified to be one carbon longer. This modification changes the binding of VCS to calcineurin. Early formulations of VCS consisted of *trans-cis* isomers.[9] The early clinical and nonclinical studies were performed using a virtually equal mixture of both isomers, i.e., ISA247 (a name used in early studies). The later clinical studies used the more potent *trans*-isomer.[10] During the different stages of development of VCS, the following names have been used to indicate the cyclic peptide, such as ISA247, ISAtx 247, LX 211, R 1524, etc.[11] VCS demonstrated good absorption after oral administration in rabbits, cats, and dogs.[10] VCS undergoes extensive first-pass metabolism, involving cytochrome P450 mediated oxidation. Fecal excretion is the primary route of elimination. VCS had comparable or even higher inhibition of lymphocyte proliferation, T cell activation, and cytokine production than that achieved by CsA in nonhuman primates.[13]

17.3 Structure–Activity Relationship (SAR)

The long-term use of CsA is limited by its side effects, notably nephrotoxicity.[14,15] The modification on multiple positions of the cyclic peptide has been attempted.[16] The biological activity is associated with a larger portion of the CsA structure and the results indicate that this probably includes amino acids 1, 2, 10, and 11 (Figure 1). In many cases, the biological activity of CsA derivatives correlates with their affinity for CypA. In general, modifications at the CypA-binding residues 1, 2, 9, 10, or 11 that diminish binding to CypA also result in diminished immunosuppressive activity, while modifications of the effector-loop residues 4, 5, or 6 can strongly affect the immunosuppressive activity without significantly affecting CypA binding. But notable exceptions have been reported too.[16–18]

Modifications of CsA amino acids have been done to improve CsA activity and stability, including (1) chemical substitution and optional deuterium substitution of amino acid 1; and (2) deuterium substitution at key sites of metabolism of CsA, such as amino acids 1, 4, and 9.[19,20] However, the substitution of deuterium in various methyl groups at the side chain of NMeBmt did not provide any advantage over the non-deuterated compound.

The chemical structure of VCS (**5**) is similar to that of CsA (**6**). The only modification is on the side-chain of amino acid NMeBmt in CsA. A single carbon was added to the terminus, changing it to *E/Z*-NMePmt in VCS. The binding affinities of the *E/Z*-ISA247 isomers were measured to be 15 and 61 nM *in vitro* using a ^{32}P-labelled calcineurin activity assay, respectively, by fluorescence spectroscopy.[13]

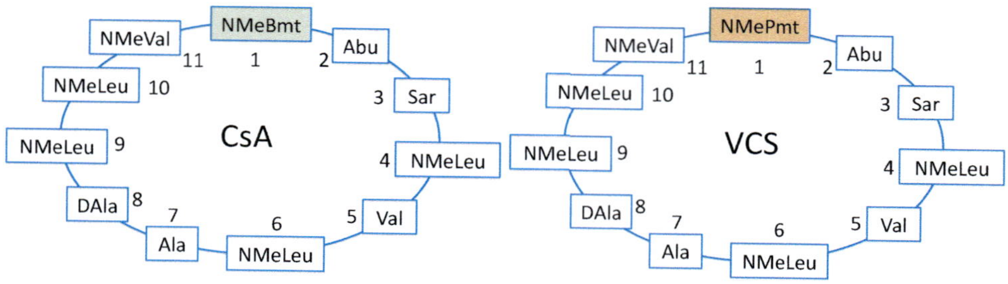

Figure 1. Amino acid composition of Cyclosporin A and Voclosporin (with the labeling of amino acid positions).

X-ray crystallographic studies have shown chemical modifications of the ISA247 isomers at residue 1 (Figure 2A−C) do not induce major re-arrangements of the CypA side chains in the drug–CypA complex and that functional differences between the two ISA247 isomers result from the structural difference at the modified NMePmt residue only.[13] This increase in carbon length alters the way the cyclophilin-VCS complex binds to a composite surface of catalytic and regulatory subunits in calcineurin. *E*-ISA247 demonstrates superior immunosuppressive activity compared to CsA.

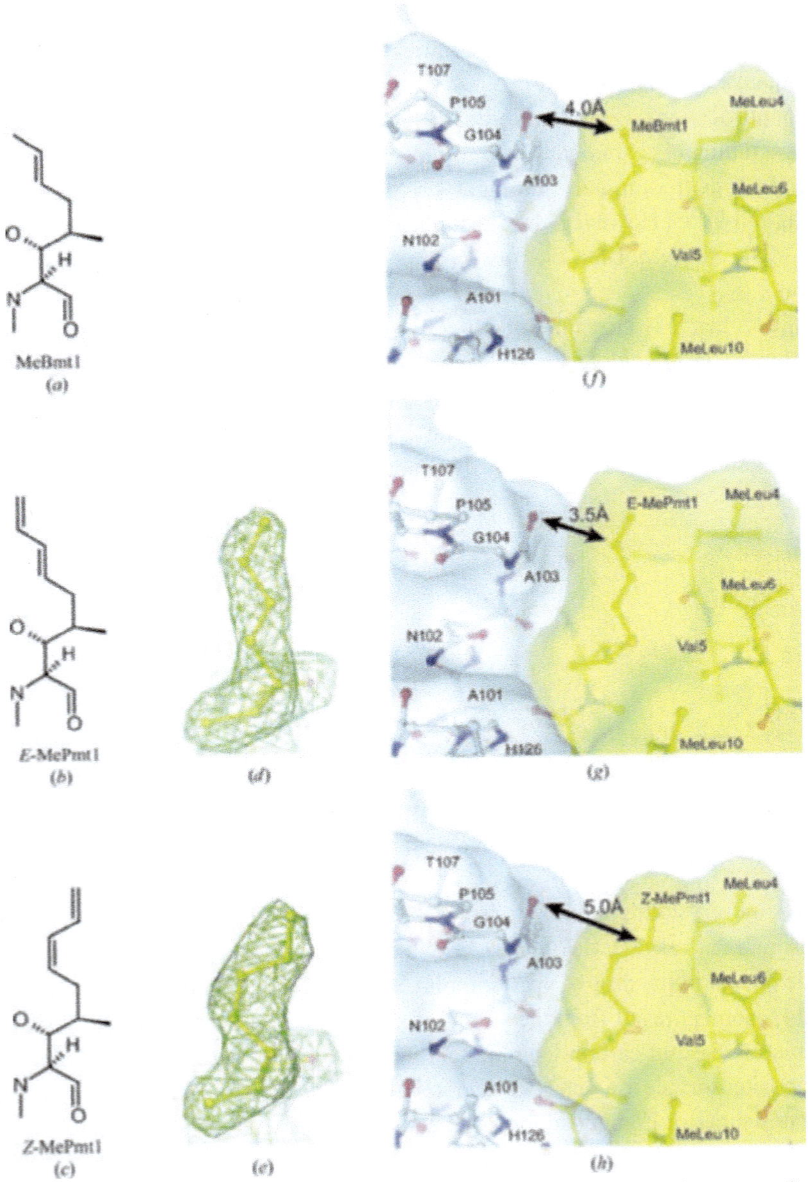

Figure 2. Residues at position 1 of CsA and the ISA247 stereoisomers. Structural formulae are displayed for (A) the cyclosporin residue NMeBmt 1, (B) the *E*-ISA247 residue *E*-NMePmt 1, (C) the *Z*-ISA247 residue *Z*-NMePmt 1, (D) *E*-NMePmt 1, (E) *Z*-NMePmt 1, (F) CsA, (G) *E*-ISA247, and (H) *Z*-ISA247 in yellow are shown. The distances between the Ala 103 backbone carbonyl of CypA and residue 1 of the cyclic peptides are marked by black arrows.

17.4 Pharmacokinetics and Drug Metabolism

VCS and mix-ISA247 were metabolized by hydroxylation and oxidative *N*-demethylation reactions in all species studied, including humans.[10,21,22] The metabolism of VCS involves cytochrome P450 oxidation of amino acids, other than amino acid 1, to yield the quantifiable metabolites, IM9, IM4, and IM4n, and also oxidation of amino acid 1 to generate the quantifiable metabolites, IM1-Diol-1, IM1c (R), and IM1w. Metabolism of VCS is similar across species, with no unique human metabolites (Table 1). The primary metabolism site of CsA is at amino acid position 1.[23] Therefore, the modification of this site has shifted the primary metabolism site of VCS to the amino acid 9 position. VCS has increased potency and a change in metabolite profile, which allows for the administration of lower doses, and less pharmacokinetic–pharmacodynamic variability. It has a potentially improved safety profile.[10,24]

Table 1: Major ISA247 metabolites observed in the various species.

	Quantity (expressed as a relative % to ISA247)			
Metabolite mass	Human (%)	Monkey (%)	Dog (%)	Rat (%)
1223 (De-methyl)	21.6	22.2	67.2	12.8
1239(De-methyl hydroxyl)	3.2	18.6	38.7	7.6
1253 (Hydroxyl)	40.7	93.5	90.3	131.1
1255 (De-hydro)	5.7	6.6	0	0
1271(ISA-AM1-diol)	38.7	21.3	10.7	7.0

Source: Based on reference 10.

A phase I study of VCS evaluated the pharmacokinetics, pharmacodynamics, and the effect of food in single ascending doses.[25] The dose range was from 0.25 through 4.5 mg/kg. Whole blood concentrations of VCS peaked within 2 hours of drug administration at all dose levels. Administration of VCS with a low-fat and high-fat meal decreased C_{max} by 29% and 53%, respectively, and AUC_{inf} by 15% and 25%, respectively. VCS inhibited calcineurin activity in a dose-related fashion with maximal inhibition peaking at 3.0 mg/kg. Administration of VCS with food resulted in diminished peak and extent of absorption and delayed peak of absorption that appeared related to the fat content of the meal.

Another phase I special population pharmacokinetic study evaluated the effect of renal or hepatic impairment on the pharmacokinetics of VCS.[26] Individual C_{max} and AUC plots indicate an almost complete overlapping range of values for mild and moderate renal impairment and normal subjects. Severe renal impairment resulted in a 1.5-fold increase in AUC without an increase in C_{max}. Mild to moderate hepatic impairment resulted in a 1.5- to 2-fold increase in VCS exposure.

Pharmacokinetic drug interactions between VCS and a CYP3A inhibitor (**7**, ketoconazole), inducer (**8**, rifampin), and substrate (**9**, midazolam) and a P-glycoprotein inhibitor (**10**, verapamil) and substrate (**11**, digoxin) were studied.[24] Co-administration of VCS with CYP3A inhibitor (**7**) increases VCS peak concentration and extent of exposure by 6.4- and 18-fold, respectively. Co-administration of VCS with CYP3A inducer (**8**) reduced VCS exposure by 90%. Co-administration of VCS with P-glycoprotein inhibitor (**10**) increased VCS C_{max}, AUC, and pre-dose concentrations by 2.1-, 2.7-, and 2.8-fold, respectively. Co-administration of VCS with P-glycoprotein substrate (**11**) resulted in a significant increase in digoxin C_{max}, AUC, and urinary excretion by 50%, 25%, and 20%, respectively, while digoxin renal clearance was unchanged.

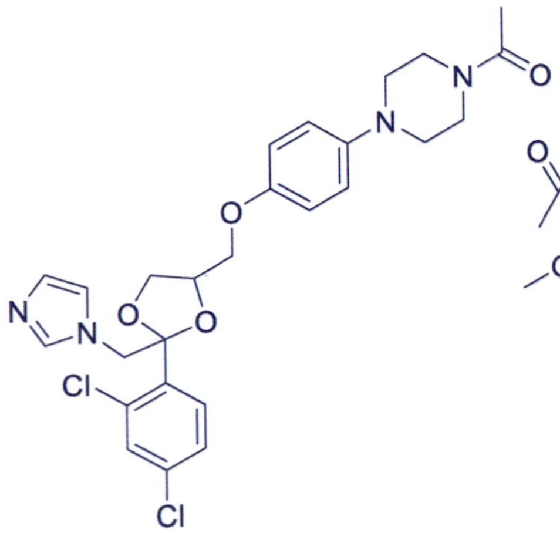

7, Ketoconazole
(CYP3A inhibitor)

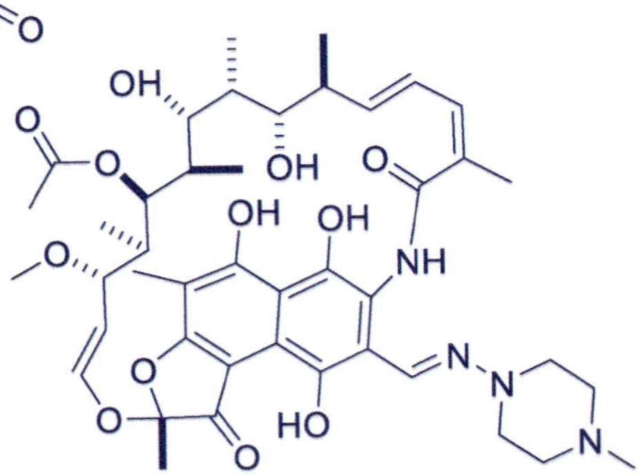

8, Rifampin
(CYP3A inducer)

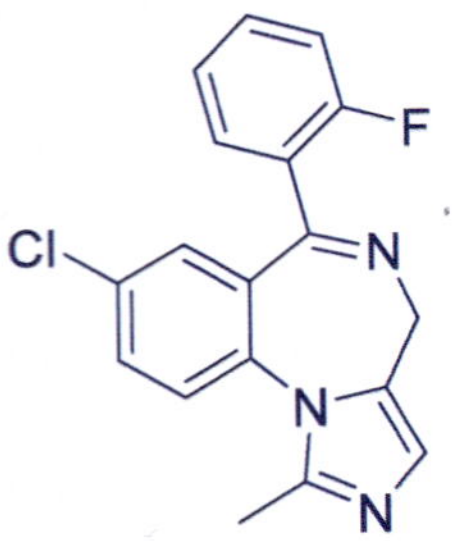

9, Midazolam
(CYP3A substrate)

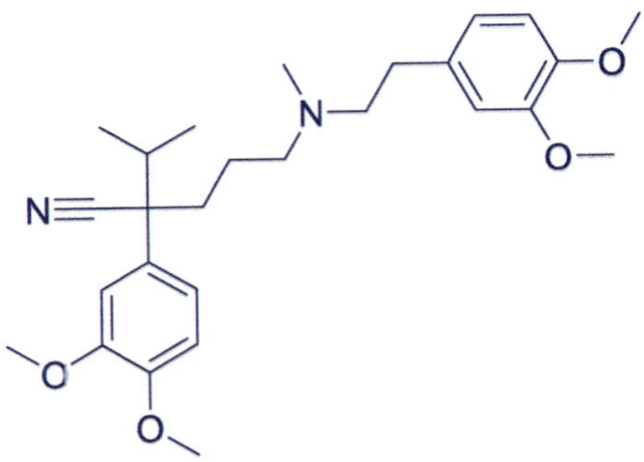

10, Verapamil
(P-glycoprotein inhibitor)

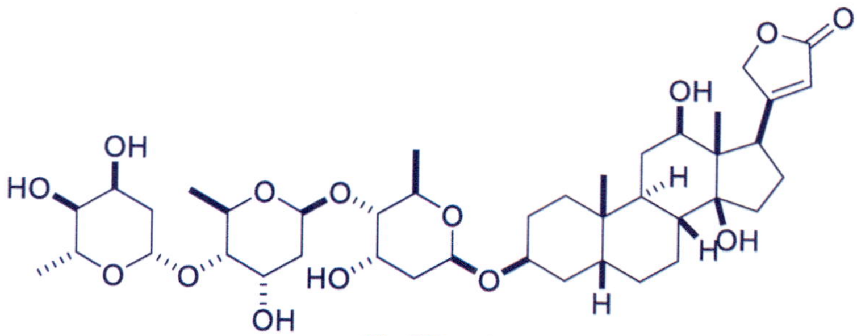

11, Digoxin
(P-glycoprotein substrate)

In summary, upon administration of VCS on an empty stomach, the median T_{max} is 1.5 hours, but it can range from 1 to 4 hours. The apparent volume of distribution (V_{ss}/F) of VCS is 2154 L. It distributes extensively into red blood cells. The distribution between whole blood and plasma is dependent on concentration and temperature. The protein binding of VCS is approximately 97%. VCS is mainly metabolized by the CYP3A4 hepatic cytochrome enzyme. Pharmacologic activity is mainly attributed to the parent molecule. The average terminal half-life of VCS is about 30 hours (24.9−36.5 hours) with a peak plasma time of 1.5 hours (empty stomach). Steady-state is reached at 6 days. The mean apparent steady-state clearance of VCS is 63.6 L/h. Hepatic and renal impairment significantly reduce its clearance.[10]

17.5 Efficacy and Safety

VCS shows greater immunosuppressive activity than CsA *in vitro*,[15] in animal models of autoimmunity[27] and transplantation.[28] A phase II, multicenter, randomized, double-blind, placebo-controlled trial demonstrated that the addition of low-dose VCS to MMF and corticosteroids for induction therapy of active LN results in a superior renal response.[29] The encouraging data of that trial led to the testing of the VCS regimen in phase III randomized placebo-controlled trial.[30] In the multicenter phase III trial, VCS in combination with MMF (**4**) and low-dose steroids led to a clinically and statistically superior complete renal response rate vs. MMF and low-dose steroids alone, with a comparable safety profile.

The phase I pharmacokinetic, single ascending dose studies of VCS were carried out in healthy volunteers.[25] The majority of adverse effects reported were associated with the gastrointestinal system. The adverse effects reported in this study were graded as mild and resolved spontaneously. It demonstrated that VCS was safe and well-tolerated after oral administration at doses ranging from 0.25 to 4.5 mg/kg.

The phase I special population pharmacokinetic studies evaluated the effect of renal or hepatic impairment on the pharmacokinetics of VCS.[26] It can be administered safely to patients with mild to moderate renal impairment without dose modification. Appropriate safety monitoring with concentration-based adjustments in transplantation is recommended for patients with severe renal impairment, and patients with hepatic impairment.

Cytochrome P450 3A and P-glycoprotein drug–drug interactions with VCS were evaluated and reported.[24] Ketoconazole (**7**), rifampin (**8**), and midazolam (**9**) are considered to be suitable models for a CYP3A inhibitor, inducer, and substrate, respectively. Verapamil (**10**) and digoxin (**11**) are considered as suitable models of P-glycoprotein inhibitor and substrate, respectively. Administration of VCS concomitantly with strong inhibitors and inducers of CYP3A should be considered contraindicated. Co-administration of VCS with midazolam (**9**), a known CYP3A substrate, did not result in significant changes to exposure of midazolam or its metabolite α-hydroxymidazolam after a single dose. Drug−drug interactions involving VCS and other CYP3A substrates are not expected.

To evaluate the safety and tolerability of the combination of VCS and MMF (**4**), an open-label phase I study was conducted.[31] It evaluated the effect of VCS following dosing with MMF on blood levels of mycophenolic acid (MPA, the active moiety of MMF, **12**) and MPA glucuronide (MPAG) (the pharmacologically inactive metabolite of MMF, **13**) in subjects with systemic lupus erythematosus. As changes in MPA (**12**) may affect efficacy and safety, these data confirm that the two drugs can be administered concomitantly without the need to adjust the dose of MMF (**4**).

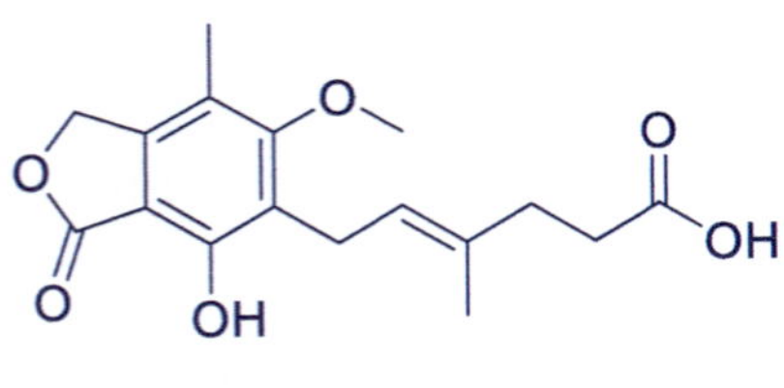

12, Mycophenolic acid (MPA)

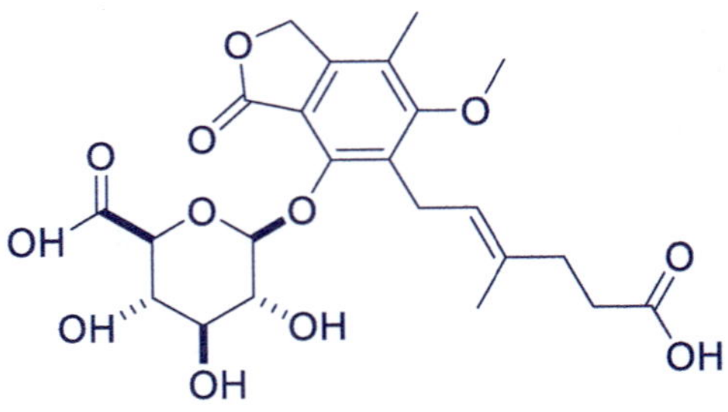

13, MPA glucuronide (MPAG)

In the multicenter phase II trial, subject deaths were observed in the low-dose group in the early stage.[29] This outcome is difficult to explain. It mandated careful surveillance when this regimen is first initiated.

17.6 Syntheses

CsA (**6**) is a cyclic, hydrophobic peptide that is composed of 11 amino acids. It contains several *N*-methyl amino acids, a DAla at position 8, and NMeBmt at position 1. The CsA sequence is cyclo-(NMeBmt-Abu-Sar-NMeLeu-Val-NMeLeu-Ala-DAla-NMeLeu-NMeLeu-NMeVal) (Figure 1). CsA can be biosynthesized by fermentation[32] or chemically synthesized using solid phase peptide synthesis (SPPS).[33–35] CsA is widely produced by submerged fermentation of aerobic fungi identified as *Trichoderma polysporum*. The general protocol for isolation and purification of CsA is explained in Figure 3. It is the main component of a group of naturally occurring cyclosporine analogs in normal fermentation broths. They usually differ in their chemical structure only at one amino acid and their biosynthesis can be directed by externally supplying the corresponding precursor.[32]

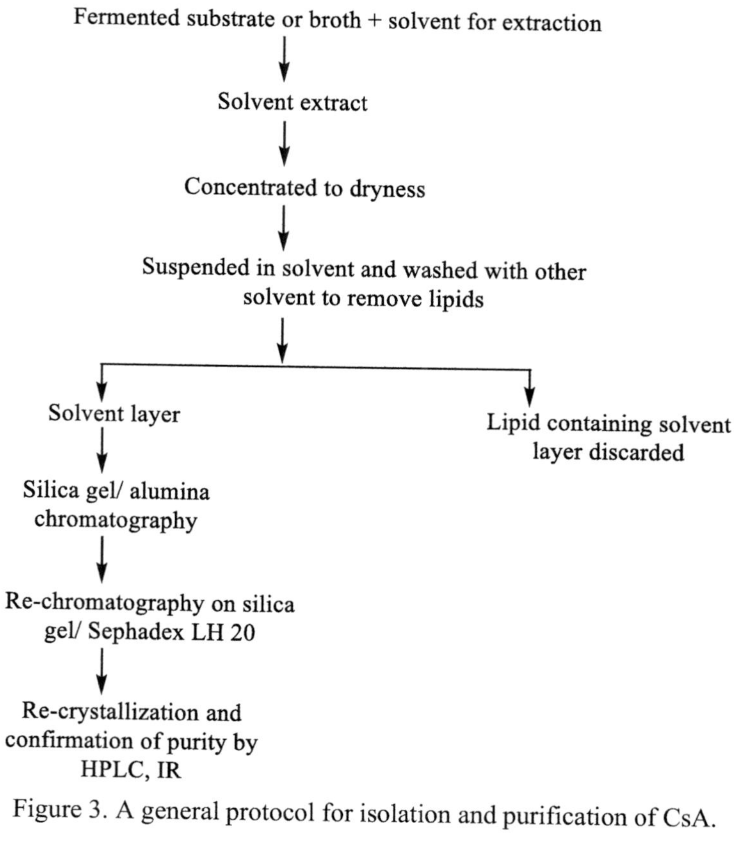

Figure 3. A general protocol for isolation and purification of CsA.

CsA also can be chemically synthesized.[33–35] The assembly of the linear undecapeptide precursor was difficult under the SPPS conditions. The coupling of sterically hindered or *N*-methylated amino acids often resulted in incomplete reactions. And the coupling of Fmoc-NMeLeu at the 9- and 10-positions and Fmoc-NMeVal at the 11-position were particularly difficult. With the development of peptide synthesis techniques, it was finally achieved using different coupling reagents (HATU, HOAt/DIPCDI, HOAt/DIC), double coupling, and longer reaction times. After cleavage, cyclization was completed in solution. The cyclization site was chosen between alanine in position 7 and the DAla in position 8 since the bond formation between non-*N*-methylated amino acids is easier.

VCS, an analog of cyclosporine, was discovered by Robert T. Foster and his team at Isotechnika in the mid-1990s. ISA247 was being co-developed as R 1524 by Isotechnika and Roche. Isotechnika and Roche entered into a co-development and commercialization agreement in April 2002, with Roche gaining the exclusive worldwide marketing rights for ISA247. The agreement was restructured in April 2004, under which Isotechnika solely managed and funded the clinical development of trans-ISA 247. Then Isotechnika merged with Aurinia Pharmaceuticals in 2013.[11]

The chemical structure of VCS (**5**) is similar to that of CsA (**6**). The only modification is on the terminal end of the side chain of amino acid NMeBmt. The terminal methyl group of this residue was converted to CH=CH_2, changing it from (4*R*)-4-[(*E*)-2-butenyl]-4, *N*-dimethyl-l-threonine (NMeBmt, **14**) in CsA to 4-[(2*E*/*Z*, 4*EZ*)-2,4-pentadienyl]-4,*N*-dimethyl-l-threonine (*E*/*Z*-NMePmt, **15**) in ISAtx 247.[36] ISAtx 247 is prepared from CsA (**6**) *via* a semisynthetic route (Scheme 1).

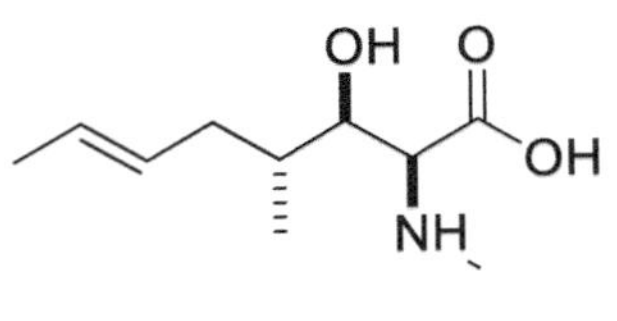

14, NMeBmt

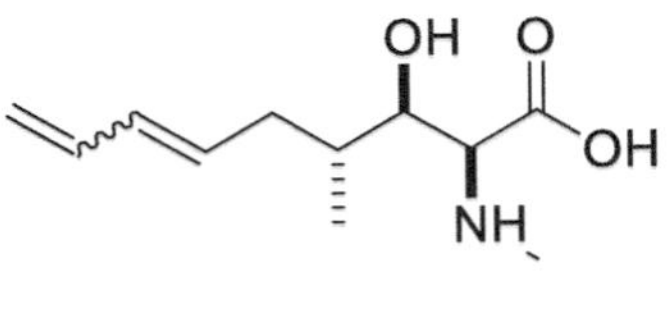

15, *E/Z*-NMePmt

ISAtx 247 is a mixture of *cis* and *trans* isomers. The *trans* (*E*)-isomer accounts for 45–50% and the *cis*-(*Z*)-isomer is from 50 to 55%. The synthesis of ISAtx 247 and enriched either *E* or *Z* isomer utilizing stereoselective reaction conditions are described in the Aurinia Pharmaceuticals patent.[36] The *trans*-isomer (*E*) of ISAtx 247 has indicated better activity on immunosuppression and improved therapeutic index compared with the *cis*-isomer.

The first step of synthesis (Scheme 1) is to protect the β-carbon hydroxyl group of CsA to obtain acetate **16**. Next, $OsO_4/NaIO_4$ is one of the reagents of choice for the transformation to aldehyde product **17**.

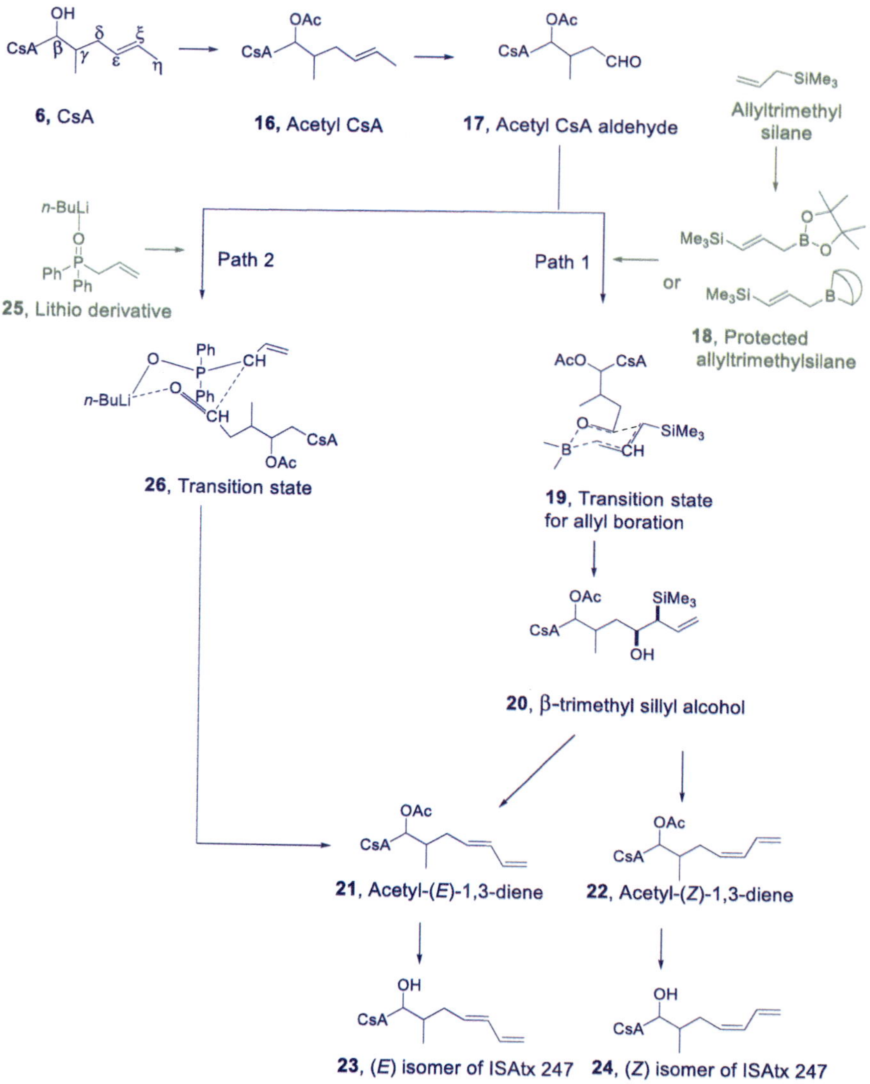

Scheme 1. Synthesis of ISAtx 247.

Subsequently **17** is treated with protected allyltrimethylsilane (**18**, Scheme 1, Path 1) and goes through a six-member ring, boron-containing transition state **19**, to form β-trimethylsilyl alcohol (**20**). The β-trimethylsilyl alcohol may then undergo a Peterson olefination to prepare an alkene (**21** or **22**). The elimination reaction can be carried out under acidic or basic conditions. Under acidic conditions an anti-elimination occurs forming the (*E*)-isomer (**23**) enriched ISAtx 247. Under basic conditions a *cis*-elimination occurs to form the (*Z*)-isomer (**24**) enriched ISAtx 247.

In another synthetic route (Scheme 1, Path 2), Wittig reagent (lithio derivative **25**) reacts with the aldehyde (**17**) to produce the lithium-containing six-membered ring transition state (**26**), which forms the 1,3-diene in *E/Z* ratios (**23**/**24**) in a range greater than approximately 75:25.

In December 2020, Aurinia Pharmaceuticals and Lonza announced an exclusive agreement for the dedicated manufacture of VCS. Lonza's expertise has helped Aurinia to cost-effectively optimize the unique and complex manufacturing process required for the synthesis of VCS. Their monoplant is estimated to be operational in 2023.[37]

17.7 References

1. Anders, H.; Saxena, R.; Zhao, M.; Parodis, I.; Salmon, J. E.; Mohan, C. Lupus nephritis. *Nat. Rev. Dis. Primers* **2020**, *6*, 1–25.
2. Parikh, S.; Almaani, S.; Brodsky, S.; Rovin, B. Update on lupus nephritis: core curriculum 2020. *AJKD* **2020**, *76*, 265–281.
3. Jaryal, A.; Vikrant, S. Current status of lupus nephritis. *Indian J. Med. Res.* **2017**, *145*, 167–178.
4. https://www.medicinenet.com/lupus_nephritis_treatment/views.htm.
5. https://www.drugs.com/history/lupkynis.html.
6. https://www.biospace.com/article/aurinia-pharmaceuticals-snags-fda-approval-for-first-oral-ln-treatment/.
7. Anglade, E.; Yatscoff, R.; Foster, R.; Grau, U. Next-generation calcineurin inhibitors for ophthalmic indications. *Expert Opin. Investig. Drugs* **2007**, *16*, 1525–1540.
8. Wenger, R.; Payne, T.; Schreier, M. Cyclosporine: chemistry, structure-activity relationships and mode of action. *Prog. Clin. Biochem. Med.* **1986**, *3*, 159–191.
9. Naicker, S.; Yatscoff, R. W.; Foster, R. T. Cyclosporine Analogue Mixtures and their Use as Immunomodulating Agents. US Patent 9765119 B2 (**2017**).
10. Multi-disciplinary review and evaluation: Voclosporin. **2020**. https://www.accessdata.fda.gov/drugsatfda_docs/nda/2021/213716Orig1s000MultidisciplineR.pdf.
11. ISA 247: trans-ISA 247, trans-R 1524, ISA(TX)247, ISAtx 247, ISATx247, LX 211, LX211, R 1524, R-1524. *Drugs R&D* **2007**, *8*, 103–112.
12. Roesel, M.; Tappeiner, C.; Heiligenhaus, A.; Heinz, C. Oral voclosporin: novel calcineurin inhibitor for treatment of noninfectious uveitis. *Clin. Ophthalmol.* **2011**, *5*, 1309–13013.
13. Kuglstatter, A.; Mueller, F.; Kusznir, E.; Gsell, B.; Stihle, M.; Thoma, R.; Benz, J.; Aspeslet, L.; Freitag, D.; Hennig, M. Structural basis for the cyclophilin A binding affinity and immunosuppressive potency of E-ISA247 (voclosporin). *Acta Cryst.* **2011**, *D67*, 119–123.
14. Burdmann, E.; Andoh, T.; Yu, L.; Bennett, W. Cyclosporine nephrotoxicity. *Semin. Nephrol.* **2003**, *23*, 465–476.
15. Birsan, T.; Dambrin, C.; Freitag, D.; Yatscoff, R.; Morris, R. The novel calcineurin inhibitor ISA247: a more potent immunosuppressant than cyclosporine in vitro. *Transpl. Int.* **2005**, *17*, 767–771.

16. Kallen, J.; Mikol, V.; Taylor, P.; Walkinshaw, M. X-ray structures and analysis of 11 cyclosporin derivatives complexed with cyclophilin A. *J. Mol. Biol.* **1998**, *283*, 435–449.
17. Gottschalk, S.; Cummins, C.; Leibfritz, D.; Christians, U.; Benet, L.; Serkova, N. Age and sex differences in the effects of the immunosuppressants cyclosporine, sirolimus and everolimus on rat brain metabolism. *Neurotoxicology.* **2011**, *31*, 50–57.
18. Taylor, P.; Husi, H.; Kontopidis, G.; Walkinshaw, M. Structural basis for the cyclophilin A binding affinity and immunosuppressive potency of E-ISA247 (voclosporin). *Prog. Biophys. Mol. Biol.* **1997**, *67*, 155–181.
19. Naicker, S.; Yatscoff, R.; Foster, R. Deuterated Cyclosporine Analogs and their Use as Immunomodulating Agents. US Patent 06605593 B1 (**2003**).
20. Hegmans, A.; Fenske, B.; Trepanier, J. Cyclosporine Analogue Molecules Modified at Amino Acid 1 and 3. US Patent 2016/0207961 A1 (**2016**).
21. FDA Approved Products: LUPKYNIS (voclosporin) capsules, for oral use. https://d1io3yog0oux5.cloudfront.net/auriniapharma/files/pages/lupkynis-prescribing-information/FPI-0011+Approved+USPI++MG.pdf; **2021**.
22. EMA Assessment Report: Luveniq (voclosporin) oral capsules. https://www.ema.europa.eu/en/documents/withdrawal-report/withdrawal-assessment-report-luveniq_en.pdf; **2021**.
23. https://go.drugbank.com/drugs/DB00091.
24. Ling, S.; Huizinga, R.; Mayo, P.; Larouche, R.; Freitag, D.; Aspeslet, L.; Foster, R. Cytochrome P450 3A and P-glycoprotein drug–drug interactions with voclosporin. *Br. J. Clin. Pharmacol.* **2013**, *77*, 1039–1050.
25. Mayo, P.; Huizinga, R.; Ling, S.; Freitag, D.; Aspeslet, L.; Foster, R. Voclosporin food effect and single oral ascending dose pharmacokinetic and pharmacodynamic studies in healthy human subjects. *J. Clin. Pharmacol.* **2013**, *53*, 819–826.
26. Ling, S.; Huizinga, R.; Mayo, P.; Freitag, D.; Aspeslet L.; Foster, R. Pharmacokinetics of voclosporin in renal impairment and hepatic impairment. *J. Clin. Pharmacol.* **2013**, *53*, 1303–1312.
27. Maksymowych, W.; Jhangri, G.; Aspeslet, L.; Abel, M.; Trepanier, D.; Naicker, S.; Freitag, D.; Cooper, B.; Foster, R.; Yatscoff, R. Amelioration of accelerated collagen induced arthritis by a novel calcineurin inhibitor, ISA_{TX}247. *J. Rheumatol.* **2002**, *29*, 1646–1652.
28. Gregory, C.; Kyles, A.; Bernsteen, L.; Wagner, G.; Tarantal, A.; Christe, K.; Brignolo, L.; Spinner, A.; Griffey, S.; Paniagua, R.; Hubble, R.; Borie, D.; Morris, R. Compared with cyclosporine, ISA_{TX}247 significantly prolongs renal-allograft survival in a nonhuman primate model. *Transplantation* **2004**, *78*, 681–685.
29. Busquea, S.; Cantarovich, M.; Mulgaonkar, S.; Gaston, R.; Gaber, A.; Mayo, P.; Ling, S.; Huizinga, R.; Meier-Kriesche H. The promise study: a phase 2b multicenter study of voclosporin (ISA247) versus tacrolimus in *De Novo* kidney transplantation. *Am. J. Transplant.* **2011**, *11*, 2675–2684.
30. Rovin, B.; Teng, K.; Ginzler, E.; Arriens, C.; Caster, D.; Romero-Diaz, J.; Gibson, K.; Kaplan, J.; Lisk, L.; Navarra, S.; Parikh, S.; Randhawa, S.;

Solomons, N.; Huizinga, R. Efficacy and safety of voclosporin versus placebo for lupus nephritis (AURORA 1): a double-blind, randomised, multicentre, placebo-controlled, phase 3 trial. *Lancet* **2021**, *397*, 2070–2080.

31. Gelder, T.; Huizinga, R.; Lisk, L.; Solomons, N. Voclosporin: a novel calcineurin inhibitor with no impact on mycophenolic acid levels in patients with SLE. *Nephrol. Dial Transpl.* **2021**, 1–6.

32. Survase, S.; Kagliwal, L.; Annapure, U.; Singhal, R. Cyclosporin A — a review on fermentative production, downstream processing and pharmacological applications. *Biotechnol. Adv.* **2011**, *29*, 418–435.

33. Maurice, R.; Wenger, M. Synthesis of cyclosporine and analogues: structural requirements for immunosuppressive activity. *Angew. Chem. Int. Ed.* **1985**, *24*, 77–138.

34. Angell, Y.; Thomas, T.; Flentke, G.; Rich, D. Solid-phase synthesis of cyclosporine peptides. *J. Am. Chem. Soc.* **1995**, *117*, 7279–7280.

35. Ko, S.; Wenger, R.; Solid-phase total synthesis of cyclosporine analogues. *Helv. Chim. Acta* **1997**, *80*, 695–705.

36. Naicker, A.; Yarcoff, W.; Foster, T. Cyclosporine Aanalogue Mixtures and their Use as Immunomodulating agents. US Patent 9765119B2 (**2017**).

37. https://ir.auriniapharma.com/press-releases/detail/207/aurinia-and-lonza-announce-exclusive-agreement-for.

18

Computer-Aided Drug Design

Jinxia N. Deng

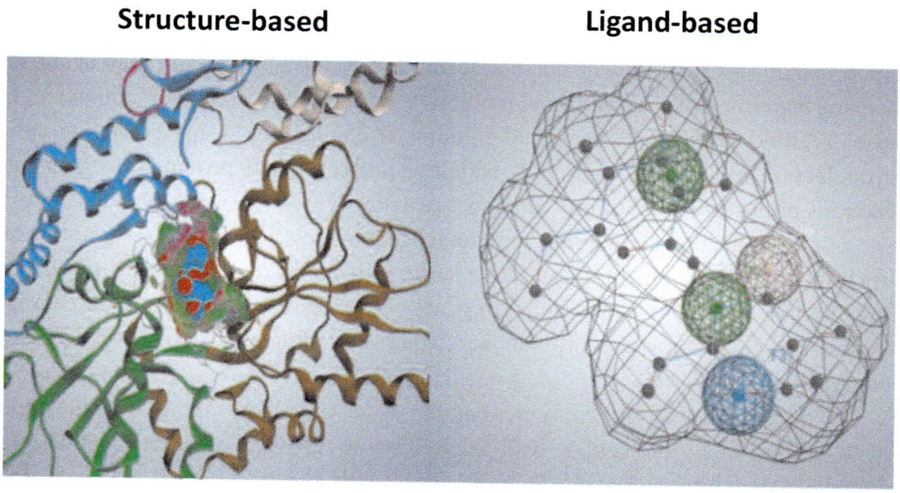

18.1 Background

The pharmaceutical industry is operating in an era where development costs are under pressure, higher percentages of drugs are demanded, and the drug-discovery process is a trial-and-error run. The profit that flows in with the discovery of new drugs has always been the motivation for the industry to keep up the pace and keep abreast with the endless demand for medicines. The process of finding a molecule that binds to the target protein using in silico tools has made computational chemistry a valuable tool in drug discovery in both academic research and pharmaceutical industry.[1]

Current Drug Synthesis, First Edition. Edited by Jie Jack Li.

The traditional drug-discovery process is expensive and can take up to 12–16 years of exhaustive research with huge financial investment before a molecule can be recognized as a drug. One of the major factors contributing to the high attrition rates is an active compound with unacceptable properties due to absorption, distribution, metabolism, excretion, and toxicity (ADMET) adverse effects. This represents approximately 50% of all costly failures in drug development and thus should be considered as early as possible in the drug–discovery process. It is evident that novel approaches to truncate the R&D time and reduce the cost involved is necessary.[1,2]

With the rapid development of both computer hardware, software, and algorithms, drug screening and design have benefited much from various computational methods, which greatly reduce the time and cost of drug development. For example, bioinformatics can help reveal the key genes from a massive amount of genomic data, thus provide possible target proteins for drug screening and design.[3] As a supplement to experiments, protein structure prediction methods, can provide protein structures with reasonable precision.[4] Most recent progress can be exemplified by the Alphafold on improved protein structure prediction from DeepMind, which can predict the 3D structure very close to experimental observation with notable accuracy just based on amino acid sequences by machine learning (ML) and artificial intelligence techniques.[5,6] Another similar example is RoseTTA fold that enables the rapid solution of challenging X-ray crystallography and cryo–electron microscopy structure modeling problems and provides insights into the functions of proteins of currently unknown structures.[7]

Biomolecular simulations with multiscale models allow for investigations of both structural and thermodynamic features of target proteins on different levels for identifying drug binding sites and elucidating drug action mechanisms.[8,9] Virtual screening then searches chemical libraries to provide possible drug candidates based on drug binding sites on target proteins.[10,11]

In addition to virtual screening, *de novo* drug design methods that generate synthesizable small molecules with high binding affinity, provide another type of computer-aided drug design (CADD) direction. Artificial intelligence (AI), e.g., ML and deep learning, is playing more and more important roles in the computational methods and thus drug development.[12–15] In brief, the many roles of computation in drug discovery have been fully appreciated in nowadays[16–18] and thus the sucessful applications have been evidented in many disease area, such as cancer drug developmet, anti-bacterial, and anti viral drugs, etc.[19,20] incuding SARS-CoV-2 novel drug deveopment.[21–24]

Usually, the CADD integrates ligand- and structure- based drug design strategies (LBDD and SBDD, respectively) with a combination of experimental techniques. Broadly used SBDD approaches, such as molecular docking, homology modeling, molecular dynamics (MD) and structure-based virtual screening, have provided relevant insights into ligand–receptor interactions.[25] Equally important, LBDD methods such as pharmacophore modeling, quantitative structure–activity relationships (QSAR), and ligand-based virtual screening have been actively used to explore small-molecule databases and produce correlations between chemical features and pharmacological activity.[26] Also, a hot-topic in LBDD, quantitative structure–property relationship (QSPR) models are central for predicting pharmacokinetics (PK) and toxicity-related characteristics.[25,27] In this chapter, we will present QSAR as a representative techinique for LBDD, while place MD simulation, molecular docking, and pharmacophore models in the section of SBDD, though pharmacophore can be developed from either SBDD or LBDD.

18.2 Structure-Based Drug Design (SBDD)

The availability of the three-dimensional structure of the therapeutic proteins and knowledge of the binding pocket lay the foundation of the SBDD.[28,29] This approach is specific and effectively fast in the identification of lead molecules and their optimization which has helped to understand disease at a molecular level.[22,30] Some of the common methods employed in SBDD as afore mentioned include structure-based virtual screening, molecular docking, and MD simulations etc. These methods find numerous applications such as assessment of binding energetics, protein–ligand interactions, and conformational changes in the receptor upon binding with a ligand. Being used by many pharmaceutical industries and medicinal chemists, SBDD has greatly helped the discovery of some drugs available in the market.[24] For example, the discovery of amprenavir as a potential inhibitor of the human immunodeficiency virus (HIV) protease using protein modeling and MD simulations,[31,32] thymidylate synthase inhibitor, raltitrexed against HIV using SBDD approach.[31] Besides, identification of topoisomerase II and IV inhibitor, norfloxacin, an antibiotic commonly used against urinary tract infection involved SBDD,[28] and the discovery of dorzolamide, a carbonic anhydrase inhibitor used against glaucoma, cystoid macular oedema using fragment-based screening,[33] as well as flurbiprofen, a nonsteroidal anti-inflammatory drug (NSAID) used against rheumatoid arthritis, osteoarthritis targeting cyclooxygenase-2 (COX-2) were discovered through molecular docking combined with other computational approaches.[34,35]

The basic steps involved in SBDD consist of the preparation of target structure, identification of the ligand binding site, compound library preparation, molecular docking and scoring functions, molecular dynamic simulation, and binding free energy calculation.[24]

18.2.1 Molecular Docking

Molecular docking has been consolidated as one of the most important methods in the molecular modeling field and thus been widely used ever since the early 1980s.[36] It has been recognized as a prominent tool in the study of protein–ligand complexes, to describe intermolecular interactions to predict poses of multiple ligands, to discover novel promising bioactive compounds.[37] Many comprehensive reviews on docking have been published in the past[38–45] and a fair amount of comparison studies were conducted to evaluate the relative performance of the programs.[46–51]

Molecular docking can be involved in multiple areas of the modern drug hunting process. It can be employed to help rationalizing ligands activity towards a target of interest and to perform structure-based virtual screening campaigns. It can also be used to identify series of targets for which the ligands present good complementarity (target fishing and profiling), some of them being potentially responsible for unexpected drug adverse reactions (off-targets prediction). Moreover, docking is also currently employed for the identification of ligands that simultaneously bind to a pool of selected targets of interest (poly-pharmacology) and for identifying novel uses for chemical compounds with already optimized safety profiles (drug repositioning).[42]

Theoretically, molecular docking is composed mainly of two stages: first is to sample conformations of the ligand in the active site of the protein; then to rank these conformations by a scoring function. Ideally, sampling algorithms should be able to reproduce the experimental binding mode and the scoring function should also rank it highest among all generated conformations. The sampling process should effectively search the conformational space described by the free energy landscape, where energy, in docking, is approximated by the scoring function. The scoring function should be able to associate the native bound conformation to the global minimum of the energy hypersurface.[38,39] Thus, from these two perspectives, a brief background of the basic docking theory is described herein.

18.2.1.1 Sampling Algorithms

Due to the translational and rotational freedom of both ligands as well as receptor, there are a huge number of possible binding modes between two molecules. However, it is expensive to computationally generate all the possible conformations. Various sampling algorithms have been published and widely used in molecular docking software. The first molecular docking algorithm was developed in the 1980s.[36] The receptor was approximated by a set of spheres filling its surface clefts, and the ligand by another set of spheres defining its volume. This is considered as rigid receptor docking and the search was made to find the best steric overlap between binding site and receptor spheres, neglecting any kind of conformational movement.[38,39]

In molecular biology, there are two major process where docking can be helpful, ligand–protein docking and protein–protein docking. Based on the level of flexibility, the sampling process can be classified into three groups, i.e., rigid docking, semi-flexible docking, and flexible docking.

Rigid Docking

Both ligand and protein are considered rigid objects, and just the three translational and three rotational degrees of freedom are considered during sampling. This approximation is analogous to the "lock–key" binding model and is still used commonly for protein-protein docking. Briefly, in these methods, the binding site and the ligand are approximated by "hot" points and the superposition of matching point is evaluated.[52]

Semi–flexible Docking

During this process, only the ligand is flexible while the protein is rigid. Thus, the conformations of the ligand are sampled and fit into the pocket of the fixed conformation of a protein. Numerous docking algorithms have been developed since the 1980s for the purpose. It is difficult to clearly classify each docking software because different algorithms may be integrated into a multiphase approach.[39,44–53] However, docking algorithms can be classified into systematic search techniques, such as that in Glide software,[54–56] and stochastic methods that potentially find optimal solution very fast, such as Monte Carlo sampling used in many docking programs, including AutoDock Vina,[57] AutoDock,[58] etc. Other well–known stochastic search is evolutionary algorithms which has been well known through GOLD docking software.[38,39,59,60] In MOE Dock, it also provides option of lowModeMD to sample the conformational space of ligands.[61]

Flexible Docking

The concept of this docking procedure is to consider the dynamic nature of both a protein and a ligand being flexible and being sampled the conformations. Over the years, lots of methods have been published, and some focused on induced fit docking and others on conformational selection. These methods take advantage of different degrees of approximation and can be divided into approaches that consider single protein or multiple protein conformations.[62]

The single conformational flexibility consideration was so often referred to as "soft docking" approach, which consists of an implicit and rough treatment of protein flexibility.[63] Several years later, side chain flexibility was introduced as a new strategy by exploiting rotamer libraries.[64] Some docking methods, such as GOLD, sample certain degrees of freedom within their own search engine. Considering side chain flexibility, huge conformational variations of the protein are neglected by these methods.

Multiple protein conformations provide more description of the dynamic nature of the receptor protein. Sometimes multiple experimental structures may be available for the same target, such as nuclear magnetic resonance (NMR) structures, or multiple X-ray structures. Computationally, an ensemble of protein conformations can be obtained *via* multiple simulation techniques, such as Monte Carlo or MD simulations. The purpose of multiple protein conformations docking is to consider all the diverse structures.

Several possible strategies have been summarized recently: One is to construct a single average grid from the ensemble of the structures, and the single average-grid can be either a simple or weighted average combination of them.[65] Another one is to create a united description of the protein. In that case, the structures do not collapse into an average grid but are used to construct the best performing "chimera" protein. This can be exemplified by the program FlexE.[66] The more common practice is using individual conformations. That is to say, the structures of the ensemble are considered as conformations that can possibly be bound by the ligand, so multiple docking runs are performed, evaluating the ligands of interest on all the target conformations.[39,53,67]

18.2.1.2 Scoring Functions

The evaluation and ranking of predicted ligand conformations are a crucial aspect of docking. It serves to not only provide the prediction of the docked orientation most accurately represents the "true" structure in the intermolecular complex, but also to rank one ligand relative to another. Therefore, the design of reliable scoring functions and schemes that can rank different poses is of fundamental importance.[44]

The scoring functions usually estimate binding energy of a complex by certain level of assumptions and simplifications to achieve as close as possible to the actual binding energy in the minimum time. Popular scoring functions have an adequate balance between accurate estimation of binding energy and computational cost in terms of time. There have been many scoring functions developed over the past many years and can be classified into three main categories—force field, empirical, and knowledge based.[68]

<u>Force Field Functions</u>

Force field scoring functions are developed based on physical atomic interactions like van der Waals interactions, electrostatic interactions and bond lengths, bond angles, and torsions. Force field functions and parameters are usually derived from both experimental data and *ab initio* quantum mechanical calculations according to the principles of physics.[69]

$$E = \sum_i \sum_j \left(\frac{A_{ij}}{r_{ij}^{12}} - \frac{B_{ij}}{r_{ij}^{12}} + \frac{q_i q_j}{\epsilon(r_{ij}) r_{ij}}\right) \tag{18.1}$$

Here, r_{ij} stands for the distance between protein atom i and ligand atom j, A_{ij} and B_{ij} are the van der Waal parameters, q_i and q_j are the atomic charges, and $\epsilon(r_{ij})$ is the distance-dependent dielectric constant.[38,70]

<u>Empirical Scoring Functions</u>

The basis of this scoring function is that the binding energies of a complex can be approximated by a sum of individual uncorrelated terms. The coefficients of the various terms involved in calculation of binding energy are obtained from regression analysis using experimentally determined binding energies or potentially from X-ray structural information. Empirical functions have simpler energy terms to evaluate when compared with force field scoring functions and thus are much faster in binding score calculations.

The first empirical scoring function developed to predict binding free energies was implemented in LUDI, credited to the pioneering work of Bohm. The energy was derived using experimental binding free energies and multiple protein–ligand crystal complexes.

$$\Delta G_{bind} = \Delta G_0 + \Delta G_{hb} \sum_{h-bond} f(\Delta R, \Delta\alpha) + \Delta G_{ionic} \sum_{ionic\ int.} f(\Delta R, \Delta\alpha) + \Delta G_{lipo}|A_{lipo}| + \Delta G_{rot} N_{rot} \quad (18.2)$$

Here, ΔG_0 is the binding energy independent of protein interactions, ΔG_{hb} describes contribution to binding energy from hydrogen bonds, ΔG_{ionic} denotes contribution to binding energy from unperturbed ionic interactions, ΔG_{lipo} considers contribution to binding energy through lipophilic interactions, while A_{lipo} is the lipophilic contact surface between the protein and the ligand, ΔG_{rot} describes the loss of binding energy due to freezing of internal degrees of freedom in the ligand while N_{rot} represents the number of rotatable bonds and $f(\Delta R, \Delta\alpha)$ is a penalty function that accounts for large deviations from ideal hydrogen bond and salt bridge geometry. As shown in Eq. (18.2), the binding free energy is modeled using hydrogen bonds, salt bridges, the hydrophobic effect, and solute entropy terms.[70,71]

Knowledge-Based Scoring Functions

Knowledge-based scoring functions have recently emerged as an alternative and very promising way of ranking protein-ligand complexes with known 3D structure according to their binding affinities. They are derived from the structural information embedded in experimentally determined atomic structures. The functions use statistical analysis on crystal structures of complexes to obtain the interatomic contact frequencies between the protein and the ligand based on the presumption that the stronger an interaction is, the greater the frequency of its occurrence will be.

The overall score is calculated by Eq. (18.3) to account for favorable contacts and repulsive interactions between each atom in the ligand and protein lying within a sphere with a specified cutoff.

$$\omega(r) = -k_B T \ln[g(r)] \quad (18.3a)$$

$$g(r) = (r)\rho(r)/\rho^*(r) \quad (18.3b)$$

Here, k_B is the Boltzmann constant, T is the absolute temperature of the system, $\rho(r)$ is the number density of the protein–ligand atom at distance r, $\rho^*(r)$ is the pair density in the reference state where interatomic interactions are zero, and $g(r)$ is pair distribution function.[72–74]

Table 1 lists the available software packages commonly used in various research groups and pharmaceutical industries. It summarizes their algorithm characteristics and applications. In these applications, for example, freely available software mainly include DOCK, AutoDock, AutoDOCKVina, 3D-DOCK, LeDock, rDock, UCSF DOCK, Surflex (for academic users), and HEX; while commercial software consists mostly of Glide, GOLD, MOE Dock, ICM-Dock, MCDOCK, Surflex-Dock, LigandFit, FlexX, etc.[45,46]

Table 1. Selected molecular docking programs, their algorithm characteristics, and applications.

Program and developing group	Algorithm characteristics	Typical applications
Autodoc, Scripps Research, https://ccsb.scripps.edu/autodock	Lamarck genetic algorithm and experience-based scoring function; the flexibilities of small molecules and some residue side chains can be fully taken into consideration[75]	Protein–small molecule
AutoDock Vina, Scripps Research, http://vina.scripps.edu/	The upgraded version of AutoDock; the success rate and calculation speed are greatly improved compared to AutoDock; simple parameter setting, easy to use, and parallel operation on multi-core machines for docking flexible ligands and flexible protein side chains[57]	Protein–small molecule
DOCK, Kuntz group at UCSF, http://dock.compbio.ucsf.edu	Step-by-step geometric matching strategy; AMBER force field experience-based scoring function. As a kind of commonly used molecular docking software, it can be used for docking between flexible small-molecule ligands and flexible proteins[76]	Protein–small molecule
FlexX, BioSolveIT, https://www.biosolveit.de	The best conformation is selected according to the size of the docking free energy, which has the advantages of fast speed, high efficiency, and easy operation. It is the representative software of the flexible docking and can also be used for the virtual screening of small molecule database[66]	Protein–small molecule

Glide, Schrodinger, https://www.schrodinger.com/products/glide	Docking program based on search algorithms, including the modes of extra precision (XP), standard precision (SP), and a high-throughput virtual filter. It is mainly used for the flexible docking of small-molecule ligands and proteins[54,55]	Protein–small molecule
PIPER, Schrodinger, https://www.schrodinger.com/products/piper	Fast Fourier Transform (FFT) search algorithm; the knowledge-based atomic statistical potential scoring function and applied to the ClusPro Server[77]	Protein–protein
MOE, Chemical computing group (CCG), https://www.chemcomp.com/	A comprehensive software system for the pharmaceutical and life science, which could fully support drug design and research through molecular simulation, protein structure analysis, small molecule database processing and protein and small molecule docking research in a unified operating environment[78]	Protein–protein/small molecule
ICM-Dock, Molsoft, http://www.molsoft.com/docking.html	User-friendly interactive image display, and the software also supports fast and accurate docking optimization[79]	Protein–protein/polypeptide/ small molecule
HADDOCK, Bonvin Lab, https://www.bonvinlab.org/software/haddock2.2/	Docking program based on experimental data (such as NMR chemical shifts and point mutations), which was invented from protein–protein docking and can also be used for protein–ligand docking[80]	Protein–protein/DNA/ RNA/ small molecule
RosettaDock, https://www.rosettacommons.org/software/servers	Monte Carlo search algorithm; the experience-based energy scoring function[81]	Protein–protein/DNA/ RNA/ small molecule

Source: modified from Reference[45].

18.2.2 Pharmacophore Model

Pharmacophore modeling is a successful yet very diverse subtopic of CADD. The concept of the pharmacophore has been widely applied to the rational novel drug design. In this section, we review the computational implementation of this concept and its common usage in the drug discovery process. The most common application of pharmacophores is virtual screening, and different strategies are possible depending on the prior knowledge. Besides, the pharmacophore concept is also useful for ADMET modeling, side effect, and off-target prediction as well as target identification. Pharmacophores are often combined with molecular docking simulations to improve the speed and efficiency of virtual screening.[82]

The original concept of the pharmacophore was developed during the late 1800s.[83] At that time, the understanding was that certain "chemical groups" or functions in a molecule were responsible for a biological effect, and molecules with similar effect had similar functions in common. The word pharmacophore was coined much later as "a molecular framework that carries (phoros) the essential features responsible for a drug's (pharmacon) biological activity." by Schueler in his book.[84] The definition of a pharmacophore was therefore no longer concerned with "chemical groups" but "patterns of abstract features."

Since 1997, the term of pharmacophore has been defined by the International Union of Pure and Applied Chemistry as: the ensemble of steric and electronic features that is necessary to ensure the optimal supramolecular interactions with a specific biological target and to trigger (or block) its biological response.[85]

Thus, pharmacophore model should be considered as the largest common denominator of the molecular interaction features shared by a set of active molecules. It does not represent a real molecule or a set of chemical groups but is an abstract concept.[82] Some recent review not only captured the successful applications of the pharmacophore modeling in current drug design, but also extend the combination of pharmacophore with other modeling techniques, such as MD simulations to investigate the dynamics during ligand and protein binding. Besides, other trends include the efficient usage of 3D pharmacophore information in ML and AI applications or freely accessible web servers for 3D pharmacophore.[86]

Herein we use HIV integrase inhibitor design to describe the concept of pharmacophore model development and its applications in novel inhibitor design.

HIV encodes three enzymes: reverse transcriptase, integrase, and protease. Integrase has been an attractive drug target due to rapid and sensitive assays exist for ascertaining enzymatic activity, and crystal and NMR structures are available for use in rational SBDD. However, integrase does not have a welled fined, buried binding pocket, but a shallow solvent-exposed one instead. In addition, it exists in a multimeric state in pre-integration complexes. These post difficulties for structure-based drug discovery.

Several approaches to develop pharmacophore models have been published, some are ligand-based,[87–89] and some are receptor-based,[90–92] leading to a series of novel categories of inhibitors have been discovered in the past. The dynamic pharmacophore method considers multiple receptor conformations that are used to generate a receptor-based pharmacophore model which is then used to search available compound databases. The protein conformations were collected from MD trajectory. The complex used for the study has an inhibitor (1QS4; 5-CITEP) in the active site of the catalytic domain.[93] The inhibitor was found to associate with an essential residue, E152, by donating a hydrogen-bond. In addition, one Mg^{2+} ion was resolved in the complex chelated with D64 and D116. The subsequent pharmacophore model development which included the Mg^{2+}, D64, D116, and E152 in the active site, as well as the nearby residues K156 and K159. The models were applied to virtual library screening, and some potent and novel chemicals were mapped out for further optimization in the HIV integrase inhibitor design. Figure 1 shows the concept of a receptor-based dynamic pharmacophore model development based on the multiple HIV integrase core domain X-ray structures. The pharmacophore model was applied to screen Enamine database and some of the hits are interesting for further optimization.

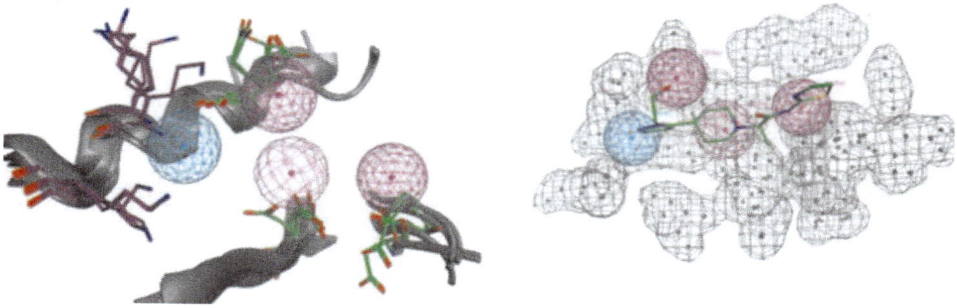

(A) (B)

Figure 1. Concept of dynamic pharmacophore model considering of flexibility of the side chains from multiple X-ray structures, and its application to map out novel inhibitor from virtual library screening. (A) Multiple HIV integrase X-ray structures superposed against 1QS4. The complementary feature is defined as pharmacophore feature. Pink is H-bond donor to complement with DDE motifs, and the light blue is H-bond acceptor to interact with K159, K156. Stick model is the side chain of the key amino acid residues in the HIV integrase. (B) One compound (stick model) selected to demonstrate the application of the pharmacophore model against Enamine database screening. The gray mesh area is so called excluded volume as part of the pharmacophore model representing the space of the integrase.

18.2.3 Molecular Dynamics (MD) Simulations

MD simulation is an established computational technique to understand the protein structure–function relationship and guide the drug design. Since the first case of MD on bovine pancreatic trypsin inhibitor in the late 1970s,[94] MD and its applications have been extended successfully in many areas. Especially in recent years, benefit from the development of modern graphics processing units (GPUs) hardware and several force fields, MD simulation has made huge impact onto life science research.[95,96] Some great progress can be exemplified by understanding the dynamics of protein conformation which is hardly available by current experimental techniques, such as the folding and aggregation of amyloid-related proteins,[97,98] and conformational transition caused by mutations, temperature, and pH values.[95–101] MD simulation plays an important role in characterizing receptor–ligand interaction (protein–protein, protein–DNA/RNA, protein–small molecule).[102–104] Besides, it helps to reveal the novel binding sites which has not been captured by NMR or X-ray crystallographic analysis. For example, novel binding sites in HIV-1 integrase,[105] cruzain,[106] Ras,[107] and five allosteric sites of human β1(β1AR) and β2 (β2AR) adrenergic receptor[108] were served as new drug targets. In addition, MD simulation provides multiple typical conformations for virtual screening to obtain more reasonable ligand binding modes.[90–92] And it can optimize the position between protein and ligand with estimation of the average binding free energy, providing the guidance of SBDD.[9,109] Plus, MD simulation have been successfully carried out in many large systems, such as the complete HIV1 capsid which is composed of 64 million atoms up to 100 ns,[110] satellite tobacco mosaic capsid of one million atoms up to 50 ns.[111] For some systems, MD simulations have been reported to study protein folding and function regulation with a simulation time of 10–100 ns.[112,113]

Recently, we published an enhanced sampling MD simulation using well-tempered meta-dynamics to estimate the conformational free energy surfaces for the 5-HT_{2B} serotonin receptor in the unliganded (APO system), lysergic acid diethylamide (LSD)-bound (LSD system), and lisuride-bound (LIS system) states. 5-HT_{2B} receptor is one of the G–protein coupled receptors (GPCRs) which constitutes the therapeutic targets of approximately 34% of Food and Drug Administration (FDA)-approved drugs with an annual market of over US $150 billion.[114]

Experimental crystal structures and structure-directed mutagenesis experiments conducted on 5-HT_{2B} showed that the differential binding poses of LSD, a prototypical agonist, and LIS, a prototypical antagonist that differs from LSD only in its stereochemistry and two extra atoms in a NH group, were responsible for the different pharmacological activity.[115] LSD, commonly known as the psychedelic drug "acid" is thought to mediate hallucinogenic effect. Therapeutic interest in LSD have emerged from its potential in treating alcoholism, depression, and anxiety in the terminally ill.[116] 5-HT_{2B} and LIS are also pharmacologically interesting for their role in cardiovascular health.[117]

The MD simulations revealed that the binding of LSD induces a large perturbation to the conformational free energy landscape of the receptor within the ligand–receptor complex to the degree that the structural ensembles of the apo and LSD-bound states become effectively disjoint (Figure 2). LSD binding shifts the global minimum of the ligand–receptor complex free energy landscape to the active state and induces a thermodynamic driving force for structural activation of ΔG is around –110 kJ/mol. We also observe the presence of a metastable inactive-like LSD-bound structure that lies ΔG is close to 25 kJ/mol higher in free energy. LIS, on the other hand, is a prototypical antagonist and its binding induces a relatively smaller perturbation of the conformational free energy landscape. The structural ensembles for the apo and LIS-bound states show a high degree of similarity, and the global minimum of the LIS-bound free energy landscape shows close structural similarity with the inactive apo form.

This research reveals that the structural conformations adopted by long 1.5 μs unbiased MD simulations do not correspond to the most stable structures identified in our enhanced sampling calculations, revealing the value of accelerated sampling, and cautioning against over interpreting unbiased molecular simulations due to the long relaxation time scales associated with these systems. From that study, the results quantify the driving forces for activation of 5-HT_{2B} by LSD binding and demonstrate the absence of this driving force under LIS binding, thereby shedding light on the molecular-level structural and thermodynamic basis for ligand-induced conformational specificity and functional selectivity. Thus, the understanding of the action of mechanism establishes a framework for high-throughput virtual screening of putative 5-HT_{2B} ligands, and a principled platform for drug design through the rational engineering of ligand-bound free energy landscapes.[119]

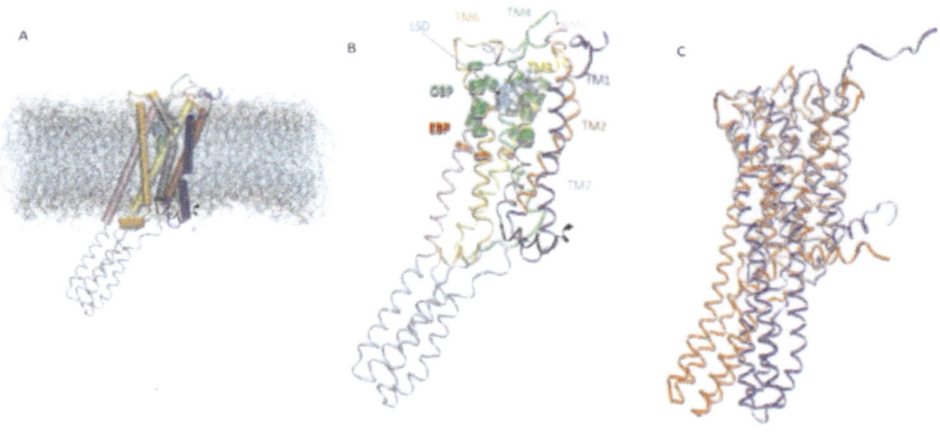

Figure 2. Molecular renderings of 5HT_{2B}-transmembering (TM) helices in GPCR. (A) A snapshot of 5HT_{2B}-TM in a membrane-bound state from LSD-bound crystal structure (PDB: 5TVN).[118]. Each helix is represented as a cylinder. (B) Structure of the LSD-bound complex, where the transmembrane helices TM1-7 are labeled by color, and the LSD molecule within the binding pocket represented as van der Waals space-filling spheres. (C) Superposition of aligned active (red) and inactive (blue) states of 5HT_{2B}-TM. The active state is distinguished by a large conformational rearrangement of transmembrane helices TM5-7 revealed by MD.

3 Ligand-based Drug Design (LBDD)

LBDD, also known as indirect drug design, depends on the information of diverse molecules that bind to the biological target of interests. Thus, LBDD approach is useful in the absence of an experimental 3D structure and the known ligands that bind to the drug target are studied to understand the structural and physio-chemical properties of the molecules that correlate with the desired pharmacological activity of those ligands. The most popular approaches for LBDD are QSAR method and pharmacophore modeling.[120, 121]

More than five–decade has passed since the QSAR modeling was initiated by Corwin Hansch et al.[115] Throughout its history, it has drawn both praise and criticism concerning its reliability, limitations, successes, and failures. In this section, we will start the fundamentals for descriptors and methods on the development of QSAR models.[122]

In fact, the analysis of Google Books (Figure 3) indicates that the continuing growth of chemical data and databases especially in the public domain has stimulated the concurrent growth in QSAR publications. QSAR modeling is widely practiced in academy, industry, and government institutions around the world.[122]

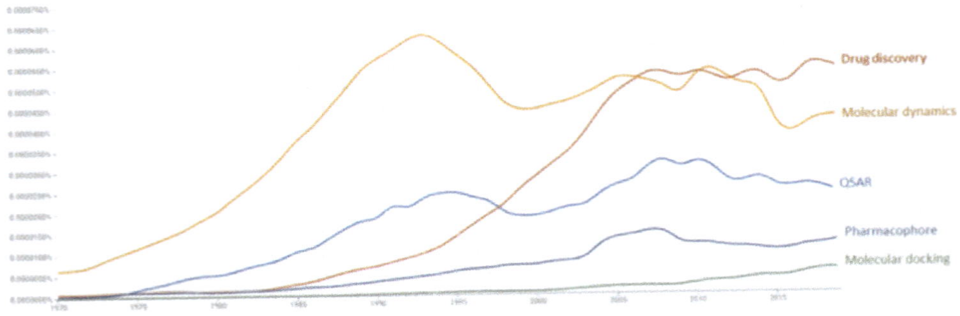

Figure 3. The growth of QSAR modeling is caused by the growth of experimental data. Chart is generated by Google Ngram Viewer (http://books.google.com/ngrams): *Y*-axis, percentage among all books in the Google Ngram database; *X*-axis, years from 1975 till 2019, the most recent year available for search at this time of written.

The goal of carrying out a QSAR study is with the help of computational methods to analyze biological data, virtual library screening or predict in silico-designed compounds aiming to reduce failure rate in the drug discovery process.[123,124]

3.1 Molecular Descriptors

Chemical descriptors are at the core of QSAR modeling, and so many different types of chemical descriptors reflecting various levels of chemical structure representation have been proposed so far. These range from molecular formula (so-called 1D) to the two-dimensional structural formula (2D), which is most popular among medicinal chemists, then to the three–dimensional which is conformation-dependent (3D) and even higher levels descriptors considering mutual orientation and time dependent dynamics of molecules (4D and higher).[124–127]

3.1.1 1D Descriptors for QSAR Modeling (1D QSAR)

Various parameters are used to select the potential molecular descriptor that defines the specific molecular properties such as electronic constraints, hydrophobic constraints, and steric constraints.

3.1.2 2D Descriptors for QSAR Modeling (2D QSAR)

The two-dimensional representation of a molecule, commonly referred to as topological representation, defines the connectivity of atoms in the molecule in terms of the presence and nature of chemical bonds. Such two-dimensional representation enables the definition of so-called molecular 2D descriptors. The main advantages of these QSAR parameters are that they (i) contain simple and useful information about the molecular structure, (ii) are invariant to molecule roto-translation, and (iii) can be calculated avoiding structure optimization. In general, 2D descriptors include constitutional, topology, total polar surface area, electrostatic and quantum chemical, geometrical and molecular fingerprints property of the chemical compound.[124]

Based on Randić, a list of attributes for topological indices include that they should (1) have structural interpretation, (2) have good correlation with at least one property, (3) preferably discriminate among isomers, (4) be able to be applied to local structure, (5) be preferably independent, (6) be simple, (7) not be based on experimental properties, (8) not be trivially related to other descriptors, (9) be possible to construct efficiently, (10) use familiar structural concepts, (11) have the correct size dependence and (12) change gradually with gradual change in structures.[128] Most of the topological descriptors have the aformentioned characteristics, which is why they have been prolifically applied in characterizing the structural similarity/dissimilarity of molecules and in QSAR/QSPR modeling.[122]

3.1.3 3D Descriptors for QSAR Modeling (3D QSAR).

3D QSAR focused broadly on all such properties of atoms in a compound that are represented as descriptors corresponding to spatial representation of a molecule. Since in many cases, 3D information of receptor molecules was not known, and in these cases the indirect method of 3D-QSAR is mostly based on information of the ligand molecule such as molecular alignment of atoms, pharmacophores, volume, or fields to generate a virtual receptor environment.[124] The major objective of a 3D-QSAR study is to improve the activity of the lead compound by optimization and structural modification.[129]

The 3D descriptors can be calculated with (dependent) or without (independent) the alignment of bioactive conformers.[130]

Alignment–Dependent Descriptor Methods

There are several methods used in the calculation of 3D descriptors focusing on molecular alignment prior to the calculation of 3D descriptors. These methods calculate the descriptor by mapping receptor atoms or ligand atoms or complexes of receptor–ligand atoms. Various alignment–dependent descriptors are Comparative Molecular Field Analysis (CoMFA), Comparative Molecular Similarity Indices Analysis (CoMSIA), Genetically Evolved Receptor Modeling (GERM), Comparative Binding Energy Analysis (CoMBINE), Adaptation of the Fields for Molecular Comparison (AFMoC), Hint Interaction field analysis (HIFA), and Comparative Residue Interaction Analysis (CoRIA).[124,130,131]

Alignment–Independent Descriptor Methods

The conventional methods based on the alignment approach have many limitations like they are time consuming and can introduce user biasness and it may affect the sensitivity of the resultant model. To overcome all these limitations, a novel class of method has been adopted, which is independent of alignment and is not affected by radiation or transformation of the molecule. The different methods belonging to this category include Comparative Molecular Moment Analysis (CoMMA), COMPASS, Holo-QSAR (HQSAR), Weighted Holistic Invariant Molecular Descriptors (WHIM), Comparative Spectral Analysis (CoSA) and Grid–Independent Descriptors (GRIND).[130]

During the last two decades, new multidimensional descriptors have been incorporated into QSAR modeling, termed 4D to 6D descriptors. These are based on structural parameters associated with the flexibility of the receptor binding site along with the topology of the ligand. Specifically, the 5D descriptors are calculated from multiple conformations, orientations, protonation states, and stereoisomers of the ligand under analysis. In the case of the 6D descriptors, it is necessary to consider the solvation scenarios of the complex, the ligand, and the interacting environment.[124,132–134]

3.2 Machine Learning on QSAR and QSAR Methods

Application of artificial intelligence (AI) in drug discovery and development has emerged as a key promising pillar. Its importance has been consolidated by the need of new strategies to overcome the high failure rate of ~90% in drug development. As such, pharmaceutical companies are beginning to explore how various AI frameworks can be integrated into the current drug discovery and development processes.[135]

ML, a branch of AI (Figure 4), is "based on the idea that systems can learn from data, identify patterns, and make decisions with minimal human intervention."[136] AI frameworks may contain several different ML methods applied together. For example, an AI framework in drug discovery may optimize drug candidates through a combination of ML models that predict favorable physicochemical characteristics (e.g., solubility and permeability), PK, safety, and efficacy.[15,133,137–140]

Wild spectrum of data analysis methods has been used by working on various statistical correlations about dependent and independent variables in a correlation study, such as activity (y-variable) and molecular descriptors (x-variable).[130]

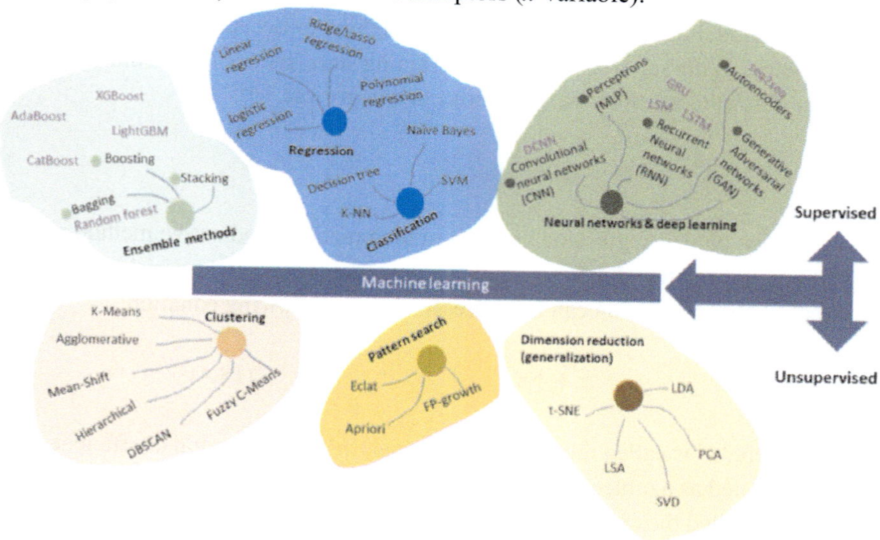

Figure 4. Overview of both supervised and unsupervised machine learning algorithms. AdaBoost: adaptive boosting; DBSCAN: density-based spatial clustering of applications with noise; DCNN: deep convolutional neural networks; Eclat: equivalence class transformation; FP-growth: frequent pattern growth; GRU, gated recurrent unit; K-NN: K-nearest neighbors; LDA: linear discriminant analysis; LightGBM: light gradient boosting machine; LSA: latent semantic analysis; LSM: liquid state machine; LSTM: long short-

term memory; MLP: multilayer perceptron; PCA: principal component analysis; seq2seq: sequence-to-sequence; SVD: singular value decomposition; SVM: support vector machine; t-SNE: t-distributed stochastic neighbor embedding; XGBoost: extreme gradient boosting[15]

Both supervised and unsupervised learning applications have been published in the realm of drug discovery, and the difference in between is the label of the data. Supervised learning relies on a labeled dataset, such as biological activity as a trainer to teach the model or the machine to eventually reach its prediction power for decision making. This can be used to do regression analysis or classification model development such as active vs. inactive prediction. Unsupervised learning identifies the associations or pattern recognition in unlabled data, such as clusting.[15] Lots of reviews summarize the principles as well as applications of unsupervised learning on drug discovery. In this chapter, we cover the supervised learning of ML on QSAR methods and model development.

As mentioned earlier, a supervised ML algorithm requires an input data set split into a "training" data set and a "test" or "validation" data set. The process of fitting (or "calibrating") the model to the training data set is called "model training." The "trained" ML model can then be validated having its predictive performance assessed using the test or validation data set. The "validated" ML model can then be applied to new data sets (i.e., not used for model development) to make predictions or decisions based on the new data set covariates. With supervised ML, the training data set contains both covariates and outcomes. The outcome may be continuous or categorical.[15,141]

3.2.1 Linear Regression Analyis (LRA)

This is the first type of regression analysis in the practical predicative method mostly used for prediction of relationship between dependent and independent variables (i.e., *x* and *y*). The simple linear regression can be expressed by the following equation:

$$y = a + bx \tag{18.4}$$

where '*a*' is called the intercept constant, '*b*' the regression coefficient, and "*x*" is depicted as a molecular descriptor, which is one or more than one in numbers and also called the explanatory variable, for example, molecular weight, or Log P etc., whereas "*y*" is called as the dependent variable, or bioactivity, and it mainly corresponds to activity in the QSAR study.[124]

3.2.2 Multiple Linear Regressions (MLR)

MLR is a widely used algorithm in supervised ML when a continuous outcome varies linearly with the independent variables or covariates. An outcome or dependent variable can be represented as Eq. (18.5):

$$y = \beta_0 + \beta_1 x_1 + \cdots + \beta_n x_n \quad (18.5)$$

where y is the outcome prediction, $\beta_{1\text{-}n}$ are the coefficients, and $x_1 - x_n$ are the covariates. Ordinary least squares based on the principle of maximum likelihood is the simplest and most common way of estimating the coefficients. This model is not designed to handle strong collinearities between covariates. Severe multicollinearity greatly reduces the precision of the estimated coefficients. In such cases, principal component analysis (PCA) may be used to reduce the number of features to a smaller set of unrelated components.

3.2.3 Logistic Regressions

Logistic regression is a widely used supervised ML method for modeling binary outcomes, such as yes/no, success/failure, and survived/died. The logistic regression model estimates the probability that the outcome belongs to a particular binary category. The probability curve (Eq. (18.6)) is sigmoid or S-shaped and constrained between 0 and 1.[15]

$$P_{(y=1)} = \frac{1}{1+e^{-(\beta_0+\beta_1 x_1+\cdots+\beta_n x_n)}} \quad (18.6)$$

3.2.4 Multivariate Data Analysis

The chemical data used in QSAR analysis are multidimensional in nature where features of a chemical compound are defined by many other data components. Thus, the multivariate techniques need specifically to reduce the multiple components within the data. Some commonly used are PCA, as previously mentioned, partial least square (PLS) analysis, genetic algorithms (GA), and genetic algorithms-partial least square (GA-PLS) analysis. The statistical analysis of these features is represented using the matrix that has row and column. Each of them represents the property and features of chemical compounds.[124,142]

Partial Least Square (PLS)

PLS is an improved QSAR model predication technique and became very frequently used technique in QSAR model analysis. It is also called the latent or projection structure method. Through this procedure, the large numbers of descriptors can be transformed into a small number of new orthogonal terms, called the latent variables. The numbers of latent variables are used to define the dependent variable. The main aim of this technique is to form a relationship between matrixes (features and property).[124]

Genetic Algorithm

The GA is inspired by the process of natural selection and belongs to the larger class of evolutionary algorithms. It works as following:

• First, the solution space is defined and corresponds to a set of models to test. The search for the best model is initialized by randomly creating an initial population, defined as a set of individuals with a genome (model structures) to identity the best model based on a goodness-of-fit statistic criterion (e.g., Akaike information criterion).

• Then, the next generation can be created by first selecting sets of "parents" also selected randomly, with replacement and probability of selection proportional to fitness.

• Next, the genomes of the parents are lined up and "crossover" occurs, i.e., some user-defined fraction of the parent sets (e.g., 60%), at a single, random location in the genome. For instance, if the genome contains four genes (model characteristics: one compartment PK model + first-order absorption + presence of lag time + proportional residual error), then a 50% split would generate the left part with "one compartment PK model + first-order absorption" and the right part with "presence of lag time + proportional residual error." Then, the left part of the first parent is combined with the right part of the second parent and *vice versa*.

• Subsequently, a mutation is randomly applied based on a prespecified small proportion. For instance, the "presence of lag time" feature is reversed to "absence of lag time."

• This process is repeated until no further improvement is seen (i.e., a model with an improved goodness-of-fit criterion is no longer obtained).[15,143]

3.2.5 Decision Trees, Random Forest, and Gradient Boosting

Decision trees are a commonly used group of nonlinear ML methods. Each model is a set of rules organized in the form of a tree. Starting at the base or "root node," the algorithm selects a "branch" based on the decision rule at the root node. The decision rule is generally based on a single covariate and a specified threshold. The algorithm then reaches the next node and follows the next decision rule. Eventually, the algorithm reaches a "leaf node," which represents a specific output decision, returned as a result. Decision trees can be used for building both classification models for making qualitative predictions and regression models for making quantitative predictions.[15,144,145]

One reason for the popularity of single decision trees is their ease of interpretation and representation. However, they are outperformed by more advanced ML algorithms, such as random forest or gradient boosting modeling.

Random forest is a method that creates a large collection of decision trees, where each decision tree makes a prediction or "vote" for a certain outcome. For each tree, typically, one third of the training data set is randomly selected and set aside, and the remaining two thirds of the training data set is used for model development. When these decision trees are being built, each time a split in a tree is considered, a random sample of predictors is chosen as split candidates from the full set of predictors. A single predictor and a corresponding decision rule threshold are identified that maximize the accuracy of the decision tree. When all the trees are completed, the model is ready to make predictions about new data. The prediction is defined by majority voting from this collection of decision trees. To evaluate the accuracy of the model, the trained forest can be used to predict the remaining one-third of the observations and calculate the out-of-bag (OOB) error.[146]

Random forest is a type of ensemble modeling because it involves combining multiple ML models to improve the overall performance of the model. Ensemble models generally combine the results by voting or taking the mode, mean, or weighted mean of the results from different models. More advanced techniques include bagging and boosting. Bagging involves creating random subsamples, or bootstrapping, of training data with replacement and building a model for each subset.[15,147]

Gradient boosting differs from bagging methods in that the trees are trained and combined sequentially. This algorithm generates models by computing a sequence of trees in which each successive tree is built from the prediction residuals of the preceding tree. A simple partitioning of the data is determined at each step in the boosting tree algorithm, and the deviations of the observed values from the respective residuals for each partition are computed. Given the preceding sequence of trees, the next tree will then be fitted to the residuals to find another partition that will further reduce the error in the model.[145] (148) A recent publication showed a novel method, named DTIGBDT, based on gradient boosting decision trees outperforms several state-of-the-art methods for drug–target interaction prediction. Besides, case studies on Quetiapine, Clozapine, Olanzapine, Aripiprazole, and Ziprasidone demonstrate the ability of DTIGBDT to discover potential drug–target interactions.[15,149]

3.2.6 Neural Networks (NNs)

NNs were widely used in QSAR in the 1990s. Due to various practical issues, such as slow on large data set, difficult to train the model, prone to overfitting etc., more robust methods like support vector machine (SVM), random forest, arose in the early 2000s. However, the most recent 10-year has witnessed an NN revival in ML community for both advances of algorithms as well as computational hardware.[150]

NNs are composed of units called artificial neurons. Each connection between neurons can transmit a signal to another neuron. All neurons have multiple inputs and one output. The receiving neuron can process the signals and then signal downstream neurons connected to it. Each input is associated with a weight that varies as learning proceeds, which can increase or decrease the strength of the signal that is sent downstream.

A general formula for the output Y of a neuron is:

$$Y = f(\sum_i W_i X_i) \qquad (18.7)$$

where f is a specified function, X_i is the ith input which can be physical chemical properties of small molecules or experiment observations, and W_i is the weight associated with the ith input. Neurons are typically organized in the following three types of layers: (1) the input layer (i.e., the bottom layer), where the predictors are entered; (2) the hidden (middle) layers; and (3) the output layer (i.e., the top layer), where the predictions are generated.[15]

Deep NN (DNN), also known as deep learning, refers to a model with more than one hidden (middle) layer. DNNs can be further categorized based on model architectures; convolution NN, recurrent NN, and long short-term memory- based recurrent NN.[15,150–153]

Though we present QSAR model as a representive technique in LBDD, it is common to combine it with other techniques to contribute to current CADD process. Among many applications, a couple of examples recently reported include that a potential novel COVID-19 drug was proposed based on 3D-QSAR followed by molecular docking and the ADMET properties,[126,154] and another showed drug repurposing by QSAR combined with docking and MD studies to find inhibitors against SARS-CoV-2 M^{pro} protease.[155]

4 Summary

The process of drug development and drug discovery is very challenging, expensive, and time consuming. It has been accelerated due to continuous development of computational tools and methods. Over the past several decades, CADD also known as *in silico* drug discovery has become a powerful technique because of its utility in various phases of drug discovery through various advanced features. This chapter focused on fundamental methods in computational chemistry to illustrate the concept of computational approaches in structural- and ligand-based drug discovery.[156]

Applications of CADD to drug discovery are like other subjects, also progressing day by day. By more interactively working with high quality data, CADD will continuously contribute its unique and irreplaceable role in the modern drug hunting by rationally design potent therapeutics with multi-targeting effects, higher efficacies, and fewer side effects, especially in terms of toxicity.

5 References

1. Arodola, O.; Soliman, M. Quantum mechanics implementation in drug design workflows: does it really help? *Drug Des. Dev. Ther.* **2017**, *11*, 2551–2564.
2. Parasuraman, S. Toxicological screening. *J. Pharmacol. Pharmacother.* **2011**, *2*, 74–79.
3. Yamanishi, Y.; Araki, M.; Gutteridge, A.; Honda, W.; et al. Prediction of drug–target interaction networks from the integration of chemical and genomic spaces. *Bioinf.* **2008**, *24*, i232–i240.
4. Moult, J.; Fidelis, K.; Kryshtafovych, A.; Schwede, T.; et al. Critical assessment of methods of protein structure prediction (CASP)—Round XII, *Proteins Struct. Funct. Bioinformatics.* **2018**, *86*, 7–15.
5. Jumper, J.; Evans, R.; Pritzel, A.; Green, T.; et al. Highly accurate protein structure prediction with AlphaFold. *Nature* **2021**, *596*, 583–594.
6. Senior, A. W.; Evans, R.; Jumper, J.; et al. Improved protein structure prediction using potentials from deep learning. *Nature* **2020**, *577*, 706–710.
7. Baek, M.; DiMaio, F.; Anishichenko, I.; Dauparas, J.; et al. Accurate prediction of protein structures and interactions using a three-track neural network. *Science* **2021**, *373*, 871–876.
8. De Vivo, M.; Masetti, M.; Bottegoni, B.; Cavalli, A.; et al. Role of molecular dynamics and related methods in drug discovery. *J. Med. Chem.* **2016**, *59*, 4035–4061.
9. Salo-Ahen, O.; Alanko, I.; Bhadane, R.; Bonvin, A.; et al. Molecular dynamics simulations in drug discovery and pharmaceutical developmnet. *Processes* **2021**, *9*, 71.
10. Shoichet, B. K. Virtual screening of chemical libraries. *Nature* **2004**, *432*, 862–865.
11. Kimber, T. B.; Chen, Y. Deep learning in virtual screening: recent applications and developments. *Int. J. Mol. Sci.* **2021**, *22*, 4435.
12. Sliwoski, G.; Kothiwale, S.; Meiler, J.; Lowe, E. W.; et al. Computational methods in drug discovery. *Pharmacol. Rev.* **2014**, *66*, 334–395.
13. Popova, M.; Isayev, O.; Tropsha, A. Deep reinforcement learning for de novo drugdesign. *Sci. Adv.* **2018**, *4*, 14

14. Vamathevan, J.; Clark, D.; Czodrowski, P.; et al. Applications of machine learning in drug discovery and development. *Nat. Rev. Drug Discov.* **2019**, *18*, 463–477.
15. Talevi, A.; Morales, J. F.; Hather, G.; et al. Machine learning in drug discovery and development part 1: a primer. *CPT Pharmacomet. Syst. Pharmacol.* **2020**, *9*, 129–142.
16. Jorgensen, W. The many roles of computation in drug discovery. *Science* **2004**, *303*, 1813–1818.
17. Wade, R.; Salo-Ahen, O. Molecular modeling in drug design. *Molecules* **2019**, *24*, 321.
18. Lin, X.; Li, X.; Lin, X. A review on applications of computational methods in drug screening and design. *Molecules* **2020**, *25*, 1375.
19. Liao, C.; Nicklaus, M. Computer tools in the discovery of HIV-I integrase inhibitors. *Future Med. Chem.* **2010**, *2*, 1123–1140.
20. Serrao, E.; Odde, S.; Ramkumar, K.; Neamati, N.; et al. Raltegravir, elvitegravir, and metoogravir: the birth of "me-too" HIV-1 integrase inhibitors. *Retrovirology* **2009**, *6*, 25.
21. Jin, Z.; Du, X.; Xu, Y.; Deng, Y.; et al. Structure of Mpro from SARS-CoV-2 and discovery of its inhibitors. *Nature* **2020**, *582*, 289–293.
22. Panda, P. K.; Arul, M. N.; Patel, P.; Verma, S.; Luo, W.; et al. Structure-based drug designing and immunoinformatics approach for SARS-CoV-2. *Sci. Adv.* **2020**, *6*, 14.
23. Muratov, E.; Amaro, R.; Andrade, C.; Brown, N.; Ekins, S.; et al. A critical overview of computational approaches. *Chem. Soc. Rev.* **2021**, *50,* 9121.
24. Gurung, A. B.; Ali, M. A.; Lee, J.; Farah, M. A.; et al. An updated review of computer-aided drug design and its application to COVID-19. *BioMed. Res. Int.* **2021**, *2021, 1–18.*
25. Ferreira, L. G.; Andricopulo, A. D. Editorial: chemoinformatics approaches to structure- and ligand-based drug design. *Front. Pharmacol.* **2018,** *9*, 1–2.
26. Lavecchia, A. Machine-learning approaches in drug discovery: methods and applications. *Drug Discov. Today* **2015**, *20*, 318–331.
27. Tao, L.; Zhang, P.; Qin, C.; Chen, S.; et al. Recent progresses in the exploration of machine learning methods as in-silico ADME prediction tools. *Adv. Drug Deliv. Rev.* **2015**, *86*, 83–100.
28. Batool, M.; Ahmad, B.; Choi, S. A structure-based drug discovery paradigm. *Intl. J. Mol. Sci.* **2019**, *20*, 2783.
29. Anderson, A. C. The process of structure-based drug design. *Chem. Biol.* **2003**, *10*, 787–797

30. Lionta, E.; Spyrou, G.; Vassilatis, D. K.; Cournia, Z.; et al. Structure-based virtual screening for drug discovery: principles, applications and recent advances. *Curr. Top. Med. Chem.* **2014**, *14*, 1923–1938.
31. Wlodawer, A.; Vondrasek, J. Inhibitors of HIV-1 Protease: A Major Success of Structure-Assisted Drug Design. *Annu. Rev. Biophys. Biomol.* **1998**, *27*, 249–284.
32. Clark, D. E. What has computer-aided molecular design ever done for drug discovery. *Expert Opin. Drug Discov.* **2006**, *1*, 103–110.
33. Grover, S.; Apushkin, M. A.; Fishman, G. A. Topical dorzolamide for the treatment of cystoid macular edema in patients with retinitis pigmentosa. *Am. J. Ophthalmol.* **2006**, *141*, 850–858.
34. Bauer, A.; Kovári, Z.; Keseru, G. M. Optimization of virtual screening protocols: FlexX based virtual screening for COX-2 inhibitors reveals the importance of tailoring screen parameters. *J. Mol. Struct. THEOCHEM.* **2001**, *676*, 1–5.
35. Miller, Z.; Kim, K. S.; Lee, D. M.; Kasam, V.; et al. Proteasome inhibitors with pyrazole scaffolds from structure-based virtual screening. *J. Med. Chem.* **2015**, *58*, 2036–2041.
36. Kuntz, I. D.; Blaney, J. M.; Oatley, S. J.; Langridge, R.; et al. A geometric approach to macromolecule-ligand interactions. *J. Mol. Biol.* **1982**, *161*, 269–288.
37. Caballero, J. The latest automated docking technologies for novel drug discovery. *Exp. Opin. Drug Discov.* **2021**, *16*, 625–645.
38. Meng, X.; Zhang, H.; Mezei, M.; Cui, M.; et al. Molecular docking: a powerful approach for structure-based drug discovery. *Curr. Comput. Aided Drug Des.* **2011**, *7,* 146–157.
39. Salmaso, V.; Moro, S. Bridging molecular docking to molecular dynamics in exploring ligand-protein recognition process: an overview. *Front. Pharmacol.* **2018**, *9,* 1–16.
40. Chaudhary, K. K.; Mishra, N. A review on molecular docking: novel tool for drug discovery. *JSM Chem.* **2016**, *4*, 3, 1029.
41. de Ruyck, J.; Brysbaert, G.; Blossey, R.; Lensink, M. F. Molecular docking as a popular tool in drug design, an in silico travel. *Adv. Appl. Bioinformat. Chem.* **2016**, *9*, 1–11.
42. Pinzi, L.; Rastelli, G. Molecular docking: shifting paradigms in drug discovery. *Int. J. Mol. Sci.,* **2019**, *20*, 4331.
43. Stanzione, F.; Giangreco, I; Cole, J. C. Chapter four - use of molecular docking computational tools in drug discovery. *Prog. Med. Chem.,* **2021**, *60*, 273–343.
44. Kitchen, D.; Decornez, H.; Furr, J.; et al. Docking and scoring in virtual screening for drug discovery: methods and applications. *Nat. Rev. Drug Discov.* **2004**, *3*, 935–949.

45. Chen, G.; Seukep, A. J.; Guo, M. Q. Recent Advances in Molecular Docking for the Research and Discovery of Potential Marine Drugs. *Marine Drugs* **2020**, *18*, 545.
46. Pagadala, N. S.; Syed, K.; Tuszynski, J. Software for molecular docking: a review. *Biophys. Rev.* **2017**, *9*, 91–102.
47. Brooijmans, N.; Kuntz, I. D. Molecular recognition and docking algorithms. *Annu. Rev. Biophys. Biomol. Struct.* **2003**, *32*, 335–373.
48. Agrawal, A.; Singh, H.; Srivastava H. K.; et al. Benchmarking of different molecular docking methods for protein-peptide. *BMC Bioinformatics* **2019**, *19*, 426.
49. Wang, Z.; Sun, H.; Yao, X.; Li, D.; et al. Comprehensive evaluation of ten docking programs on a diverse set of protein–ligand complexes: the prediction accuracy of sampling power and scoring power. *Phys. Chem. Chem. Phys.* **2016**, *18*, 12964–12975.
50. Castro-Alvarez, A.; Costa, A. M.; Vilarrasa, J. The performance of several docking programs at reproducing protein–macrolide-like crystal structures. *Molecules* **2017**, *22*, 136.
51. Boittier, E. D.; Tang, Y. T.; Buckley, M.; et al. Assessing molecular docking tools to guide targeted drug discovery of CD38 inhibitors. *Int. J. Mol. Sci.* **2020**, *21*, 5183.
52. Taylor, R. D.; Jewsbury, P. J.; Essex, J. W. A review of protein-small molecule docking methods. *J. Comput. Aided Mol. Des.* **2002**, *16*, 151–166.
53. Huang, S.Y.; Zou, X. Advances and challenges in protein-ligand docking. *Int. J. Mol. Sci.* **2010**, *11*, 3016–3034.
54. Friesner, R. A.; Banks, J. L.; Murphy, R. B.; Halgren, T. A.; et al. Glide: a new approach for rapid, accurate docking and scoring. 1. Method and assessment of docking accuracy. *J. Med. Chem.* **2004**, *47*, 1739–1749.
55. Halgren, T. A.; Murphy, R. B.; Friesner, R. A.; Beard, H. S.; et al. Glide: a new approach for rapid, accurate docking and scoring. 2. Enrichment factors in database screening. *J. Med. Chem.* **2004**, *47*, 1750–1759.
56. Friesner, R. A.; Murphy, R. B.; Repasky, M. P.; Frye, L.; et al. Extra precision glide: docking and scoring incorporating a model of hydrophobic enclosure for protein-ligand complexes. *J. Med. Chem.* **2006**, *49*, 6177–6196.
57. Trott, O.; Olson, A. J. AutoDock Vina: improving the speed and accuracy of docking with a new scoring function, efficient optimization, and multithreading. *J. Comput. Chem.* **2010**, *31*, 455–461.
58. Goodsell, D. S.; Olson, A. J. Automated docking of substrates to proteins by simulated annealing. *Proteins* **1990**, *8*, 195–202.

59. Jones, G.; Willett, P.; Glen, R. C. Molecular recognition of receptor sites using a genetic algorithm with a description of desolvation. *J. Mol. Biol.* **1995**, *245*, 43–53.
60. Jones, G.; Willett, P.; Glen, R. C. Development and validation of a genetic algorithm for flexible docking. *J. Mol. Biol.* **1997**, *267*, 727–748.
61. Labute, P. LowModeMD—implicit low-mode velocity filtering applied to conformational search of macrocycles and protein loops. *J. Chem. Inf. Model.* **2010**, *50*, 792–800.
62. Alonso, H.; Bliznyuk, A.; Gready, J. E. Combining docking and molecular dynamic simulations in drug design, *Med. Res. Rev.* **2006**, *26*, 531–568.
63. Jiang, F.; Kim, S. H. "Soft docking": matching of molecular surface cubes. *J. Mol. Biol.* **1991**, *219*, 79–102.
64. Leach, A. R. Ligand docking to proteins with discrete side-chain flexibility. *J. Mol. Biol.* **1994**, *235*, 345–356.
65. Knegtel, R. M.; Kuntz, I. D.; Oshiro, C. M. Molecular docking to ensembles of protein structures. *J. Mol. Biol.* **1997**, *266*, 424–440.
66. Rarey, M.; Kramer, B.; Lengauer, T.; Klebe, G. A fast flexible docking method using an incremental construction algorithm. *J. Mol. Biol.* **1996**, *3*, 261, 470–489.
67. Huang, S.; Zou, X. Ensemble docking of multiple protein structures: considering protein structural variations in molecular docking. *Proteins* **2007**, *66*, 399–421.
68. Liao, C.; Sitzmann, M.; Pugliese, A.; Nicklaus, M. C. Software and resources for computational medicinal chemistry. *Future Med. Chem.* **2011**, *3*, 1057–1085.
69. Huang, N.; Kalyanaraman, C.; Irwin, J. J.; Jacobson, M. P.; et al. Physics-based scoring of protein-ligand complexes: Eenrichment of known inhibitors in large-scale virtual screening. *J. Chem. Inf. Model.* **2006**, *46*, 243–253.
70. Sethi, A.; Joshi, K.; Sasikala, K.; Alvala, M. Molecular docking in modern drug discovery: principles and recent applications. *IntechOpen* **2019**, *21*.
71. Böhm, H. J. *LUDI:* Rule-based automatic design of new substituents for enzyme inhibitor leads. *J. Comput.-Aided Mol. Des.* **1992**, *6*, 593–606.
72. Muegge, I.; Martin, Y. C. A general and fast scoring function for protein–ligand interactions: a simplified potential approach. *J. Med. Chem.* **1999**, *42*, 791–804.
73. Ishchenko, A. V.; Shakhnovich, E. I. Small molecule growth 2001 (SMoG2001): an improved knowledge-based scoring function for protein ligand interactions. *J. Med. Chem.* **2002**, *45*, 2770–2780.
74. Muegge, I. A. knowledge-based scoring function for protein-ligand interactions: Probing the reference state. *Perspect. Drug Discov. Des.* **2000**, *20*, 99–114.
75. Morris, G. M.; Huey, R.; Lindstrom, W.; Sanner, M. F.; et al. AutoDock4 and AutoDockTools4: automated docking with selective receptor flexibility. *J. Comput. Chem.* **2009**, *30*, 2785–2791.

76. Allen, W. J.; Balius, T. E.; Mukherjee, S.; Brozell, S. R.; et al. DOCK 6: Impact of new features and current docking performance. *J. Comput. Chem.* **2015**, *36*, 1132–1156.
77. Kozakov, D.; Brenke, R.; Comeau, S. R.; Vajda, S. PIPER: An FFT-based protein docking program with pairwise potentials. *Proteins Struct. Funct. Bioinf.* **2006**, *65*, 392–406.
78. Vilar, S.; Cozza, G.; Moro, S. Medicinal chemistry and the molecular operating environment (MOE): application of QSAR and molecular docking to drug discovery. *Curr. Top. Med. Chem.* **2008**, *8*, 1555–1572.
79. Abagyan, R.; Totrov, M.; Kuznetsov, M. ICM—A new method for protein modeling and design: Applications to docking and structure prediction from the distorted native conformation. *Comput. Chem.* **1994**, *15*, 488–506.
80. Dominguez, C.; Boelens, R.; Bonvin, A. M. HADDOCK: A protein-protein docking approach based on biochemical or biophysical information. *J. Am. Chem. Soc.* **2003**, *125*, 1731–1737.
81. Gray, J. J.; Moughon, S.; Wang, C.; Schueler-Furman, O.; et al. Protein-protein docking with simultaneous optimization of rigid-body displacement and side-chain conformations. *J. Mol. Biol.* **2003**, *331*, 281–299.
82. Qing, X.; Lee, X.; Raeymaeker, J.; Tame, J.; et al. Pharmacophore modeling: advances, limitations, and current utility in drug discovery. *J. Recept. Ligand Channel Res.* **2014**, *7*, 81–92.
83. Ehrlich, P. Über den jetzigen Stand der Chemotherapie. *Ber. Dtsch. Chem. Ges.,* **1909**, *42*, 17–47.
84. Schueler, F. W. *Chemobiodynamics and Drug Design.* New York: McGrawHill Book Co., Inc., **1960**. p. 638.
85. Wermuth, C. G.; Ganellin, C. R.; Lindberg, P.; Mitscher, L. A. Glossary of terms used in medicinal chemistry (IUPAC recommendations 1998). *Pure Appl. Chem.* **1998**, *70*, 1129–1143.
86. Schaller, D.; Šribar, D.; Noonan, T.; Deng, L.; et al. Next generation 3D pharmacophore modeling. *WIREs Comput. Mol. Sci.* **2020**, *10*, 20.
87. Barreca, M. L.; Ferro, S.; Rao, A.; Luca, L. D.; et al. Pharmacophore-based design of HIV-1 integrase strand-transfer inhibitors. *J. Med. Chem.* **2005**, *48*, 7084–7088.
88. Dayam, R.; Sanchez, T.; Neamati, N. Diketo acid pharmacophore. 2. Discovery of structurally diverse inhibitors of HIV-1 integrase. *J. Med. Chem.* **2005**, *48*, 8009–8015.
89. Hong, H.; Neamati, N.; Wang, S.; Nicklaus, M. C. et al. Discovery of HIV-1 integrase inhibitors by pharmacophore searching. *J. Med. Chem.* **1997**, *40*, 930–936.
90. Carlson, H. A.; Masukawa, K. M.; and Rubins, K.; et al. Developing a dynamic pharmacophore model for HIV-1 integrase. *J. Med. Chem.* **2000**, *43*, 2100–2114.

91. Deng, J. X.; Lee, K. W.; Sanchez, T.; Cui, M.; Neamati, N.; Briggs, J. M. Dynamic receptor-based pharmacophore model development and its application in designing novel HIV-1 integrase inhibitors. *J. Med. Chem.* **2005**, *48*, 1496–1505.
92. Deng, J. X.; Sanchez, T.; Neamati, N.; Briggs, J. M. Dynamic pharmacophore model optimization: identification of novel HIV-1 integrase inhibitors. *J. Med. Chem.* **2006**, *49*, 1684–1692.
93. Goldgur, Y.; Craigie, R.; Cohen, G. H.; Fujiwara, T.; et al. Structure of the HIV-1 integrase catalytic domain complexed with an inhibitor: a platform for antiviral drug design. *Proc. Natl. Acad. Sci. U.S.A.* **1999**, *6*, 13040–13043.
94. McCammon, J. A.; Gelin, B. R.; Karplus, M. Dynamics of folded proteins. *Nature* **1977**, *267*, 585–590.
95. Karplus, M.; McCammon, J. Molecular dynamics simulations of biomolecules. *Nat. Struct. Mol. Biol.* **2002**, *9*, 646–652.
96. Hollingsworth, S. A.; Dror, R. O. Molecular Dynamics Simulation for All. *Neuron* **2018**, *99*, 1129–1143.
97. Urbanc, B.; Betnel, M.; Cruz, L.; Bitan, G.; et al. Elucidation of amyloid beta-protein oligomerization mechanisms: discrete molecular dynamics study. *J. Am. Chem. Soc.* **2010**, *132*, 4266–4280.
98. Gsponer, J.; Haberthür, U.; Caflisch, A. The role of side-chain interactions in the early steps of aggregation: molecular dynamics simulations of an amyloid-forming Peptide from the yeast prion Sup35. *Proc. Natl. Acad. Sci. U.S.A.* **2003**, *100*, 5154–5159.
99. Woods, C. J.; Malaisree, M.; Pattarapongdilok, N.; Sompornpisut, P.; et al. Long time scale GPU dynamics reveal the mechanism of drug resistance of the dual mutant I223R/H275Y neuraminidase from H1N1-2009 influenza virus. *Biochemistry.* **2012**, *51*, 4364–4375.
100. Phanich, J.; Rungrotmongkol, T.; Kungwan, N.; Hannongbua, S. Role of R292K mutation in influenza H7N9 neuraminidase toward oseltamivir susceptibility: MD and MM/PB(GB)SA study. *J. Comput. Aided Mol. Des.* **2016**, *30*, 917–926.
101. Campos, S. R.; Machuqueiro, M.; Baptista, A. M. Constant-pH molecular dynamics simulations reveal a β-rich form of the human prion protein. *J. Phys. Chem. B* **2010**, *114*, 12692–12700.
102. Sousa, S. F.; Tamames, B.; Fernandes, P. A.; Ramos, M. J. Detailed atomistic analysis of the HIV-1 protease interface. *J. Phys. Chem. B* **2011**, *115*, 7045–7057.
103. Viricel, C.; Ahmed, M.; Barakat, K. Human PD-1 binds differently to its human ligands: a comprehensive modeling study. *J. Mol. Graph. Model.* **2015**, *57*, 131–142.
104. Etheve, L.; Martin, J.; Lavery, R. Protein-DNA interfaces: a molecular dynamics analysis of time-dependent recognition processes for three transcription factors. *Nucleic Acids Res.* **2016**, *44*, 9990–10002.

105. Schames, J. R.; Henchman, R. H.; Siegel, J. S.; Sotriffer, C. A.; et al. Discovery of a novel binding trench in HIV integrase. *J. Med. Chem.* **2004**, *47*, 1879–1881.
106. Durrant, J. D.; Keränen, H.; Wilson, B. A.; McCammon, J. A.; et al. Computational identification of uncharacterized cruzain binding sites. *PLoS Negl. Trop. Dis.* **2010**, *4,* 1-11.
107. Grant, B. J.; Lukman, S.; Hocker, H. J.; Sayyah, J.; et al. Novel allosteric sites on Ras for lead generation. *PLoS One* **2011**, *6,* 1-10.
108. Ivetac, A.; McCammon, J. A. Mapping the druggable allosteric space of G-protein coupled receptors: a fragment-based molecular dynamics approach. *Chem. Biol. Drug Des.* **2010**, *76*, 201–217.
109. Hucke, O.; Coulombe, R.; Bonneau, P.; Bertrand-Laperle, M.; et al. Molecular dynamics simulations and structure-based rational design lead to allosteric HCV NS5B polymerase thumb pocket 2 inhibitor with picomolar cellular replicon potency. *J. Med. Chem.* **2014**, *57*, 1932–1943.
110. Zhao, G.; Perilla, J.; Yufenyuy, E.; Meng, X.; et al. Mature HIV-1 capsid structure by cryo-electron microscopy and all-atom molecular dynamics. *Nature* **2013**, *497*, 643–646.
111. Freddolino, P. L.; Arkhipov, A. S.; Larson, S. B.; McPherson, A.; et al. Molecular dynamics simulations of the complete satellite tobacco mosaic virus. *Structure* **2006**, *14*, 437–449.
112. Freddolino, P. L.; Liu, F.; Gruebele, M.; Schulten, K. Ten-microsecond molecular dynamics simulation of a fast-folding WW domain. *Biophys. J.* **2008**, *94*, 157–177.
113. Klepeis, J. L.; Lindorff-Larsen, K.; Dror, R. O.; Shaw, D. E. Long-timescale molecular dynamics simulations of protein structure and function. *Curr. Opin. Struct. Biol.* **2009**, *19*, 120–127.
114. Hauser, A. S.; Chavali, S.; Masuho, I.; Jahn, L. J.; et al. Pharmacogenomics of GPCR drug targets. *Cell* **2018***, 172*, 41–54.
115. McCorvy, J. D.; Wacker, D.; Wang, S.; Agegnehu, B.; et al. Structural determinants of 5-HT2b receptor activation and biased agonism. *Nat. Struct. Mol. Biol.* **2018**, *25*, 787–796.
116. Chen, Q.; Tesmer, J. J. A receptor on acid. *Cell* **2017**, *168*, 339–341.
117. Hofmann, C.; Penner, U.; Dorow, R.; Pertz, H.; et al. Lisuride, a dopamine receptor agonist with 5-HT2b receptor antagonist properties: absence of cardiac valvulopathy adverse drug reaction reports supports the concept of a crucial role for 5-HT2B receptor agonism in cardiac valvular fibrosis. *Clin. Neuropharmacol.* **2006**, *29*, 80–86.
118. Wacker, D.; Wang, S.; McCorvy, J. D.; Betz, R. M.; Venkatakrishnan, A. J.; et al. Crystal structure of an LSD-bound human serotonin receptor. *Cell* **2017**, *168*, 377–389.

119. Peters B.; Deng, J.; Ferguson, A. Free energy calculations of the functional selectivityof 5-HT2BG protein-coupled receptor. *PLoS One* **2020**, *15*.1-21.

120. Acharya, C.; Coop, A.; Polli, J. E.; MacKerell Jr, A. D. Recent advances in ligand-based drug design: relevance and utility of the conformationally sampled pharmacophore approach. *Curr. Comput. Aided Drug Des.* **2011**, *7*, 10–22.

121. Hansch, C.; Maloney, P.; Fujita, T.; Muir, R. Correlation of biological activity of phenoxyacetic acids with Hammett substituent constants and partition coefficients. *Nature* **1962**, *194*, 178–180.

122. Cherkasov, A.; Muratov, E. N.; Fourches, D.; Varnek, A.; et al. QSAR modeling: where have you been? where are you going to? *J. Med. Chem.* **2014**, *57*, 4977–5010.

123. Liew, C. Y.; Yap, C. W. Statistical Modelling of Molecular Descriptors in QSAR/QSPR **2012,** *2*, 1–436 .

124. Damale, M. G.; Harke, S. N.; Khan, F.; Shinde, D. B.; et al. Recent advances in multidimensional QSAR (4D-6D): A Critical Review. *Mini-Rev. Med. Chem.* **2014**, *14*, 35–55.

125. Polanski, J. Receptor dependent multidimensional QSAR for modeling drug–receptor interactions. *Curr. Med. Chem.* **2009**, *16*, 3243–3257.

126. Ishola, A. A.; Adedirin, O.; Joshi, T. Chandra, S. QSAR modeling and pharmacoinformatics of SARS coronavirus 3C-like protease inhibitors. *Comput. Biol. Med.* **2021**, *134,* 1–16.

127. Kuz'min, V. E.; Artemenko, A. G.; Polischuk, P. G.; Muratov, E. N. Hierarchic system of QSAR models (1D – 4D) on the base of simplex representation of molecular structure. *J. Mol. Model* **2005**, *11,* 457–467.

128. Randić, M. Generalized molecular descriptors. *J. Math. Chem.* **1991**, 7, 155–168.

129. Verma, J.; Khedkar, V. M.; Coutinho, E. C. 3D-QSAR in drug design-a review. *Curr. Top. Med. Chem.* **2010**, *10*, 95–115.

130. Dudek, A. Z.; Arodz, T.; Galvez, J. Computational methods in developing quantitative structure-activity relationships (QSAR): a review. *Combin. Chem. High Throughput Screen.* **2006**, *9*, 213–228.

131. Lemmen, C.; Lengauer, T. Computational methods for the structural alignment of molecules. *J. Comput. Aided Mol. Des.* **2000**, *14*, 215–232.

132. Wang, T.; Yuan, X. S.; Wu, M. B.; Lin, J. P.; et al. The advancement of multidimensional qsar for novel drug discovery-where are we headed? *Exp. Opin. Drug Discov.* **2017**, *12*, 769–784.

133. Carracedo-Reboredo, P.; Liñares-Blanco, J.; Rodríguez-Fernández, N.; Cedrón, F.; et al. A review on machine learning approaches and trends in drug discovery. *Comput. Struct. Biotechnol. J.* **2021**, *19*, 4538–4558.

134. Bak, A. Two decades of 4D-QSAR: a dying art or staging a comeback? *Int. J. Mol. Sci.* **2021**, *22*, 5212.
135. Fleming, N. How artificial intelligence is changing drug discovery. *Nature* **2018**, *557*, S55–S57.
136. SAS Institute. *Machine learning: what it is and why it matters.* [Online]. https://www.sas.com/en_us/insights/analytics/machine-learning.html.
137. Palmer, D. S.; O'Boyle, N. M.; Glen, R. C.; Mitchell, J. Random forest models to predict aqueous solubility. *J. Chem. Inf. Model.* **2007**, *47*, 150–158.
138. Yamashita, F.; et al. An evolutionary search algorithm for covariate models in population. *J. Pharm. Sci.* **2017**, *106*, 2407–2411.
139. Gayvert, K. M.; Madhukar, N. S.; Elemento, O. A data-driven approach to predicting successes and failures of clinical trials. *Cell Chem. Biol.* **2016**, *23*, 1294–1301.
140. Korolev, D.; et al. Modeling of human cytochrome p450-mediated drug metabolism using unsupervised machine learning approach. *J. Med. Chem.* **2003**, *46*, 3631–3643.
141. Kotsiantis, S. B. Supervised machine learning: a review of classification techniques. *Informatica* **2007**, *31*, 249–268.
142. Rogers, D.; Hopfinger, A. J. Application of genetic function approximation to quantitative structure-activity relationships and quantitative structure-property relationships. *J. Chem. Inf. Comput.* **1994**, *34*, 854–866.
143. Bies, R. R.; et al. A genetic algorithm-based, hybrid machine learning approach to model selection. *J. Pharmacokinet. Pharmacodyn.* **2006**, *33*, 195–221.
144. Schwaighofer, A.; Schroeter, T.; Mika, S.; Blanchard, G. How wrong can we get? A review of machine learning approaches and error bars. *Comb. Chem. High Throughput Screen.* **2009**, *12*, 453–468.
145. Schöning, V.; Hammann, F. How far have decision tree models come for data mining in drug discovery? *Expert Opin. Drug Discov.* **2018**, *13*, 1067–1069.
146. Priya, N.; Shobana, G. Application of machine learning models in drug discovery: a review. *Int. J. Emerg. Technol.* **2019**, *10*, 268–275.
147. Breiman, L. Random forests. *Mach. Learn.* **2001**, *45*, 5–32.
148. Friedman, J. H. Greedy function approximation: a gradient boosting machine. *Ann. Stat.* **2001**, *29*, 1189–1232.
149. Xuan, P.; Sun, C.; Zhang, T.; Ye, Y.; et al. Decision tree-based method for predicting interactions between target genes and drugs. *Front. Genet.* **2019**, *10*, 1-11.
150. Ma, J.; Sheridan, R.; Liaw, A.; Dahl, G.; et al. Deep neural nets as a method for quantitative structure–activity relationships. *J. Chem. Inf. Model.* **2015**, *55*, 263–274.

151. Lipinski, C. F.; Maltarollo, V. G.; Oliveira, P. R.; da Silva, A. B. F.; et al. Advances and perspectives in applying deep learning for drug design and discovery. *Front. Robot. AI* **2019**, *6,* 1–6.

152. Hinton, G.; et al. Deep neural networks for acoustic modeling in speech recognition: the shared views of four research groups. *IEEE Signal Process. Mag.* **2012**, *29*, 82–97.

153. Han, Y.; et al. Deep learning with domain adaptation for accelerated projection-reconstruction MR. *Magn. Reson. Med.* **2018**, *80*, 1189–1205.

154. Khaldan, A.; Bouamrane, S.; Fatima En-Nahli, F.; El-mernissi, R.; et al. Prediction of potential inhibitors of SARS-CoV-2 using 3D-QSAR, moleculardocking modeling and ADMET properties. *Heliyon* **2021**, *7, 1–15.*

155. Tejera, E.; Munteanu, C. R. Drugs repurposing using QSAR, docking and molecular dynamics for possible inhibitors of the SARS-CoV-2 M^{Pro} protease. *Molecules* **2020**, *25*, 16.

156. Baig, M. H.; Ahmad, K.; Rabbani, G.; et al. Computer aided drug design and its application to the development of potential drugs for neurodegenerative disorders. *Curr. Neuropharmacol.* **2018**, *16*, 740–748.

Index

Current Drug Synthesis, First Edition. Edited by Jie Jack Li.

Printed and bound by CPI Group (UK) Ltd, Croydon, CR0 4YY
17/09/2026
14964258-0001